3rd International Virtual Conference of Chemistry (IVCC 2021)

Journal of Physics: Conference Series Volume 2063

Online
14-18 February 2021

Editors:

Athir M. Haddad
Zaki N. Kadhim
Salah S. Hashim

ISBN: 978-1-7138-5034-2
ISSN: 1742-6588

Printed from e-media with permission by:

Curran Associates, Inc.
57 Morehouse Lane
Red Hook, NY 12571

Some format issues inherent in the e-media version may also appear in this print version.

Additional copies of this publication are available from:

Curran Associates, Inc.
57 Morehouse Lane
Red Hook, NY 12571 USA
Phone: 845-758-0400
Fax: 845-758-2633
Email: curran@proceedings.com
Web: www.proceedings.com

TABLE OF CONTENTS

Journal of Physics: Conference Series **2063** (2021) 011001 doi:10.1088/1742-6596/2063/1/011001

The Conference took place virtually due to covid-19 outbreak and travel restrictions. The organisers were from different cities in Iraq but most of them from Basrah. There were five plenary speakers and these speakers were from UK, Australia, South Africa, Jordan and Iraq. Each speaker took approximately an hour with Q&A session. In overall, there were 11 sessions that were delivered by 82 speakers, who were from different countries, each speaker spent 15 min to deliver their talks. All the discussions and Q&A time were live through Zoom rooms and FCC. The overall attendance number was more than 500 people during 5 days. Laptops, phones, Zoom and free conference call were needed to deliver the meeting successfully. The only technical difficulty that we faced sometimes is the weak WiFi signals. The drawbacks in the delivery of the conference virtually, was we couldn't welcome people in person and couldn't have Gala dinner. Other than that everything went very well.

The idea of the conference is for the chemist, scientists, scholars, engineers, and students from Iraq and all around the world and the industry to present ongoing research activities, and hence to foster research relations between the Universities and the industry. We think it an ideal platform for people to share views and experiences in different aspects of chemistry and its relation with other sciences.

The conference program consists of invited sessions and technical workshops and discussions with eminent speakers covering a wide range of topics in chemistry. This rich program provides all attendees with the opportunities to meet online and interact with one another. We hope your experience with the 3rd VICC is a fruitful and long-lasting one.

We would like to thank the organization staff, the members of the program committees, and reviewers. They have worked very hard in reviewing papers and making valuable suggestions for the authors to improve their work. We also would like to express our gratitude to the external reviewers, for providing extra help in the review process, and the authors for contributing their research results to the conference. Special thanks go to IOP Publishing-Journal of Physics Conference Series for their generous assistance and cooperation.

Prof.Dr.Athir M. Haddad

Prof.Dr. Zaki N. Kadhim

Prof.Dr.Salah S. Hashim

Editors of Conference proceedings, 3VICC-2021

Organizing Committee

Assist. Prof. Dr. Muwafaq F. Al- Shahwan	Chairperson	Basrah University
Prof. Dr. Zaki N. Kadhim	Vice Chairperson	Basrah University
Prof. Dr. Hadi S. Al- Lami	Member	Basrah University
Prof. Dr. Widad S. Hannosh	Member	Basrah University
Prof. Dr. Majeed Y. A-Luaibi	Member	Basrah University
Prof. Dr. Dawood S. Ali	Member	Basrah University
Prof. Dr. Hadi Z. Mohammed	Member	Basrah University
Prof. Dr. Ammar K. Al- Baage	Member	Basrah University
Prof. Dr. Hassan T. Abdulsahib	Member	Basrah University
Assist.Prof. Dr. Dakhil Z. Alkremawi	Member	
Assist. Prof. Dr. Bushra Al- Salami	Member	Basrah University
Assist. Prof. Dr. Abdellah Hussein	Member	
Prof. Dr. Athir M. Haddad	Conference secretary	Basrah University

Scientific Committee

Prof. Dr. Salah Shaker	Chairperson	Basrah University
Prof. Dr. Nadhum A Awad	Member	Basrah University
Prof. Dr. Hanan Abduljaleel	Member	Basrah University
Prof. Dr. Iqbal J. Al-Asadi	Member	Basrah University
Prof. Dr. Abass Dawas	Member	
Prof. Dr. Adel A. Al-Fregie	Member	Basrah University
Prof. Dr Abbas F. Abbas	Member	Basrah University
Assist. Prof. Dr. Ali Al-Riyahee	Member	Basrah University
Assist. Prof. Dr. Salih H. Abbas	Member	Basrah University

Journal of Physics: Conference Series **2063** (2021) 011001 doi:10.1088/1742-6596/2063/1/011001

Technical committee

Prof. Dr. Athir M. Haddad	Chairperson	Basrah University
Assist. Prof. Dr. Mohammed J. Al- Asadi	Member	Basrah University
Prof. Dr. Ali J. Hameed	Member	Basrah University
Prof. Dr. Basil A. Saleh	Member	Basrah University
Assist. Prof. Dr. Mohammed A. Abed	Member	Basrah University
Assist. Prof. Dr. Raed K. Al-Tameemi	Member	Basrah University
Assist. Prof. Dr. Baqer A. Al-Mayyahi	Member	Basrah University
Assist. Prof. Dr. Zainab S. Al-Al	Member	Basrah University
Assist. Prof. Dr. Ali A. Abdulwahid	Member	Basrah University
Assist. Prof. Dr. Roza A. Al-Aqar	Member	Basrah University
Assist. Prof. Dr . Ibrahim M. Al-Naiema	Member	Basrah University
Lecturer Dr. Aula A. Alwattar	Member	Basrah University

Journal of Physics: Conference Series **2063** (2021) 011002 doi:10.1088/1742-6596/2063/1/011002

Peer review declaration

Abstract
All papers published in this volume of Journal of Physics: Conference Series have been peer reviewed through processes administered by the Editors. Reviews were conducted by expert referees to the professional and scientific standards expected of a proceedings journal published by IOP Publishing.

- **Type of peer review: Double-blind**

- **Conference submission management system: google website**
  https://sites.google.com/uobasrah.edu.iq/3rd-ivccbasrah/english
  Online Conference Email: athir.haddad@uobasrah.edu.iq

- **Number of submissions received: 65**

- **Number of submissions sent for review: 47**

- **Number of submissions accepted: 31**

- **Acceptance Rate (Number of Submissions Accepted / Number of Submissions Received X 100): 47.7 %**

- **Average number of reviews per paper: 2-3 Reviewers**

- **Total number of reviewers involved: 35**

- **Any additional info on review process:** The papers were received was evaluated using Turnitin software at Basrah University. Then the papers were sent to the reviewers of related research areas for peer-review process. The referees gave their decision (acceptance or acceptance after revisions or rejection). Rejected papers were sent back to the authors. The accepted papers by two reviewers were sent for the publications.

- **Contact person for queries:**

Prof. Dr. Salah Shaker Hashim,

Dept. of Chemistry,

College of Science,

University of Basrah,

salah.hashim@uobasrah.edu.iq

Mobile No.: 00964 774 070 8188

Extraction of phenolic compounds from Iraqi *Coriandrum Sativum* L. and loaded on copolymeric hydrogels and examine there as drug delivery system and antioxidant

Ula Ali Sadoun, Zainab Shakir Al-Ali* **and Athir Mahmood Haddad**
Department of Chemistry, College of Science, University of Basrah, karmat Ali, Basrah, Iraq

*zainab.abdulah@uobasrah.edu.iq

Abstract. The phenolic extracts for leaves and stems (L+S) parts and leaves (L) part of Iraqi *Coriandrum Sativum* L. and their total phenols, total tannins and total flavonoids are described. Three copolymeric hydrogels prepared and loaded with phenolic extract 4 (U1-U3). The HPLC results show three phenolic compounds, while the GC-Mass results show one phenolic compound and four non-phenolic compounds. Gained results showed that there are significant ($P < 0.05$) variations in total phenols (9.822 ± 0.634–4.015 ± 0.118 mg GAE/g DW), total flavonoids (8.112 ± 0.115–2.811 ± 0.371 mg QE/g DW) and total condensed tannin (4.245 ± 0.276–1.135 ± 0.091 mg QE/g DW) contents for all phenolic extracts. The swelling rate for (U1-U3) in distilled water, the SGF, and the SIF was estimated. The maximum swelling was observed in copolymeric hydrogels at pH 6.9 in distilled water. The IC_{50} values of radical scavenging activity of the phenolic extracts 4, 8 and phenolic extract 4 released from copolymeric hydrogels (U1-U3) show varied significantly ($P < 0.05$). Our results indicated that Iraqi *Coriandrum Sativum* L. could constitute a rich and novel source of natural antioxidants. When it loaded on, copolymeric hydrogels could be used as a drug delivery system.

1. Introduction

Medicinal plants are necessary due to having bioactive compounds like phenolic compounds applied in the manufacture and development of drugs [1]. Phenolic compounds in plants already proved beneficial effects in cardiovascular diseases, diabetes, cancer, anti-inflammatory effects [2], and antioxidant activities. They have been proved to be more potent antioxidants than vitamins E and C and carotenoids [3] [4, 5]. Antioxidants can inhibit the oxidation processes called oxidative stress caused by oxygen or reactive oxygen species (ROS). The main problems of oxidative stress in cells were damaging proteins, lipids, and DNA [6]. Phenolic compounds have high antioxidant activity due to their phenoxyl radicals that could stabilize the whole structure through resonance when reacting with oxidants [7].

Coriandrum sativum L. is an annual medicinal herb belonging to the Apiaceae family. It uses many issues such as to treat loss of appetite, convulsion, anxiety, and insomnia [8], antioxidant [9-11], antifungal [12, 13], antibacterial [14, 15], diuretic [16], antidiabetic [17, 18], cancer chemopreventive activities [19], exhibits anxiolytic properties and a sedative effect [20, 21], and provides significant protection from heart failure [22].

Hydrogels are a class of three-dimensional (3D) hydrophilic polymeric network. Polymeric structures can swell in water and biological fluid and keep water and biological fluids in the swelling state due to hydrophilic groups' presence in gel structure (e.g., $CONH_2$, -COOH, -OH, $-SO_3H$) that absorbs water and keeps it inside the hydrogel chains. They are having many applications, such as drug delivery [23], gene delivery [24], tissue engineering [25], and wound healing [26]. The oral drug delivery system is the most appropriate method in many clinical trials. The hydrogels as drug delivery are excellent due to swelling in water, permeable of the solute through the pores, the controlled release systems and reducing side effects and toxicity within the organism's body [23]. To date, there is a good idea to develop an integrated approach for developing bioapplications using polymers hydrogels to carry the natural bioactive compounds extracted from plants. In the present study, eight phenolic extracts for leaves and stems (L+S) parts and leaves (L) part of Iraqi *Coriandrum Sativum* L. were prepared. Then, the total phenols, total tannins, total flavonoids and antioxidant capacities were examined. Three copolymeric hydrogels loaded with phenol extract 4 of *Coriandrum Sativum* L. (U1-U3) *via* redox polymerization were prepared. It is best to know that there are no reports on the loaded phenol extract on hydrogels and studies it as a drug delivery system and antioxidant.

2. Materials and Methods

2.1. Chemicals

Chemicals used in this study were obtained from different sources. Absolute ethanol and methanol were supplied from V.W.R. Acryl amide, 2-Acrylamido-2-methylpropane sulfonic acid, DMSO, and sodium chloride was supplied from Merck, Germany. BHT, DPPH, quercetin, N, N\-Methylenebis(acrylamide), and Potassium hydroxide were supplied from Sigma Aldrich, Germany. Sodium carbonate, Vanillin was delivered from Alpha. Sodium nitrite, sodium persulfate, sulfuric acid, N, N, N\,N\Tetramethyleneethylendiamine were supplied from Fluka. Potassium dihydrogen phosphate was supplied from H.W. Sodium hydroxide, Folin–Ciocalteu reagent were supplied from HIM. Aluminium chloride was supplied from B.D.H. Gallic acid was supplied from M.N.B.

2.2. Plant material and extracts preparation

The fresh leaves and stems of *Coriandrum Stavium* L. were collected from a local market in Basrah city, Iraq, in September 2019. They were classified in the Department of Biology, Faculty of Science, University of Basrah, Basrah, Iraq. These parts were manually divided into two groups; the leaves (L) part and the mixture of leaves and stems (L+S) part, cleaned, washed with distilled water, dried away by the direct sunlight, and then ground into powder an electrical grinder. The powder was kept in a closed container at 7°C until the time of use. Eight phenolic extracts were prepared, two groups were divided, and four groups of leaves and steams mixture (L + S) and four groups of leaves (L) were used with both ethanol and methanol solvent and stirred at different times. The extracts were prepared by the modified method [27]. Three g of dry powder sample was separately extracted by stirring with 30 mL of solvent for a specific time, and the extracts were kept for 24 h at 4°C, filtered through a Whatman paper No.6, then dried and stored until use. See table 1.

2.3. Total phenols content

Total phenols contents were tested using the Folin–Ciocalteu reagent by following slightly modified literature Singleton's method [28]. An aliquot 0.125 mL of convenient diluted phenolic extracts were taken in a test tube and added to 0.5 mL of deionized water and 0.125 mL of the Folin–Ciocalteu reagent. The mixture was shaken and allowed to stand for 6 min before adding 1.25 mL of 7% Na_2CO_3 solution. The

Journal of Physics: Conference Series **2063** (2021) 012001 doi:10.1088/1742-6596/2063/1/012001

solution was then adjusted with deionized water to a final volume of 3 mL and mixed thoroughly. After incubation for 90 min at 23°C, the absorbance versus prepared blank was read at 760 nm. Gallic acid was used as the standard. The total phenol amount was calculated using the standard curve of gallic acid drawn within a concentration range of 0.2 to 1 mg/mL had an R^2 value of 0.994. It was expressed as mg gallic acid equivalents g^{-1} (mg GAE g^{-1}) leaves and stems. All samples were performed in duplicate.

Table 1. Preparation of the extracts of *Coriandrum Stavium* L leaves and stems mixture (L+S) and leaves (L) using methanol and ethanol solvent.

No. of extract	Type of part used from the plant	Solvent	Time of stirrer
1	Leaves and stems (L+S)	Methanol	30 min
2	Leaves and stems (L+S)	Methanol	24 h
3	Leaves and stems (L+S)	Ethanol	30 min
4	Leaves and stems ((L+S	Ethanol	24 h
5	Leaves (L)	Methanol	30 min
6	Leaves (L)	Methanol	24 h
7	Leaves (L)	Ethanol	30 min
8	Leaves (L)	Ethanol	24 h

2.4. Total flavonoid content

Total flavonoid contents were measured in line with the slightly modified literature method described by [29]. A 250 µL of phenolic extracts was briefly taken in a test tube mixed with 75 µL $NaNO_2$ (5%). After 6 min, 150 µL of 10% $AlCl_3$ and 500 µL of NaOH (1M) were added to the mixture. Finally, the mixture was adjusted to 2.5 mL with distilled water. The absorbance versus prepared blank was read at 510 nm on a UV spectrophotometer. Total flavonoid contents of all extracts (two replicates per treatment) were expressed as mg quercetin equivalents per gram (mg QE/g) using a standard curve of quercetin drawn within a concentration range of 0.2 to 1 mg/mL had an R^2 value of 0.986.

2.5. Condensed tannin content

In concentrated H_2SO_4, condensed tannins were turned by vanillin's reaction into anthocyanidols by following a slightly modified literature method [30]. Fifty µL of the phenolic extract appropriately diluted was taken in a test tube and mixed with 3 mL of 4% methanol vanillin solution and 1.5 mL of H_2SO_4. After 15 min, the absorbance was measured at 500 nm. Condensed tannin contents of all extracts (two replicates per treatment) were expressed as mg quercetin equivalents per gram (mg QE/g) using a standard curve of quercetin drawn within a concentration range of 0.2 to 1 mg/mL had an R^2 value of 0.987. duplicate measurements were taken for all samples.

2.6. Hydrolysis and identification of phenolic compounds using HPLC and GC-MS

The two dried samples of leaves and stems mixture (L+S) and leaves (L) for *Coriandrum Sativum* L. were acidic hydrolyzed according to modified literature method [31] to show the phenolic composition by GC/MS and HPLC. The acidic hydrolysis was followed to remove the aglycones. That will help identify the samples

Journal of Physics: Conference Series **2063** (2021) 012001 doi:10.1088/1742-6596/2063/1/012001

(L+S) and (L) simply due to the plants' phenolic compounds found as glycosides esters or bound to the cell wall. They are rarely present as free forms [32]. Twenty mL of methanol containing BHT (1 g/L) was added to 0.5 g of a dried sample. Then 10 mL of 1 M HCl was added. The mixture was stirred carefully and sonicated for 15 min and refluxed in a water bath at 90°C for 2 h. The obtained mixture was injected into HPLC and GC-MS.

2.6.1. Gas Chromatography-Mass Spectroscopy. Gas Chromatography-Mass Spectrometry (GC-MS) analysis was performed in Tehran University, Iran, by an Agilent 7890A\GCMS (USA). An instrument using a Hp 25 column and 1μL of each dried samples dissolved in ethanol was injected in the following conditions: injector temperature, 300°C; carrier gas, helium; pressure, 11.962 psi. Compounds were identified based on their mass spectral data.

2.6.2. HPLC analysis. Analysis of phenolic compounds was carried out in Tehran Unversity, Iran using A waters liquid chromatography apparatus consisting of a Separations module: waters 2695 (USA) and a PDA Detector waters 996 (USA). Data acquisition and integration were performed with Millennium 32 software. The injection was an autosampler injector equipped with the chromatographic. The assay was performed on a 15 cm×4.6 mm with pre-column, Eurospher 100-5 C18 analytical column provided by waters (Sunfire) reversed-phase matrix (3.5 μm) (Waters). And elution was carried out in a gradient system with methanol as the organic phase (solvent A) and distilled water (solvent B). The gradient programme was as follows: 15%A/85%B 0–12 min, 40%A/60%B 12–14 min, 60%A/40%B 14–18 min, 80%A/20%B 18–20 min, 90%A/10%B 20–24 min, 100%A 24–28 min [33]. The flow-rate was kept at 1 mL min^{-1}. Peaks were monitored at 280 nm wavelength. The injection volume was 20 μL, and the temperature was maintained at 25°C. Phenolic compounds were identified according to their retention times and their peaks' spectral characteristics against those of standards.

2.7. Preparation of copolymeric hydrogels U1-U3

Copolymeric hydrogel U1 loaded with *Coriandrum Sativum* L. extract salt was prepared by dissolving 2 gm acryl amide and 0.2 gm of the cross-linked agent (N, N\-methylene bisacrylamide) in 5 mL of distilled water with stirring. Then, 1 mL of *Coriandrum Sativum* L. extract salt (0.05 mg/mL) was added. 2 g of 2-Acrylamido-2-methylpropane sulfonic acid (AMPS) was dissolved in 5 mL of water and added to the acryl amide solution with constant stirring at 30-35°C. Then 1 mL ammonium persulfate (10% W/V) was added as an initiator of the polymerization reaction and then add 5 drops of N,N,N\,N\tetramethylenediamine (TMEDA) as an accelerating agent for initiator dissociation with good mixing for three minutes until the polymerization, and crosslinking process is complete. The gel was cut off and left to dry at laboratory temperature. See Scheme 1.

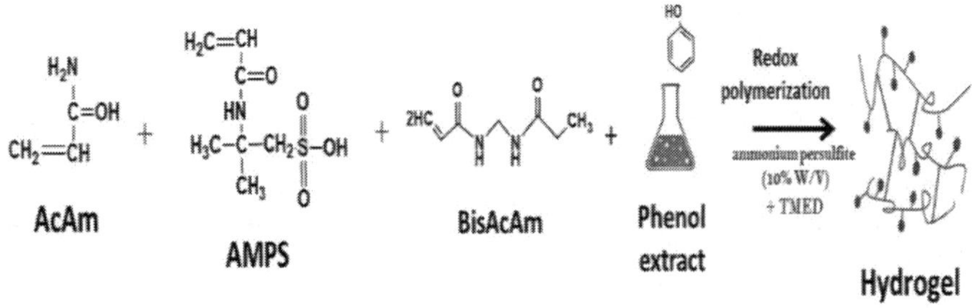

Scheme 1. The reaction of preparing the copolymeric hydrogels (U1-U3).

The U2 and U3 hydrogels were prepared at the same protocol with different monomers ratio, as shown in table 2.

Table 2. Composition of materials for the preparation of hydrogels U1-U3.

Hydrogels	AcAm (g)	Bis AcAm (g)	AMPS (g)	Extract salt (mL)	Total polymer weight (g)	Ratio AMPS:AcAm
U1	3	1.2	3	1	7.2	1:1
U2	6	1.8	3	1	10.8	1:2
U3	3	1.8	6	1	10.8	2:1

2.8. Preparation of Simulated Gastric Fluid (SGF) and Simulated Intestine Fluid (SIF)
The Simulated Gastric Fluid (SGF) and Simulated Intestinal Fluid (SIF) were prepared. According to the American Medicines Encyclopedia [34].

2.9. Scanning electron microscope (SEM) of (U1-U3) hydrogels
The prepared copolymeric hydrogels (U1-U3) were dipped in distilled water. After reaching the equilibrium swelling ratio, the hydrogels were removed from the distilled water, freezing and carried out under freeze-drying for 3 hours at (-48°C) to preserve their porous structure. Then, they are examined using a scanning electron microscope SEM.

2.10. Swelling and release studies of prepared hydrogels
The swelling ratio and release rate of the copolymeric hydrogels was studied in different pH at 37°C. The copolymeric hydrogels species were immersed in distilled water (pH=6.9), SGF (pH=2.1) and SIF (pH=8.2) solutions and at different time intervals. The swelling ratio was calculated according to the following equation [35].

$$Swelling\ ratio = \frac{Ws-Wd}{Wd} *100\%$$

Where Wd= dry weight of hydrogel and Ws = the weight of the swollen hydrogels. The release of the phenolic extract was followed using a UV-vis spectrophotometer with time intervals.

2.11. Set standard calibration curves for Coriandrum Sativum L. extract salt
The λ_{max} of the *Coriandrum Stavium* L. extract salt was determined using ultraviolet spectroscopy, where the extracted salt showed absorbance at the wavelength (636 nm) in distilled water. Simulated Gastric and Intestine Fluids showed absorbance at the wavelength (655 nm), (665 nm), respectively.

2.12. DPPH radical scavenging activity of phenolic extracts 4, 8, and the extracts released from the polymers
The obtained extracts electron donation ability was tested by bleaching the purple-coloured solution of 1,1-diphenyl-2 picrylhydrazyl radical (DPPH) in line with the method [36]. One mL different concentrations of extracts (1, 10, 100 and 200 μg/mL) prepared in ethanol were added to 1 mL of a 0.2 mmol/L DPPH ethanolic solution. The mixture was shaken vigorously and left standing at room temperature for 30 min. The absorbance of the resulting solution was then measured at 517 nm after 30 min. The antiradical was expressed as IC_{50} (mg/mL), the concentration required to cause a 50% DPPH inhibition. The ability to scavenge the DPPH radical was calculated using the following equation:

Journal of Physics: Conference Series **2063** (2021) 012001 doi:10.1088/1742-6596/2063/1/012001

$$\text{DPPH scavenging effect (\%)} = \{[A_0 - A_1] / A_0\} \times 100$$

A_0 is the absorbance of the control at 30 min, and A_1 is the absorbance of the sample at 30 min. BHT was used as a positive control.

2.13. Statistical analysis

Phenolic extracts were assays in duplicate, and results were expressed based on dry matter weight. Data are expressed as mean ± SD. The means were compared using the one-way and multivariate analysis of variance (ANOVA). The differences between individual means were deemed to be significant at $p < 0.05$. All analyses were performed by using the "Minitab 19" software.

3. Results and discussion

3.1. Total phenolic, total flavonoid and total tannin contents

Total phenolic contents in methanolic and ethanolic extracts of *Coriandrum Sativum* L. is shown in table 3. Total phenolic content expressed as mg gallic acid equivalent per gram dry weight. The extracts exhibited high phenol content with 9.822, 9.364 and 8.373 mg GAE/g DW in extract 7, 4 and 3, respectively. Thus, the order of the total phenolic content of all extracts determined by the Folin -Ciocalteu method is as follows: (L) ethanolic extracts > (L+S) ethanolic extracts > methanolic extracts for both (L) and (L+S).

Table 3. Total phenolic, flavonoid and tannin contents of *Coriandrum Sativum* L. parts.

No. of extracts	Type of Extracts	Total phenol content mg GAE/g DW	Total flavonoids content mg QE/g DW	Total tannin content mg QE/g DW
1	(L+S) in methanol at 30 min	5.921 ± 0.361	4.472 ± 0.121	2.650 ± 0.070
2	(L+S) in methanol at 24 h	4.689 ± 0.187	3.793 ± 0.928	4.245 ± 0.276
3	(L+S) in ethanol at 30 min	8.373 ± 0.168	3.422 ± 0.044	1.935 ± 0.077
4	(L+S) in ethanol at 24 h	9.364 ± 0.012	7.909 ± 0.160	1.840 ± 0.141
5	(L) in methanol at 30 min	7.308 ± 0.218	5.213 ± 0.186	1.480 ± 0.070
6	(L) in methanol at 24 h	4.015 ± 0.118	2.811 ± 0.371	1.135 ± 0.091
7	(L) in ethanol at 30 min	9.822 ± 0.634	7.957 ± 0.014	1.855 ± 0.077
8	(L) in ethanol at 24 h	7.726 ± 0.124	8.112 ± 0.115	1.855 ± 0.106

Values are duplicate ± standard deviation; data were compared statistically by one-way ANOVA at $p = 0.000$

In India, Jangra *et al.* analyzed the total phenols found in the ethanolic extract of *Coriandrum Sativum* L. leave ranged from 6.14-7.27 mg GAE/g DW of total phenol. And this is close to our results, while the extract with acetone and water was ranged from 5.06-6.18 mg GAE/g DW, 6.97-8.97) mg GAE/g DW respectively [37]. The other work demonstrated that the ethanolic extract of *Coriandrum Sativum* L. leaves 0.36 mg GAE/g DW of total phenol [38] and this is less than our present study. On the other hand, a study from Udaipur (Agrawal *et al.* 2016) found that the extraction solvents significantly affected the extract's polyphenol content. Flavonoids are one of the phenolic compounds that widely spread in leaves, seeds, bark, and plant flowers, and they are responsible for the colours of plants and the blue and red colours of berries and wines [39]. Quercetin dihydrate was taken as standard flavonoid, and results were represented as mg quercetin equivalent per gram dry weight (mg QE/g DW).

The flavonoid contents varied according to the fractions of *Coriandrum Sativum* L.and the solvent used. Extract 8 showed maximum flavonoids content of 8.112 mg QE/g DW, followed by extract 7 and 4 with mean value 7.957 and7.909 QE/g DW respectively, as shown in table 3. Ethanolic extract shows higher TFC than methanolic extracts. The leaves (L) extracts in ethanol and methanol showed more flavonoids substances than leaves and stems mixture (L+S) extracts. Jangra SS, Madan V, Singh I Analyzed the total flavonoids in India's study; they found in the ethanolic extract of *Coriandrum Sativum* L. leaves ranged from 8.35-9.41 mg QE/g DW [37]. These results were close to the present study results.

Total tannin content expressed as mg Quercetin equivalent per gram dry weight. In general, the leaves and stems mixture (L+S) extracts in ethanol and methanol showed more tannins substances than the leaves extracts (L). extract 2 showed maximum tannin content of 4.245 mg QE/g DW. There are no previous studies to estimate tannins in *Coriandrum Sativum* L. leaves and stems mixture or leaves.

3.2. Phenolic identification by HPLC and GC-MS

The results of GC-MS revealed that the (L+S) and (L) of *Coriandrum Sativum* L. have one phenolic compounds *O*-tert-Butyl-*p*-cresol contributing to about 21.27% in (L+S) and approximately 16.47% in (L). See Figure 1.

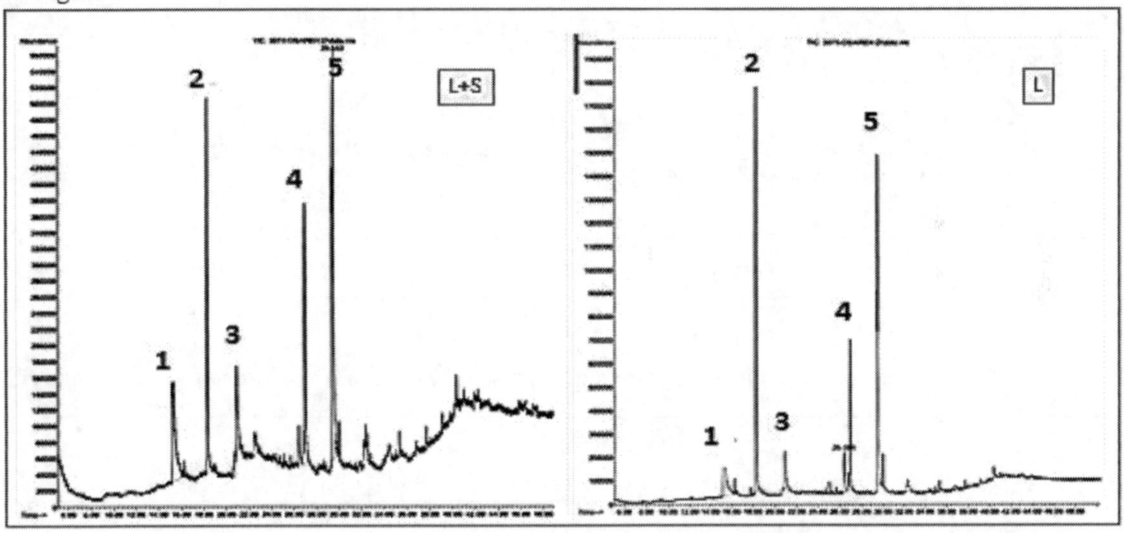

Figure 1. Gas chromatography and mass spectroscopy (GC-MS) for (L+S) and (L) dried samples Peaks numbers corresponding to: 1. *O*-tert-Butyl-*p*-cresol, 2. Butylated hydroxytoluene, 3. Citroflex 2, 4. Palmitic acid, ethyl ester, 5. Linolenic acid, ethyl ester. Peaks without numbers were not identified.

In addition, non-phenolic compounds in the *Coriandrum Sativum* L. were identified by comparing their mass spectra with the database of the spectrum of known components stored in the Gas Chromatography-Mass Spectrometry library, as shown in table 4.

Table 4. Phytocomponents identified for (L+S) and (L) dried samples of *Coriandrum Sativum* L. using GC-MS.

PK. No.	RT (min)	Area%		Name of compound
		(L+S)	**(L)**	
1	15.251	21.27	16.47	*O-tert*-Butyl-*p*-cresol
2	18.261	15.78	22.16	Butylated hydroxytoluene (BHT)
3	20.871	10.81	5.19	Citroflex 2
4	26.936	14.83	11.66	Palmitic acid, ethyl ester
5	29.453	37.30	40.28	Linolenic acid, ethyl ester

On the other hands, HPLC (L+S) and (L) dried samples revealed the presence of three phenolic compounds in both samples (L+S) and (L) as shown in figure 2.

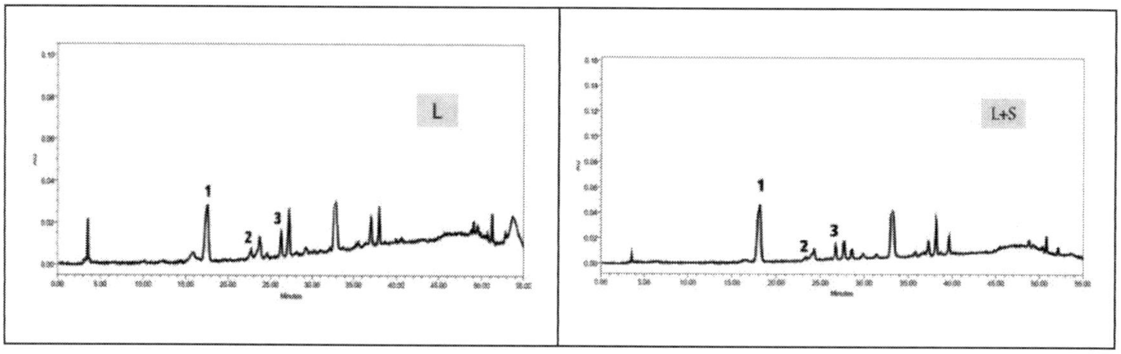

Figure 2. HPLC chromatogram of Coriandrum Sativum L. (L+S) and (L), the signal was collected at λ 195-400 nm. Peaks numbers corresponding to 1. Caffeic acid, 2. *p*-coumaric acid, 3. Chicoric acid. Peaks without numbers were not identified.

Table 5. HPLC chromatogram of Phenolic compounds present in *Coriandrum Sativum L.* (L+S) and (L).

PK.No.	Peak name	Retention Time (Min)		μg/5mg DW		Area [%]	
		(L+S)	(L)	(L+S)	(L)	(L+S)	(L)
1	Caffeic acid	18.150	17.555	16.99	10.59	86.01	75.23
2	*p*-coumaric acid	23.309	22.690	0.2	0.08	4.77	5.86
3	Chicoric acid	26.752	26.256	3.05	4.16	9.20	18.90

Caffeic acid contributing for about 86.016% (16.99 mg/100 g DW) in (L+S) and about 75.232% (10.59 mg/100 g DW) in (L). *P*–coumaric acid (4.779%, 0.2 mg/100 g DW) in (L+S) and (5.863%, 0.08mg/100 g DW) in (L), and chicoric acid (9.203%, 3.05mg/100 g DW) in (L+S) and about (18.903%, 4.16 mg/100 g

DW) in (L) were identified by comparison of their retention times with authentic standards analyzed under the same conditions, as shown in table 5.

3.3. Morphology of the prepared hydrogels

The copolymeric hydrogels were analysis by scanning electron microscopy. Figure 3 of prepared copolymeric hydrogels (U1-U3) find that an unregular shape with many holes results from their swelling. As the number of acrylamide decreases, the pore size decreases when compared between U1 and U3. The surface roughness becomes less in the case of U3 hydrogel due to the small pore size. Therefore, the U1 hydrogel gave a higher swelling ratio than the U3 hydrogel. SEM image of the U2 hydrogel in a shrinking state shows that the shrinking gel does not have any pores, so the surface appears regular and smooth except for some small bumps in nano size.

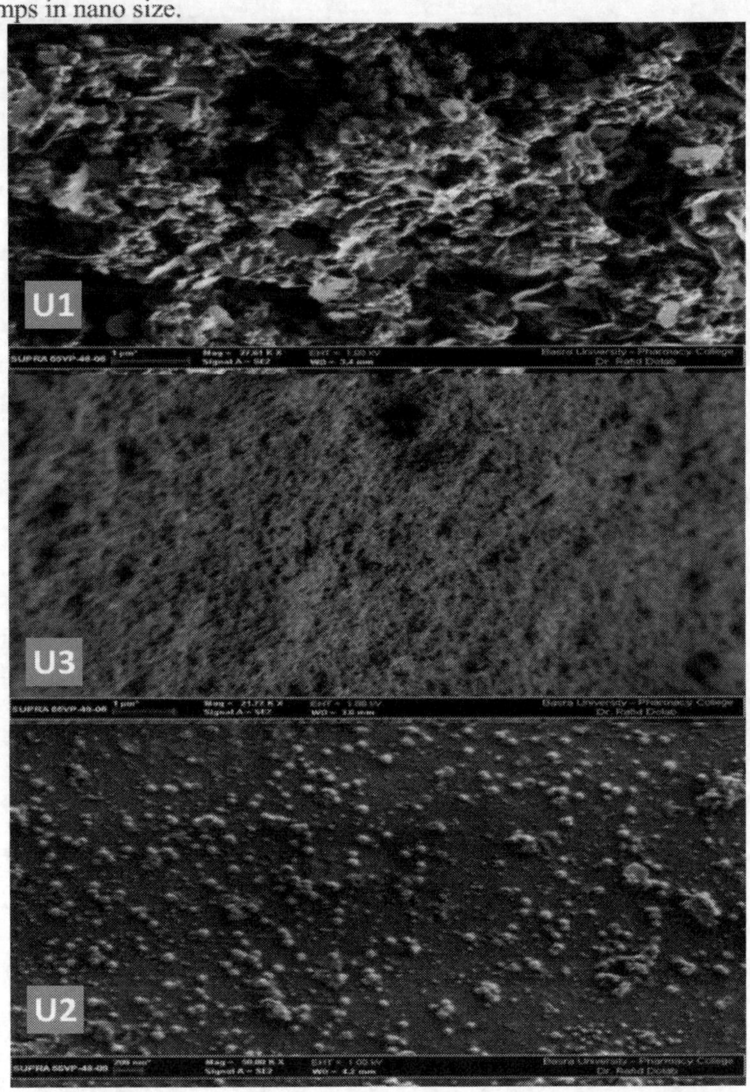

Figure 3. SEM images of U1, U3 in a swell state and U3 in a shrinking state.

Journal of Physics: Conference Series **2063** (2021) 012001 doi:10.1088/1742-6596/2063/1/012001

3.4. Bioapplications of Iraqi Coriandrum Sativum L. extracts and the copolymeric hydrogels loaded with extract salt

3.4.1. drug delivery system. The swelling ratio of the prepared copolymeric hydrogels was determined as a function of time in distilled water, SFG and SIF at 37°C and six hours. The pH of the swelling medium is one of the crucial factors affecting a hydrogel's swelling behaviour [40]. Scheme 2 shows the results of the swelling ratios of copolymeric hydrogels U1-U3.

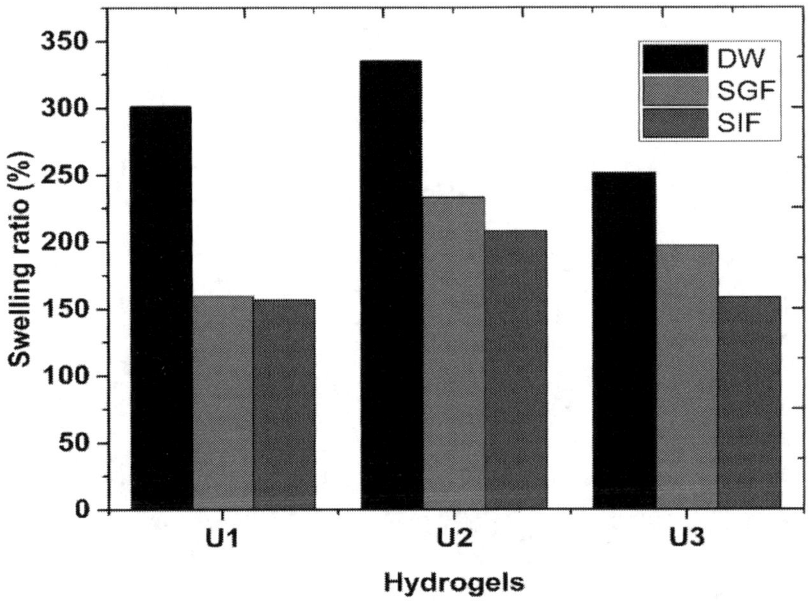

Scheme 2. **The swelling ratio for hydrogels (U1-U3) at SIF, SGF and distilled water**

The maximum swelling of copolymeric hydrogel was observed at pH 6.9 (in distilled water) due to ionization of $-SO_3H$ groups in AMPS. These sulfonate groups can be ionized quite easily, even in neutral pH, and it allures the polar water molecules to enter into the crosslinked network. The cage structure has been formed from the intermolecular H-bonding between the amide's carbonyl group and the AMPS unit acid moieties. It may lead to designs that can hold many water molecules [41]. Also, the presence of hydrophilic groups ($CONH_2$ secondary amine) that are present in the copolymeric hydrogels (AcAm-co-AMPS) have an impact on the swelling behavior [42], as the amide group enhances the hydrogen bonding between copolymeric structures and water molecules [43].

In an acidic medium (SGF, pH<7), the amino groups may protonate. Thus, the acidic protons of sulfonyl groups of AMPS interact with the amide group's nitrogen or oxygen from acrylamide, which prevents polymer water interaction. And the formation of the desirable H-bond responsible for accommodating water molecules and thus may be responsible for reducing equilibrium swelling [44, 45]. An increase in protonation amide groups will generate electrostatic repulsion between neighbouring ions in the hydrogel's polymer structure and the ions in the hypothetical gastric solution quaternary ammonium groups responsible for polymer swelling [46].

In a simulated intestinal fluid (pH=8.2), the acidic group's ionisation increases due to the deprotonation process. Many SO_3^- ions are formed, which repel each other, leading to the cage structure's disturbance and reduce swelling [41]. On further increases in AMPS content (beyond 10 mol%), the content of the strong ionic group-SO_3H will increase in the polymer network. Thus, maybe expected to create ionic repulsion between the similar charges leads to the increased distance between the functional groups, resulting in a gradual decrease in hydrogen bonds forming, which prevents the water molecule penetration inside the network [44]. Thus the swelling decreases gradually with an increasing amount of AMPS. While increasing the amount of AM, the hydrogels' swelling ratios are found to grow.

The released percentage of phenolic extract salt with time was studied at room temperature (37°C) in distilled water, SGF and SIF for nine hours. Figure 4 shows there are no significant differences in the rate of phenolic extract salt in SGF solution. So, there is no effect of the acidic medium on the release rate of phenolic extract from the copolymeric hydrogel.

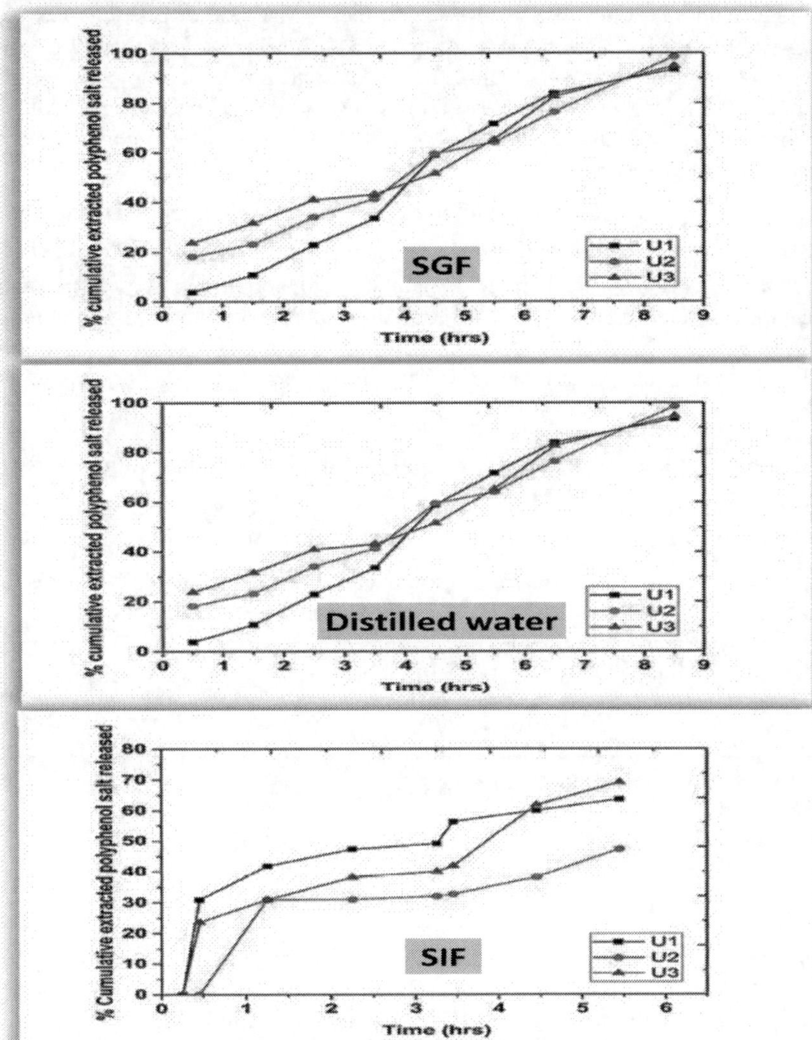

Figure 4. The release percentage of cumulative extracted phenol extract salt from copolymeric hydrogels U1-U3 in SGF, distilled water, and SIF.

Figure 4 indicates that by increasing the amount of acryl amide in the gel structure (U2), the phenolic extract's release rate is high distilled water. That maybe due to the high swelling ratio of the hydrogel U2 in distilled water compared to the other two hydrogels U1 and U3. Polyacrylamide has a very high swelling rate when immersed in water [47]. It leads to the formation of large pore size quickly, which explains that the highest release rate of phenolic extract salt from the hydrogel U1 compares with U3, which has the least amount of polyacrylamide.

The hydrogel U3 has a higher amount of AMPS. Thus, the possibility of forming hydrogen bonds between the gel and the phenolic extract salt will remain high due to the presence of non-ionized sulfonic acid groups and amine groups, so the rate of phenolic extract salt release decreases dramatically from within the hydrogel.

On the contrary, there are no noticeable differences in the release rate of phenolic extract salt from the U1 and U3 copolymeric hydrogels in the basic medium (SIF). Simultaneously, the U2 gel gives the lowest percentage of phenolic extract salt release because the sulfonic groups are completely ionized in the basic medium, which reduces the hydrogen bonds between the gel and the phenolic extract salt. Due to a more significant amount of acrylamide in the U2 hydrogel, maintaining the hydrogen bonds between the phenols and carbonyl groups remains possible.

Thus, the drug release depends on the composition of the hydrogel polymer and the active materials with hydrogels via physical interactions, the swelling ratio, and the pH of the environment [48].

The Higuchi release equation is the simplest mathematical equations. It was used to theoretically determine the percentage of released phenol extract salts as a function of time-based on the practical results obtained from a slow release of phenolic extract salt from all prepared copolymeric hydrogels in distilled water, SGF and SIF. Higuchi equation can be represented in the form:

$$Q_{(t)}=K_H \times t^{1/2}$$

$Q_{(t)}$ is the cumulative percentage phenol release, and K_H is the Higuchi dissolution constant [49].

Hence, if the correlation coefficient (R^2) is high for this plot, the release mechanism follows a diffusion control release mechanism [50, 51]. See table 6. Thus, the Higuchi model confirmed the release mechanism.

Table 6. Higuchi dissolution constant and correlation coefficient (R^2) for released phenolic extract salts from copolymeric hydrogel (U1-U3) in distilled water, SGF and SIF.

Hydrogel	Distilled Water		SGF		SIF	
	K_H	R^2	K_H	R^2	K_H	R^2
U1	68.816	0.9895	48.872	0.9530	19.825	0.9553
U2	32.189	0.9919	44.511	0.8799	27.776	0.9952
U3	46.534	0.9444	44.189	0.9750	35.283	0.9410

3.4.2. Antioxidant activity. Total antioxidant capacities of different parts of *Coriandrum Sativum* L. were significantly various. See table 7. The assay is based on reducing purple DPPH to a yellow-coloured diphenyl picrylhydrazine and the remaining DPPH. This ability was higher in extract 4 (77.46 mg GAE/g DW) and lowered in extract 8 (61.41 mg GAE/g DW). However, the DPPH scavenging abilities of the ethanolic extracts (L+S) and (L) were lower than those of synthetic antioxidant BHT.

Table 7. DPPH Radical Scavenging % at different concentration of *Coriandrum Sativum* L. phenolic extracts.

Phenolic extracts	DPPH Radical Scavenging % at Different Concentration of *C. Sativum* L. ethanol Extracts and BHT					IC_{50} µg/mL
	25 µg/mL	50 µg/mL	100 µg/mL	200 µg/mL	300 µg/mL	
Extract 4 (L+S)	16.3%	28%	36%	40%	59%	241.585 ± 0.007
Extract 8 (L)	3.2%	5%	14.2%	23.3%	32.4%	455.150 ± 0.071
BHT	38%	47%	59%	61%	65%	88.115 ± 0.163

Values are duplicate ± standard deviation; data were compared statistically by one-way ANOVA at p = 0.000

The high free radical scavenging activity of extract 4 is directly related to its high phenolic and flavonoid content (9.364 mg GAE/g DW) and (7.909 mg QE/g DW) where such a relationship has also been noted in other studies [9, 52].

There are no reports on the antioxidant capacities of *Coriandrum Sativum* L. for the composition of phenolic fraction present in the leaves and stems (L+S) and leaves (L) cultivated in Iraq.

The DPPH assay results for phenolic extracts released from the polymers (U1-U3) in distilled water, SGF, and SIF were presented in table 8.

Table 8. DPPH radical scavenging % activity for the phenolic extracts released from the copolymeric hydrogels.

sample	DPPH Radical Scavenging % of samples		
	in SGF	in SIF	in distilled water
Phenolic extracts released from U1	11.515 ± 0.276 at 88 µg/mL	17.1450 ± 0.0778 at 93 µg/mL	17.3150 ± 0.0212 at 93 µg/mL
Phenolic extracts released from U2	24.455 ± 0.488 at 100 µg/mL	8.3650 ± 0.0636 at 84 µg/mL	12.3900 ± 0.0141 at 88 µg/mL
Phenolic extracts released from U3	10.300 ± 0.283 at 85 µg/mL	14.5550 ± 0.0636 at 90 µg/mL	12.9050 ± 0.0212 at 88 µg/mL

Values are duplicate ± standard deviation; data were compared statistically by one-way ANOVA at p = 0.000

The scavenging activity on DPPH radicals for copolymeric hydrogels (U1-U3) ranged from 10.300 - 24.455% in SGF, from 8.3650-17.1450% in SIF, and from 12.3900-17.3150% in distilled water. These results are consistent with the results we obtained from the release study as the active extract is released from the copolymeric hydrogels. Thus, their radical scavenging activity increases.

4. Conclusion

The extraction methods play an essential role in extracting phenolic compounds, as it notices that different extraction methods give different in, yield, phenol, flavonoid, and tannin were obtained. The phytochemical, HPLC, and GC-MS analysis results showed phenolic compounds in Iraqi *Coriandrum Sativum* L. extracts. Hence, this enhances being a cure for many diseases. Ethanolic extracts of (L+S) and (L) showed more phenols and flavonoids content than methanolic extracts for both (L) and (L+S). Leaves and stems mixture (L+S) extracts in ethanol and methanol showed more tannins substances than the leaves extracts (L). SEM images of the prepared copolymeric hydrogels showed large pores that increased the speed of drug release. As AM increases, the hydrogels' swelling rates increase due to the hydrogen bonds between polymeric structures and water molecules. The ethanolic extract 4 (L+S) showed higher scavenging activity on DPPH radicals than extract 8 (L). The high free radical scavenging activity of extract 4 is directly related to its high phenolic and flavonoid content. Also, the study showed the scavenging activity of extracts released from the prepared copolymeric hydrogels.

References

[1] Jamshidi-Kia F, Lorigooini Z, Amini-Khoei H 2018, Medicinal plants: Past history and future perspective. *Journal of herbmed pharmacology*.**7**.

[2] Kumar N, Goel N 2019, Phenolic acids: Natural versatile molecules with promising therapeutic applications. *Biotechnology Reports*.**24**:e00370.

[3] Zhao Y, Yue Z, Zhong X, Lei J, Tao P, Li B 2020, Distribution of primary and secondary metabolites among the leaf layers of headed cabbage (Brassica oleracea var. capitata). *Food chemistry*.**312**:126028.

[4] Kiokias S, Proestos C, Oreopoulou V 2020, Phenolic acids of plant origin—a review on their antioxidant activity in vitro (O/W emulsion systems) along with their in vivo health biochemical properties. *Foods*.**9**:534.

[5] Fan Z, Wang Y, Yang M, Cao J, Khan A, Cheng G 2020, UHPLC-ESI-HRMS/MS analysis on phenolic compositions of different E Se tea extracts and their antioxidant and cytoprotective activities. *Food chemistry*.**318**:126512.

[6] Zińczuk J, Maciejczyk M, Zaręba K, Pryczynicz A, Dymicka-Piekarska V, Kamińska J, *et al* 2020, Pro-Oxidant Enzymes, Redox Balance and Oxidative Damage to Proteins, Lipids and DNA in Colorectal Cancer Tissue. Is Oxidative Stress Dependent on Tumour Budding and Inflammatory Infiltration? *Cancers*.**12**:1636.

[7] de Morais JS, Sant'Ana AS, Dantas AM, Silva BS, Lima MS, Borges GC, *et al* 2020, Antioxidant activity and bioaccessibility of phenolic compounds in white, red, blue, purple, yellow and orange edible flowers through a simulated intestinal barrier. *Food Research International*.**131**:109046.

[8] Önder A 2018, Coriander and its phytoconstituents for the beneficial effects. *Potential of essential oils*.165-85.

[9] Demir S, Korukluoglu M 2020, A comparative study about antioxidant activity and phenolic composition of cumin (*Cuminum cyminum* L.) and coriander (*Coriandrum sativum* L.). *Indian Journal of Traditional Knowledge (IJTK)*. **19**:383-93.

[10] Imankulova G, Moldabayeva Z, Mammadov R, Kassenov A, Utegenova A 2020, Nutritional value and antioxidant activity of brassica oleracea and *coriandrum sativum* vegetable crops. *EurAsian Journal of BioSciences*. **14**:901-6.

[11] Jeya K, Veerapagu M, Sangeetha V 2019, Antimicrobial and antioxidant properties of *Coriandrum sativum* L. seed essential oil. *American Journal of Essential Oils and Natural Products*. **7**:06-10.

[12] Kačániová M, Galovičová L, Ivanišová E, Vukovic NL, Štefániková J, Valková V, *et al* 2020, Antioxidant, antimicrobial and antibiofilm activity of coriander (*Coriandrum sativum* L.) essential oil for its application in foods. *Foods*.;**9**:282.

[13] Trifan A, BOSTĂNARU A-C, Luca SV, GRĂDINARU AC, JITĂREANU A, Aprotosoaie AC, *et al* 2020, Antifungal potential of *Pimpinella anisum*, *Carum carvi* and *Coriandrum sativum* extracts. A comparative study with focus on the phenolic composition. *Farmacia.*;**68**:22-7.
[14] Ahmed FBM, Alla AE-MJ 2019, Effectiveness of *Coriandrum sativum* Extract as Antimicrobial by Applying on Six Bacterial Strains. *Journal of Diseases and Medicinal Plants*. **5**:103-6.
[15] Verma A, Agarwal D, Dhanik J, Arya N, Nand V 2018, Antibacterial Efficacy and Cluster Analysis of Genotypic Extracts of Coriander Leaves and Seeds against Human Pathogenic Bacterial Strains. *Int J Curr Microbiol App Sci*. **7**:2727-36.
[16] Thuraisingam S, Sunilson JAJ, Kumari A, Anandarajagopal K 2019, Preliminary Phytochemical Analysis and Diuretic Activity of the Extracts of *Coriandrum Sativum* Leaves in Wistar Albino Rats. *Int Res J Pharm Med Sci*. **3**:1-3.
[17] Yella SST, Kumar RN, Ayyanna C, Varghese AM, Amaravathi P, Vangoori Y 2019, The combined effect of Trigonella foenum seeds and *Coriandrum sativum* leaf extracts in alloxan-induced diabetes mellitus wistar albino rats. *Bioinformation*. **15**:716.
[18] Sá-Caputo D, Dionello C, Morel D, Paineiras-Domingos L 2016, Possible Benefits of the Coriandrum sativum in the Management of Diabetes in Animal Model: A Systematic Review. *ARCHIVOS DE MEDICINA*. **2**:10.
[19] Tamura H, Maeyama K, Yoshida E, Kori M 2013, Aroma character of coriander (*Coriandrum sativum* L.) leaves: limited odor unit method and sensory perception in preference. *Nutrition, Functional and Sensory Properties of Foods*. p. 66-73.
[20] Sahoo S, Brijesh S 2020, Anxiolytic activity of Coriandrum sativum seeds aqueous extract on chronic restraint stressed mice and effect on brain neurotransmitters. *Journal of Functional Foods*.**68**:103884.
[21] Sakurai J, Izumo N, Wanadabe Y 2019, Effect of Coriandrum Sativum L. leaf extract on the brain GABA neurons in mice. *J Nutr Health Food Sci*. **7**:1-7.
[22] Dhyani N, Parveen A, Siddiqi A, Hussain ME, Fahim M 2020, Cardioprotective Efficacy of Coriandrum sativum (L.) seed extract in heart failure rats through modulation of endothelin receptors and antioxidant potential. *Journal of dietary supplements*. **17**:13-26.
[23] Narayanaswamy R, Torchilin VP 2019, Hydrogels and their applications in targeted drug delivery. *Molecules*. **24**:603.
[24] Ziminska M, Wilson JJ, McErlean E, Dunne N, McCarthy HO 2020, Synthesis and Evaluation of a Thermoresponsive Degradable Chitosan-Grafted PNIPAAm Hydrogel as a "Smart" Gene Delivery System. *Materials*.**13**:2530.
[25] Spicer CD 2020, Hydrogel scaffolds for tissue engineering: the importance of polymer choice. *Polymer Chemistry*.**11**:184-219.
[26] Liang Y, Chen B, Li M, He J, Yin Z, Guo B 2020, Injectable antimicrobial conductive hydrogels for wound disinfection and infectious wound healing. *Biomacromolecules*. **21**:1841-52.
[27] Mau J-L, Chao G-R, Wu K-T 2001, Antioxidant properties of methanolic extracts from several ear mushrooms. *Journal of Agricultural and Food Chemistry*. **49**:5461-7.
[28] Bettaieb I, Bourgou S, Wannes WA, Hamrouni I, Limam F, Marzouk B 2010, Essential oils, phenolics, and antioxidant activities of different parts of cumin (*Cuminum cyminum* L.). *Journal of agricultural and food chemistry*. **58**:10410-8.
[29] Dewanto V, Wu X, Adom KK, Liu RH 2002, Thermal processing enhances the nutritional value of tomatoes by increasing total antioxidant activity. *Journal of agricultural and food chemistry*. **50**:3010-4.
[30] Sun B, Ricardo-da-Silva JM, Spranger I 1998, Critical factors of vanillin assay for catechins and proanthocyanidins. *Journal of agricultural and food chemistry*.**46**:4267-74.

[31] Proestos C, Boziaris I, Nychas G-J, Komaitis M 2006, Analysis of flavonoids and phenolic acids in Greek aromatic plants: Investigation of their antioxidant capacity and antimicrobial activity. *Food chemistry*.**95**:664-71.

[32] Nuutila A-M, Kammiovirta K, Oksman-Caldentey K-M 2002, Comparison of methods for the hydrolysis of flavonoids and phenolic acids from onion and spinach for HPLC analysis. *Food chemistry*.**76**:519-25.

[33] Bourgou S, Ksouri R, Bellila A, Skandrani I, Falleh H, Marzouk B 2008, Phenolic composition and biological activities of Tunisian *Nigella sativa* L. shoots and roots. *Comptes Rendus Biologies*.**331**:48-55.

[34] Europe. Co 2005, *European pharmacopoeia*. Print book : English : 5th ed ed: Strasbourg : Council of Europe, 2005.

[35] Nogueira CCdS, Cabral LM, Santos TCd, Marucci A, Alhaique F 2003, Evaluation of new polysaccharides networks for extended-release purposes: mesquite seed gum (MSG), xanthan gum and chitosan. *Revista Brasileira de Ciências Farmacêuticas*.**39**:273-88.

[36] Shimada K, Fujikawa K, Yahara K, Nakamura T 1992, Antioxidative properties of xanthan on the autoxidation of soybean oil in cyclodextrin emulsion. *Journal of agricultural and food chemistry*. **40**:945-8.

[37] Jangra SS, Madan V, Singh I 2018, Comparative Analysis of Phytochemical Profile and Antioxidant Activity of Coriander (*Coriandrum sativum* L.). *Asian Journal of Chemistry*.**30**.

[38] Wangensteen H, Samuelsen AB, Malterud KE 2004, Antioxidant activity in extracts from coriander. *Food chemistry*.**88**:293-7.

[39] Iwashina T 2015, Contribution to flower colors of flavonoids including anthocyanins: a review. *Natural product communications*.**10**:1934578X1501000335.

[40] Rizwan M, Yahya R, Hassan A, Yar M, Azzahari AD, Selvanathan V, *et al* 2017, pH sensitive hydrogels in drug delivery: Brief history, properties, swelling, and release mechanism, material selection and applications. *Polymers*.**9**:137.

[41] Bera R, Dey A, Chakrabarty D 2017, Tuning of the swelling and dye removal efficacy of poly (acrylamide-AMPS)-based smart hydrogel. *Separation Science and Technology*.**52**:743-55.

[42] Chen J, Liu M, Liu H, Ma L, Gao C, Zhu S, *et al* 2010, Synthesis and properties of thermo-and pH-sensitive poly (diallyldimethylammonium chloride)/poly (N, N-diethylacrylamide) semi-IPN hydrogel. *Chemical Engineering Journal*.**159**:247-56.

[43] Bhadani R, Mitra UK 2016, Synthesis and studies on water swelling behaviour of polyacrylamide hydrogels. *Macromolecular Symposia*: *Wiley Online Library*.**369**: p. 30-4.

[44] Devi N, Narzary A 2012, Release dynamics of brufen from a drug-loaded polymer hydrogel containing polyvinyl alcohol, 2-acrylamide-2-methylpropane sulfonic acid and acrylamide. *International Journal of Polymeric Materials*.**61**:821-33.

[45] Anirudhan TS, Rejeena SR 2014, Poly (acrylic acid-co-acrylamide-co-2-acrylamido-2-methyl-1-propanesulfonic acid)-grafted nanocellulose/poly (vinyl alcohol) composite for the in vitro gastrointestinal release of amoxicillin. *Journal of Applied Polymer Science*.**131**.

[46] Varaprasad K, Reddy NN, Ravindra S, Vimala K, Raju KM 2011, Synthesis and characterizations of macroporous poly (acrylamide-2-acrylamido-2-methyl-1-propanesulfonic acid) hydrogels for in vitro drug release of ranitidine hydrochloride. *International Journal of Polymeric Materials*. **60**:490-503.

[47] Jayaramudu T, Ko H-U, Kim HC, Kim JW, Kim J 2019, Swelling behavior of polyacrylamide–cellulose nanocrystal hydrogels: swelling kinetics, temperature, and pH effects. *Materials*.**12**:2080.

[48] Zaid A, Athir M 2018, Preparation of Polyelectrolyte Hydrogels and Study Their Controlled Release of Gabapentin. *Drug Des Int Prop Int J* **1 (2)**. DDIPIJ MS ID. 108.

[49] Paarakh MP, Jose PA, Setty C, Christoper G 2018, Release kinetics–concepts and applications. *Int J Pharm Res Tech*.**8**:12-20.

[50] Paul D 2011, Elaborations on the Higuchi model for drug delivery. International journal of pharmaceutics. **418**:13-7.
[51] Gouda R, Baishya H, Qing Z 2017, Application of mathematical models in drug release kinetics of carbidopa and levodopa ER tablets. *J Dev Drugs*.**6**:1-8.
[52] Mitrevska K, Grigorakis S, Loupassaki S, Calokerinos AC 2020, Antioxidant Activity and Polyphenolic Content of North Macedonian Wines. *Applied Sciences*.**10**:2010.

Journal of Physics: Conference Series **2063** (2021) 012002 doi:10.1088/1742-6596/2063/1/012002

Simulation Study of Structural and Electronic Properties for Adducts complexes of Bis(Acetylacetonato)oxoVandium (IV) with 4-(*Para*-substituted phenyl)-1,2,3-Selenadiazole

Dalal H Alsawad, Ali A Al-Riyahee, Ali J Hameed*

Department of Chemistry, College of Science, University of Basrah, Basrah, Iraq

Email: ali.alhamad@uobasrah.edu.iq

Abstract. A series of 4-(*para*-substituted phenyl)-1,2,3-selenadiazole adducts of [$VO(acac)_2$] were studied by density functional theory (DFT) calculations. The 4-(*para*-substituted phenyl)-1,2,3-selenadiazole molecules have been selected to be bound with vanadium atom in [$VO(acac)_2$] through Se, N2 and N3. The resulting adducts have been investigated in two geometries (*cis* and *trans*) in order to show the effect of such structural change on the electronic properties of the studied adducts. The optimized geometries, (binding and reorganization) energies and the spatial distribution of the highest molecular orbital (HOMO) and lowest unoccupied molecular orbital (LUMO) of the adducts are presented and discussed.

1. Introduction

Coordination complexes are of great significance in different fields due to their presence in many shapes and structures. There are numerous examples of coordination complexes in biochemistry field. For example, the iron, magnesium, cobalt and copper coordination complexes are demonstrated by hemoglobin, chlorophyll, Vitamin B-12 and hemocyanin respectively [1-4].

The Vanadyl moiety in coordination complexes has been extensively studied. Bis(acetylacetonato)oxidovanadium (IV), or [$VO(acac)_2$], is one of the most important vanadium complexes [5]. X-ray diffraction studies on the [$VO(acac)_2$] complex clearly show the existence and stability of VO^{+2} entity in the solid state. The geometry of complex demonstrates as an almost square pyramidal geometry [6]. Many applications have been demonstrated for the [$VO(acac)_2$] complex. It has been used as the vanadium precursor for the preparation of different vanadium compounds [7-9] and also it has been used as a catalyst precursor in organic reactions [10-12]. Furthermore, Vanadium complexes show a wide variety of biological applications, and numerous complexes have been investigated as anti-parasitic, spermicidal, antiviral, anti-HIV, anti-tuberculosis, and antitumor agents [13].

Organo heterocyclic compounds of selenium which containing nitrogen such as selenirenes, selenophenes, selenadiazoles, selenatriazoles and benzisoselanazolones can have many biological effects. Among such features they are active immunostimulants, inhibitors of enzymes, antioxidants, anti-inflammatory, antitumor, antiviral and antimicrobial agents [14].

Bis(acetylacetonato)oxidovanadium(IV), or [$VO(acac)_2$] is known to interact with various ligand containing donor atoms like N, P, O, S and Se to form adduct compounds. Although some vanadium-chalcogene complexes have been synthesized and characterized, only very few examples with analogous organo selenium derivatives as ligands have been described [15-18].

In this work, a series of 4-(*para*-substituted phenyl)-1,2,3-selenadiazole adducts of [$VO(acac)_2$] have been investigated theoretically in order to show the effect of such structural change on the structural and electronic properties of the adducts.

2. Computational Methods

The calculations presented in this paper have been carried out by using hyperchem program 7.5 [19]. The geometry of the 4-(*para*-substituted phenyl)-1,2,3-selenadiazole ligands were optimized by carrying out the semiempirical molecular orbital theory at the PM3 level [20], using the restricted Hartree–Fock (RHF) procedure [21]. The Polak–Ribier algorithm [22] was used for the optimization, with the termination condition being a root mean square (RMS) of <0.001 kcal/mol. Further, Geometry optimization was done by performing the B3LYP/3-21G theory method [23-25]. Concerning the adduct molecules, the initial geometry optimization carried out with the molecular mechanics (MMC) force field [26-27], where the lowest energy conformations are obtained. Further, Geometry optimization was done was done by performing the B3LYP/STOG and B3LYP/3-21G theory methods. All calculations have been done on Pc, Intel(R) Core(TM) i3-3220 CPU @ 3.30GHz. Due to computational limitations and the large size of the studied adduct molecules, we were unable to achieve higher level optimizations on the adducts molecules.

3. Result and discussion

Scheme 1 shows the 4-(*para*-substituted phenyl)-1,2,3-Selenadiazole ligands. These ligands have been investigated theoretically to study their structural and electronic properties. Table 1 shows the calculated proton affinities of the substituted ligands according to semiemperical PM3 and BLYP/3-21G levels, the highest occupied and the lowest unoccupied molecular orbital energies (HOMO and LUMO respectively), their calculated hardness (1/2 the HOMO-LUMO gap). The values of hardness can give an information about the molecules if they are soft or hard, where the molecules which have a large HOMO-LUMO gap are hard while the molecules which have a small HOMO-LUMO gap are soft [28]. It can be seen clearly that electron-donating and electron-releasing groups have small effects on the hardness values for the calculations methods. For the PM3 calculation the trend of hardness of the substituted 1,2,3-selenadiazole ligands decrease in the order $CF_3>H>Br>F>CN>CH_3>Cl>OCH_3>Ph>NH_2$; while for BLYP/3-21G the hardness decrease in the order $H>CF_3>CN>F>CH_3>Cl>Br>OCH_3>Ph>NH_2$.

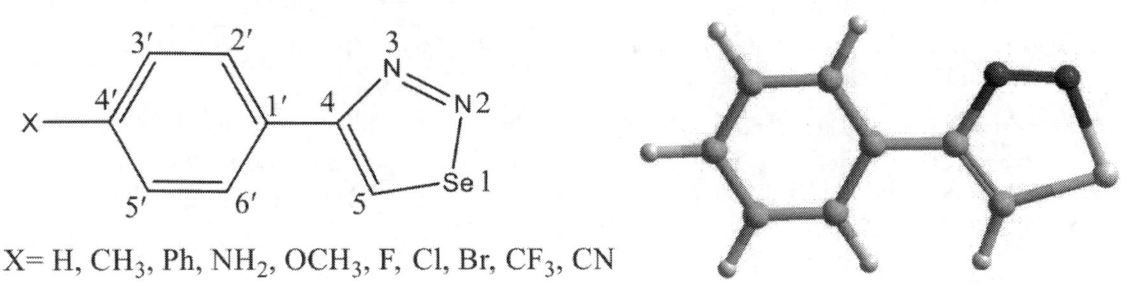

Scheme 1. The general structure of 4-(*para*-substituted phenyl)-1,2,3-selenadiazole molecules (left) and optimized molecular structure of 4- phenyl-1,2,3-selenadiazole molecules (right)

The values of partial charges of selenium and nitrogen atoms are also given in Table 1. For both type of calculations, selenium atom indicates a positive charge. The Positive charge on the selenium atom should considerably decrease their base strengths electrostatically. Nitrogen is a better donor atom than Se. N2 atom possess negative charge vary from -0.092 to -0.085 and from -0.302 to -0.297 according to PM3 and BLYP/3-21G calculations respectively. N3 atom possess positive charge vary from 0.097 to 0.092 and negative charge vary from -0.294 to -0.272 according to PM3 and BLYP/3-21G calculations respectively.

Journal of Physics: Conference Series **2063** (2021) 012002 doi:10.1088/1742-6596/2063/1/012002

Table 1. Selected calculated energy, hardness and parial charges of the 4-(*para*-substituted phenyl)-1,2,3-selenadiazole ligands according to semiempirical PM3 and BLYP/3-21G calculation.

X	Methods	E_{HOMO}	E_{LUMO}	Hardness	Mulliken charge			PA[a] (kcal/mol)	
		eV	eV	eV	Se	N2	N3	N2	N3
H	PM3	-9.357	-1.280	4.039	0.158	-0.090	0.097	172.90	178.20
	3-21G	-6.222	-1.455	2.384	0.671	-0.296	-0.271	197.92	215.29
CH_3	PM3	-9.198	-1.251	3.974	0.156	-0.090	0.097	173.82	179.24
	3-21G	-6.009	-1.379	2.315	0.667	-0.297	-0.273	199.88	217.47
Ph	PM3	-8.921	-1.347	3.787	0.158	-0.090	0.096	174.21	181.78
	3-21G	-5.802	-1.750	2.026	0.652	-0.297	-0.289	201.11	220.32
NH_2	PM3	-8.226	-1.038	3.594	0.146	-0.092	0.092	178.57	185.30
	3-21G	-4.990	-1.327	1.832	0.630	-0.302	-0.294	194.33	209.03
OCH_3	PM3	-9.00	-1.221	3.890	0.155	-0.090	0.095	174.63	180.08
	3-21G	-5.655	-1.316	2.170	0.664	-0.297	-0.274	194.19	211.77
F	PM3	-9.444	-1.433	4.006	0.165	-0.087	0.095	170.76	175.57
	3-21G	-6.166	-1.529	2.319	0.675	-0.296	-0.273	195.86	212.71
Cl	PM3	-9.250	-1.395	3.928	0.163	-0.088	0.096	171.95	177.07
	3-21G	-6.253	-1.638	2.308	0.679	-0.294	-0.272	198.64	215.42
Br	PM3	-9.470	-1.433	4.019	0.165	-0.088	0.097	171.09	176.15
	3-21G	-6.141	-1.595	2.273	0.679	-0.295	-0273	197.83	214.86
CF_3	PM3	-9.810	-1.674	4.068	0.175	-0.085	0.098	167.68	172.28
	3-21G	-6.608	-1.884	2.362	o.690	-0.293	-0.269	195.46	212.17
CN	PM3	-9.654	-1.656	3.999	0.173	-0.086	0.097	168.50	173.19
	3-21G	-6.668	-2.031	2.319	0.710	-0.305	-0.282	190.62	207.06

[a] PA are the values of ΔH and ΔE for PM3 and 3-21G calculations respectively for the reaction:
$BH^+ \longrightarrow B + H^+$(B selenadiazole)

The electron-releasing and electron-withdrawing effects of substituted groups are not reflected very well on the charges of both nitrogen and selenium atoms. The values of calculated proton affinity for N2 and N3 atoms are listed in Table 1. According to the PM3 and BLYP/3-21G calculations, N3 atom indicated the greatest values of proton affinity which means that N3 is more basicity than N2.

A few studies have been achieved to study the coordination chemistry of 1,2,3-selenadiazole ligands [29]. These studies indicated that the coordination between metal and 1,2,3-selenadiazole ring might be through N2 in 4-methyl-1,2,3-selenadiazole [30-32] or through N3 in 4-(2-pyridyl)-1,2,3-selenadiazole [33]. Connecting this N3 atom into coordination becomes more likely because of the metal chelation with the pyridinyl nitrogen and the nearby selenadiazole-N3.

In an attempt to detect the donor atom on 1,2,3-selenadiazole ring which is coordinating with $[VO(acac)_2]$, various 4-(*para*-substituted phenyl)1,2,3-selenadiazole molecules have been selected to be bound with vanadium atom in $[VO(acac)_2]$ through Se, N2 and N3. The resulting molecules have been selected to adopt *cis* or *trans* isomer.

Scheme 2 denotes the structures formula of *cis* and *trans* $[VO(acac)_2L]$ adducts. The donor atom (Se, N2 and N3) of 4-(*para*-substituted phenyl)-1,2,3-selenadiazole may be coordinate *cis* or *tans* to the O-oxido. The studied molecules have been optimized by performing the molecular mechanics force field, where the lowest energy conformations are obtained.

Journal of Physics: Conference Series **2063** (2021) 012002 doi:10.1088/1742-6596/2063/1/012002

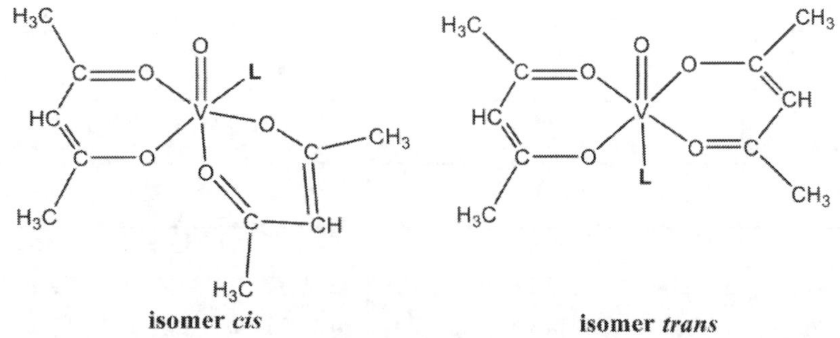

L= *para*-Substituted phenyl -1,2,3-Selenadiazole

Scheme 2. Structures formula of *cis* and *trans* [VO(acac)$_2$**L**] adducts, the donor atoms (Se, N2, N3) in 4-(*para*-substituted phenyl)-1,2,3-Selenadiazole may coordinate ***cis*** or ***tans*** to the O-oxido.

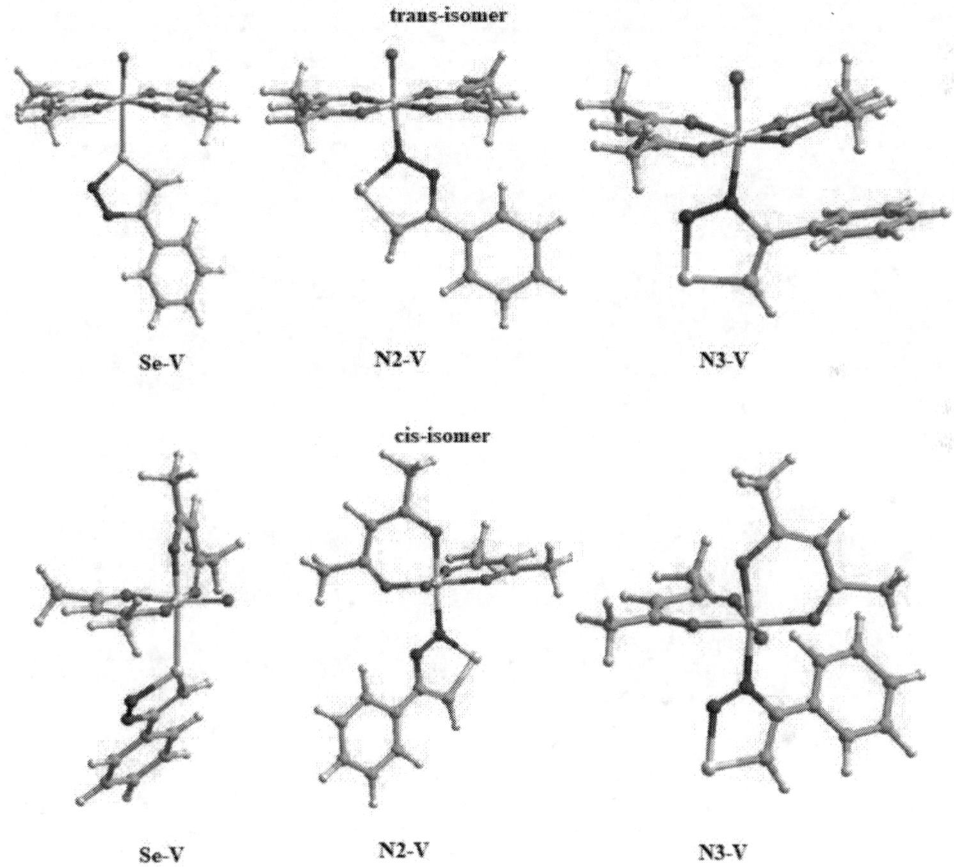

Figure1.Optimized structures formula of ***cis*** and ***trans***-[VO(acac)$_2$L] adducts isomers, the donor atoms (Se, N2, N3) in 4-(*para*-substituted phenyl)-1,2,3-selenadiazole may coordinate *cis* or *tans* to the O-oxido.

Figure 1 presents the optimized structures formula of *cis* and *trans* [VO(acac)$_2$L] adduct. L is 4-phenyl-1,2,3-selenadiazole.

Table 2. Total energy of conformations (in kcal/mol) of *cis* and *trans* [VO(acac)$_2$**L**] adducts, the donor atom (Se, N2 or N3) of 4-(*para*-substituted phenyl)-1,2,3-selenadiazole may coordinate *cis* or *tans* to the O-oxido.

X	***trans*-isomer**			***cis*-isomer**		
	Se-V	N2-V	N3-V	Se-V	N2-V	N3-V
H	50.377	47.993	50.862	51.046	23.988	39.657
CH_3	52.283	49.719	52.020	50.777	22.3659	39.122
Ph	60.881	58.343	60.010	49.901	23.306	37,867
NH_2	51.365	48.871	50.953	55.923	28.912	45.028
OCH_3	52.640	50.188	52.170	55.062	28.398	43.160
F	50.286	47.791	49.988	50.988	23.978	39.934
Cl	50.257	47.784	50.139	51.379	24.363	40.198
Br	50.050	47.549	49.512	51.512	24.480	40.237
CF_3	52.701	50.153	52.338	55.711	28.618	43.479
CN	53.185	49.524	51.415	51.566	24.286	39.468

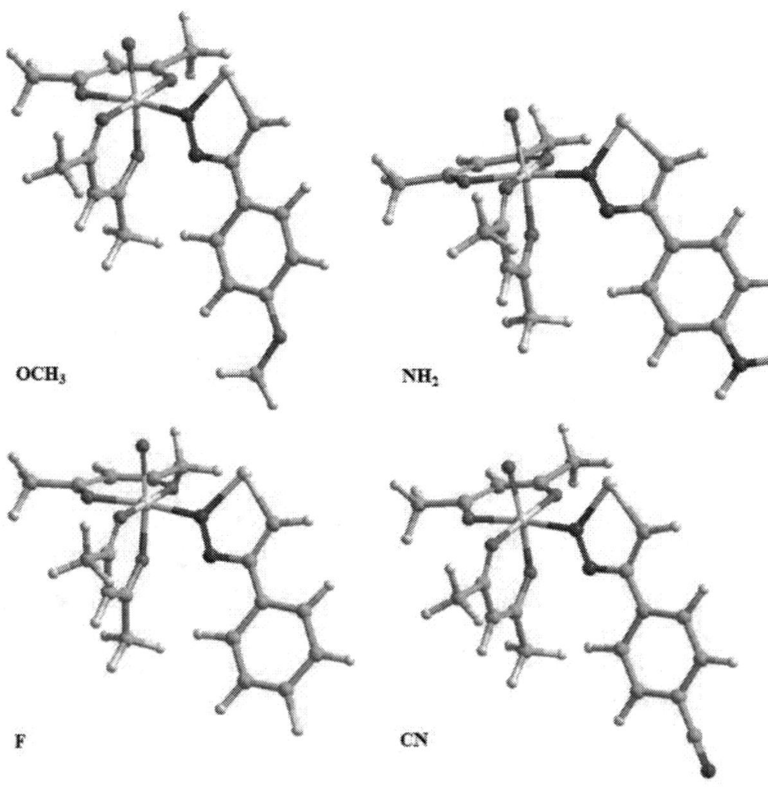

Figure 2. Optimized structures formula of ***cis***-[VO(acac)$_2$L] adduct. L is 4-(*para*-substituted phenyl)-1,2,3-selenadiazole.

Table 2 listed the total energy of *cis* and *trans* [VO(acac)$_2$**L**] adducts conformations. The results of the computed adducts revealed that all isomers of *cis* and *trans* adopted distorted octahedral. It can be seen clearly form Table 2 that the *cis*-adduct isomers have the lowest total energy in comparison with *trans*-adduct isomers. It's means that these molecules may be more stable compared with other studied molecules. Furthermore, the *cis*-adducts isomers with N2-V binding have the lowest energy in comparison with the other *cis*-adducts isomer with Se-V or N3-V binding. Thus, the coordination of $[VO(acac)_2]$ with 4-(*para*-substituted phenyl)1,2,3-selenadiazole molecules is more likely to be through N2 atom on the selenadiazole ring.

Table 3. Selected geometrical parameters of *cis*-$[VO(acac)_2\mathbf{L}]$ adducts, L is 4-(*para*-substituted phenyl)-1,2,3-selenadiazole calculated using BLYP/3-21G calculation. N2 atom is coordinated *cis* to the O-oxido.

X	H	CH_3	Ph	NH_2	OCH_3	F	Cl	Br	CF_3	CN
Bond distance (Å)										
V=O (oxo)	1.789	1.789	1.789	1.789	1.789	1.789	1.789	1.789	1.789	1.789
V-O [a] (ketonic)	1.854	1.854	1.854	1.855	1.854	1.854	1.854	1.855	1.854	1.854
V-O [b] (enolic)	1.850	1.850	1. 854	1. 850	1. 851	1. 850	1. 850	1.850	1. 854	1. 850
V-N2	1.917	1.917	1.920	1.917	1.917	1.917	1.917	1.918	1.919	1.917
Se-N2	1.834	1.834	1.835	1.833	1.834	1.834	1.833	1.834	1.834	1.834
N2=N3	1.252	1.253	1.253	1.252	1.252	1.252	1.252	1.252	1.253	1.252
N3-C4	1.267	1.267	1.267	1.267	1.267	1.267	1.267	1.267	1.267	1.267
C4=C5	1.342	1.342	1.342	1.342	1.342	1.342	1.342	1.342	1.342	1.342
C5-Se	1.879	1.879	1.879	1.880	1.879	1.880	1.880	1.879	1.879	1.879
Bond angle (°)										
O=V-O	170.08	169.85	177.90	170.00	169.95	170.28	170.19	170.83	176.09	169.72
O-V-O	96.09	96.09	80.23	95.97	96.03	95.98	95.91	95.81	95.78	90.03
O=V-N2	98.20	98.28	91.25	98.26	98.46	98.10	98.16	91.27	90.69	98.34
V-N2-Se	121.85	121.71	122.43	121.89	121.89	121.86	121.88	122.35	122.53	121.89
V-N2=N3	120.12	120.25	119.98	120.10	120.08	120.13	120.10	119.76	119.77	120.07
Se-N2=N3	118.03	118.04	117.59	118.01	118.04	120.13	118.01	117.89	117.71	118.03
N2-Se-C5	78.88	78.89	79.13	78.88	78.89	78.89	78.87	78.96	70.07	78.89
Se-C5=C4	113.34	113.34	113.19	113.36	113.31	113.34	113.37	113.32	113.23	113.32
C5=C4-N3	113.95	113.97	113.96	113.91	114.00	113.94	113.91	113.93	113.96	113.98
Bond dihedral (°)										
N3-C4′-C1′-C2′	-0.17	-0.36	0.10	-0.16	0.15	-0.09	-0.10	-0.07	0.48	-0.30
C5-C4-C1′-C6′	-0.13	-0.21	0.00	-0.05	0.00	-0,09	-0.05	0.03	-0.23	-0.21

Further geometry optimization has been carried out for the *cis*-$[VO(acac)_2L]$ adducts at the B3LYP/STOG and B3LYP/3-21G levels of theory in order to study the structural and electronic properties of resulting adducts. Figure 2 presents the optimized structures formula of *cis*-$[VO(acac)_2L]$ adduct. Table 3 shows the selected geometrical parameters of *cis*-$[VO(acac)_2L]$ adducts, calculated using BLYP/3-21G method. The experimental data of crystallography of $[VO(acac)_2]$ and 4-(4-Chlorophenyl)-1,2,3-selenadiazole moieties have reported [34,35]. It is very useful to compare these data with the theoretical data that calculated by B3LYP/3-21G method.

The optimized structures of the studied molecules indicate to have a distorted octahedral geometry. The bonds angle of O=V-O and O-V-O gave values in the range of 169.72-177.9 and 80.23-96.09

respectively. As shown in Table 3, the length of V=O (oxo) bond was 1.789 Å compared with 1.586 Å in free [VO(acac)$_2$] [25]. The calculated bond length of V-O (ketonic) and V-O (enolic) bonds in the adducts vary from 1.850 to 1.855 Å compared with 1.973 Å and 1.967 Å respectively in the free [VO(acac)$_2$]. Concerning 1,2,3- selenadiazole ring, the experimental bonds length of Se-N2, N2=N3, N3-C4, C4=C5 and C5-Se were 1.857, 1.265, 1.381 and 1.353 Å respectively in 4-(4-Chlorophenyl)-1,2,3-selenadiazole [26]. On the other hand, the experimental bonds angle of Se-N2=N3, N2-Se-C5, Se-C5=C4 and C5=C4-N3 were 111, 86.42, 111.9, and 113.1 (°) respectively in 4-(4-Chlorophenyl)-1,2,3-selenadiazole [26]. All these values are in good agreement with the calculated values, see Table 3. Finally, there is no clear trend for the variation of electronic donating or withdrawing groups on the structural properties of the studied adducts.

Some of the calculated energies of *cis*-[VO(acac)$_2$L] adducts using BLYP/STOG and BLYP/3-21G calculations are given in Table 4. The Binding energy of adducts formation and the reorganization energy of 4-(*para*-substituted phenyl)-1,2,3-Selenadiazole and [vo(acac)$_2$] in the adducts using BLYP/3-21G calculation are given in Table 5.

Table 4. Some of the calculated energies of *cis*-[VO(acac)$_2$**L**] adducts, L is 4-(*para*-substituted phenyl)-1,2,3-selenadiazole calculated using BLYP/STOG and BLYP/3-21G calculations. N2 atom is coordinated *cis* to the O-oxido.

X	Methods	Binding energy kcal/mol	E_{HOMO} eV	E_{LUMO} eV
H	STOG	-2736528.57	-5.661	-3.884
	3-21G	-2506056.73	-9.762	-7.882
CH_3	STOG	-2760219.78	-5.444	-3.642
	3-21G	-2061255.01	-9.545	-7.578
Ph	STOG	-2372471.90	-5.345	-3.467
	3-21G	-2484322.38	-9.657	-7.802
NH_2	STOG	-2769768.20	-5.469	-3.6
	3-21G	-2098283.03	-9.834	-7.732
OCH_3	STOG	-2777009.67	-5.499	-3.622
	3-21G	-2151226.07	-9.945	-7.711
F	STOG	-2796070.93	-5.696	-3.679
	3-21G	-2148539.39	-9.563	-7.43
Cl	STOG	-2976746.76	-5.36	-3.34
	3-21G	-2244784.61	-9.472	-7.23
Br	STOG	-4295115.76	-5.803	-3.742
	3-21G	-4105436.87	-9.603	-7.216
CF_3	STOG	-2616844.67	-5.106	-3.121
	3-21G	-1971597.43	-9.754	-7.301
CN	STOG	-2515070.12	-5.14	-3.162
	3-21G	-1957462.17	-9.876	-7.233

The binding energy of adduct formation reactions has been calculated by the difference between the binding energy of *cis*-[VO(acac)$_2$L] adduct and that of individual [VO(acac)$_2$] and 4-(*para*-substituted phenyl)-1,2,3-selenadiazole ligand. Also, the reorganization energy of ligands and [VO(acac)$_2$] that is needed to change the geometry of them into that present in the final [VO(acac)$_2$L] adduct has been calculated [36-37].

Journal of Physics: Conference Series **2063** (2021) 012002 doi:10.1088/1742-6596/2063/1/012002

As shown in Table 5, the values of binding energy are somewhat large values (vary from 1066083.30 to 321355.55 kcal/mol) these may be due to needing to overcome steric interactions of [$VO(acac)_2$] and 4-(*para*-substituted phenyl)-1,2,3-selenadiazole ligands.

The organization energy of [$VO(acac)_2$] represents the required energy to convert the geometry of vanadyl complex from a square pyramidal to distorted octahedral, while the organization energy of ligands represents the required energy to change the bonds lengths and angles of free ligand into that present in adduct. It can be seen clearly form Table 5 that the required energy to convert the geometry of vanadyl complex from a square pyramidal to distorted octahedral is almost similar (vary from 29.94-30.66 kcal/mol). On the other hand, the organization energy of ligand was different values for each ligand (vary form 20.44-23.90 kcal/mol).

Figure 3 shows the spatial distribution of the highest molecular orbital (HOMO) and lowest unoccupied molecular orbital of selected adducts. The negative region is showed with violet color, while the positive region is showed with green color. In general, HOMOs orbitals are located mainly on the nitrogen atom in diazole ring and on [$VO(acac)_2$] moiety. In contrast, LUMOs orbitals are located mainly on[$VO(acac)_2$] moiety and on different atoms (Se, N and C) in selenadiazole moiety. In other words, from the diagrams of HOMO and LUMO, it can be concluding existence two types of interaction between nitrogen and vanadium atoms in the adducts. First type of interaction is pπ-dπ, where the nitrogen atom donates pairs of electrons to vanadium atom to form σ covalent bond. The second type of interaction is dπ-pπ, where the vanadium atom donates single electron to nitrogen atom to form π bond. These two types of interactions are illustrated in Figure 4.

Table 5. The Binding energy of adducts formation and the reorganization energy of 4-(*para*-substituted phenyl)-1,2,3-Selenadiazole and [$vo(acac)_2$] in the adducts using BLYP/3-21G calculation.

X	ΔE [a] kcal/mol	E_{reorg} of L [b] kcal/mol	E_{reorg} of [$vo(acac)_2$] [c] kcal/mol
H	321355.55	23.50	30.14
CH_3	790676.75	23.17	30.45
Ph	487183.07	20.89	29.94
NH_2	763650.40	22.36	30.05
OCH_3	747617.61	23.72	30.43
F	740787.81	23.90	30.55
Cl	869602.35	20.44	30.12
Br	329069.26	21.23	30.66
CF_3	1066083.36	20.69	30.01
CN	927484.28	29.85	30.00

[a] The binding energy of adduct formation which is calculated by the difference between the energy of the adduct and that of the individual 4-(*para*-Substituted phenyl)-1,2,3-selenadiazole and [$vo(acac)_2$] moieties.

[b] Reorganization energy of 4-(*para*-substituted phenyl)-1,2,3-selenadiazole. [c] Reorganization energy of [$vo(acac)_2$].

4. Conclusions

In the present investigation we have investigated the adducts formed between [$Vo(acac)_2$] and 4-(*para*-substituted phenyl)-1,2,3-selenadiazole ligands.

The studied ligands have more than one donor atoms (Se, N2 and N3) that may coordinate with

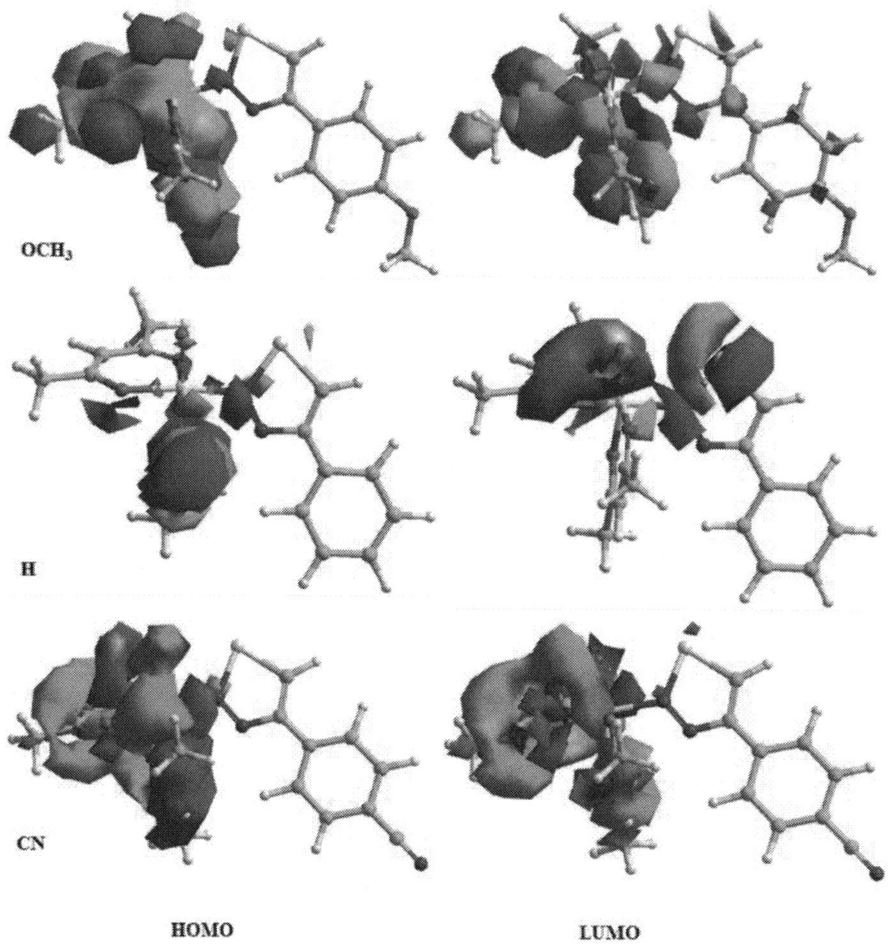

Figure 3. Molecular orbital spatial distribution for the HOMO and LUMO of ***cis***-[VO(acac)$_2$L] adduct. L is 4-(*para*-substituted phenyl)-1,2,3-Selenadiazole.

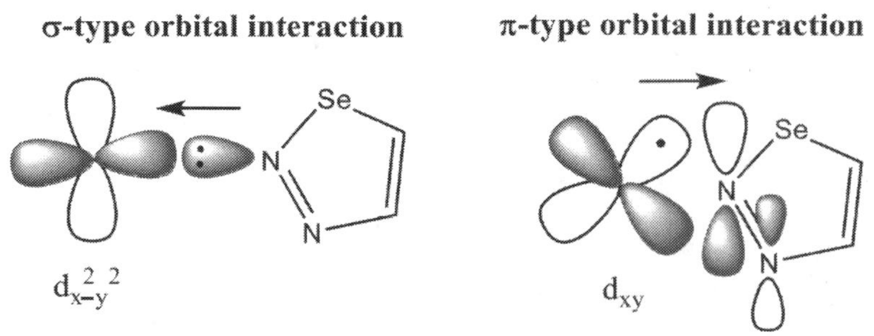

Figure 4: interaction types of selenadiazole ring with d orbitals of vanadium atom

vanadium atom on the [Vo(acac)$_2$]. In an attempt to detect the donor atom on 1,2,3-selenadiazole ring which is coordinating with [VO(acac)$_2$], the 4-(*para*-substituted phenyl)1,2,3-selenadiazole molecules have been selected to be bound with vanadium atom in [VO(acac)$_2$] through Se, N2 and N3. The resulting adducts have been investigated in two geometries (*cis* and *trans*) in order to show the effect of such structural change on the electronic properties of the studied adducts. Based on the electron density of nitrogen atoms, calculations of basicity of nitrogen atom and calculations of lowest energy conformation, the suggested structures of adducts of [VO(acac)$_2$] with 4-(*para*-substituted phenyl)1,2,3-selenadiazole molecules have been detected.

The resulting structure of adduct have been studied at two levels of DFT theory to investigated their structural and electronic properties. The structural properties (i.e. bond lengths and angles) of the studied [VO(acac)2] and 4-(*para*-substituted phenyl)1,2,3-selenadiazole moieties indicated a good agreement with experimental data of free molecules.

The organization energy of both [VO(acac)$_2$] and 4-(*para*-substituted phenyl)1,2,3-selenadiazole moieties into that present in adducts have been calculated, where the required energy to convert the geometry of vanadyl complex from a square pyramidal to distorted octahedral was almost 30 kcal/mol. The HOM and LUMO spatial distribution indicated interesting interactions between nitrogen and vanadium atoms.

The variation of electronic donating or withdrawing groups on the structural properties of the studied adducts have no interesting effects on the structural and electronic properties of the studied adducts. Finally, the new adducts species have interesting properties and deserve to be investigated experimentally.

References

[1] Buchler, J. W. (1978) hemoglobin-an inspiration for research in coordination chemistry. Angew. Chem. In. Ed. Engl. 17:407-423.

[2] Senge, M. O., Ryan, A. A., Letchford K. A., MacGowan S. A. and Mielke, T. (2014) chlorophylls, symmetry, chirality, and photosynthesis. Symmetry. 6: 781–843.

[3] Randaccio, L. Geremia, S. Demitri N. and J. Wuerges (2010) vitamin B12: unique metalorganic compounds and the most complex vitamins. Molecules.15: 3228–3259.

[4] Van Holde, K. E., Miller K. I. and H. Decker (2001) hemocyanins and invertebrate evolution. J. Biol. Chem. 276:15563–15566.

[5] Dodge, R. P., Zalkin, A., and Templeton, D. H. (1961) crystal structure of vanadyl bisacetylacetonate. geometry of vanadium in fivefold coordination. J. Chem. Phys.35:55-67.

[6] Fedorova, E. V. , Rybakov, V. B., Senyavin, V. M., Anisimov, A. V. and Aslanov , L. A. (2005) synthesis and structure of oxovanadium(IV) complexes [VO(Acac)2] and [VO(Sal: L-alanine)(H2O)]. Crystallogr. Rep., 50:224–229.

[7] Maurya, M. R., Uprety B., Avecilla, F. and Adão, P. (2015) vanadium(v) complexes of a tripodal ligand, their characterisation and biological implications. J. C. Pessoa, Dalton Trans. 44:17736-17755.

[8] Maurya, M. R., Chaudhary, N., Avecilla, F., Adão, P. (2015) oxidovanadium(iv) and dioxidovanadium(v) complexes of hydrazones of 2-benzoylpyridine and their catalytic applications. J. C. Pessoa, Dalton Trans.44:1211-1232.

[9] Maurya, M. R. (2003) Development of the coordination chemistry of vanadium through bis(acetylacetonato)oxovanadium(IV): synthesis, reactivity and structural aspects. Coord. Chem. Rev. 237:163-181.

[10] Liu, H. B., Wang, M., Wang, Y., Sun, H., and Sun, L. C. (2009) asymmetric oxidation of sulfides with hydrogen peroxide catalyzed by a vanadium complex of a new chiral NOO-ligand. Catal. Commun.11:294-297.

[11] Hirao, T. (1997) vanadium in modern organic synthesis. Chem. Rev.97:2707-2724.

[12] Pereira, C., Silva, J. F., Pereira, A. M., Araujo, J. P., Blanco, G., Pintado, J. M., and Freire, C. (2011) [VO(acac)2] hybrid catalyst: from complex immobilization onto silicananoparticles to catalytic application in the epoxidation of geraniol. Catal. Sc. Techn. 1:784-793.

[13] Pessoa, J. C., and Tomaz, I. (2010) transport of therapeutic vanadium and ruthenium complexes by blood plasma components. Curr. Med. Chem.17:3701–3738.

[14] Jain, V. K., Priyadarsini K. I. Organoselenium Compounds in Biology and Medicine: Synthesis, Biological and Therapeutic Treatments,2017.
[15] Tetteh, S., and Zugle, R. (2017) theoretical study of terminal vanadium(v) chalcogenido complexes bearing chlorido and methoxido ligands. J. Chem.2017:1-8.
[16] Molter, A., Bill, E., and Mohr, F. (2012) synthesis, structures and reactivity of two oxidovanadium(IV) and dioxidovanadium(V) selenosemicarbazonato complexes. Inorg. Chem. Comm.17:124–127.
[17] Ying Luo, H., Zhou J. and Zou , H. (2019) a series of new vanadium(iii) chalcogenido-antimonates: an unusual seven-coordinate nitro-selenidovanadium(iii) complex Dalton Trans.48: 3090-3097.
[18] Pisk, J., Daran, J.-C., Poli, R. and Agustin, D. (2015) Pyridoxal based ONS and ONO vanadium(V) complexes: Structural analysis and catalytic application in organic solvent free epoxidation. J. of Mol. Cata. A-Chemical.403:52–63.
[19] HyperchemTM Release 7.52, Windows Molecular Modeling System, hypercube, Inc. 2002.
[20] Stewart, J.J.P. (1989) Optimization of parameters for semiempirical methods II. Applications. J. Comput. Chem.10:221-261.
[21] Roothaan, C.C. (1951) New Developments in Molecular Orbital Theory. J. Rev. Mod. Phys. 23 69-89.
[22] Gill, P.E., Murray, W., Wright, M.H. Practical Optimization, Academic Press, New York, 1981.
[23] Lee, C., Yang, W., and Parr, R.G. Phys. (1988) Development of the Colle-Salvetti correlation-energy formula into a functional of the electron density. Rev. B 37: 785-789.
[24] Becke, A.D. (1993) density-functional thermochemistry. III. The role of exact exchange J. Chem. Phys. 98:5648-5652.
[25] Keith, T.A., and Bader, R.F.W. (1992) calculation of magnetic response properties using atoms in molecules Chem. Phys. Lett.194:1-8.
[26] Burkert, U., and Allinger, N.L. Molecular Mechanics ACS Monograph,American Chemical Society, Washington, DC, 1982, p. 177.
[27] Li, J.-H., Gallion, S. Bender, C., Wikström, H., Allinger, N.L., Flurchick, K.M., and Teeter M.M. (1989) molecular mechanics (MM2) calculations on peptides and on the protein crambin using the CYBER 205. J. Comput. Chem.10:503-513.
[28] Pearson, R. G. (1988) absolute electronegativity and hardness: application to inorganic chemistry. Inorg. Chem.27:734–740.
[29] Arsenyan, P., Rubina, K., Shestakova, I. and Domracheva, I. (2007b) 4-Methyl-1,2,3-selenadiazole-5-carboxylic acid amides: Antitumor action and cytotoxic effect correlation.Eur. J. Med. Chem. 42:635-640.
[30] Todres, Z. V. Chalcogenadiazoles: Chemistry and Applications. CRC Press, 2011.
[31] Baetzel, V. and Boese, R. (1981) synthese und struktur von 1.2.3-Thiadiazol- und 1.2.3-Selenadiazol-Pentaearbonyl-Komplexen der Elemente Chrom und Wolfram / synthesis and structure of 1,2,3-thiadiazol- and 1,2,3-selenadiazol-pentacarbonyl-complexes of the elements chromium and tungsten. Zeitschr. Naturforsch.36B:172-179.
[32] Richardson, Ch. and Steel, P.J. (2002) metal complexes of 2-[1,2,3-thia(and selena)diazol-4-yl]pyridine and an unusual silver-induced selenadiazole rearrangement. Austral. J. Chem.55 783-788.
[33] Pannell, K. H., Mayr, A. J., Hoggard, R., McKennis, J.S., and Dawson, J.C. (1983) (cycloalkeno-1,2,3-selenadiazol- und -thiadiazol)carbonyl-(6b)-metall-komplexe. Chem. Ber. 116: 230-237.
[34] Ravichandran, K., Ranjith, S., Sankari, S., Beemaraoa M. and Ponnuswamy M. N. (2018) 4-(4-Chloro-phen-yl)-1,2,3-selena-diazole. IUCrData.3: 80462-80465.
[35]Oshi , R., Abdalla , S., and Springborg, M. (2017) study of the influence of functionalization on the reorganization energy of naphthalene using DFT. Comput. Theoret. Chem.1099 209-215.
[36] D. F. Shriver, P. Atkins, Inorganic Chemistry, 5th ed.; Freeman: New York, 2010.
[37] B. E. Douglas, McDaniel, D. H.; Alexander, J. J. Concepts and Models of Inorganic Chemistry, 3rd ed.; Wiley: New York,1994.

Synthesis, characterization and biological studies of some new organometallic compounds containing mercury, selenium and tellurium based on *p*-aminobenzoic acid

Ali Z Al-Rubaie*[1], Shaker A S Al-Jadaan[2], Anwar T Abd Al-Wahed[1] and Ibraheem A Raadah[2]

[1]Department of Chemistry, College of Science, University of Basrah, Basrah, Iraq
[2]Department of Chemistry, College of Pharmacy, University of Basrah, Basrah, Iraq

* Corresponding author; e.mail: alrubaie49@yahoo.com

Abstract. Ten chalcogen and mercury bearing compounds based on 4-aminobenzoic acid (*i.e.*, (2-amino-5-(ethoxycarbonyl)phenyl)mercury(II) chloride (**1**), (2-amino-5-(ethoxycarbonyl)phenyl) phenyl selenide (**2**), (2-amino-5-(ethoxycarbonyl)phenyl) phenyl telluride (**3**), (4-carboxyphenyl)mercury(II) chloride (**4**), 4-selenocyanatobenzoic acid (**5**), 4-tellurocyanatobenzoic acid (**6**), bis(4-carboxyphenyl) diselenide (**7**) bis(4-carboxyphenyl) ditelluride (**8**), bis(4-carboxyphenyl) selenide (**9**) bis(4-carboxyphenyl) telluride (**10**) were prepared and characterized by various spectroscopic techniques. All compounds were screened for antibacterial activity against Gram-positive bacterial strains of *Staphylococcus aureus* and the Gram-negative bacteria *Escherichia coli* by using the disk diffusion technique. The antibacterial activity of these compounds was dependent on the molecular structure of the compounds, and the bacterial strain under consideration.

1. Introduction

p-Aminobenzoic acid (PABA) and its derivatives showed interesting pharmacological properties [1,2]. PABA is a chemical found in the folic acid vitamin and also in several foods including grains, eggs, milk, and meat. Furthermore, this compound is a water soluble, non-poisonous drug and is easily absorbed by intestinal treatment. Some of its derivatives are used as acetyl cholinesterase inhibitors for the treatment of Alzheimer's disease [3]. p-Aminobenzoic acid analogs such as procaine and lidocaine are anesthetics [4]. p-Aminobenzoic acid is also a member of the B vitamins and is part of the folic acid molecule which showed antioxidant properties [5]. It is well known that PABA is important to skin, hair pigment, and intestinal health. It is used as a sunscreen; it also can protect against the development of sunburn and skin cancer from excess ultraviolet light exposure [6]. On the other hand, the interest in the field of synthesis and reactivity of organoselenium and organotellurium compounds has increased; particularly, because of the observations that they can exhibit important biological activities [7-9].

The present work describes the synthesis of several compounds based on p-aminobenzoic acid-containing mercury, selenium, and tellurium and investigates their biological activity.

2. Experimental

2.1. Physical measurements

All melting points were determined on a Gallenkamp melting point apparatus and are uncorrected.
FT-IR spectra were recorded on FT-IR spectrophotometer -8400s Shimadza as KBr disc. The 1H and

13C NMR spectra were measured on a Bruker spectrometer at 300 (1H NMR) MHz and 75 (13C NMR) MHz or 200/50 MHz on an AC-200 NMR spectrometer. Spectra were recorded in CDCl3 solutions. Chemical shifts are reported in ppm, referenced to the solvent peak of CDCl3 or tetramethylsilane (TMS) as the external reference. Elemental analyses were determined on an MT-3 elemental analyzer within ±5 % of the theoretical values. Mass spectra were recorded on a HP-598 8A MS instrument at 70 eV.

2.2. Synthesis

All reactions were carried out under nitrogen or argon atmosphere and monitored by the conventional TLC method. All organic solvents were dried prior to use according to standard methods. Ethyl 4-aminobenzoate (Benzocaine) was prepared by the reaction of p-aminobenzoic acid with absolute ethanol in presence of conc. H_2SO_4 [10]. Phenylselenium trichloride and phenyltellurium tribromide were prepared according to literature methods [11, 12].

2.1.1. (2-Amino-5-(ethoxycarbonyl)phenyl)mercury(II) chloride (**1**)

Ethyl 4-aminobenzoate (5.0g; 30 mmol) was treated with mercuric acetate by a similar method used for the preparation of 2-amino-5-methylphenyl mercury chloride [13].

2.1.2. (2-Amino-5-(ethoxycarbonyl)phenyl) phenyl selenide (**2**)

(2-Amino-5-(ethoxycarbonyl)phenyl)mercury(II) chloride (4.83 g, 12.5 mmol) and phenylselenium trichloride (3.28 g, 12.5 mmol) in dry dioxane (100 cm3) was refluxed for 12h. The hot solution was filtered hot. The filtrate was cooled to room temperature. The HgClBr.2diox complex was removed by filtration as white plates. The filtrate was evaporated to dryness on a rotary evaporator at 40°C. A dark brown precipitate was obtained. The product was dissolved in CH_2Cl_2 (100 cm^3) and shaken vigorously in a separating funnel for 30 min with an aqueous solution (50 cm^3) of $Na_2S_2O_5$ (1.73g; 9 mmol). The yellow organic phase was separated, washed with water, and dried over calcium chloride. Evaporation and recrystallization from petroleum ether (60-80°C) gave a brown-yellow precipitate.

2.1.3. (2-Amino-5-(ethoxycarbonyl)phenyl) phenyl telluride (**3**)

This compound was prepared by the same above method by reaction of phenyltellurium tribromide(3.55 g; 8 mmol)) with (2-Amino-5-(ethoxy-carbonyl)phenyl)mercury(II) chloride (3.10 g, 8 mmol). A brown precipitate was obtained.

2.1.4. (4-Carboxyphenyl)mercury(II) chloride (**4**)

Compound (4-carboxyphenyl) mercury (II) chloride was prepared according to a literature method [14] by heating the double salt of p-toluenediazonium chloride and mercuric chloride in acetone with copper powder, followed by oxidation of p-tolylmercuric chloride by $KMnO_4$[14], m.p. 287°C (dec.).

2.1.5. 4-Selenocyanatobenzoic acid (**5**)

A stirred solution containing *p*-aminobenzoic acid (4.7 g, 34.91 mmol), 20 cm^3 of conc. hydrochloric acid and 20 cm^3 of water in an ice bath (0 - 5°C), was added dropwise to sodium nitrite (9.0 g, 134.31 mmol) in 25 cm^3 of water. A solution of p-carboxybenzodiazonium chloride was formed. The resulting solution was added dropwise to a stirred solution of potassium selenocyanate (5.0g; 34.91 mmol) in ethanol (25 cm^3) while a brisk stream of dry nitrogen was passed into the system. The mixture was stirred at room temperature for an additional 2 h. The mixture was diluted with CH_2Cl_2 (250 cm^3), filtered from the dark insoluble material and washed several times with H_2O. The solution was dried ($CaCl_2$) and evaporated, giving a solid, which was dissolved in CH_2Cl_2 and filtered through a short SiO_2 column.

2.1.6. 4-Tellurocyanatobenzoic acid (***6***)

Finely ground tellurium powder (1.6 g, 12.5 mmol) and KCN (0.82 g, 12.5 mmol) in dry DMSO (20 cm^3) were heated at 100°C, for 1 h under dry nitrogen. After all the tellurium dissolved, the mixture was cooled in an ice bath until most of the solvent has solidified. The p-carboxybenzodiazonium chloride

(12.5 mmol) was added rapidly while a brisk stream of nitrogen is passed into the system. When the initial violent gas evolution has ceased, the ice bath was removed and stirring was continued at room temperature for 6 h. The mixture was diluted with CH_2Cl_2 (200 cm^3), filtered from the dark insoluble material and washed several times with H_2O. The solution was dried over anhydrous calcium chloride. Evaporation of solvent gave a yellow solid which was dissolved in dichloromethane and filtered through a short SiO_2 column. A yellow solid was obtained in 58% yield.

2.1.7. Bis(4-carboxyphenyl) diselenide (**7**)
This compound was prepared by two methods and as follow:
a) Compound **5** (2.26g; 10 mmol) was acidified with aqueous HCl (10%, 25 cm^3), gently heated with mixing and left at room temperature overnight. The solid was filtered off, washed with hot water several times, and air-dried. The crude product was recrystallized from hot dioxane to give compound **7** as a pale brick powder. Yield: 37%; m.p. 321–323°C.
b) $NaBH_4$ (0.22g, 6 mmol) was added in small portions with caution to a solution of 4-selenocyanatobenzoic acid (1.36g; 6 mmol) in absolute ethanol (40 cm^3). The mixture was stirred at room temperature for 2 h. The solvents were removed under vacuum by a rotary evaporator and the residue was treated with water. The mixture was extracted with dichloromethane. The organic phase was washed with water and dried with anhydrous Na_2SO_4. The dichloromethane was removed under vacuum and the residue product was recrystallized from hot dioxane to give compound **7** in 26% yield.

2.1.8. Bis(4-carboxyphenyl) ditelluride (**8**)
Compound **6** (2.75 g; 10 mmol) was treated with aqueous hydrochloric acid (10%, 25 cm^3) in a similar above method (method b) described for compound **7.** A yellow precipitate was obtained in 41% yield; m.p. 288–290°C.

2.1.9. Bis(4-carboxyphenyl) selenide (**9**)
This compound was prepared by using previously reported method [15] with a slight modification. A solution of bis(4-carboxyphenyl) diselenide (**7**) (2 g; 5 mmol) in dioxane (50 cm^3) was refluxed in presence of copper powder (0.70g; 11 mmol) for 12 h. The hot solution was filtered and the filtrate was evaporated to give compound **9** in almost quantitative yield, m.p. 198-199°C.

2.1.10. Bis(4-carboxyphenyl) telluride (**10**)
Compound **8** was converted into bis(4-carboxyphenyl) telluride (**10**) by treatment with copper powder in refluxing dioxane by the same above method. Compound 10 was obtained in 87% yield, m.p. 135-137°C.

2.2. Antibacterial Activity
Compounds **1-10** were tested for their antibacterial activity by using cup plate agar diffusion method [16] and the inhibition zones were measured in millimetre (mm). The nutrient agar broth prepared by the usual method, was inoculated aseptically with 0.5 cm^3 of 24 h old subculture of *S. aureus* and *E. coli* in separate conical flasks at 40-50°C and mixed well by gentle shaking. About 25 cm^3 of the contents of the flask were poured and evenly spread in petridish (90 mm in diameter) and allowed to set for 2h. The cups (10 mm in diameter) were formed by the help of borer in agar medium and filled with 0.02 cm^3 (20 μg/ cm^3) and 0.04 cm^3 (40 μg/ cm^3) solution of sample in DMSO. The plates were incubated at 37°C for 24 h and the control was also maintained with 0.04 cm^3 of DMSO similarly and the zones of inhibition of the bacterial growth were measured in millimetre and recorded in Table 3.

3. Results and Discussion

3.1. Synthesis
Organoselenium and organotellurium compounds are fast gaining contemporary interest due to their indispensable applications in the electronic industry [17], as organic conductors [18] and precursors for

Journal of Physics: Conference Series 2063 (2021) 012003 doi:10.1088/1742-6596/2063/1/012003

semiconducting materials [19], in biology [20], and medical imaging. They also finding renewed interest as synthetic reagents in organic synthesis [21, 22]. Thus, in the present work, a new series of new organometallic compounds containing mercury, selenium, and tellurium based on *p*-aminobenzoic acid were prepared.

Compound (2-Amino-5-(ethoxycarbonyl)phenyl)mercury(II) chloride (**1**) can be easily prepared from the reaction of ethyl 4-aminobenzoate with mercuric acetate followed by addition of lithium chloride, Scheme 1. Compound **1** was reacted with phenylselenium trichloride and phenyltellurium tribromide followed by reduction by $Na_2S_2O_5$ to give compounds **2** and **3**, respectively in good yield, Scheme 1, Table 1. It is worth noting that direct mercuration of p-aminobenzoic acid gave an unidentified product although a previous work [23] reported the formation of (2-amino-5-carboxyphenyl) mercury (II) chloride by direct mercuration of p-aminobenzoic acid.

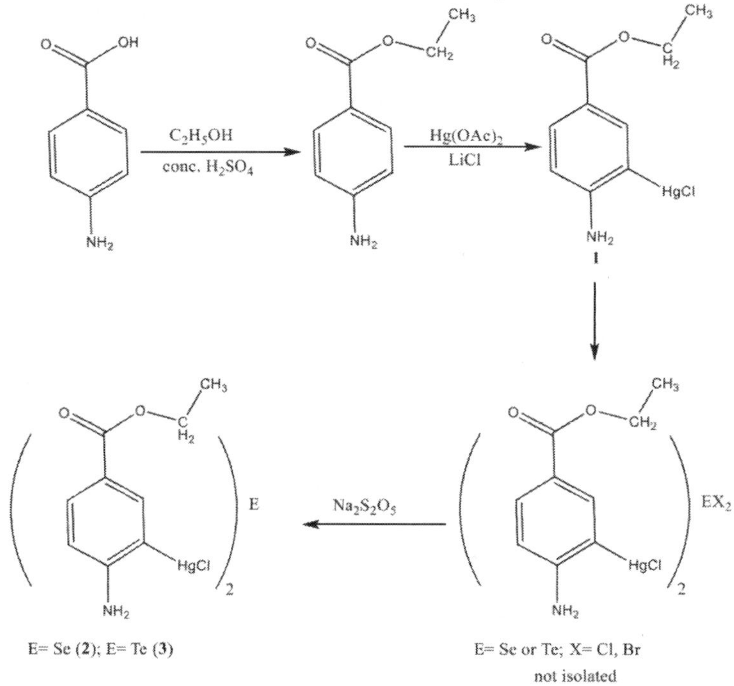

Scheme 1. Preparative methods for compounds **1**, **2** and **3**.

Compound **4** (4-carboxyphenyl)mercury(II) chloride) was prepared by heating the double salt of *p*-toluenediazonium chloride and mercuric chloride in acetone with copper powder, followed by oxidation of *p*-tolylmercuric chloride by $KMnO_4$[14], Scheme 2. Compounds **5** and **6** were prepared by reaction of *p*-carboxybenzodiazonium chloride with KSeCN and KTeCN, respectively, Scheme 2. Acid hydrolysis of compounds **5** and **6** gave bis(4-carboxyphenyl) diselenide (**7**) and bis(4-carboxyphenyl) ditelluride (**8**), respectively. Compound **7** was also prepared in low yield by reaction of 4-carboxybenzodiazonium chloride with $NaBH_4$, Scheme 2. The subsequent reaction of Compounds **7** and **8** with copper powder in refluxing dioxane afforded bis(4-carboxyphenyl) selenide (**9**) and Bis(4-carboxyphenyl) telluride (**10**), Scheme 2.

Journal of Physics: Conference Series **2063** (2021) 012003 doi:10.1088/1742-6596/2063/1/012003

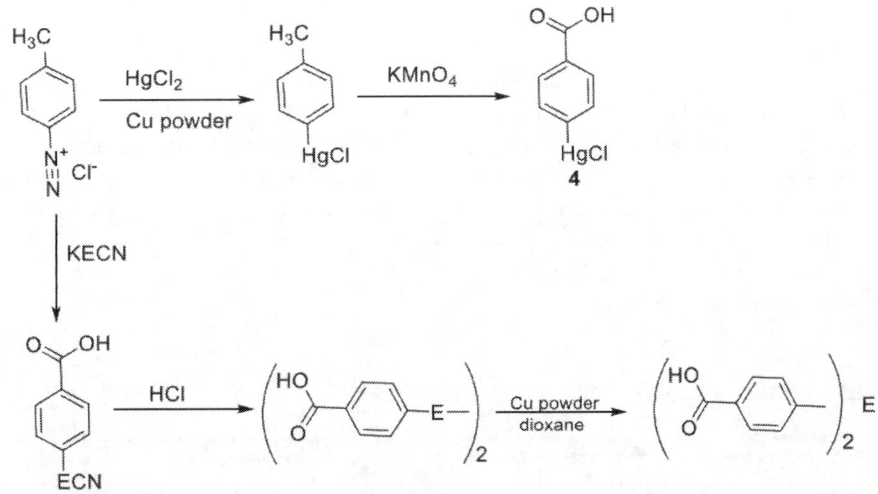

5: E = Se; 6: E= Te   7: E = Se; 8: E= Te   9: E = Se; 10: E= Te

Scheme 2. Preparative methods for compounds **4** – **10**

All compounds are brown to white solids and mostly soluble in DMF and DMSO. Isolated yields, melting points, colours, and analytical data for compounds **1**- **10** are shown in Table 1.

Table 1. Isolated yield, analytical and IR data for compounds **1-10**

Compd.	Yield (%)	M. p. (°C)	Found (calc.) (%) C	H	N	IR (cm^{-1}, KBr)
1	86	234	26.76 (27.01)	2.33 (2.52)	3.44 (3.50)	3560(OH), 3340, 3285(NH), 3047 (C-H str.), 2931 (C-H str.), 1699 (C=O stretch), 1600(NH), 1581 (C=C), 1365 (CH_3), 1045 (CO str.), 775 (C-H).
2	74	204-206	55.84 (56.26)	4.56 (4.72)	4.13 (4.37)	3562(OH), 3328, 3265(NH), 3034 (C-H str.), 2925 (C-H str.), 1706 (C=O), 1602(NH), 1585 (C=C), 1313 (CH_3), 1024 (C-O str.), 815 (C-H).
3	68	175-177	48.47 (48.84)	3.83 (4.10)	3.52 (3.80)	3556(OH), 3315, 3257(NH), 3032 (C-H str.), 2943 (C-H str.), 1705 (C=O),1608 (NH), 1578(C=C),1280 (CH_3), 1103 (C-O str.), 871 (C-H)
4	59	282-284 (dec.) *	23.22 (23.54)	1.38 (1.41)	-	3532(OH), 3012(CH), 1686(CO).
5	75	193	41.83 (42.50)	2.12 (2.23)	5.86 (6.20)	3455(OH), 3105(CH), 2065(νCN), 1686(CO), 690(νC-Se), 430(δ(NCSe)),
6	58	104	34.77 (34.98)	1.58 (1.83)	4.86 (5.10)	3502 (OH), 3095(CH), 2172((νCN), 1692(CO), 540(νC-Te), 415(δ(NCTe)),
7	37	278-280	41.76 (42.02)	2.33 (2.52)	-	3478 (OH), 3078(C-H), 1678(CO).
8	41	188-190	34.21 (33.80)	1.79 (2.03)	-	3490 (OH), 3011(C-H), 1674(CO).
9	95	198-199	52.41 (52.35)	2.78 (3.14)	-	3504(OH), 3060(C-H), 1664(CO).
10	87	135-137	44.87 (45.47)	2.53 (2.73)	-	3493(OH), 3088(C-H), 1667(CO).

* Lit. 287°C[14]

The IR spectra of compounds **1-3** the compounds showed medium bands due to N-H asymmetric and symmetric stretching (ν_{as} and ν_s) between 3560 and 3315 cm^{-1}, respectively. The spectra showed strong bands at 1600 cm^{-1} due to the internal deformation mode of N-H. A comparison of the ν(N-H) bands in compounds **1-3** with ethyl 4-aminobenzoate indicates that the NH_2 frequencies in the new compounds are shifted to a lower position. This may indicate that the NH_2 group coordinates to the mercury, selenium, or tellurium atom by donation of the lone pair of nitrogen, as was observed for bis(2-aminophenyl) ditelluride [24]. Furthermore, a band around 1700 cm^{-1} may be attributed to carbonyl group. The IR spectra of compounds **4** and **5** show strong bands at 2078, and 2160 cm^{-1}, which are attributed to ν(Se-CN) and ν(Te-CN), respectively [25, 26] together with a strong band at 1695 and cm^{-1}, respectively are due to carbonyl group.

Table 2. NMR spectroscopic data for compounds **1-10**

Compound	^{1}H NMR (δ, 300 MHz, $CDCl_3$); TMS= 0 ppm	^{13}C NMR
1	8.03 (d, *J* = 1.5 Hz, 1H), 7.73 (dd, *J* = 7.4, 1.5 Hz, 1H), 6.71 (d, *J* = 7.5 Hz, 1H), 4.65 (s, br, 2H), 4.34 (q, *J* = 8.0 Hz, 2H), 1.39 (t, *J* = 8.0 Hz, 3H).	166.2, 153.0, 138.1, 135.1, 129.8, 121.8, 116.3, 60.9, 14.4.
2	7.68 (d, *J* = 2.0 Hz, 1H), 7.60 (dd, *J* = 7.46, 2.0 Hz, 1H), 7.29 (s, 5H), 6.72 (d, *J* = 7.5 Hz, 1H), 4.48 (s, 2H), 4.32(q, 4H), 1.34 (t, *J* = 8.0 Hz, 3H).	166.2, 150.4, 138.5, 133.7 ,132.8, 131.7, 129.3, 129.0, 127.2, 122.8, 117.5, 61.3, 14.42.
3	7.80 (d, *J* = 2.0 Hz, 1H), 7.46 (ddd, *J* = 9.8, 7.6, 2.0 Hz, 3H), 7.38 – 7.24 (m, 2H), 7.15 (ddt, *J* = 7.5, 6.1, 2.1 Hz, 1H), 6.58 (d, *J* = 7.5 Hz, 1H), 4.38 – 4.24 (m, 4H), 1.34 (t, *J* = 8.0 Hz, 3H).	165.4, 149.2, 136.1, 135.2, 130.8, 129.15, 127.7, 127.5, 126.3, 117.3, 116.3, 60.8, 14.3.
4	12.54 (s, 1H, OH), 8.15 (d, *J* = 7.8 Hz, 2H), 7.52 (d, *J* = 7.8 Hz, 2H).	170.0, 155.6, 136.7, 128.4, 131.1
5	8.12 – 8.02 (m, 1H), 7.55 – 7.45 (m, 1H).	172.2, 132.9, 128.5, 127.1, 124.1, 103.0.
6	11.8 (s, 1H, OH), 7.73 (d, *J* = 8.1 Hz, 2H), 6.80 (d, *J* = 7.9 Hz, 2H).,	169.6, 136.5, 129.4, 127.4, 115.6, 103.3.
7	12.0 (s, 1H, OH), 8.12 (d, *J* = 7.9 Hz, 2H), 7.55 – 7.45 (m, 2H).	168.8, 138.7, 130.4, 129.1, 124.7.
8	11.7 (s, 2H, OH), 7.9 (d, 4H), 7.6 (d, *J* = 8.4 Hz, 4H).	167.5, 135.1, 130.0, 127.2, 113.2.
9	11.9(s, 2H, OH), 8.08(d, *J* = 7.5 Hz, 4H), 7.44(d, *J* = 8.0 Hz, 4H).	172.3, 133.8, 128.4, 127.3, 101.9
10	11.5(s, 2H, OH), 8.27(d, *J* = 8.5 Hz, 4H), 7.64(d, *J* = 8.2 Hz, 4H).	169.8, 134.3, 126.4, 125.3, 108.8.

Compounds **4-10** showed one or two broad strong bands in the range 3495 -3220 cm^{-1} due to ν(OH). All protons in compounds **1-10** were identified, and the total number of protons calculated from the integration curve was tallied with the expected molecular formula in the ^{1}H NMR spectra, Table 2. Thus, the methylene protons ($COCH_2$-CH_3) appeared as a quartet at around 4.40 ppm for **1**, **2** and **3**, while for *CH_3* proton appeared as triplets, Table 2. ^{13}C-NMR spectra of compounds 1, **2**, and **3** revealed the presence of CO at around 170.0 ppm.

Treatment of compounds **5** and **6** with aqueous hydrochloric acid (10%) is the most obvious way to prepare the diselenide (**7**) and ditelluride (**8)** respectively, Scheme 1. Compounds **7** and **8** were obtained as red solids. These new compounds were obtained in good yields and quite stable to handle at room temperature (Scheme 1). IR and NMR data together with elemental analyses support the formation of these new compounds. IR spectra of **7** and **8** indicate the presence of νCO at 1678 and 1674 cm^{-1}, respectively as a strong band. ^{13}C-NMR spectra of **7** and **8** showed the presence of CO signal at 168.8 and 167.5 ppm, respectively, Tables 1 and 2.

Refluxing compounds **7** and **8** in dry dioxane in presence of a threefold molar excess of activated copper gave compounds **9** and **10**, respectively as yellow solids in good yields, Table 1, Scheme 1. The structure of the obtained compounds was assessed according to IR, ^{1}H NMR, ^{13}C NMR spectroscopic data and their element analysis, Tables 1 and 2.

3.2. Antibacterial Activity

The in vitro antimicrobial activity of compounds **1-10** were screened against Gram-positive bacterial strains (*S. aureus*) and Gram-negative bacterial strains (*E. coli*). Inhibition zones demonstrated that compounds **1**and **4** showed significant high activity which may be due to the presence of mercury and N in the form of the amine group. Compounds **2** and **3** showed high activity. The presence of the cyano group in compounds **5** and **6**, is probably responsible for the high activity against bacterial strains than compounds **2** and **3,** Table 3. This may be due to an increase in lipophilicity and easier penetration of the compounds into the outer cell wall of the microorganisms, which causes death due to cell membrane rupture. Compounds **7**–**10** were also effective against both bacterial strains, Table 3.

Table 3: Antibacterial activity of compounds **1**–**10** expressed as inhibition zone against *S. aureus* and *E. coli* at different concentration

Compound	Diameter of inhibition zone (mm)*			
	S. aureus		*E. coli*	
	20 μg/ml	40 μg/ml	20 μg/ml	40 μg/ml
1	30	43	38	44
2	24	28	20	28
3	26	30	25	32
4	24	39	35	39
5	18	22	15	19
6	15	18	13	15
7	22	25	17	22
8	18	21	15	19
9	25	28	24	30
10	23	27	17	22
Ampicillin	25	30	24	29

* the diameter (mm) of the inhibition zone, produced around each disc, are average of three separate experiments; DMSO was used as a control and ampicillin as standard drug.

In general, the presence of NH_2 and COOH groups, as well as Hg, Se or Te in the studied compounds, leads to higher activity.

The results presented in Table 3 indicate that compounds **9** and **10** are higher activity against both bacterial strains compared to compounds 7 and 8. This may be attributed to the structures of diselenide (**9**) and ditelluride (**10**). This agrees with the previous work [27-30] described that diorganyl dichalcogenides have higher biological activities in comparison with other types of organoselenium and organotellurium compounds [27-30].

Journal of Physics: Conference Series **2063** (2021) 012003 doi:10.1088/1742-6596/2063/1/012003

4. CONCLUSION

In this study, several new compounds were prepared and characterized by various spectroscopic techniques and elemental analysis. All prepared compounds (**1-10**) exhibited high antimicrobial activity against *Staphylococcus aureus* and *Escherchia coli*. The combination of an organochalcogen group with nitrogen and/or oxygen in the same molecule increases the antibacterial capabilities of these compounds.

ACKNOWLEDGEMENTS

The financial support of this work by the Ministry of Higher Education and Scientific Research/ Research and Development is hereby gratefully acknowledged.

References

[1] Krátký M, Konečná K, Janoušek J, Brablíková M, Janďourek O, Trejtnar F, Jiřina Stolaříková J and Vinšová J 2020 *Biomolecules* **10**, 9.

[2] Patel H M, Bhardwaj V, Sharma P, Noolvi M N, Lohan S, Bansal S, and Sharma A. 2019 *J. Mol. Struct.* **1184**, 562.

[3] a) Anand P, Singh B. 2013 *Arch Pharm Res.* **36(4)** 375-99.
b) Shrivastava S K, Sinha S K, Srivastava P, Tripathi P N, Sharma P, Tripathi M K, Tripathi A, Choubey P K, Waiker D K, Aggarwal L M, Dixit M, Kheruka S C, Gambhir S, Shankar S and Srivastava R K 2019 *Bioorg Chem.* **82** 211.

[4] Jenerowicz D, Polańska A, Glińska O, Czarnecka-Operacz M, Schwartz Ra. 2014 *Adv. Dermatol. Allergol./Postępy Dermatologii i Alergologii.* **31** 134.

[5] Basset GJC, Quinlivan EP, Ravanel S, Rebeille F, Nichols BP 2004 *Proc Natl Acad Sci USA* **101** 1496.

[6] Sowinska M, Morawiak M, Bochyńska-Czyż M, Lipkowski A W, Ziemińska E, Zabłocka, B and Urbanczyk-Lipkowska Z. 2019 *Biomolecules* **9** 89

[7] a) Nogueira CW, Barbosa NV, and Rocha JBT. 2021 *Arch Toxicol* **95** 1179.
b) Nogueira CW, and Rocha JBT. 2012 *Organoselenium and organotellurium compounds: Toxicology and pharmacology, in Organic Selenium and Tellurium Compounds*. John Wiley & Sons, USA.

[8] Edward R. T. Tiekink E R T, 2012 *Dalton Trans.* **41** 6390.

[9] Zhang Z, Zhang J, and Xiao J. 2014 *Biochimica et Biophysica Acta.* **1840**, 3246.

[10] Pavia D L, Lampman G M, Kriz G S, and Engel R G. *F C* 1999 *Introduction to Organic Laboratory Techniques. A Microscale Approach.* 3° ed., Saunders: Fort Worth, pp 362-364.

[11] Engman L,1987 *J. Org. Chem.* **52**, 4086.

[12] Petragnani N. 1994 *Tellurium in Organic Synthesis*, Academic Press: London, p. 65.

[13] a) Junk T, and Irgolic K J, 1988 *Phosph. and Sulfur* **38**, 121.
b) Al-Rubaie A Z, Al-Salim N I, and Al-Jadaan S A N, 1993 *J. Organomet. Chem.* **443**, 67 and references therein.

[14] Whitmore F C, and G. E. Woodward G E, Organic Syntheses, Coll. Vol. 1, p.159 (1941); Vol. 7, p.18 (1927).

[15] Sadekov T D, Bushkov AY, and Minkin, VT. 1973 *J. Gen. Chem. USSR.* **43**, 815.

[16] Doughari J H, El-mahmood A M, and Manzara S. 2007 *African J Microbiol Res.***3**, 37.

[17] Singh H B, and Sudha N,1996 *Polyhedron* **15**, 745.

[18] Siato G and Kagoshima S, Eds., 1990 *The Physics and Chemistry of Organic Superconductors*, Springer, Berlin, Germany.

[19] Somasundaram S, Chenthamarakshan CR, Tacconi N R De, Ming Y, and Rajeshwar K. 2004 *Chemistry of Materials* **16**, 3846.

[20] Hatfield D L, Ed., 2001 *Selenium: Its Molecular Biology and Role in Human Health*, Kluwer Academic, Boston, Mass, USA.

Journal of Physics: Conference Series **2063** (2021) 012003 doi:10.1088/1742-6596/2063/1/012003

[21] Wirth T, Ed., 2000 *Organoselenium Chemistry, Modern Development in Organic Synthesis*, Springer, New York, NY, USA.
[22] a) Irgolic K J, *Organotellurium Compounds Houben-Weyl, Methods of Organic Chemistry*, George Thieme, Stuttgart, Germany.
b) Petragnani N and Stefani H A, 2007 *Tellurium in Organic Synthesis*, Elsevier, The Netherlands.
[23] Al-Asadi R H, Al-Masoudi W A and Abdual-Rassol K S, 2016 *Asian J. Chem.* **28**, 1171.
[24] Junk T and Irgolic K J. 1988 *Phosph. and Sulfur* **38** 121.
[25] a) Falcon SJ, Cava MP. 1980 *J. Org. Chem.* **45**, 1044.
b) Guillemin J-C, 2011 *Current Org. Chem.* **15**, 1670.
[26] Engman L, 1983 *J. Org. Chem.* **48**, 2920.
[27] Nascimento V, Ferreira NL, Canto RF, Schott KL, Waczuk EP, Sancineto L, Santi C, and Rocha JB. 2014 *Eur. J. Med. Chem.* **87**, 131.
[28] Shaaban S, Negm A, Sobh MA, and Wessjohann LA, 2015 *Eur. J. Med. Chem.* **97** 190.
[29] Sartori G, Jardim NS, Marcondes Sari MH, Dobrachinski F, Pesarico AP, Rodrigues LC , Cargnelutti J, Flores EF, Prigol M, and Nogueira CW. 2016 *J. Cell. Biochem.* **117** 1638.
[30] Ba LA, Doring M, Jamier V and Jaco C. 2010 *Org. Biomol. Chem.*, **8** 4203.

Journal of Physics: Conference Series **2063** (2021) 012004 doi:10.1088/1742-6596/2063/1/012004

Synthesis , Characterization and Thermal Studies of Fe(II) , Co(II) , Ni(II) and Cu(II) Complexes with bis-(4-benzylidenamino-phenyl)-methane Schiff base ligand

Atared F. Hassan , Hiba H. Sabah and Afaq A. Turki

Department of chemistry, College of Science, University of Basrah

Atared.hassan@uobasrah.edu.iq , heba.hani@uobasrah.edu.iq,

afaq.turki@uobasrah.edu.iq

Abstract. A new series of Co(II), Fe(II) , Cu(II) and Ni(II) complexes of Bis –(4-benzylidenamino-phenyl)-methane (L) have been prepared and characterized . The complexes conform to general formula $[CoL(H_2O)_3]\ Cl_2$, $[FeL(H_2O)_5]SO_4.2H_2O$, $[CuL(H_2O)_5]\ NO_3$ and $[NiL(H_2O)_3Cl_2]$, Infra red spectra and elemental analysis reveal uni-dentate coordination of Schiff base to metal ions. Molar conductivity measurements show that complexes (I , III) behave as 2:1 electrolyte while complex (II) behave as 1:1 electrolyte. The thermal studies for complexes shows that the complexes are thermal stable.

1. Introduction

Schiff bases are an important organic compounds which carrying imin or azomethane group . Schiff bases are versatile compounds in many fields including biological , analytical , medical and pharmaceutical such as Anti tubercular [1,2] , Anti depressant [3] , Anticonvulsant [4,5] and anti-inflammatory [6]. Schiff bases easily formed high thermal stability complexes with almost all metal ions were useful for uses as catalysis in highly temperature for both homogeneous and heterogeneous reaction such as polymerization [7,8] , For reduction of thionyl chloride [9] and oxidation of organic compounds [10].

These organometallic compounds exhibit larger biological activity than free Schiff base themselves such as anti bacterial [11] , Anti fungal [12-14], Anti cancer [15-17] and antiviral activity [18].

Journal of Physics: Conference Series **2063** (2021) 012004 doi:10.1088/1742-6596/2063/1/012004

The aim of this work , the preparation , spectroscopic studies of Fe(II) ,Co(II) ,Ni(II) and Cu(II) complexes with mono-dentate bis-(4-benzylidenamino-phenyl)-methane ligand and we have study thermal behaviors of these complexes .

2. Material and methods

All chemicals and solvents used in this study were supplied (Merck and Fluka) company . The ^{1}H NMR spectra were recorded with (Bruker DPX-300)Spectrometer(Iran) . The elemental analysis was done by using (CE-440)Elemental analyzer(Iran) . FT-Infrared spectra were recorded as KBr pellets with (FT-IP 8400S) Shimadzu instrument(Univercity of BAsrah) . Melting point of complexes were determined on (Gallnkamp melting point apparatus). Thermal analysis were carried out under (N_2) atmosphere using α -Al_2O_3 as reference(Iran) . The ionic conductivity was measured by using (Konduktoskop E 365B Metrohm Herisau) conductivity instrument at room temperature.

2.1. Synthesis:

2.1.2. Preparation of Schiff base complexes:

To a hot solution of $CoCl_2.6H_2O$, $FeSO_4.7H_2O$, $Cu(NO_3)_2.3H_2O$,and $NiCl_2.6H_2O$ in (15 ml)ethyl alcohol (2.5mmol)was added (2.5mmol) solution of Schiff base (L) are mentioned in literature [19] in 25ml ethyl alcohol drop wise with continues stirring and gentle heating for 1hr. The volume of solution reduced to half, colored solid compound was formed were filtered off , washed several times with hot deionized water and then with ethyl alcohol and dried under vacuum.

2.2. Computational methods

The theoretical study in work was performed using the computationals method by hyper chem. 8.02. Geometry optimization of the studied , compounds by performing the semi-empirical molecular orbital theory at the level PM3[20].

Result and discussion

The C.H.N , molar conductunce data and physical characteristic of new complexes are listed in table (1).

Table(1). Physical properties , molar conductivity and C.H.N data of prepared complexes.

No.	Empirical formula	Ω^{-1} cm^2 mol^{-1}	%	Color	m.p ^oC	Elemental analysis found(calc) C	H	N	S
I	$[CoL(H_2O)_3]\ Cl_2$	243.7	%60	Dark green	> 320	58.61(57.97)	5.39(5.04)	5.11(5.01)	
II	$[FeL(H_2O)_5]SO_4,2H_2O$	108.6	%70	Golden yellow	126-128	48.89(49.70)	5,23(5.56)	4.20(4.29)	4.33(4.90)
III	$[CuL(H_2O)_5]\ NO_3.H_2O$	226	%65	Dark brown	260-262	48.33(48.29)	4.80(5.01)	9.01(8.36)	
IV	$[NiL(H_2O)_3Cl_2]$	-	%62	Green	233-235	60.60(60.15)	5.71(4.67)	5.73(5.19)	

Journal of Physics: Conference Series **2063** (2021) 012004 doi:10.1088/1742-6596/2063/1/012004

The assignment bands of(IR) spectra of Schiff base ligand and their metal complexes are listed in table(2).

In The Infrared spectra of Schiff base ligand (L) , The νC=N stretching frequency for azomethane group is observed at 1625 cm^{-1} . The value of these band was shifted to lower wave number in the spectra of all metal complexes in the range (1589-1618)cm^{-1} which due to coordination the nitrogen of azomethane with the central metal ions [21,22]. On the other hand all complexes exhibit a broad bands at (3315-3454) and bands at (910 -960)cm^{-1} which due to stretching modes of O-H group and rocking mode of coordinated water to metal ion respectively [23,24]. The ν(C-N) stretching modes appeared around(1172-1244) cm^{-1} and the ν(C=C)stretching frequency of aromatic ring appears a round(1508-1591) cm^{-1}, While the bands in the (3020-3059) cm^{-1} and (2860-2924) cm^{-1} attributed to ν(C-H) stretching of aromatic rings and ν(C-H) aliphatic respectivily. The new bands appears in the spectral of all metal complexes a round (509-580) and (428-518)cm^{-1}due to Metal-Oxigen(M-O) and Metal-Nitrogen(M-N)stretching mode[25]

Table2. The assignment IR bands of schiff base ligand and their complexes:

Com	υC=N (cm^{-1})	υO-H (cm^{-1})	υC-H aro. (cm^{-1})	υC-H ali. (cm^{-1})	υC-N (cm^{-1})	υM-N (cm^{-1})	υM-O (cm^{-1})
L	1625		3030	2950	1192		
Co(II)	1589	(3348-3444)	3036	2924	1222	428	516
Fe(II)	1599	(3358-3454)	3020	2916	1244	456	520
Cu(II)	1600	(3352-3444)	3059	2920	1172	432	509
Ni(ii)	1618	(3315-3446)	3030	2860	1234	518	580

Figures 1 and 2 show the typical ^{1}H NMR spectrum of compound(I) and (III) respectively , The N M R data for all complexes are summarized in the table(3). All the ^{1}H NMR spectrum of new complexes were studied [26,27] by the presence of singlet signals in the range (7.73-8.56) ppm which due to proton of azomethane group in (1) position. The aromatic protons showed multiplet signals in the range (6.18-8.22) ppm due to the protons in (4,5,6,7 and 8) positions in the aromatic rings. While the aliphatic proton of CH_2 group (proton in 2 position) showed singlet signals within the range (1.19-2.08) ppm . The OH protons showed singlet signal in the region (3.81-4.68) ppm which appeared to coordinated water with metals. While the peak of solvent (DMSO) showed singlet signal within the range 2.50 ppm and 3.30 ppm

Table (3). The ^{1}H NMR spectral data of new complexes

Complex	δ ppm CH=N	δ ppm aromatic protons	δ ppm $-CH_2-$	δ ppm O-H
Co(II)	8.56	(6.35-6.92)	2.04	4.68
Fe(II)	7.73	(7.01-7.20)	2.08	3.81
Cu(II)	8.23	(6.25-7.52)	1.20	4.14
Ni(IV)	8.56	(6.87-8.22)	1.19	3.92

Journal of Physics: Conference Series **2063** (2021) 012004 doi:10.1088/1742-6596/2063/1/012004

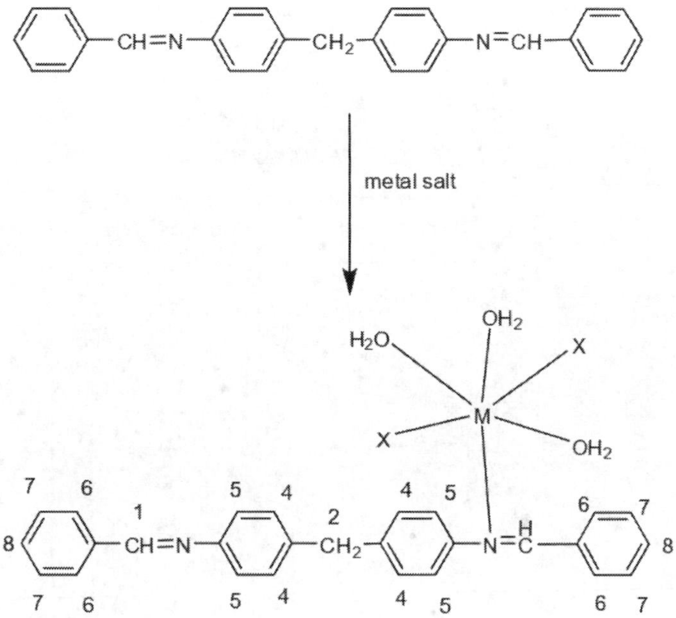

X = Cl in Co S1 and Ni S1 compounds
X = OH_2 in FeS1 and Cu S1 compounds

Scheme (1)

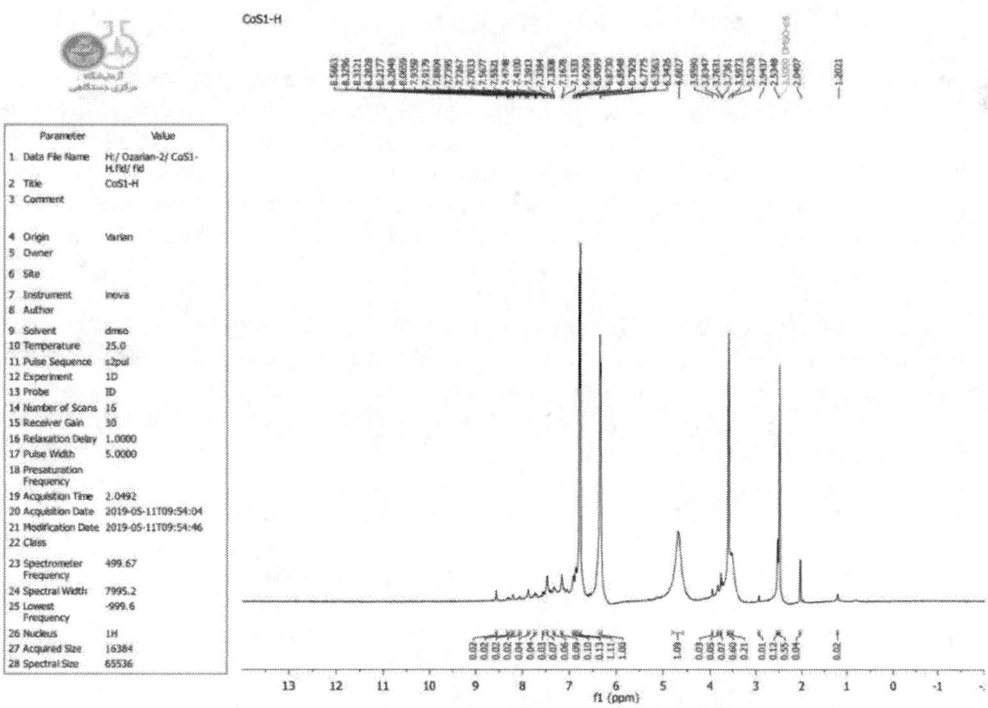

Figure (1) ^{1}HNMR spectrum of (I)

Journal of Physics: Conference Series **2063** (2021) 012004 doi:10.1088/1742-6596/2063/1/012004

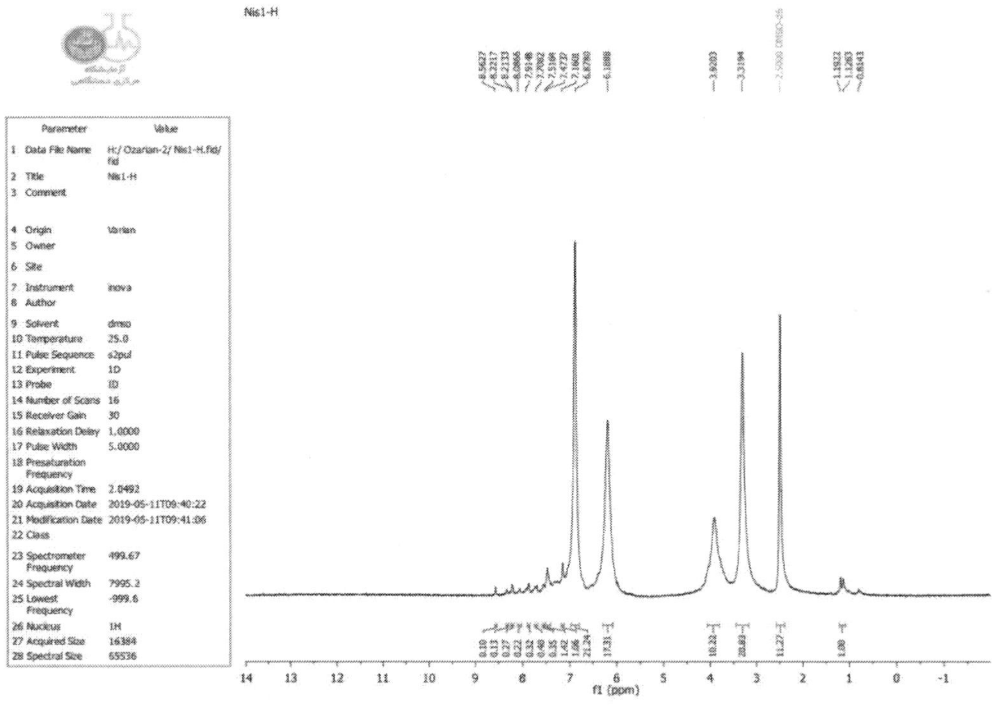

Figure (2) ^{1}HNMR spectrum of (III)

TG/DTA analysis of synthesized complexes was done from ambient temperature to 817 °C under (N_2) atmosphere. The TG/DTA data for prepared complexes are listed in table (4), and the thermo grams are given in figurs (3-6). The TG/DTA curve of Co(II) complex (I) exhibit one exothermic decomposition step occur in the range 334-360 °C with weight loss 36.17% calc.(36.60 21% found) indicate to removal the three coordinated water ,chlorine and non-coordinated ligand moiety [29-30].The mass residue at 801°C is 29.12%(30.44%) may be attributed to formation C_7H_6NCo.

The TG/DTA curve of Fe (II) complex (II) exhibit three exothermic decomposition step . The first step occur in the range 116-140 °C with mass loss of 5.517%(5.14%) due to removal the two crystalline water molecules. A small peak at 126°C indicate to melting point .The second step occur between 280 -340 °C with weight loss of 20.23%(20.23%) due to removal sulfate ions . In the third step which occur in the range 480-510 °C a peak exhibit at 498°C with mass loss of 54.88% (54.31%) attributed to loss the five coordinated water and non-coordinated part of Schiff base ligand[29,30]. The mass residue at 750°C is 35.48% (35.51%) may be attributed to formation Fe_3O_4.

The TG/DTA curve of Cu(II) complex(II) exhibit three endothermic decomposition step .The mass loss 2.69% (2.63%) in the first step occur at 100°C due to loss crystalline water molecules . The second step between 165-260 °C with mass loss of 16.12% (16.12%) attributed to loss coordinated water molecules. In the third stage which occur in the range 430-500 °C with mass loss of 34.62% (34. 63%) due to loss tow nitrate ions .At 817°C the mass loss 46.128% (46.52%) due to loss the non-coordinated part of Schiff base ligand .

In the TG/DTA curve of Ni(II) complex (IV)exhibit two exothermic decomposition step . The first step consist of two peaks at 128 and 198°C with mass loss of 3.33% (3.35%) due to elimination of one water molecule . A small peak at 235 °C indicate to its melting point . Next stage occur in the

Journal of Physics: Conference Series **2063** (2021) 012004 doi:10.1088/1742-6596/2063/1/012004

range 341 to 428 °C with mass loss of 54.82% (54.42) attributed to loss coordinated water , two chloride ions and non-coordinated ligand moiety .

Table(4). TG/DTA of synthesis complexes:

Complex	Step	Type of molecule loss	Temp. of TGA (°C)	Temp. of DTA°C	Mass residue%		Mass loss%	Temp. of 50% Mass loss°C
					Calc.	Found		
Co (II)	I	$3H_2O+2Cl^-$	334-	344			36.17	
	Last	$+C_6H_5$	360	801	29.3	30.44	36.21	387
		Formation						
		C_7H_6NCo						
Fe (II)	I	$2H_2O$	116-140	119			5.547	
	m.p			126			5.136	489
	II	SO_4^{2-}	280-340	298				
	III	$5H_2O$ +	480-510	498			20.23	
	Last	$2C_6H_5$		750	35.48	35.51	19.24	
		Formation of					54.88	
		F_3O_4					54.31	
Cu (II)	I	H_2O		100			2.69	>
	II	$5H_2O$	165-260	255			2.63	700
	III	$2NO_3^{-1}$	430-500	450			16.12	
	Last	C_6H_5		817			16.12	
							34.62	
							34.63	
							46.128	
							46.52	
Ni (II)	I	H_2O	123-265	128,198			3.22	
	m.p			235			3.35	403
	II	$2H_2O+2Cl^-$	341-428	411			54.82	
		$+C_{13}H_{11}N$					54.42	

Concerning the computational Study, All optimised structure of this study molecules are shown in Figures (7) and (8). The (PM3) geometry optimizations yields planar structure for the new compounds. The generals geometry of molecules all this compounds are very similar. The [the dipole moment, μ (in Debyes), highest occupied and the lowest unoccupied molecular orbital (HOMO and LUMO, respectively) energies, total energy, and the energy band gap (LUMO–HOMO energy difference) and for the studied molecules are given in Table (5)]. The dipole moment calculated indicates the molecules are pole. This means that these molecules may, especially other polar molecules.The spatial distributions of(HOMO LUMO) are shown in Figure. (9). In general the all molecules gave similar (HOMO) and (LUMO) orbitals.

Conclusion:

Co(II), Fe(II) ,Cu(II) and Ni(II) complexes with bis-(4-benzyllidenamino-phenyl)-methane have been prepared and characterized by different physical and chemical methods. The analysis confirmed the structures of complexes . All synthesized complexes are thermally stable Based on the thermal studies and most of the complexes in this studies shows similar behaviors and the Cu(II) complexes are more stable . Molar conductrivity measurements show that complexes (I , III) behave as 2:1 electrolyte while complex (II) behave as 1:1 electrolyte.

Table(5). The Total energy, MO energy of the lowest, highest, (HOMO) , (LUMO) , levels, and the dipole moment, μ (in Debyes) for the studied molecules

Mol.	Method	Total energy (Kcal/mol)	HOMO (ev)	LUMO (ev)	μ (D)
II	PM3	-1.38871	-6.503585	-0.4283036	6.287
IV	PM3	-1.50249	-9.076046	-1.378737	8.481

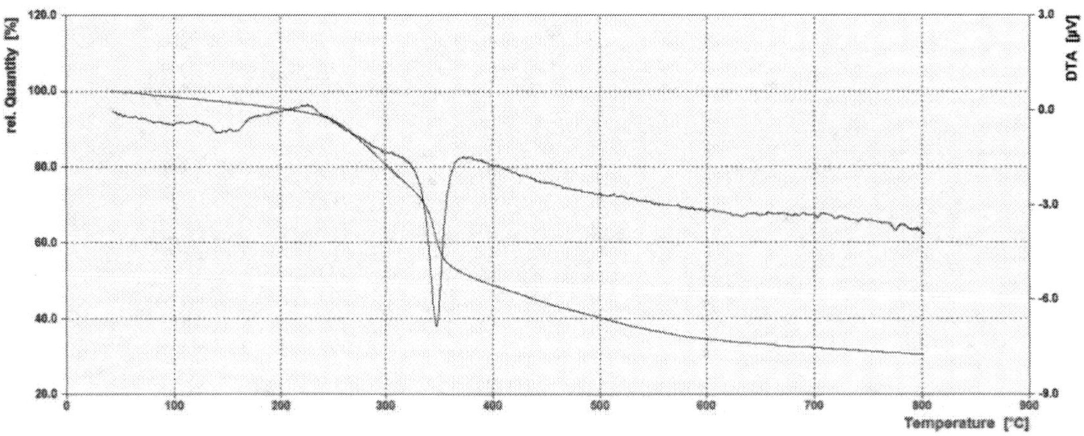

Figure (3) TG Curve of (I)

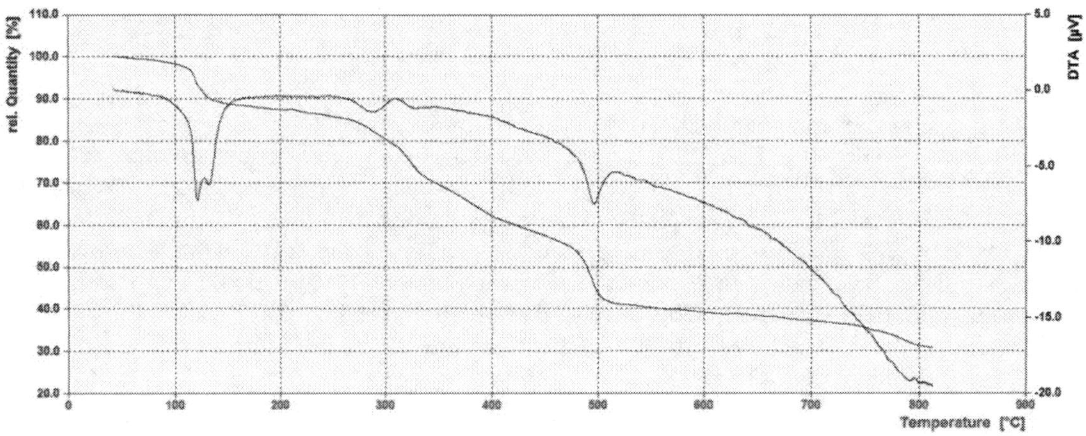

Figure (4) TG curve of (II)

Journal of Physics: Conference Series **2063** (2021) 012004 doi:10.1088/1742-6596/2063/1/012004

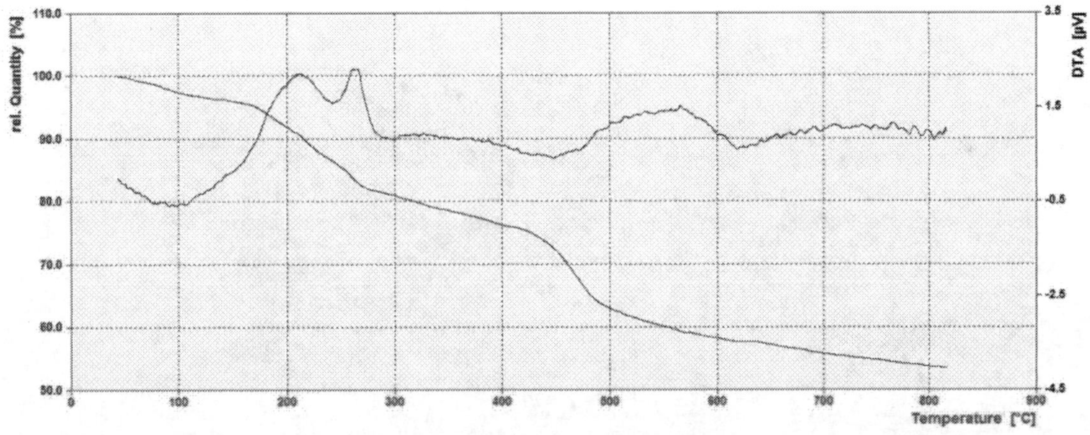

Figure (5) TG curve of (III)

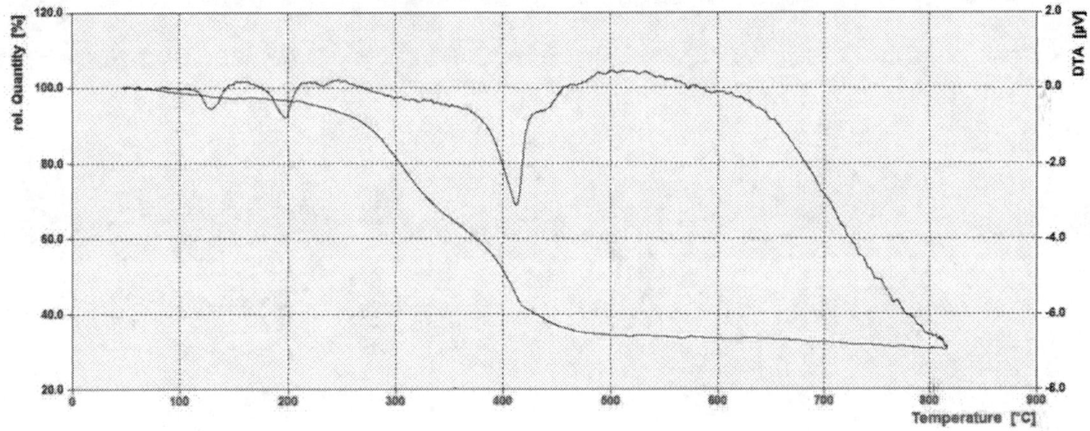

Figure (6) TG curve of (IV)

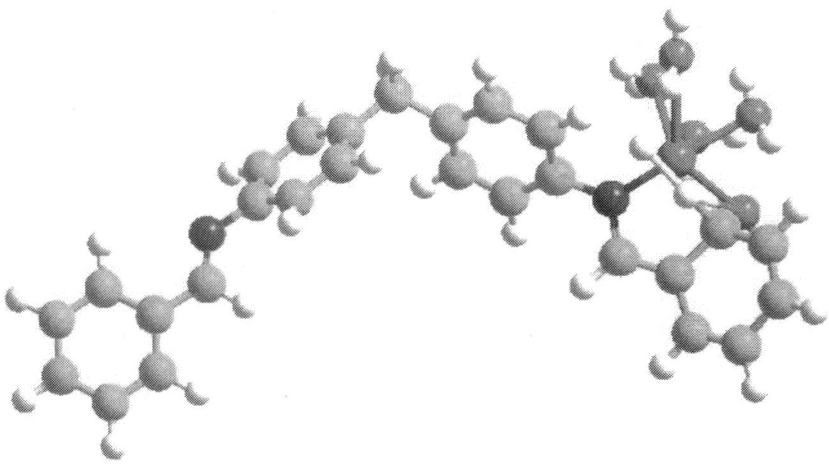

Figure (7) Balls and cylinders model of (II)

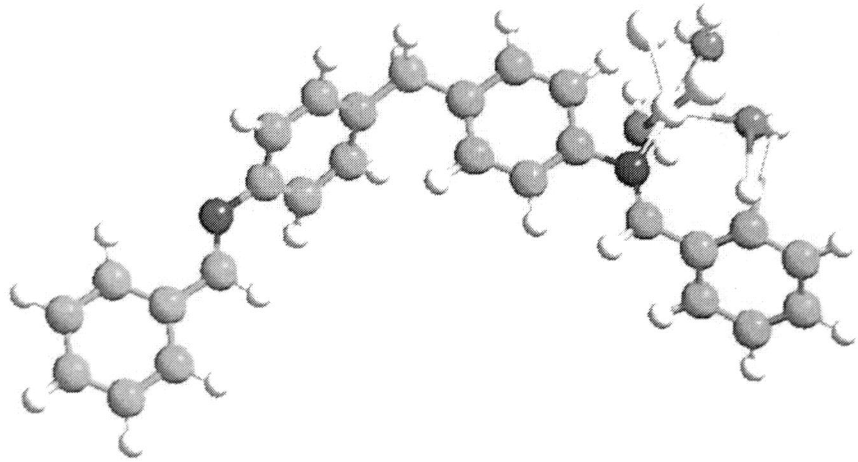

Figure (8) Balls and cylinders model of (IV)

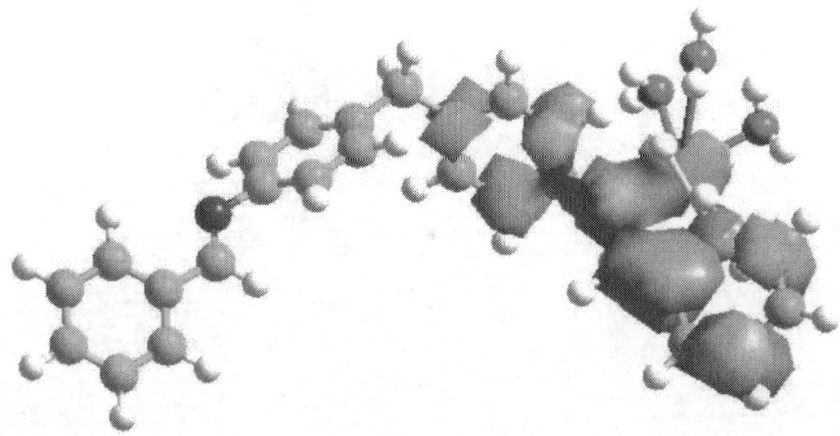

HOMO (II)

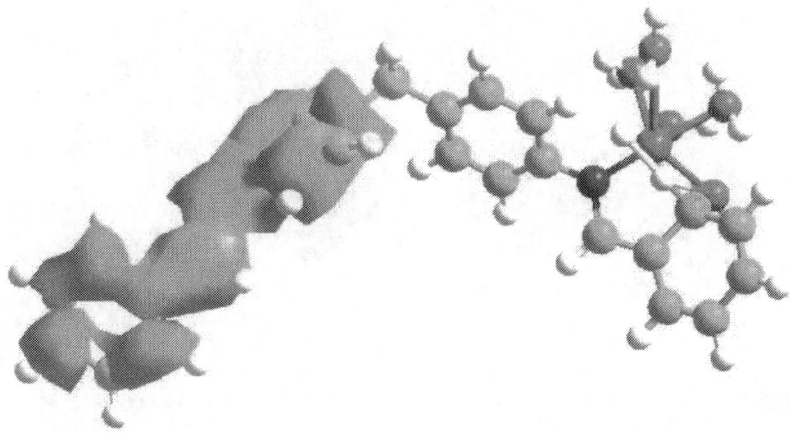

LUMO(II)

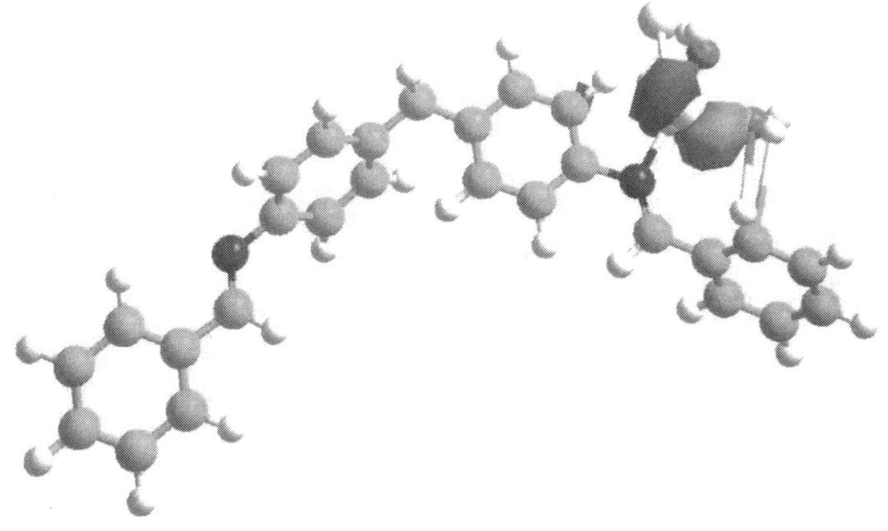

HOMO (IV)

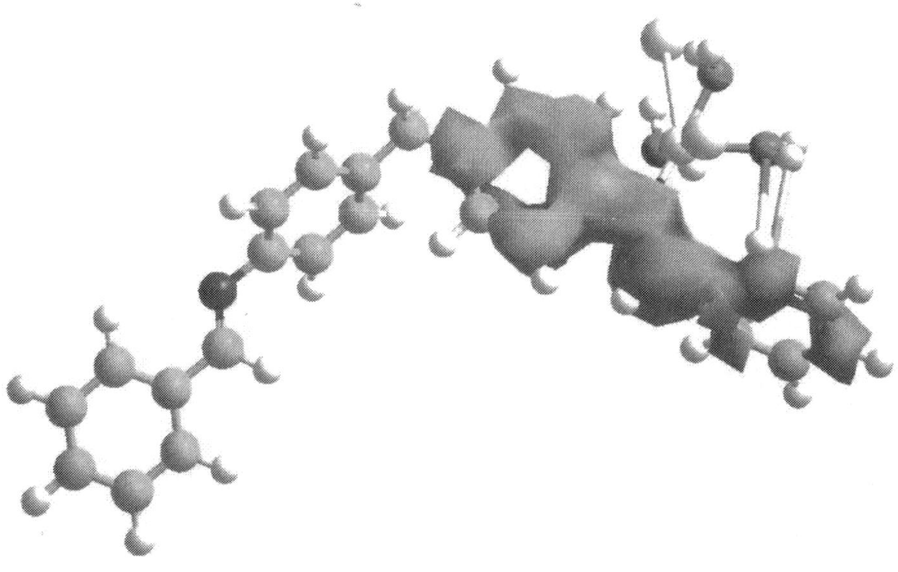

LUMO (IV)

Figure (9) The optimized structure of the studied molecules optimization has been performed by PM3 method.

References:

[1] T. Aboul-Fadl , F A S Bin-Jubair and O. Aboul-Wafa 2010 *European Journal of Medicinal Chemistry*. **45** 4578-4586 .

[2] M D I Ferreira , T R A Vasconcelos and E M. de Carvalho 2009 *Carbohydrate Research*. **334** 2042-2047.

[3] A B Thomas , R K Nanda , L P Kothapalli and S C Hamane 2011 *Arabian Journal of Chemistry* .

[4] M A Bhat and M A Al-Omar 2011 *Acta Poloniac pharmaceutica* **68** 375-380 .

[5] A A Kulkarni , S B Wankhede , N D Dhawale , P B Yadav , V V Deore and I D Gonjar 2012 *Arabian Journal of Chemistry* .

[6] Nithinchandra, B Kalluraya, S Aamir , and A R Shabaraya 2012 *European Journal of Medicinal Chemistry* **54** 597-604.

[7] Gupta K G ,Sutar AK 2008 *Gatalytic activity of Schiff base complexes coord. Chem. Rev* **25a** 1420-1450.

[8] M S Alam , J Choi and D Lee 2012 *Bioorganic Medical Chemistry* **20** 4103-4108.

[9] Lin Che T ,Chang Gao Q , She zhao J and Zhang G 2008 *Chin. Journal Cem* **26** 1079-1084 .

[10] MM Tamizh, K Mereiter, K Kurchner and R Karvembu 2012 *J Organomet Chem* **700** 194-201.

[11] AS Gaballa , MS Asker , AS Barakat and SM Teleb 2007 *J Spectrochim Acta* **67** A 114-121.

[12] N Raman, A Kulandaisamy , A Shanmugasundaram and K Jeyasubramanian 2001 *J Trans Met Chem* **26** .

[13] Y Prashanthi, K Kiranmai , Subhashini NJP and Shivaraj S 2008 *J Spectrochim Acta* **70** A 30-35.

[14] S. Santhi and CG. Radhakrishnan Namboori 2013 *Int J Chem Tech Res* **4** 1750-1755 .

[15] M. Hajrezaie, M. Paydar , SZ. Moghadamtous, P. Hassan Darvish and M. Zahedifard 2014 *Sci World J* ***1***.

[16] Li YF , Liu ZQ 2011 *Eur J Pharm Sci* **44** 158-163.

[17] TS. Kamatchi , N. Chitrapriya , H. Lee, CF. Fronczek , FR. Fronczek and K. Natrajan 2012 *Dalton Trans J* 2066-2077.

[18] KS. Kumar , S. Gangly, R. Veerasamy and E. De Clercq 2010 *Eur J Med Chem col.* **45** 5474-5479.

[19] H. Heba -Sabah 2014 *J .Derpharma chemical*, **6** 38-41.

[20] M. Yoko, M. Mai , T. Yuriko , S. Mitsuaki , W. Hidetsugu and T. Hiroyuki 2010 *J Tetrahydron,* **66** 605-611.

[21] D. Fatih , U. Mahmut , F. Omar and Ozturk 2009 *J Therm Anal Calorim* **98** 785-792 .

[22] P. Yokeswari Nithya , P. Ananthi and D. Shanmuga priya 2019 *International J of Advanced Scientific Research and Management , Special ISSU* **4** 77-81.

[23] W.J Geary 1971 *The use of conductivity measurements inorganic solvent for the characterization of coordination compound Coot. Chem. Rev.* **7** 81.

[24] Atared F Hassan , H. Hayat Abass and A. Nadhim Abul-Hussein 2016 *International Research Journal of Natural Sciences* **4** .

[25] JR. Ferraro (1971) *Loww Frequency Vibrations of inorganic and coordination compounds* 2nd. Edn..(Wiley , New York).

[26] R M Silverstien, F X Webster and D J Kiemle 2005 *Spectrometric Identification of Organic Compounds* 6th ed (John Wiley and Sons, New Yourk, USA).

[27] R L Shriner and C K Hermann 2004 *Spectroscopic Techniques for Organic Chemistry* (John Wiley and Sons, New Yourk, USA).

[28] J.W. Cooper 1980 *Spectroscopic Techniques for Organic Chemistry* (John Wiley and Sons, New Yourk, USA).

[29] N Kavitha , P V Anatha Lakshmi 2017 *Journal of Saudi chemical Society* **21** S456-S466.

[30] S M. Abdalah , MA. Zayed and G. Gehad 2010 *Arabian J Chem.* **3** 103-113 .

Synthesis, characterization and Stydy thermal stabilization effect of calcium complex of anthocyanin on the poly (vinyl chloride)

Hikmat A Ali[1,*], Hussein A Shnawa[2], Dhiaa A Abidalimam[2]

[1] Department of polymer and petrochemical engineering, Basrah university for oil and gas, Basrah,Iraq
[2] Polymer Research Center, University of Basrah, Basrah, Iraq

Corresponding author's e-mail: hikmatali42@gmail.com

Abstract. In this study, calcium complex of anthocyanin was synthesized from anthocyanin, a flavonoid type natural phenolic product, which was extracted from eggplant peel. The structure of Ca-anthocyanin was characterized by FTIR spectroscopy. Its efficiency as bio-based thermal stabilizer to stabilize poly (vinyl chloride) was investigated and compared to that of Reapak B-NT/7060, a Ca/Zn-based commercial stabilizer. The influence of Ca-anthocyanin complex on the thermal degradation reaction of PVC was monitored by differential scanning calorimetry (DSC). The results indicated that the Ca-anthocyanin complex is an efficient thermal stabilizer and it reduces the rate of dehydrochlorination reaction of PVC already at a concentration as small as 2 phr, part per hundred of PVC resin. The thermal degradation reaction of PVC through the first degradation stage is clearly limited by the addition of Ca-anthocyanin complex as single (primary) stabilizer. Its efficiency is similar to that of Reapak B-NT/7060 used as reference stabilizer and it can enhance the performance of commercial stabilizer when used as co-stabilizer at mixing ratio (1:1). This study has allowed verifying and validating the usefulness of environmental friendly thermal stabilizer for PVC with very evident stabilization effect.

1. Introduction

It is well known that the thermal stabilizers are essential additives especially for PVC, one of the most important polymers intensively used as thermoplastics. Both flexible and rigid PVC products are highly dependent of thermal stabilizers and cannot be found at marking scale without these essential additives [1] [2]. Thermal stabilizers in addition to plasticizers play crucial roles in the convertible of PVC formulations from raw material to commercial products in addition to their final quality and physical characters. They have been used for: (i) increase the resistance of this polymer against any thermal degradation which can easily initiated by heat or by another agent such as light (photo-degradation) and reduce its fragmentation, (ii) improve its processing and get various indoor, outdoor and food packaging with high quality applications and (iii) for long-term thermal, light and mechanical stabilities.

PVC is highly sensitive to whole the processing temperature ranges and undergoes thermal degradation when exposed to heat both during extruding, molding and recycle steps [3][4]. The thermal instability behavior of PVC which is a complex phenomenon leads to irreversible changes in most of its desirable physical and mechanical properties. Various thermal stabilizers types at average concentrations ranged from 2 phr to 5 phr, part per hundred of PVC resin, have been widely used in a large range of PVC commercial formulations to enhance its thermal stability especially through thermal process. The most common thermal stabilizer types include metal soaps which consist of long chain carboxylic acid salts, lead salts and organo-tin compounds [5][6] [7] and other types (auxiliary or co-thermal stabilizers). In particular, many PVC thermal stabilizers that based on heavy metals ions such as tin and lead, has been forbidden or at limited uses in many countries because of their deleterious impacts on human and environmental systems or because their high price. New thermal stabilizers or co-stabilizers with efficient stabilization and plasticization effects can be synthesized from various types of bio-based raw materials[8][9][10]. A large number of studies were carried out to synthesize and evaluation of new compounds with different chemical structures as PVC thermal stabilizes with or without another activity such as photostablization, plasticization or even with any synergistic effects with all polymer formulations components. Mei Li and co-workers [11] synthesized thermal stabilizers from dipentene and gum rosin using two step; Diels–Alder addition used to prepare maleated dipentene and acrylopimaric acid in the first step; subsequently in the second one, the researchers converted these derivatives to the corresponding zinc salts and calcium salts. Johan Steenwijk et al. [12] reported a systematic investigation on the uses of natural polyols such as sorbitol and xylitol as PVC long-term thermal stabilizers . Other natural compounds, vegetable oils, have a potential to be modified and used as PVC thermal stabilizers or co-stabilizers with plasticization effects. In a recent studies, the stabilization effect of plant oils based epoxy compounds was investigated in detail [13][14][15].

The general results of those studies revealed that the stabilization efficiency of epoxidized vegetable oils is based on ability of epoxy group to accept the evolved HCl and to formation of ester bonds with reactive chlorine atoms. Taghizadeh and co-workers [16] studied the synthesis of epoxy derivative from commercial sunflower oil. The new tannins based thermal stabilizers were synthesized by complexation with a base, Ca(OH)2 for example, through it phenolic hydroxyl groups using simple reaction conditions or by epoxidization reaction with epichlorohydrin [17][18]. Another naturally occurring phenolic compounds, anthocyanins, which might potentially be used for the stabilization of PVC. Anthocyanins are the main class of flavonoids which give different color (red and/or purple color) to fruits and many colored vegetables parts of plants. Anthocyanins are natural phenolic monomer which can be isolated from many plant parts such as eggplant peel, figs peel and other [19][20][21][22]23].

In present article, anthocynains were extracted from eggplant peels obtained as waste of local food consumption. For that purpose, it had been converted to calcium complex and evaluated as a primary/single stabilizer as well as in combination with a Ca/Zn based commercial thermal stabilizer (co-stabilization effect). Differential scanning calorimetry (DSC) (dynamic mode) was used to characterize the stabilization efficiency of this complex.

2- Experimental

2.1. Materials

PVC (a suspension polymerization type with a k-value of 65 and dioctyl phthalate (DOP) used as plasticizer were kindly provided by state company of petrochemical industries, Basrah, Iraq. HCl was supplied by Biosolve BV, Netherlands. Calcium hydroxide was purchased from Hi-Media laboratories, India. Sodium hydroxide was purchased from Thomas Baker, India. Reapak B-NT/7060 (Ca/Zn-based powder), which used reference thermal stabilizer was supplied by Reagens Co. (Italy).

All reagents and polymer involved in this work were used as received without further purification

2.2. Isolation of anthocyanins

Eggplant (*Solanum melongena* L.) was purchased from local market. Its peel was manually isolated by sharp knife and dried at 50°C for one week. The dried peel was crushed using coffee grinder. 30 g of this powder was extracted with 250 ml of 10% sodium hydroxide aqueous solution at reflux for 24hr. After that, the crude extract was filtered, acidified with 2% HCl solution to pH of 4-3, washed with distillation water and dried in oven at 60°C for 24 hr.

2.3. Calcium-anthocyanins complex synthesis

The Ca-anthocyanin complex used in this work was synthesized from anthocyanins as crude phenolic mixture and calcium hydroxide .The synthesis began with dissolved of 5g of anthocyanins in 250 ml of water using heat and continuous stirring to form homogenous solution. After complete dissolving of anthocyanins, filtered and saturated aqueous solution of calcium hydroxide was added slowly with contentious stirring. The addition of calcium hydroxide solution was contentious until the pH rise to about 10-12. The solution was allowed to stand for 24hr, then a brown thick precipitate was

Journal of Physics: Conference Series **2063** (2021) 012005 doi:10.1088/1742-6596/2063/1/012005

obtained which was filtered and washed using distillation water for several times and dried at 60oC for 24hr. The chemical structure of anthocyanin-calcium complex was confirmed using FTIR analysis.

2.4. Preparation of stabilized PVC samples

One series of PVC formulations was prepared from PVC, anthocyanin-calcium complex as thermal stabilizer and DOP as plasticizer. The DOP content was fixed at 10 phr (parts per hundred PVC resin) in all PVC blends. One PVC sample was fabricated by mixing 2phr of commercial thermal stabilizer and used for comparison in this study. PVC was thoroughly mixed with varying proportions of this derivative and constant amount of DOP in a mortar. The composition of prepared PVC samples is reported in table 1.

Table 1, Samples designations and their corresponding compositions

Ingredients	PVC	DOP	B-NT/7060 phr	Ca-Anthocyanin
Pure PVC	100	10	0	0
PVC-BNT/2phr	100	10	2	0
PVC/Ca-anth/2phr	100	10	0	2
PVC/Ca-anth: B-NT/1:1	100	10	1	1

2.5.Fourier Transform Infrared Spectroscopy (FTIR)

The chemical structures of anthocyanins and its derivative were conferred by using FTIR spectroscopy (FTIR 4200, Jasco Corporation, Japan). The anthocyanisns and its salt were mixed with dry KBr and pressed to form the pellet. Mixtures of these compounds with KBr were scanned in the range of 400-4000cm^{-1}.

2.6.Evaluation the stabilizing efficiency

Thermal properties of PVC samples were analyzed with a DSC-60 (Shimadzu Cooperation, Kyoto, Japan) for evaluation of stabilizing efficiency on synthesized salt.

A sample of 8-10 mg of each PVC samples was weighted and loaded into aluminum open pan, and the DSC scans were conducted from room temperature to 350°C. All DSC measurements were carried out at a heating rate 20°C/min in a N_2 atmosphere flow at 20 mL/min flowing rate.

3. Results and Discussion

Herein, the anthocyanins were isolated from the eggplant peel in alkaline aqueous medium then it used to prepare a new bio-based thermal stabilizer. Ca-Anthocyanin salt or complex was synthesized through precipitation process with calcium hydroxide. FTIR analysis were conducted to confirm the phenolic property of anthocyanins and to investigate its interaction with calcium ions. The FTIR spectrophotometry analysis results of anthocyanins before and after modification are shown in figure 1' and figure 2' respectively.

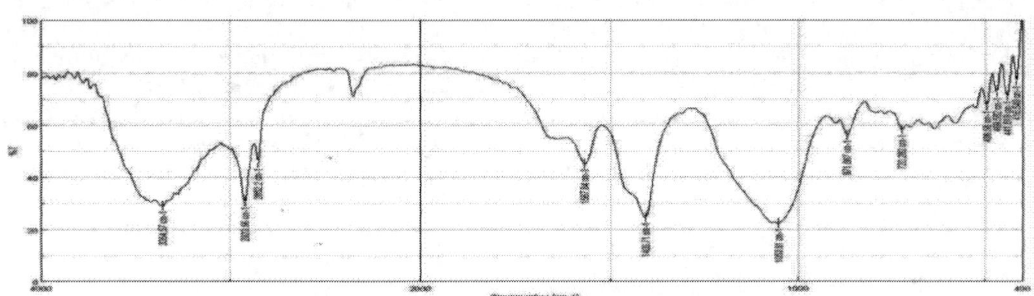

Figure 1. FTIR spectrum of anthocyanins obtained from eggplant peel by solid-liquid extraction.

Journal of Physics: Conference Series **2063** (2021) 012005 doi:10.1088/1742-6596/2063/1/012005

As can be seen in figure1' a wide and intense band appears at around 3350cm^{-1} related to the stretching vibration of phenolic –OH groups. Other characteristic peaks such as that for phenolic hydroxyl group in plane bending and C-O stretching at 1409cm^{-1} and 1050cm^{-1}, respectively, also can be seen in this figure. The skeletal vibration which absorb at 1560cm^{-1} and 1600 cm^{-1} related to C=C axial deformation are a characteristic peaks of aromatic carbon-carbon double bonds. The two bands obtained at 2920 and 2850 cm^{-1} that corresponds to symmetric and asymmetric C-H vibration respectively. The absorption peaks at 1050cm^{-1} which was due to the symmetric stretching vibration of phenolic -OH almost disappeared in the spectrum of the calcium salt of anthocyanin indicated that the hydroxyl groups of anthocyanins are coordinated to metal ions.

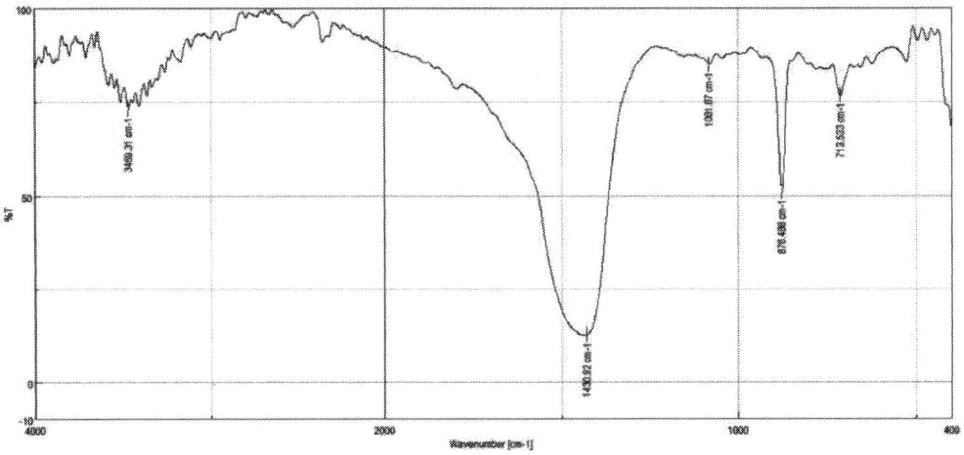

Figure 2. FTIR spectrum of anthocyanin-calcium salt.

Around region of 720-770cm-1, it was observed a series of weak combination and less intense bands related to C-H asymmetric vibration of aromatic rings. The bonding between anthocyanins as ligand and calcium ion can be inferred by comparing absorption peaks of the anthocynin-calcium complex with that of free anthocyanins. The FTIR spectrum of the calcium-anthocyanin salt is shown in Figure 2' This spectrum is very diverse to that of free anthocyanins, many characteristic peaks were disappeared or shifted and some new peaks appear. The combination of anthocyanins with calcium ion leads to weaken many bonds such as -O-H, -C-O- bonds so their position will appears at lower wave number value and intensity per that free or that combined with hydrogen because of bonding with low electron density ion. The strong absorption peaks of hydroxyl groups which appear at 3350cm-1 in the FTIR spectra of anthocyanins nearly transferred to 3470cm-1 with lower intensity after generating calcium salt, indicating to incorporation of most of hydroxyl groups of anthocyanins in the reaction with calcium hydroxide to produce anthocyanin-calcium complex. This spectrum also showed a new absorption peak at 1430 cm-1 attributed to the symmetric stretching vibration of carbon-calcium bonds in anthocyanin-Ca complex. As can be seen in figure 2' this peak has relativity high intensity due to the high amount of –O-Ca-O- group in this complex. At wave number value 876 cm-1, it was observed a sharp band related to O-Ca ionic bond [17]. These changes in the stretching vibration of OH group after complexation can be considered as an indicating to the involvement of this group in coordination with calcium ion and formation Ca-O bond.

The FTIR analysis suggests that the hydroxyl groups of anthocyanin were successfully converted to corresponding salt by direct reaction between calcium hydroxide and anthocyanins. However, there was still a small fraction of unreacted hydroxyl groups, which was not connected with calcium ion and still free.

3.1.DSC analysis

Alternatively to TGA, DSC technique has proven to be one of the most practical and powerful technique to study the thermal properties and degradation reaction of PVC and get enough information to evaluate the performance of an additive as thermal stabilizer to this polymer. In order to study the impact of Ca-anthocyanin on the thermal degradation behavior of PVC. DSC technique was used to analysis and monitored the degradation reactions of PVC mixed with appropriate amount from this product. Dynamic DSC scans of PVC samples from room temperature to 350°C are shown in Figuers. 3-6.

According to the literatures data, the thermal degradation reaction of PVC consist of two main distinct reactions, the first reaction is evolving of hydrogen chloride from PVC chains which named dehydrochlorination (DHC) reaction and forming a conjugated double bonds structure named polyenes sequence and fracture or cyclization of the double bonds structures to form other cyclic or cross-linked structures. The second one is polymer fragmentation and chains scission to low molecular weight or gaseous products. The second stage of PVC thermal degradation or decomposition usually takes place at a temperature degree higher than 400°C [24] which is higher than of the DSC scales used in this study. Figure 3 show the dynamic calorimetric curve of PVC sample without any additive (non-stabilized PVC). It can be seen that the DSC curve shows two endothermic degradation peaks of PVC degradation, the first peak located about 250°C and the second peak located at 305°C. Both two peaks associated with the first degradation stage (DHC reaction and some other reactions). Because a huge amount of literature resources about PVC thermal degradation process and mechanisms showed that the second degradation stage which consist of loss of low molecular weight fragments usually take place at temperature region higher than 400°C, the two endothermic peaks attributed to the first degradation stage.

The first peak is corresponding to the main degradation reaction which is evolving of HCl with auto catalytic effect by HCl while the second one is corresponding to secondary reactions such as crosslinking or cyclization reactions. The reason for splitting of this stage into two peaks or two steps is unclear and may be due to the formation of polyenes sequences in the first step and formation of crosslinking and cyclic structures in the second step. In order to monitor the thermal degradation reaction of stabilized PVC, a PVC samples with and without additives were typically heated to a certain temperature and then the temperature regions and absorbing or releasing heat related to thermal degradation reactions were recorded by DSC apparatus. All the DSC thermal data for the thermal degradation of PVC samples were summarized in Table 2. As mentioned above, the first stage of PVC thermal degradation occurs by complex process or chain-stripping process through two reactions. DSC curves of all PVC samples containing thermal stabilizer, in the first degradation stage, can be divided into two peaks. One can be seen that the first reaction appears as endothermic process and started in the range of 200-220°C and the second one started at about 275-320°C. Before this temperature region, PVC sample do not exhibit any thermal transition except the glass transition relaxation that seen at about 88-89°C. This temperature was specific for pure PVC without any thermal treatment and depends on its native physical and chemical properties and on the amount of the plasticizer. The Tg relaxation of PVC sample followed by relatively strain line or thermal stability phenomenon up to 200°C. After that, a broad endothermic peak appears at 250°C in the DSC curve with peak area or enthalpy equal to (-35.55 J/g) which is associated with the first type of degradation reaction (DHC reaction) for non-stabilized PVC.

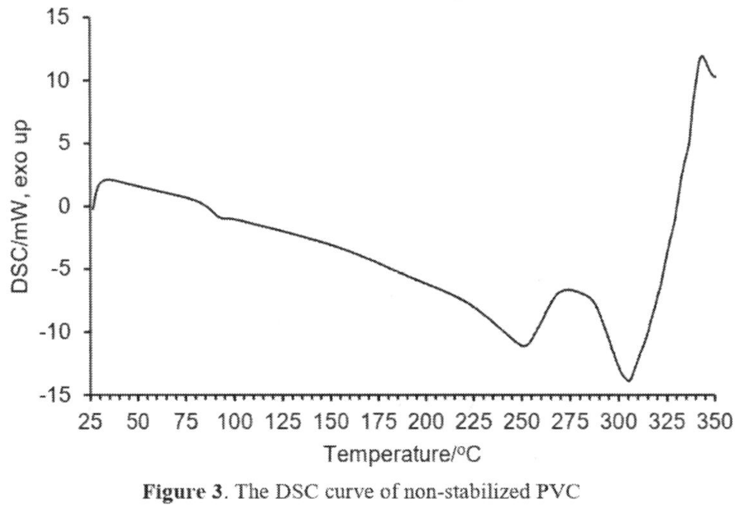

Figure 3. The DSC curve of non-stabilized PVC

During the course of PVC degradation, PVC eliminates HCl with the formation of alternating double bonds with different longs across the polymer chains. The HCl elimination reaction is endothermic process because the released energy from this reaction has been used to production of less stable double bonds [25]. The result unsaturated polymeric chains containing double bonds split up into lower molecules such as benzene and toluene or enter in additional reaction with itself or other molecules in the polymer system, crosslinking and chain scission for examples [26][27].

Attention should be paid to the fact that the main action of a primary thermal stabilizer should be in the first stage (both in the frist and second steps) of PVC degradation in which all HCl molecules release from the polymer backbone and other secondary reactions. Therefore, much attention must be directed toward the two peaks in the DSC scans. In the second step of PVC thermal degradation which appears as endothermic peak at 305°C with reaction enthalpy -74 J/g followed directly by high exothermic transition. Complex series of reactions occur simultaneously during the second stage. However, this cannot be taken as an indication of HCl scavenging ability by thermal stabilizers because it takes place at a temperature at which the additive is thermally degraded and do not consist of HCl releasing. But this process would account for the ability of stabilizer or its fragments to work with unstable groups and its retardation effects to the crosslinking and cyclization reactions as results to its action in the first stage.

By knowing the total enthalpy of totally non-stabilized PVC sample, the performance of any thermal stabilizer can be calculated by DSC. Differently to most of thermoplastics, PVC thermal degradation reaction especially the DHC reaction is endothermic process therefore it can be detected by DSC. Clearly, the total degradation enthalpy of PVC was lower in the presence of commercial thermal stabilizer compared to the pure sample. B-NT was efficient stabilizer for PVC since it can retard the degradation reaction of PVC due to its action as an effective HCl scavenger and possibly by other activities, due to the presence of metal carboxylates groups. The DSC curve obtained for PVC stabilized with 2 phr B-NT scanned in the temperature range 25-350°C is shown in Figuee 4' From this curve, it is evident that the PVC sample has more thermal stability in comparison with pure sample which was shown in Fig 3. And the thermal degradation peaks and enthalpies become smaller per that of previous sample.

Both the thermal data of the two samples as presented in Table 2 and DSC curves show that the degradation reaction of stabilized PVC start relatively at nearby temperature of pure sample but it contentious for higher temperature.This observation may explained by the effect of thermal stabilizer which leads to decrease the rate of thermal degradation reaction in in this stage. As the first step of PVC thermal degradation reaction appears as endothermic peak, the second step shows several small peaks expand to 350°C. This may be a result of accompany between the thermal degradation process of PVC and that of thermal stabilizer. It can be observed from the DSC curve in Fig 4. that the initial degradation of PVC sample occurred at 200°C, while PVC mixed with commercial thermal stabilizer displayed initial degradation at 216°C.

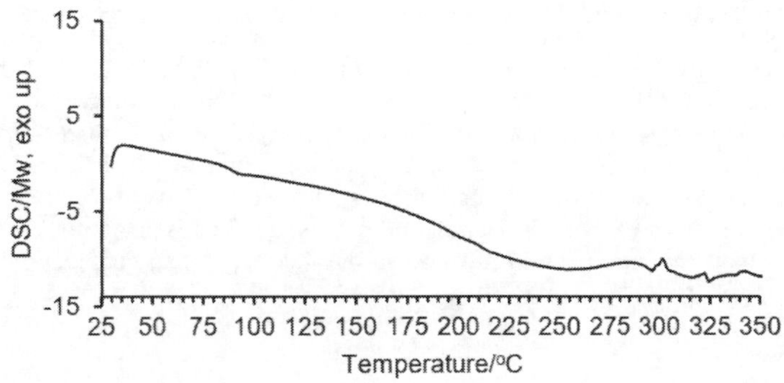

Figure 4. The DSC curve of PVC stabilized with 2 phr B-NT

It is observed that the DSC curve of PVC stabilized with 2 phr of ca-anthocyanin complex shows an evident endothermic peak at around 320°C corresponding to the second step of PVC thermal degradation, whose decomposition peak similar in its shape to that of un-stabilized PVC. In the second step of PVC thermal degradation, the absorbed energy (reaction enthalpy) equal to -2.56 J/g which is very lower than that obtained for pure PVC. By analysis these results, it was observed that the Ca-anthocyanin complex showed thermal stabilization activity closed to that of commercial one only in the first degradation step. The addition of Ca-anthocyanin complex to PVC leads to stabilize this polymer in the first stage of thermal degradation process whereas most of PVC thermal process and molding are carried out at below or around this temperature range (below 200°C). For all PVC samples with pure Ca-anthocyanin complex, or with commercial thermal stabilizer, or with mixture of the two additives, Fig. 5, this stage appears as small and broad endothermic peak in the range of 225-280°C with reaction enthalpy -2.56 J/g, Table 2.

Table 2, The DSC results of pure PVC and PVC stabilized with various thermal stabilizers.

Samples	$T_{start}{}^{a}$/ °C	$T_{max}{}^{a}$/ °C	$T_{end}{}^{a}$/ °C	Deg. rate[b]/ mW min^{-1}	Deg. heat[c]/J g^{-1}
PVC neat	222	251	270	8.71	-32.55
PVC-2phr B-NT	216	254	280	3.28	-15.44
PVC-2phr Ca-Antha	226.6	246	276	1.14	-2.56
PVC-1phr BN-T/7060:1phr Ca-Antha	212	245	255	1.23	-2.02

[a] start, maximum and end temperatures of first stage of PVC degradation. [b] Curing rate of PVC degradation was calculated from the first derivative of degradation peak at first stage. [c] Degradation heat was calculated for the first degradation stage.

The thermal degradation reaction of PVC became slower in the presence of this derivative per that of pure PVC. These results confirm that the complex of anthocyanin with calcium increase the thermal stability of PVC which exerted more thermal stabilization behavior at less by HCl scavenging

Journal of Physics: Conference Series **2063** (2021) 012005 doi:10.1088/1742-6596/2063/1/012005

mechanism. Compared the DSC results of the DSC scan of PVC mixed with Ca-eggplant anthocyanin complex with the DSC of non-stabilized PVC sample, the temperature of initiation (Ti) and temperature of maximum rate (Tmax) of degradation reaction increased from 200 to 225 °C and from 305 to 320°C, respectively, which is due to stabilization effects of this additive. The performance of this derivative derived from its ability to scavenge HCl from the polymer system and retard its thermal degradation reaction at this temperature range was shown by scheme 1. During thermal scanning of PVC/Ca-anthocyanin complex sample, not only changes in the total enthalpies and rate of degradation reaction were noticeable, but also the peak temperature of the secondary degradation reaction which consists of polyenes reactions appears at higher temperature region. Based on the changes in the positions and peaks enthalpies of the two degradation reactions of PVC thermal degradation process, it can be present the Ca-anthocycnin complex act as efficient primary thermal stabilizer for PVC. This derivative can stabilize PVC polymer through at least by one mechanism which is including direct absorption of HCl from the polymer system during thermal process or at DSC test. In other words, in spite of that the primary products of thermal degradation process such as isolated double bonds, HCl and free radical groups have the a dominant influence on the thermal ageing degradation reaction of PVC.

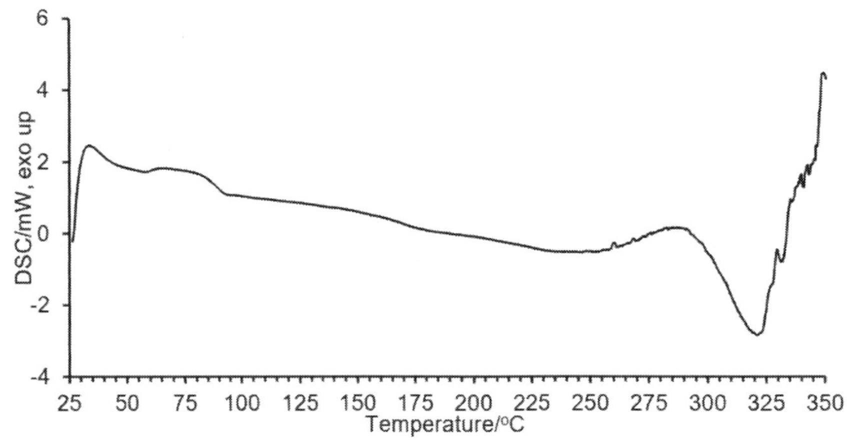

Figure 5. The DSC curve of PVC stabilized with 2 phr Ca-anthocyanin Complex.

Therefore, the thermal stabilization of PVC samples could take place as a result to one of chemical interaction between one these products and thermal stabilizer. The chemical properties of this derivative such as its coordination bonds with calcium ion and its anti-oxidation feature of phenolic groups give it its stabilization ability. The free anthocyanin reproduced by stabilization mechanism has the advantage of being an active antioxidant and therefore its phenolic hydroxyl groups could act as free radical quencher. But the salt segment or basic group of this derivative clearly had the predominant thermal stabilization effect by replacement of metal ions by relatively more chemically stable sites Scheme 1

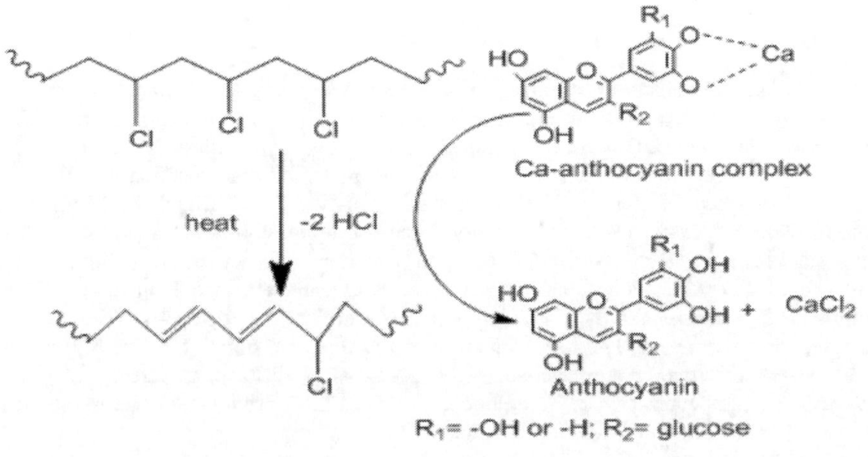

Scheme1. Possible thermal stabilization of PVC through direct removal of HCl by Ca-anthocyanin complex.

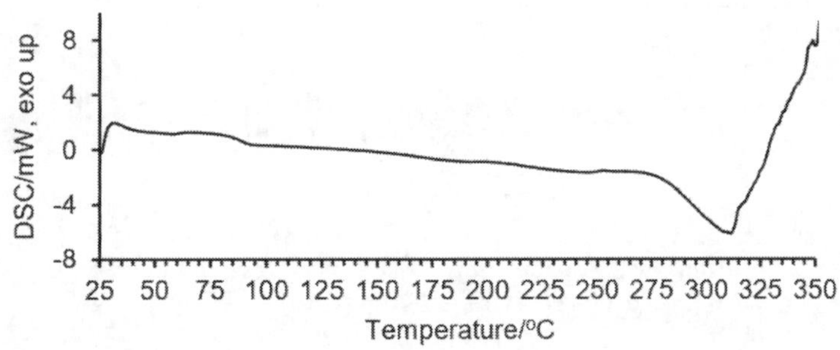

Figure 6. The DSC curve of PVC stabilized with 1 phr Ca-anthocyanin complex and 1 ph B-NT at weight ratio 1:1

From the DSC whole results, it can be observed that the calcium complex with anthocyanin has a significant property in term of thermal stabilization of PVC and its ability is comparable to that of commercial thermal stabilizer. In order to study the synergistic of the synthesized stabilizer with commercial one, B-NT/7060, the calcium complex with anthocyanin was examined as co-stabilizer with that used as reference in this study. The mixing ratio was done used equal amount from both additives (1:1), Table 1. The total amount of the mixture stabilizers in PVC was kept constant at 2phr based on the polymer amount. The result of DSC test for PVC mixed with Ca-anthocyanin/B-NT stabilizer is shown if Figure 6. It can be observed from the DSC curve that the first stage of PVC thermal degradation take place at the same temperature range but became very small reaction enthalpy (-2.02 J/g) in comparison with that for PVC stabilized with single of these additives. As shown by Table 2, the rate of dehydrochlorination (DHC) reaction of PVC stabilized with mixture of the two stabilizers is dramatically reduced by the cooperative effect of these additives. Interestingly, the significant reducing of PVC degradation enthalpy by about 16.1 times means that the thermal stability of PVC was efficaciously improved by this mixture of stabilizers. The combination of commercial stabilizer with Ca-anthocyanin complex reveals a true synergism effect. By comparing the thermal data in table 2, it can be confirmed that the mixture of anthocyanin based thermal stabilizer and Ca/Zn based commercial stabilizer displays a better stabilization behavior for PVC than that obtained from each one. The observed enhanced in the thermal resistance of PVC sample is considered to results from the varied in stabilization mechanisms and increase the active functional groups in each stabilizer type.

4. Conclusion

Calcium-anthocyanin complex was employed in this work to improve the thermal stability of PVC. It has been investigated to act as thermal stabilizer and co-stabilizer. The results obtained through thermal analysis clearly reveal that the Ca-anthocyanin complex. The addition of this derivative alone or with another stabilizer lowered the rate and enthalpy of degradation reaction of PVC. A critical analysis of DSC revealed an increase in thermal stability of the PVC through HCl removal process mainly due to the presence of -O-Ca coordination bonds. One possible mechanism was suggested to explain the stabilization effect of this derivative in which it act as thermal stabilizer. Mixing of anthocyanin based derivative with equal proportion of reference stabilizer, Reapak B-NT/7060 (1:1) greatly improves the thermal stability of PVC by synergism effect between the two additives.

Good stabilization performance can be obtained by using this derivative at a ratio similar to that of commercial stabilizer used as reference. The potential stabilization effect, renewable and environmentally friendly characters of Ca-anthocyanin, gives this derivative an important added value to submit it as PVC stabilizer in many applications.

5. Acknowledgement

Authors want to acknowledge Dr. Zaid Tariq Falak and Dr. Wisam Abidalhasn Radi, academic researchers in polymer research center/university of Basrah, Iraq for their help in the carried out of DSC and FTIR analysis.

References

[1] Allsopp MW and Vianello G 2012 Poly(vinyl chloride). In: Ullmann's encyclopedia of industrial chemistry. Wiley-VCH Verlag GmbH & Co. KGaA

[2] Ye F, Ye Q Zhan H, Ge Y, Ma X, Xu Y, Wang X 2019 Synthesis and Study of Zinc Orotate and Its Synergistic Effect with Commercial Stabilizers for Stabilizing Poly(Vinyl Chloride). Polymers; 11, 194. doi:10.3390/polym11020194 [CrossRef]

[3] Braun D 1981 Thermal degradation of poly (vinyl chloride). In: Grassie N, editor. Developments in polymer degradation. London, Applied Science Publishers.

[4] Starnes WH 2002 Structural and mechanistic aspects of the thermal degradation of poly(vinyl chloride). Prog Polym Sci.;27:2133–70. https://doi.org/10.1016/S0079-6700(02)00063-1.

[5] Liu YB, Liu WQ, Hou MH 2007 Metal dicarboxylates as thermal stabilizers for PVC. Polym Degrad Stab;92:1565–1571.

[6] Radka K, Miroslava N, Zdenek V 2004 Investigation of thermal stabilization of poly(vinyl chloride) by lead stearate and its combination with synthetic hydrotalcite. Polym Degrad Stab;85:903–909.

[7] Arkis E, Balkose D 2005 Thermal stabilization of poly (vinyl chloride) by organotin compounds. Polym Degrad Stab;88:46–51.

[8] Bueno-Ferrer C, Garrigo´s MC, Jime´nez A (2010) Characterization and thermal stability of poly (vinyl chloride) plasticized with epoxidized soybean oil for food packaging. Polym Degrad Stab;30:1–6.

[9] Benaniba MT, Belhaneche-Bensemra N, Gelbard G 2003 Stabilization of PVC by epoxidized sun flower oil in the presence of zinc and calcium stearates. Polym Degrad Stab 82:245–249.

[10] Karmalm P, Hjertberg T, Jansson A, Dahl R 2009 Thermal stability of poly (vinyl chloride) with epoxidised soybean oil as primary plasticizer. Polym Degrad Stab 94:2275–2281

[11] Mei Li, Jinwen Zhang, Kun Huang Shouhai Li, Jianchun Jiang, Jianling Xi 2014 Mixed calcium and zinc salts of dicarboxylic acids derived from rosin and dipentene: preparation and thermal stabilization for PVC. RSC Adv;4, 63576-63585.

[12] Johan Steenwijk, Rik Langerock, Daan S van Es, Jacco van Haveren, John W Geus, Leonardus W Jenneskens 2006 Long-term heat stabilisation by (natural) polyols in heavy metal- and zinc-free poly(vinyl chloride). Polym Degrad Stab; 91:52-59.

[13] Bueno-Ferrer C, Garrigós MC, Jiménez A 2010 Characterization and thermal stability of poly (vinyl chloride) plasticized with epoxidized soybean oil for food packaging. Polym Degrad Stab; doi:10.1016/j.polymdegradstab.2010.01.027.

[14] Benaniba MT, Belhaneche-Bensemra N, Gelbard G 2001 Stabilizing effect of epoxidized sunflower oil on the thermal degradation of poly(vinyl chloride). Polymer Degrad Stab;74:501-505.

[15] Qiufeng Ye, Xiaotao Ma, Bobin Li, Zhe Jin, Yingying Xu, Cheng Fang, Xiaoya Zhou, Yeqian Ge, Feng Ye 2019 Development and investigation of lanthanum sulfadiazine with calcium stearate and epoxidised soyabean oil as complex thermal stabilizers for stabilizing poly(vinyl chloride). Polymers;11:531. doi:10.3390/polym11030531.

[16]Taghizadeh MT, Nalbandi N, Bahadori A 2008 Stabilizing effect of epoxidized sunflower oil as a secondary stabilizer for Ca/Hg stabilized PVC. eXPRESS Polymer Letters;2:65–76.

[17] Shnawa HA, Khalaf MN, Jahani Y, Taobi AAH 2015 Efficient thermal stabilization of polyvinyl chloride with tannin-Ca complex as bio-based thermal stabilizer. Materials sciences and applications;6:360-372. http://dx.doi.org/10.4236/msa.2015.65042.

[18] Shnawa AS, Khalaf MN, Jahani Y (2018) Thermal degradation, dynamic mechanical and morphological properties of PVC stabilized with natural polyphenol-based epoxy resin. Polym Bull;75:3473–3498.

[19] Singh AP, Luthria D, Wilson T, Vorsa N, Singh V, Banuelos GS, Pasakdee S 2009 Polyphenols content and antioxidant capacity of eggplant pulp. *Food Chemistry*, *114*(3): 955-961. http:// dx.doi.org/10.1016/j.foodchem.2008.10.048.

[20] Todaro A, Cimino F, Rapisarda P, Catalano AE, Barbagallo RN, Spagna G 2009 Recovery of anthocyanins from eggplant peel. *Food Chemistry*, *114*(2), 434-439. http://dx.doi.org/10.1016/j. foodchem.2008.09.102.

[21] Concellón A, Zaro MJ, Chaves AR, Vicente AR 2012 Changes in quality and phenolic antioxidants in dark purple american eggplant (Solanum melongena L. cv. Lucía) as affected by storage at 0 °C and 10 °C. *Postharvest Biology and Technology*, *66*(0), 35-41. http://dx.doi.org/10.1016/j.postharvbio.2011.12.003.].

[22] Vallejo F, Marín JG, Tomás-Barberán FA 2012 Phenolic compound content of fresh and dried figs (Ficus carica L.). Food Chemistry, 130(3), 485–492.

[23] Wang Z, Cui Y, Vainstein A, Chen S, Ma H 2017 Regulation of fig (Ficus carica L.) fruit color: Metabolomic and Transcriptomic analyses of the flavonoid biosynthetic pathway. Frontiers in Plant Science, 8, 1–15.

[24] Sogbaike CE, Okieimen FE, Eromosele OC 2005 Thermo-oxidative degradation of poly(vinyl chloride) in the presence of derivatives of Khaya seed oil. Polym Degrad Stab; 88;175-181.

[25] Woo L, Ling MTK, Chan E 1991 The oxidative induction test applied to medical PVC and other polymers. J Vinyl and Additive Technology, 13, 199-203. http://dx.doi.org/10.1002/vnl.730130408.

[26] Simon P 1992 Kinetics of polymer degradation involving the splitting off of small molecules: Part 7- thermooxidative dehydrochlorination of PVC. Polym Degrad Stab;36:85-89.

[27] Radhakrishnan Nair MN, Thomas GV, Gopinathan Nair MR 2007 Thermogravimetric analysis of PVC/ELNR blends. Polym Degrad Stab;92(2): 189-96.

Light scattering detector based on light-emitting diodes-Solar cells for a flow analysis of Warfarin in pure form and pharmaceutical formulations

Nagham S Turkey [1, a] **and Jalal N Jeber** [1, b*]

[1] University of Baghdad, College of Science, Department of Chemistry, Baghdad, Al-Jadriya, Iraq

[a] **E-mail:** nagam@yahoo.com, [b*] **E-mail:** jalal.n@sc.uobaghdad.edu.iq

Abstract. Continuous turbidimetric analysis (CTA) for a distinctive analytical application by employing a homemade analyser (NAG Dual & Solo 0-180°) which contained two consecutive detection zones (measuring cells 1 & 2) is described. The analyser works based on light-emitting diodes as a light source and a set of solar cells as a light detector for turbidity measurements without needing further fibres or lenses. Formation of a turbid precipitated product with yellow colour due to the reaction between the warfarin and the precipitation reagent (Potassium dichromate) is what the developed method is based on. The CTA method was applied to determine the warfarin in pure form and pharmaceutical formulations in the concentration range from 2.0-16& 0.7-16 mmol/L with 0.58 and 0.55 mmol/L of the limit of detections. The correlation coefficients (r) of the developed method were 0.9977 and 0.9981 for cell 1 and 2 respectively. For validation of proposed method, the ICH guidelines were followed. The developed method was successfully applied for the determination of Warfarin in pure and pharmaceutical preparations. In addition, the method can be considered as a quality control method and conveniently used for routine analysis in laboratories since the method permits quantitatively determination of 60 samples/h.

1. Introduction

Chemically, warfarin is [3-(alpha-acetonylbenzyl)-4-hydroxycoumarin)], (Figure 1). In early 1950, It was approved in the US for being used as an anticoagulant (preventing thrombosis and embolism), and has remained the most popular and prescribed anticoagulant drug ever since [1]. In spite of its effectiveness, warfarin interacts with many widely used medications and some foods. Thus, it has several shortcomings and must be accurately monitored in the blood to ensure the safe dosages are taken. Decreasing the blood coagulation by warfarin is performed by the inhibit vitamin k1 reductase, this enzyme recycles the oxidized form of vitamin k1 which participates in the carboxylation of many blood coagulation proteins like factor VII. and prothrombin to its reduced form. For this reason, warfarin is also used as a function of vitamin K1 antagonist [2]. In the literature, several analytical methods have been reported for qualitative determination of warfarin in the pharmaceutical preparation and biological fluids. For example, LC-MS-MS [3], UPLC-MS-MS [4], micellar electro kinetic mass spectrometry [5], chiral HPLC [6], and dissolution methods [7] have been applied for warfarin analysis. The determination of warfarin in plasma has also been reported by using direct HPLC methods [8-10], indirect methods where a chiral stationary phase is used [11-15]. Nowadays, the most common method is the direct injection of plasma into the reverse phase [16]. Table 1

Journal of Physics: Conference Series **2063** (2021) 012006 doi:10.1088/1742-6596/2063/1/012006

summarizes different types of HPLC methods for the quantification of warfarin which found in the literature. Majority of the methods used fluorescence, MS and UV absorbance as detectors.

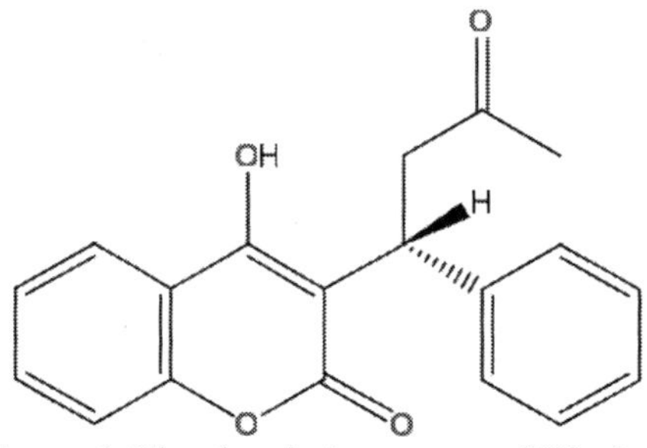

Figure 1. The chemical structure of Warfarin

Table 1. Shows previously reported HPLC methods for determination of Warfarin in tablets

Detection	Internal standard	Type	Column (stationary phase)	Ref
Fluorescence	Benoxaprofen	Chiral RP	BSA – ISRP	[11]
Fluorescence	Naproxen	Chiral RP	AGP	[14]
UV absorbance	Oxybenzone	Chiral RP	β-Cyclodextrin	[17]
UV absorbance	Naproxen	Achiral RP	C18	[18]
Mass spectrometry (LC/MS)	deuterium-labeled 7- hydroxywarfarin	Achiral RP	Luna C18	[19]
GC-MS	p-chloro warfarin	Achiral RP	β-Cyclodextrin	[20]

The turbidity concept is an analytical technique which has a wide range application in the determination of lots of active ingredient in pharmaceutical formulation. A great number of drugs have been determined using turbidity technique like Ephedrine Hydrochloride [21], Ciprofloxacin HCl [22], Cyproheptadine hydrochloride [23] and Theophylline [24]. However, the determination of warfarin using turbidity method has never been reported in the literature. The main objective of the proposed method was to develop and validate a new approach that combines the flow analysis technique with turbidimetric detection allowing the simultaneous quantitative determination of warfarin in pharmaceutical formulations for the same injected sample. Therefore, in the present work, a new, two lines manifold system for flow injection analysis is developed in which two consecutive detection zones (cells 1 & 2) for the same injected sample were applied using NAG dual & solo (0-180°) detector. This type of manifold is used for the chemical reactions when the warfarin is injected into the carrier streamline (distilled water) prior to mixing with the precipitation reagent line (sodium persulfate). The two lines are merged at Y junction point to form a yellow ion association complex due to the reaction between the warfarin and the potassium dichromate, the proposed reaction is shown in Figure 2. The intensity of the complex is detected by two consecutive detection zones (cells 1 & 2). Thus, two peaks will be obtained for any injected sample. Therefore, a new manifold system for the flow injection spectrophotometric analysis was successfully presented and applied for the quantitative determination of warfarin in different commercial tablets. The proposed method is rapid, simple, economic and more sensitive than papers reported earlier.

Journal of Physics: Conference Series **2063** (2021) 012006 doi:10.1088/1742-6596/2063/1/012006

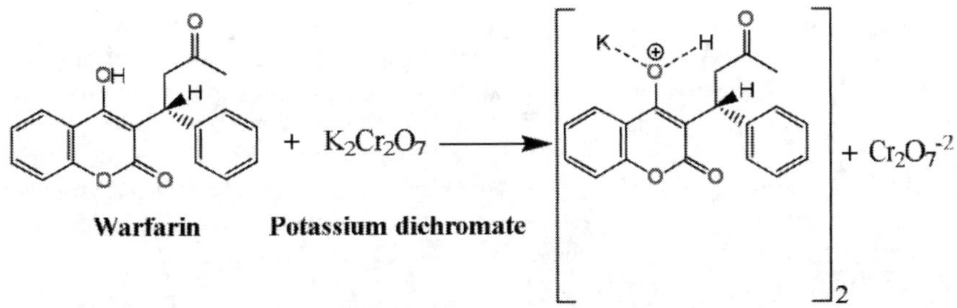

Figure 2. The proposed reaction between the warfarin and potassium dichromate

2. Materials and Methods

2.1. Material

All the standards materials of Potassium dichromate, Potassium bromide, Potassium nitrate, Potassium nitrate, Sodium nitrite, Sodium chloride, ammonium chloride and the solvents such as sulfuric acid, nitric acid, acetic acid and hydrochloric acid were with purity higher than 98% and procured from Sigma Aldrich. Double-distilled water was used in all the preparations of aqueous solutions. Samara company for pharmaceutical industries, Ltd. Iraq supplied this study with the pure material of standard warfarin, while the warfarin tablets under their brand names were procured from the local market.

2.2. Preparation of experiment solutions (Standard and sample)

Warfarin stock solution 20 mmol/L was prepared by dissolving of 1.54 g in 250 ml volumetric flask and filled to the label with double distilled water. An amber bottle was used to keep the solution away from the light and use it in a further experiment. A series of warfarin standard solution were prepared by dilution appropriate volumes of stock solution with a double distilled water to obtain standard solutions ranging from 0.25-25 mmol/L. UV spectrophotometry measurements were performed to the stock solution before each experiment in order to monitor the stability of the warfarin since some drugs become unstable in their solutions. The results showed there is no decomposition (did not break into other substance which means the warfarin has high stability) in the warfarin during the proposed method.

2.3. Preparation of samples solutions (Tablets)

In this study, two of warfarin commercial companies (Warfarin ®, Bristol, UK) and (Orfarin ®, Orion, Finland) were investigated. Twenty tablets were precisely weighed, powdered and smoothly mixed. The average weight of the tablet was accurately dissolved in double distilled water, mechanically shaken for 30 min and filtered. The residue from the filtration was washed four times with the distilled water and transferred to a volumetric flask and diluted to the mark with distilled water. The prepared solution was used for further experiments by further dilution of previously prepared warfarin standard solution. Each of the commercial tablet types has followed the above procedure for sample solution preparation.

2.4. Apparatus

All the turbidimetric measurements were performed using a homemade optoelectronic detector integrating a flow injection system [25, 26]. The detector contains two identical twin cells i.e., cell no. 1 and cell no. 2, each one of them has 100 mm length, and in between of them there is a 20 mm without any detection as shown in Fig 3. The light source in the detector unit is white snow led, each measuring cell contains 10 white snow LED (WSLED) as an irradiation source (blue-violet 42.4%, Green 56.73% and 1.15% Red). Each one of the WSLED irradiates a circle spot of 3mm diameter for the flow cell Oφ 4mm (Outer diameter of the flow cell) and Iφ 2mm (Internal diameter of the flow cell) of the active total distance of the cell. The inlet distance for each cell unit (100 mm) is covered by

3-solar cells with (37.8mm (L)x10mm (W)x1mm, thickness) as cell dimensions. Identical twins' solar cells are used with minimum 2.5VDC at ambient light. The integrated manifold flow system composed of two-channel Ismatec (Switzerland) model 796 peristaltic pump supplied with Tygon pump tubing was used for the propulsion of the fluids. Different lengths of Teflon tubing (0.8 i.d.) were used to connect the manifold parts together, mix and carry the flow streams. Sample injection was performed by a 6-way selection valve (Type Upchurch Scientific® Medium Pressure Valves) that supplied with an external sample loop style. The reactants were mixed at the Y junction point with 0.8 i.d. which was made from methyl methacrylate. The obtained profiles of the S/N against the time from the detector are shown in a chart (Figure 4).

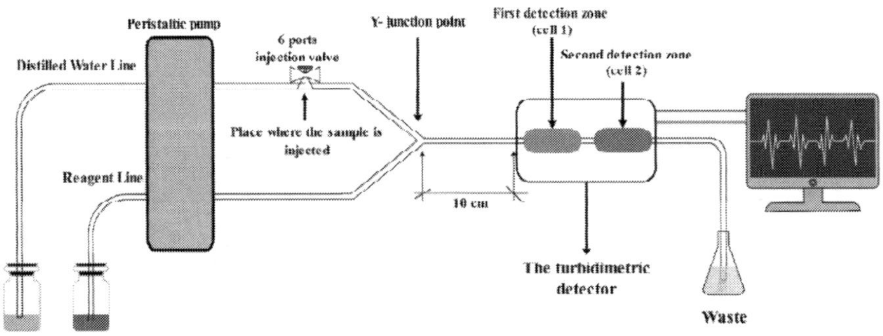

Figure 3. The configuration of the manifold system for the flow injection analysis

2.5. Recommended Procedure
Warfarin determination was performed using a manifold analysis system contains two consecutive detection zones (cells 1 & 2), which allows simultaneous detection for the same injected sample. A series of physical and chemical parameters like salts effect, precipitation reagent concentration, flow rates, light intensity, mixing coil, the volume of the sample and purge time were optimized under flow injection conditions. The proposed method based on the forming of an ion association complex, this product is white and turbid and can be determined by the detectors. The formation of the product is conducted by mixing the sample with the presentation reagent. In the beginning, 150 µL of sample solution contains 20 mmol/L of warfarin is injected into the injection valve at a place called an injection loop. Then, the direction of the injection valve is changed to the injection mode. The water stream is carried out the injected sample to the junction point where is mixed with the reagent and a yellow precipitate turbid solution is formed. The formed colored solution is transported to the detection units by the stream. The carrier streamlines (distilled water) and the reagent line solutions (6.0 mmol/L) were propelled using a two-channel peristaltic pump at flow rates 2.8 ml/min and 3.4 ml/min for the carrier stream and precipitation reagent lines respectively. The absorbance of the formed yellow product is continuously recorded on the chart paper represented (first peak for detection cell1 and the second peak for the detection cell 2). These two peaks are responsible for the determination of injected sample concentration by treated them mathematically. Figure 4 shows the shape of the peaks obtained from the recorder. The concentration of warfarin is determined by plotting the calibration graph (peaks height 1, 2).

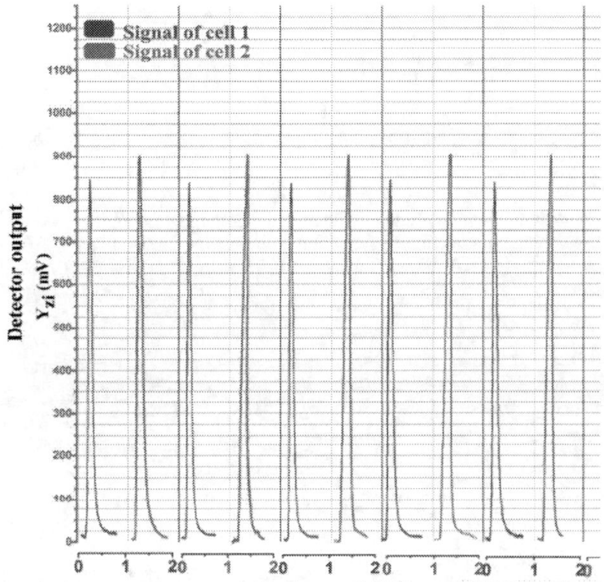

Figure 4. The signals profiles of the S/N against the time obtained for warfarin at 4.0 mmol/L

3. Results and discussion

The warfarin is found to be forming an ion association complex with potassium dichromate, Fig 2. illustrates the proposed reaction of the product. The product is a yellow, turbid and can be detected using the optoelectronic detector. The analysis system working under flow injection conditions; therefore, a series of physical and chemical parameters will be optimized.

3.1. Optimization of chemical parameters

3.1.1. Potassium dichromate concentration (precipitation reagent)

The stock solution of potassium dichromate was prepared by dissolving potassium dichromate in distilled water to obtain the concentration of 25 mmol/L. Further dilution by distilled water to the previously prepared stock solution is conducted to obtain the concentrations range from (0.3-8.0 mmol/L). The experimental conditions of the measurements were: warfarin 20 mmol/L, sample volume 150µL, light intensity (I=1, 4 for cell 1 and cell 2 respectively), flow rate of 3.2 ml.min^{-1} for carrier stream and 4.3 ml.min^{-1} flow rate for reagent and open valve mode. The obtained results showed there is a gradual increase in the peak height during increasing in the reagent concentration reaching 6.0 mmol/L of potassium dichromate. However, above 6.0 mmol/L there was insignificantly increased in the peak height which probably indicates there is no increase in the concentration of formed product at these concentrations. Therefore, the 6.0 mmol/L of potassium dichromate (precipitation reagent) was chosen to be the optimum concentration for further experiment as shown in Figure 5.

Journal of Physics: Conference Series **2063** (2021) 012006 doi:10.1088/1742-6596/2063/1/012006

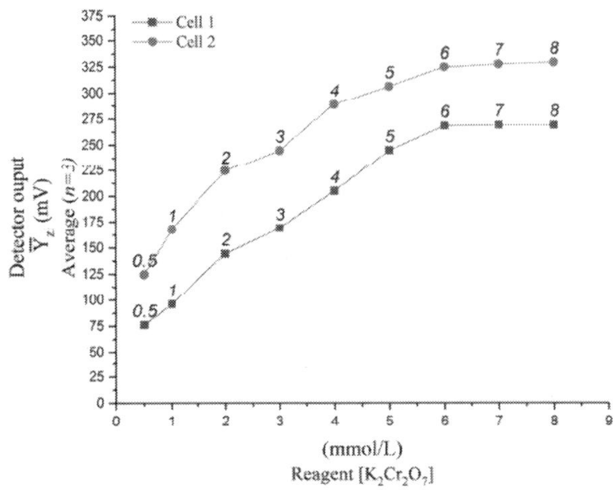

Figure 5. The effect of potassium dichromate Conc. [mmol/L]

3.1.2. Effect of aqueous salt solutions

Using different aqueous salts solutions (H_2O, NH_4Cl, NaCl, KBr, CH_3COONH_4, $NaNO_2$, KNO_3, HCl, HNO_3 and CH_3COOH) instead of the distilled water as a carrier stream were carried out to investigate whether using these solutions can enhance the concentration of the formed product. The experimental conditions of the experiment were: potassium dichromate 6.0 mmol/L, the flow rate of the carrier stream 3.3 $ml.min^{-1}$, while 4.3 $ml.min^{-1}$ flow rate for reagent with sample volume 150μL, light intensity (I=1, 4 for cell 1 and cell 2 respectively), and open valve mode. From the results, it can be observed that there is an increase in sensitivity of detector during using the distilled water as a carrier stream compared with these aqueous salts solutions which are caused to decrease in the detector sensitivity; and this might be attributed to disperse of precipitate particulate in the presence of salts. Therefore, distilled water was chosen as the optimum medium (carrier stream) (Figures 6 A and B).

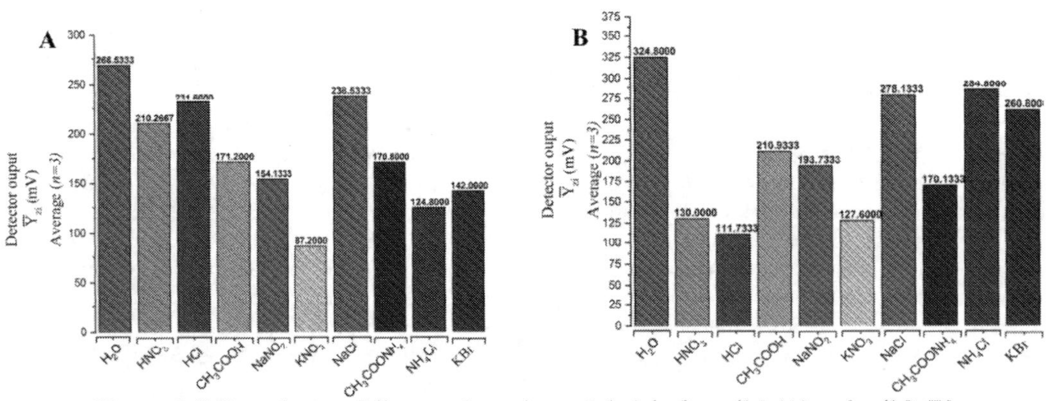

Figure 6. Effect of using different salt on the peak height for cell 1 (A) and cell 2 (B)

3.2. Optimization of physical parameters

3.2.1. The effect of using variables intensities of white snow light emitting diodes (WSLEDs)

Each cell in the detector has 5 levels of light intensity ranging from 1-5 (lowest-highest) in addition to the off position for both cells which are individually controlled which means each one of the instrument cells has its own selector switch of the light source. In order to determine the optimum light intensity for each cell (1 & 2), variables intensities of the incident light of cell 1 and 2 were carried out. Under applied following experimental conditions: 20 mmol/L of warfarin, 6.0 mmol/L of potassium dichromate, 3.3 &4.3 ml.min^{-1} flow rates for carrier stream (distilled water) and reagent respectively with sample volume 150 µL and open valve mode, the light intensity experiment was performed. From the obtained results it was observed that, for the cell 1, the favored light intensity was I=1, and this is may be attributed to the nature of the formed precipitate particles which means the well-distributed particles with small size can leave very short spaces between them during passing through the cell 1, therefore, using a low intensity of light source can detect these particles and obtaining a signal instead of using high intensity of the light source which cannot detect these small sizes of particles. On the other hand, cell 2 preferred intensity of light I= 5, and this is can be explained by the same fact, during passing through the cell 1, the particles will start to growing up and suffering from conglomeration and as a result of that, the particle size will be increased. At this point using a low intensity of light source cannot be useful therefore, the light intensity of the light source must be high. Thus, the intensity of light (I=1, 5) was chosen to be the optimum intensity for further experiment for cell1 and 2 respectively as shown in Figure 7.

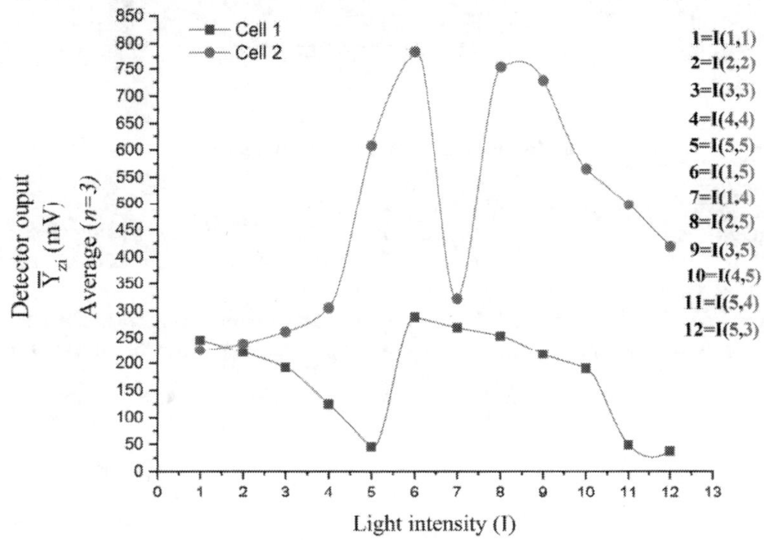

Figure 7. The effect of light intensities

3.2.2. *Effect of flow rate*

In order to choose the optimum flow rate for both of the carrier stream-line and the precipitation reagent line 6.0 mmol/L potassium dichromate, ranges of flow rates (0.5-7.0 ml min^{-1}) have been applied for the carrier stream and the precipitation reagent lines. While the other experimental conditions i.e.; warfarin 20 mmol/L, 150µL sample volume, light intensity (I=1, 5 for cell 1 and cell 2 respectively) and open valve mode are kept constant. The obtained results have shown that at a low flow rate for both of the carrier stream-line and the reagent line there is a gradual increase in detector signal for cell 1 & 2 up to (2.8 & 3.4 ml min^{-1}) for stream-line and reagent line respectively and the bases of these responses are wide. At a high flow rate (i.e.; more than 2.8 and 3.4 for both lines carrier stream and precipitation reagent respectively), the detector signal of these flow rates have decreased and became very sharp with the base of these peaks became narrow, because the precipitated particulates are moved faster and take only a very short time to passing in front of the measuring cells. Therefore, the flow rates 2.8 & 3.4 ml/min for stream-line and reagent line respectively were chosen to be the optimum flow rates (Figure 8).

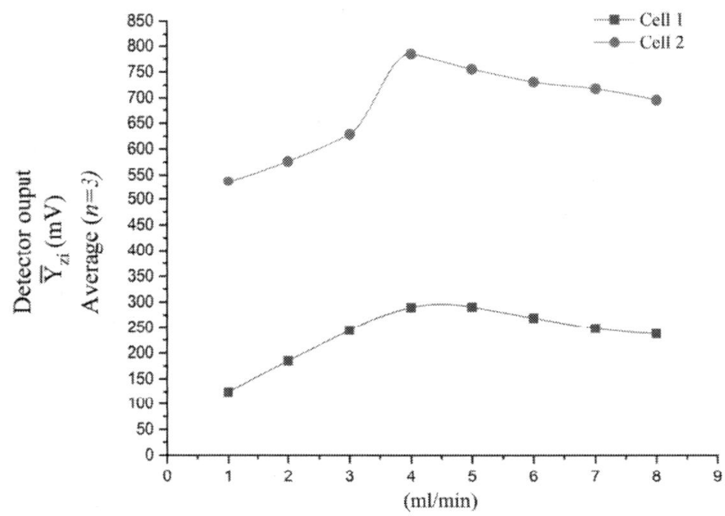

Figure 8. The effect of flow rates

3.2.3. The effect of the sample volume in CFIA

In this experiment, variable of sample volumes ranges from (32-250 μL) using Teflon tubes (4.07-38.17 cm) with 0.5 mm as a dimeter were used. The results showed there is an increase in the peak high during the increase in the sample volume up to 150 μL, at this point, the maxima peak high was recorded as sample volume 150 μL (22.45 cm as a length of sample segment). Therefore, 150 μL has chosen to be the optimum sample volume and use it in further experiments as shown in Figure 9.

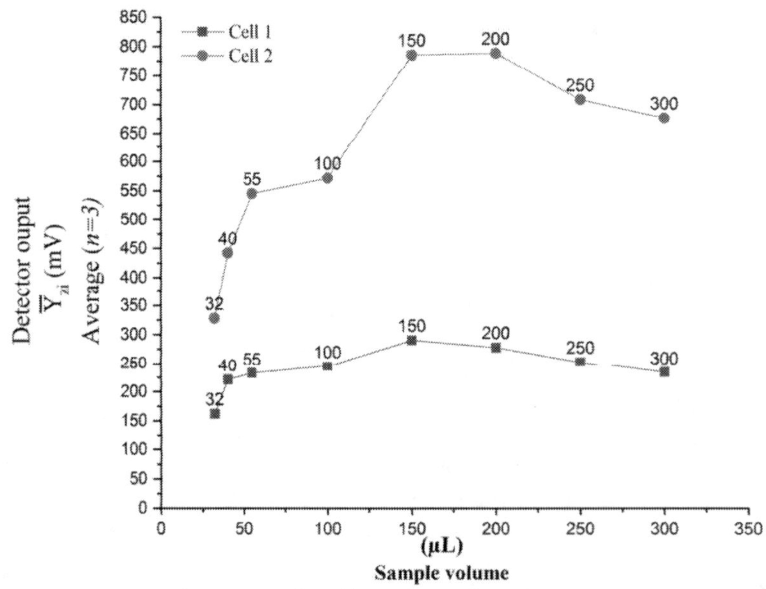

Figure 9. The effect of sample volumes

Journal of Physics: Conference Series 2063 (2021) 012006 doi:10.1088/1742-6596/2063/1/012006

3.2.4. The effect of mixing coil

In order to check whether the flow system requires adding a mixing coil to ensure a maximum reaction between the warfarin and potassium dichromate or not, under optimization conditions, mixing coils (made from Teflon, 0.5 mm i.d.) with different length of 10, 15, and 20 cm were used. The results showed there is no improvement in the sensitivity of the peak height has recorded in comparison with the system without using the mixing coil which indicates that the reaction between the warfarin and the potassium dichromate is already complete and the manifold system does not need any mixing coil. (See Figure 10).

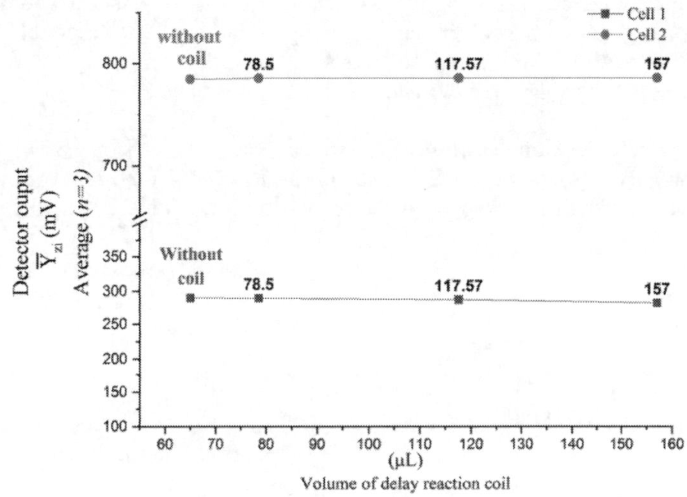

Figure 10. The effect of mixing coil

3.2.5. Purge time of the sample segment

The purge time by definition, is the required time that the sample segment needs to leave from the injection valve, mixing with the reagent and reaching the detection units. Therefore, periods of times ranged from (3-30 sec and open valve mode) were studied. The results have shown that 10 sec is the time that the sample segment requires to leave from the injection valve towards the Y-junction. Therefore, from 10- open valve mode considered as the optimum purge time of the sample segment and it can choose any one of those times that fall in that range as shown in Figure 11.

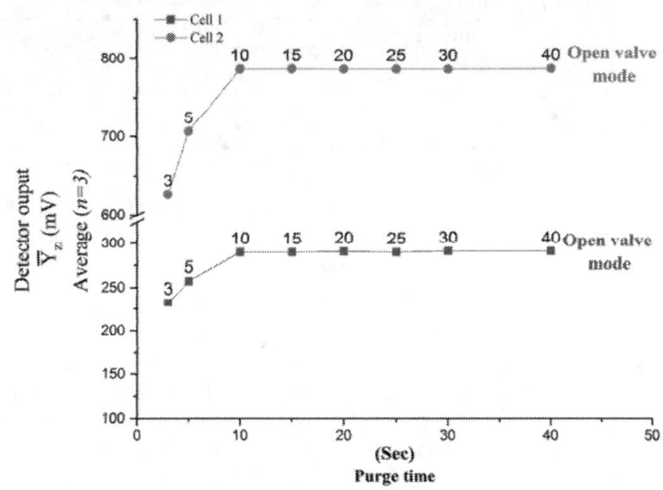

Figure 11. The effect of purge time

Journal of Physics: Conference Series **2063** (2021) 012006 doi:10.1088/1742-6596/2063/1/012006

3.3. Validation parameters of proposed methods

The proposed method has been validated based on ICH guidelines [27]. Therefore, the developed method was validated for precision, accuracy, specificity, linearity, and LOD.

3.3.1. Calibration curve and the limit of detection (LOD)

It is a general procedure which is used for determine the concentration of a chemical substance in an unknown sample by establishment a set of known concentration and comparing with the unknown. By applying all of the optimized analytical parameters of the proposed method, the calibration curve was constructed by recording the attenuated light peak heights for various concentrations of the warfarin. The linear region of the curve for warfarin has shown in the range of 2.0-16 mmol/L, 0.7-16 mmol/L with (R^2=0.9963 & 0.9956) for both cell 1 & 2 respectively as shown in Fig 12. While the limits of detection were calculated based on the following formula:

$\mathrm{LOD} = 3.3\, S_a\, /b$

Where: S_a = the standard deviation (S_a) of the Y-intercept and b = the slope which can be estimated from the calibration curve. Note that, the S_a is usually meant (RMSE) the root mean squared error or SD of the residuals which can be taken from the regression line. Table 2 describes the summary of linear regression lines.

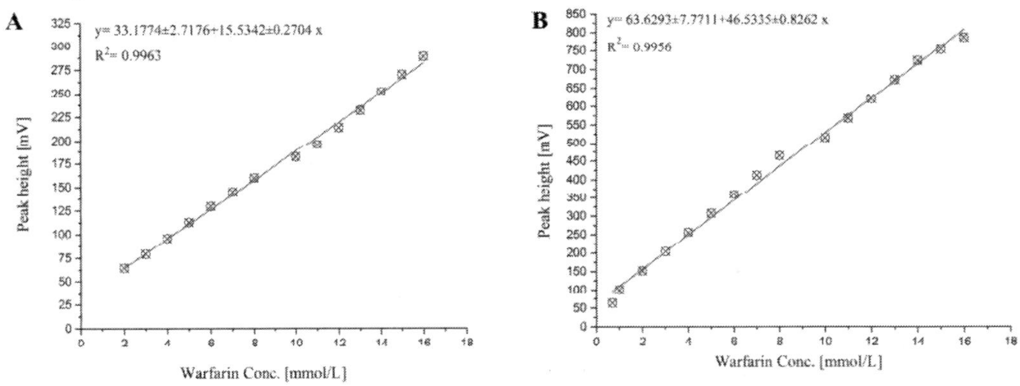

Figure 12. The linear calibration curves for determination of warfarin for measuring cell 1& 2 (A&B) respectively.

Table 2. The summary of linear regression lines of the proposed method

Parameter	Obtained Value	
	Proposed method (cell 1)	**Proposed method (cell 2)**
Linearity (mmol/L)	2.0-16	0.7-16
Regression equation	33.177+15.534 [War.](mmol/L)	63.629+46.534[War.](mmol/L)
Slope	15.534	46.534
Intercept	33.177	63.629
Correlation coefficient, r	0.9981	0.9977
coefficient of determination, R^2	0.9963	0.9956
LOD (mmol/L)	0.5773	0.5511

3.3.2. Specificity & interferences

Based on the official guideline of the ICH Q2 in method validation, the specificity can be defined as the ability of the method to assess the analyte in its matrix in the other words, in the presence of other components. The effect of interference of excipients on the analysis of warfarin in the tablet's dosage was carried out by the analysis of sample prepared with all potential the excipients that present in the tablets (a mixture of excipients) such as Lactose, Sucrose, Starch, Gelatin and Magnesium Stearate but without the warfarin. The response of the instrument (peak height) did not show any potential interference of the tablet's excipients (See table 3).

Table 3. The effect of the interferences on the determination of warfarin (5.0 mg)

Excipients	Using 8.0 mmol/L of warfarin (measuring cell 1)			Using 8.0 mmol/L of warfarin (measuring cell 2)		
	Fold added (mmol/L)	warfarin determined in (mmol/L±SD)	% E	Added (mmo/L)	warfarin determined in (mmol/L±SD)	% E
Gelatin	2.0	8.02±0.14	0.33	2.0	8.08±0.10	1.33
Magnesium stearate	2.0	8.05±0.12	0.83	2.0	8.02±0.18	0.33
Lactose	2.0	8.10±0.18	1.66	2.0	8.10±0.19	1.66
Sucrose	2.0	7.99±0.15	-0.16	2.0	8.12±0.18	2.0
Starch	2.0	7.98±0.02	-1.5	2.0	7.95±0.03	-0.83
Calcium carbonate	2.0	8.10±0.18	1.66	2.0	8.10±0.19	1.66
All above	2.0	8.05±0.11	1.16	2.0	8.13±0.06	2.16

3.3.3. Precision and accuracy

The accuracy can be defined as the closeness of the agreement in the values between the found value and an accepted reference or true value in an analytical procedure. While the precision knows as the closeness of the agreement between the values obtained from multiple measurements of the same homogeneous sample using the prescribed conditions. For the proposed method, three sets of diluted concentrations (6.0 and 8.0 mmol/L) were prepared and each one of them was injected and determined on 5 consecutive days. These values represented the inter-day precision, while the intra-day precision was evaluated by determination of three concentrations of each drug on five successive occasions. The RSD values of both inter-day and intra-day have been used for expressing the precision. RSD % of the obtained data have found to be 0.14-1.95 % and 0.24-1.61 for inter and intraday precision respectively. The error % was used for expressing the accuracy which was ranged from 1.5-3.16 % and 1.63-3.16 % for intra-day and inter-day respectively. as shown in Table 4.

Table 4. The inter- and intra-day assay of precision and accuracy of the proposed method for the determination of warfarin

Taken of warfarin (mmol/L)	Intra-day (n=5)			Inter-day (n=5)		
	Found [Warfarin ±SD] (mmol/L)	RSD %	Accuracy as E %	Found [Warfarin ±SD] (mmol/L)	RSD %	Accuracy as E%
			Cell 1 Cell 2			
6.0	6.07±0.005	0.14	1.91	6.08	0.27	2.10
	6.09±0.02	0.56	2.33	6.09	0.40	2.33
8.0	8.09±0.01	0.16	1.50	8.09	0.24	1.63
	8.10±0.02	0.43	1.66	8.10	0.33	1.66

3.4. Application of the proposed method to determine of warfarin in tablets

The proposed method was successfully applied for the determination of warfarin in its commercial tablets of two different companies (Warfarin ®, Bristol, UK) and (Orfarin ®, Orion, Finland) which comprise (5.0 mg of warfarin). The standard addition method was followed by injection of six samples and the results were satisfied and in the excellent an agreement with the label claims and with the official USP method as shown in Tables 5. The one-sample t-test and F-test were carried out for the obtained results. The t-test analysis has shown that there is no a significant difference were found between the results obtained by proposed method and the label claims for the same batch. The given values by the proposed method were also compared with the official method using the F-test. The statistical evaluation (F-test) has shown that there was no significant difference between the methods used as shown in Table 5.

Table 5. Determination of commercial tablets containing 5.0 mg warfarin using the proposed and official methods

Brand name (tablets)	Nominal amount, mg/tablet	t^a	F^a	Found [b] *(n=3)* Official method [28] UV-spectrophotometry	developed method 1st cell	developed method 2nd cell
Warfarin ® (Bristol, UK)	5.0	0.160	1.245	100.04±0.03	100.63±0.12	101.36±0.26
Orfarin ® (Orion, Finland)	5.0	0.010	1.885	101.12±0.24	103.39±0.03	103.49±0.06

[a] 2.056 and 2.57 are the tabulated values of both t and F at P=0.05 (95%C.I) respectively [29].
[b] mean found ± SD

4. Conclusion

A simple and sensitive turbidimetric method was developed and validated for the determination of warfarin. The warfarin has been determined in tablets using two consecutive detection units (cell 1 & 2). The detection units have given accuracy and trust ability to the method since each of injected sample is determined twice at the same time. The proposed method is rapid, accurate, economics and does not require any pre-treatment in comparison with the reported methods. By virtue of its rapidity and simplicity, the proposed method can be applied for routine analysis in a laboratory as a quality control method since the method permits quantitively determination of warfarin (60 samples/hr.). The method also does not require any toxic chemicals or organic solvents, free from the interference, and its procedure meets the requirements of green chemistry. Generally, the proposed method should be used for the determination of warfarin in tablets.

Acknowledgements

The authors would like to thank Prof Dr Issam M. A. Shakir for granting permission to use his analyser (NAG 0-180°) for the project and his help for typing the manuscript.

References

[1] Verhovsek M, Motlagh B, Crowther MA, Kennedy C, Dolovich L, Campbell G, et al., (2008) Quality of anticoagulation and use of warfarin-interacting medications in long-term care: a chart review. BMC geriatrics. **1:** 1-13.
[2] Ram VR, Dubal KL, Dave P, Joshi HS, (2019) Development and Validation of a Stability Indicating HPLC Assay Method for Determination of Warfarin Sodium in Tablet Formulation. Chemical Methodologies. **3,** 322-339.
[3] Park SM, Lee J-K, Chun SI, Lee HI, Kwon SU, Kang D-W, et al., (2013) VKORC1 and CYP2C9 genotype variations in relation to warfarin dosing in Korean stroke patients. Journal of stroke. **2:** 107-115.
[4] Krishna DK, Gopal DS, Parasuraman S, Rajan S, Balachander J, Sai BC, et al., (2013) Estimation of plasma levels of warfarin and 7-hydroxy warfarin by high performance liquid chromatography in patients receiving warfarin therapy. Journal of young pharmacists: JYP. **1:** 7-13.
[5] Zheng J, Shamsi SA, (2006) Simultaneous enantioseparation and sensitive detection of eight β-blockers using capillary electrochromatography-electrospray ionization-mass spectrometry. Electrophoresis. **11:** 2139-2151.
[6] Cao X, Yang X, Liu Z, Jiao H, Liu S, Liu L, et al., (2017) Rapid Simultaneous Screening and Detection of 12 Anticoagulant Rodenticides in Food by Ultra-performance Liquid Chromatography-Triple Quadrupole/Linear Ion Trap Tandem Mass Spectrometry. Food analytical methods. **11:** 3538-3547.
[7] Gordon K, Balasubramanian S, (1999) Recent advances in solid-phase chemical methodologies. Current opinion in drug discovery & development. **4:** 342-349.

[8] Banfield C, Rowland M, (1983) Stereospecific high-performance liquid chromatographic analysis of warfarin in plasma. Journal of pharmaceutical sciences. **8:** 921-924.

[9] Banfield C, Rowlandx M, (1984) Stereospecific fluorescence high-performance liquid chromatographic analysis of warfarin and its metabolites in plasma and urine. Journal of pharmaceutical sciences. **10:** 1392-1396.

[10] Carter SR, Duke CC, Cutler DJ, Holder GM, (1992) Sensitive stereospecific assay of warfarin in plasma: reversed-phase high-performance liquid chromatographic separation using diastereoisomeric esters of (−)-(1S, 2R, 4R)-endo-1, 4, 5, 6, 7, 7-hexachlorobicyclo [2.2. 1]-hept-5-ene-2-carboxylic acid. Journal of Chromatography B: Biomedical Sciences and Applications. **1:** 77-83.

[11] Chu Y-Q, Wainer IW, (1988) The measurement of warfarin enantiomers in serum using coupled achiral/chiral, high-performance liquid chromatography (HPLC). Pharmaceutical research. **10:** 680-683.

[12] Uno T, Niioka T, Hayakari M, Sugawara K, Tateishi T, (2007) Simultaneous determination of warfarin enantiomers and its metabolite in human plasma by column-switching high-performance liquid chromatography with chiral separation. Therapeutic drug monitoring. **3:** 333-339.

[13] Andersen C, Balmér K, Lagerström P-O, (1993) Enantioselective assay of warfarin in blood plasma by liquid chromatography on Chiralcel OC. Journal of Chromatography B: Biomedical Sciences and Applications. **1:** 159-163.

[14] McAleer SD, Chrystyn H, Foondun AS, (1992) Measurement of the (R)-and (S)-isomers of warfarin in patients undergoing anticoagulant therapy. Chirality. **8:** pp.488-93.

[15] Pinkerton TC, (1991) High-performance liquid chromatography packing materials for the analysis of small molecules in biological matrices by direct injection. Journal of Chromatography A. **5**: 13-23.

[16] Boppana VK, Miller-Stein C, Schaefer WH, (1996) Direct plasma liquid chromatographic-tandem mass spectrometric analysis of granisetron and its 7-hydroxy metabolite utilizing internal surface reversed-phase guard columns and automated column switching devices. Journal of Chromatography B: Biomedical Sciences and Applications. **2:** 227-236.

[17] Ring PR, Bostick JM, (2000) Validation of a method for the determination of (R)-warfarin and (S)-warfarin in human plasma using LC with UV detection. Journal of pharmaceutical and biomedical analysis. **3:** 573-581.

[18] Sun S, Wang M, Su L, Li J, Li H, Gu D, (2006) Study on warfarin plasma concentration and its correlation with international normalized ratio. Journal of pharmaceutical and biomedical analysis. **2:** 218-222.

[19] Ufer M, Kammerer B, Kirchheiner J, Rane A, Svensson J-O, (2004) Determination of phenprocoumon, warfarin and their monohydroxylated metabolites in human plasma and urine by liquid chromatography–mass spectrometry after solid-phase extraction. Journal of Chromatography B. **2:** 217-226.

[20] Naidong W, Ring PR, Midtlien C, Jiang X, (2001) Development and validation of a sensitive and robust LC–tandem MS method for the analysis of warfarin enantiomers in human plasma. Journal of pharmaceutical and biomedical analysis. **2:** 219-226.

[21] Jeber JN, (2019) Quantitative Determination of Ephedrine Hydrochloride in Pharmaceutical Injections by Highly Sensitive Turbidimetric and Reversed-Phase Combined with UFLC Methods. Chemistry & Chemical Technology. **2(13):** 269-274.

[22] Hammood MK, Jeber JN, Muhamad YH, (2016) Two Techniques (Spectrophotometric and Turbidimetric) for Determination of Ciprofloxacin HCl in Pharmaceutical Drugs with Comparison between the Techniques. Iraqi Journal of Science. **3A:** 1620-1628.

[23] Jeber JN, Turkey NS, (2020) A turbidimetric method for the quantitative determination of cyproheptadine hydrochloride in tablets using an optoelectronic detector based on the LEDs array. International Journal of Pharmaceutical Research. **4:** 2911-2294.

[24] Jeber Jalal N, Hassan Raed F, Hammood Mohammad K, (2019) Solid Phase Extraction of Theophylline in Aqueous Solutions by Modified Magnetic Iron Oxide Nanoparticles as an Extractor Material and Spectrophotometry Technique for the Determination. **23:** 94-100.
[25] Turkey NS, Jeber JN, (2021) A flow analysis system integrating an optoelectronic detector for the quantitative determination of active ingredients in pharmaceutical formulations. Microchemical Journal. **160:** 105710-105719
[26] Jeber JN, Turkey NS, (2021) An optoelectronic flow-through detectors for active ingredients determination in the pharmaceutical formulations. Journal of Pharmaceutical and Biomedical Analysis. **201:** 114128-114140
[27] Guideline IHTJQ, (2005) Validation of analytical procedures: text and methodology. 1-15.
[28] Rockville M, (2007) The United States Pharmacopoeia 30, the National Formulary 25 US Pharmacopeial Convention. Electronic version. 2287-8.
[29] Miller J, Miller JC, (2018) Statistics and chemometrics for analytical chemistry: Pearson education.

Journal of Physics: Conference Series **2063** (2021) 012007 doi:10.1088/1742-6596/2063/1/012007

Preparation of Ethanol-Free Hand Sanitizers Gels and Studying its Sterile Efficacy

Mahmood M. Fahad[1, *], Mohammed G. A. Al-Khuzaie[2]

[1, *] Medical Laboratory Techniques Department, Kufa Techincal Institute, Al-Furat Al-Awsat Technical University, Kufa, Iraq.
mahmood.mohy.iku@atu.edu.iq
[2]Department of Nursing Techniques, Technical Institute of Al-Diwaniyah, AL-Furat AL-Awsat Technical University (ATU), Al-Diwaniyah, Iraq.
mohammed.alkhuzaie@atu.edu.iq

[*]Corresponding author: mahmood.mohy.iku@atu.edu.iq

Abstract. Due to the spread of the Corona pandemic, the demand for hand sanitizers has increased dramatically. This led to a global scarcity of sterilizers. Therefore, this study was performed to produce and evaluate the effectiveness of new commercially available, cheap and ethanol-free hand sanitizers from various sources. The new hand sanitizers were prepared by mixing (glycerin) as a moisturizer and (carbomer) as thickening agents to form the gel with (tri-ethanolamine base) as pH neutralizer and (Tea tree oil, Aloe vera extract or Povidone 10%) respectively as antiseptic agents. All prepared hand sanitizers were examined with four kinds of bacteria (Enterococcus faecalis, E. coli, Pseudomonas aeruginosa and Staphylococcus aureus) And two kinds of fungi (Candida albicans and Aspergillus flavus). The prepared hand sanitizer gels showed good sterilization efficacy and similar to commercially prepared hand sanitizers (70% ethanol).

1. Introduction

Due to the spread of the Corona pandemic in the world beginning with the Chinese city of Wuhan at the end of 2019, the global demand for many preventive medical products increased, causing an acute shortage of these products. Among these items is hand sanitizer. In fact, there are two types of hand sanitizer gels. The first type is based on alcohols as an active sterilization substance, such as ethanol and isopropanol. But this type of sterilizer causes many health problems for the user, such as skin irritation. As for the second type of sterilizers, it does not depend on alcohol, but rather on other materials such as (benzene alkonium chloride and quaternary ammonium chloride) [1-2].

As part of contributing to the global efforts to confront this pandemic and stope the virus, our aim is to find alternative ethanol-free hand sanitizers that can be used as commercial, safe, inexpensive and widely available such as tea tree oil, aloe vera and povidone solution.

Aloe vera is a succulent herb belonging to the Aloe family [3]. It is an evergreen perennial that hails from the Arabian Peninsula but thrives in tropical, semitropical, and arid climates all over the world. It is grown for both agricultural and medicinal purposes such as anti-ulcer, anti-hypercholesterolemic, antioxidant, antibacterial activity, antiviral activity, antifungal activity, anti-acne, nutraceutical,

Journal of Physics: Conference Series **2063** (2021) 012007 doi:10.1088/1742-6596/2063/1/012007

humectant, skin protection against UV-A and UV-B, wound healing properties, the prevention of type II diabetes and cancer [4].

Tea tree oil is the common name for the essential oil extracted from Melaleuca alternifolia trees' leaves and terminal branches. Tea tree oil is extracted by blowing steam through the fresh leaves and terminal roots. The oil extracts from the water after cooling and can be collected. Melaleuca alternifolia essential oil is a mixture of aldehydes, terpenes, alcohols, esters, and other chemical molecules, and therefore has been used in many medical applications for its antioxidant, antimicrobial and digestive stimulant properties [5-6].

2. Experimental

The raw materials (Carbomer, Sodium methylparaben, Glycerine, Propylene glycol, and Triethanol-amine, povidone solution 10%) were purchased from chemical centres, Baghdad city. While, tea tree oil from herbs centres, Najaf city. All solvents in the experiment were supplied by Fluke, Merck, and BDH company.

2.1. Chemical Methods

2.1.1. Preparation of Povidone hand sanitizers gels (A)

The first brown product A, Ethanol-free, was prepared by using (2%) of Povidone solution 10% as disinfectant material. (0.3%) of Carbomer material 940 type as a thickening agent was mixed with the amount of water in a beaker for one day. The preservative (Sodium methyl paraben) (0.2%) was dissolved in another beaker with a small amount of water. The povidone and preservative were added to the carbomer mixture, After the components are stabilized, (0.1%) of triethanolamine base as pH neutralizer was added to the mixture in droplets to form the gel.

2.1.2. Preparation of tea tree oil hand sanitizers gels (B)

As in the first product, but here tea tree oil is used. The blue second product B. was prepared by using (0.07%) of tea tree oil as disinfectant material was added to the mixture of (0.3%) Carbomer material 940, (99%) of D. water, (0.2%) Sodium methyl paraben as preservative material and (0.05%) of Glycerine. After (24 hrs), (0.1%) of triethanolamine base was added to the mixture drop by drop to manufacture the gel [7].

2.1.3. Extraction of aloe vera extract

Aloe Vera extract was extracted from mixing (25 %) of aloe Vera plant after removing the leaves with (75%) of distilled water. After mixing and faltering, (60%) of the aloe Vera juice was treated with (30%) isopropanol and (10%) glycerin. After two weeks, the product is purified and used to prepare hand sanitizer gel [7].

2.1.4. Preparation of Aloe Vera hand sanitizers gels (C)

The last yellow product C was prepared using (2%) of aloe Vera extract, (0.05%) of Glycerine and (0.2%) of Sodium methyl paraben, the carbomer (0.3%) and triethanolamine (0.1%) were responsible for forming the yellow gel [7], Table 1.

2.2. Biological activity test [8-9]

Studies of antibacterial and antifungal activity were performed using a disc diffusion technique with (Muller Hinton) Agar plates for bacteria and (potato dextrose) agar plates for fungi. In testing the products against bacteria and fungi, the bacteria and fungi kinds (*Enterococcus feacalis*, *E. coli*, *Pseudomonas aeruginosa,* and *Staphylococcus aureus*) and (*Candida albicans* and *Aspergillus flavus*) were spread and activated on the Muller Hinton agar and potato dextrose agar respectively. The thin samples (1:1, water: gel), heavy samples, in addition to the standard product were injected for each disc.

Journal of Physics: Conference Series **2063** (2021) 012007 doi:10.1088/1742-6596/2063/1/012007

The bacterial and fungal plates were incubated at 27°C and 24°C for 24 and 72 hrs. respectively. The results were recorded based on the diameter of the inhibition zone in mm.

Table 1. Percentages of prepared hand sanitizers gels

No	Ingredients	(Stander)	A	B	C
1	Ethyl Alcohol	70%	-	-	-
2	Povidone 10%	-	2%	-	-
3	Tea tree oil	-	-	0.07%	-
4	Aloe Vera extract	-	-	-	2%
5	Glycerin	0.7%	0.05%	0.05%	0.05%
6	Propylene glycol	0.2%	-	-	-
7	Carbomer	0.47%	0.3%	0.3%	0.3%
8	Triethanolamine	0.33%	0.1%	0.1%	0.1%
9	D.water	28.77%	97.3%	99.229%	97.299
10	Sodium methyl paraben	-	0.2%	0.2%	0.2%
11	Color	-	-	0.001%	0.001%
12	Perfume	-	0.05%	0.05%	0.05%

3. Results and Discussion

Industrial companies and factories resorted to the manufacture of sterile products profusely, which led to an increase in the prices of raw materials and their lack in non-industrialized countries. The most important of these raw materials are ethanol and isopropanol and Hydrogen peroxide. That's why, we've prepared a group of hand sanitizers gels using alternative materials for ethanol material such as povidone 10%, tea tree oil, and aloe Vera extract.

The commercially colorless product (S) was prepared as a stander product to compare it with new products in terms of effectiveness against germs. The product was prepared using 70% of ethyl alcohol; Ethanol has been used as an effective, preservative and deadly substance for all kinds of bacteria and viruses as (COVID-19) according to recent studies [10-12].

The three new hand sanitizers gels (A, B, and C) were prepared by using povidone 10%, tea tree oil (Melaleuca alternifolia), and aloe Vera extract as sterile raw materials[13-15]. Glycerin was used as an emollient agent for the hands [16], while sodium methylparapene or (Sodium 4-(methoxycarbonyl) phenolate) was used as a preservative for the product. The acid-based polymer (Polyacrylic acid or carbomer) was used as a chelating agent for the gel [17]. Triethanolamine as the final step was added to neutralize the acidity and form the gel. It is a high safety compound that is widely used in many products such as thickeners, emulsifiers, detergents, wetting agents, shampoos, and cosmetics [18], Figure 1.

Journal of Physics: Conference Series **2063** (2021) 012007 doi:10.1088/1742-6596/2063/1/012007

Figure 1. Prepared hand sanitizer gels

The physical stability of the three prepared products (A, B and C) was examined after 10 weeks of storage, the color, odor and homogeneity were stable at 25 ° C and 10 ° C. The pH value as chemical property was stable at the mentioned temperatures for the three products, the pH value was within the normal values of the stander hand sanitizer gel to avoid health problems for skin such as itching, irritation, and scaling. The pH was ranged between (6.2-6.6), Table 2.

Table 2. Some physical and chemical properties of prepared products

No	Properties	After 10 w	(Stander)	A	B	C
1	State	Gel	Clear	Turbid	Semi turbid	Clear
2	Color	Stable	Silvery	Brown	Blue	Yellow
3	Odor	Stable	Alcohol	Pine	Pine	Pine
4	Homogeneity	Stable	Homogeneous	Homogeneous	Homogeneous	Homogeneous
5	pH	Stable	6.5	6.2	6.4	6.6

As for the gel volatilization time with rubbing hands, the time was calculated in seconds, the first product (A) remained on the skin for (13) seconds, the second product (B) remained for (15) seconds, the third product (C) was stayed for (13) seconds, while the standard product (S) was volatilized for (7) seconds. The reason is that the products contain a high percentage of water (97.299- 99.229%) and that the products do not contain a fast volatile substance such as ethanol as in the standard product. The gel volatilization time in the prepared compounds is approximately twice that of the standard gel, the gel staying time on the hands for a longer time, gives a good result in sterilization, Figure 2.

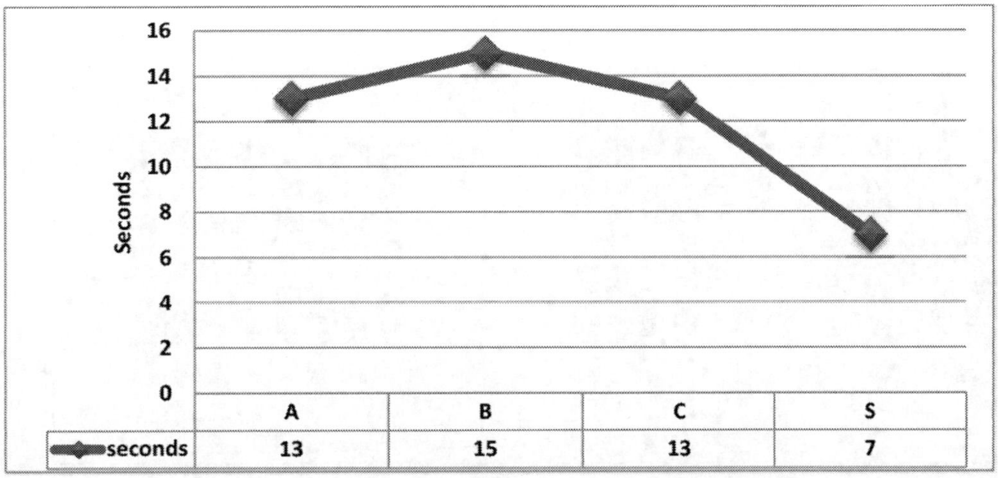

Figure 2. Gel volatilization time of products

The second part of the research is to study the microbial activity of the prepared products and compare the results with the prepared standard product. The prepared products were examined in the Central Laboratory /Biology Science / University of Baghdad. By using Muller Hinton Agar medium for bacteria and Potato Dextrose agar for Fungal, the prepared products (A, B, and C) were examined against four kinds (Enterococcus feacalis, E. coli, Pseudomonas aeruginosa, and Staphylococcus aureus) with two kinds of fungal (Candida albicans and Aspergillus flavus). In Table 3 & Figures 3,4, All prepared products gave similar results to the standard product in the thin and heavy products, the product (C) exhibited high antibacterial activity against all bacteria types. The prepared compounds and the standard gel in the thin state gave inactive results against bacteria (E. Coli and P. aeruginosa).

Table 3. Antibacterial activity for heavy products

Sample	**Inhibition zone diameter (mm\mg sample)**							
	E. feacalis		***E. Coli***		***P. aeruginosa***		***S.aureus***	
	Heavy	Thin	Heavy	Thin	Heavy	Thin	Heavy	Thin
A	13	11	11	R	13	R	12	11
B	12	11	11	R	12	R	13	12
C	14	11	12	R	13	R	16	12
S	12	11	11	R	12	R	11	11

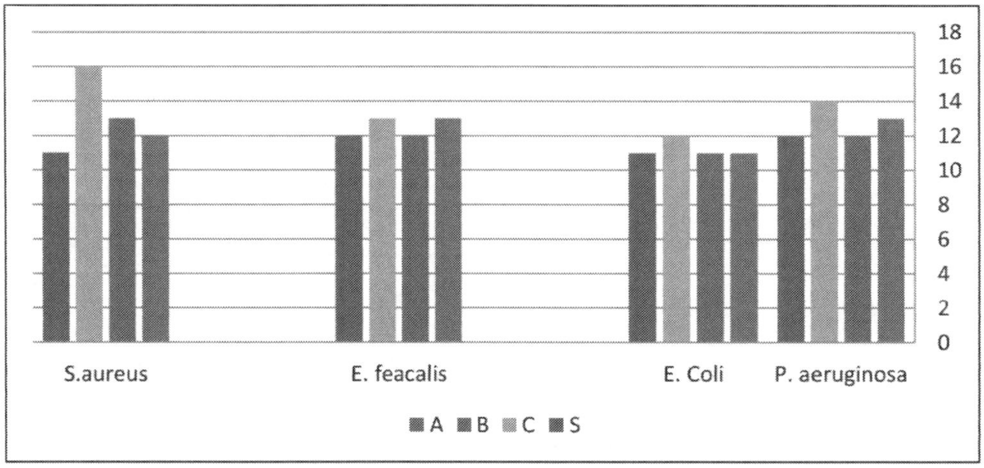

Figure 3. Antibacterial activity for heavy products

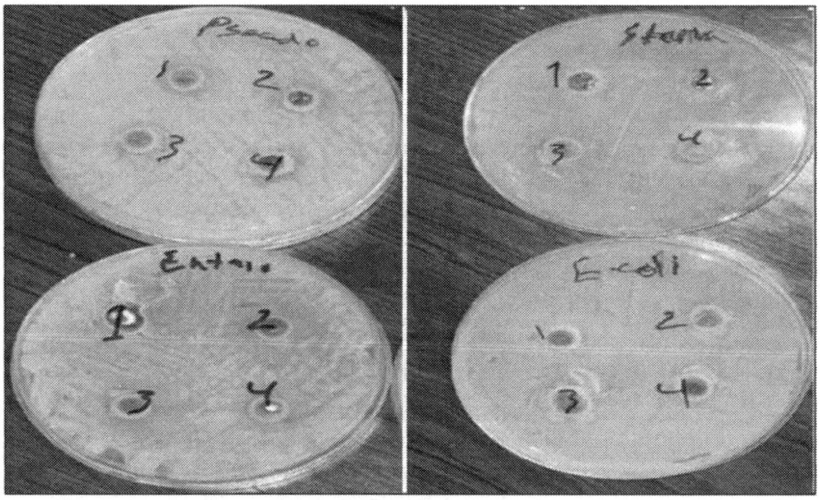

Figure 4. Antibacterial activity of heavy products against all bacteria

In the antifungal activity, the prepared products (A, B, and C) were tested in the Central Laboratory /Biology Science / University of Kufa. The products were also tested against two kinds of fungal (Candida albicans and Aspergillus flavus). The results of the fungi test differed significantly from the bacterial test. The thin and heavy prepared products gave high activity against fungi. In Candida albicans, the results ranged between (30-25) mm\mg in heavy products, whereas in the thin products between (20-10) mm\mg, as the C product gave highly active compared with other products (A and B) and similar with standard product. While in Aspergillus flavus, the prepared product A was showed high activity compared with other products (B and C) and nearer than standard product, Table 4 & Figure 5,6.

Table 4. Antifungal activity for heavy and thin products

Sample	**Inhibition zone diameter (mm\mg sample)**			
	C. albicans		***A. flavus***	
	Heavy	Thin	Heavy	Thin
A	25	20	28	10
B	28	15	20	15
C	30	10	10	10
S	30	20	30	20

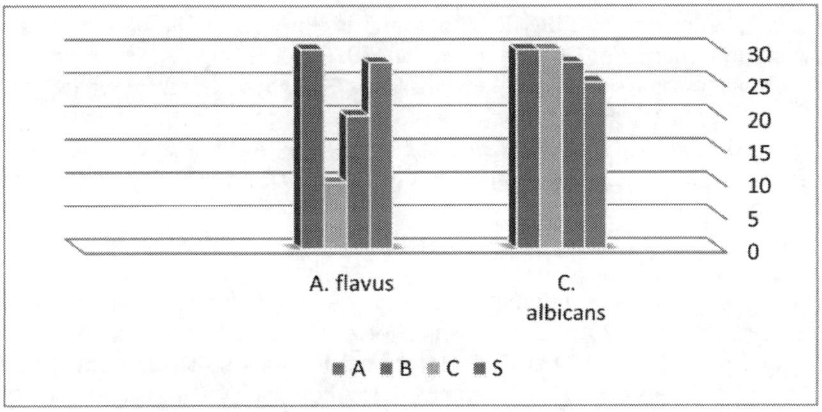

Figure 5. Antifungal activity for heavy products

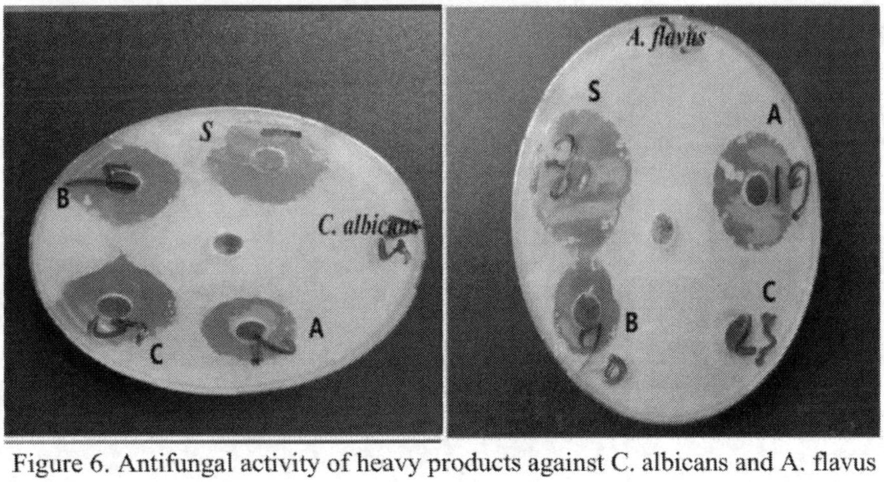

Figure 6. Antifungal activity of heavy products against C. albicans and A. flavus

4. Conclusion

Using active raw materials instead of ethanol 70%, such as povidone 10%, tea tree oil, and aloe Vera extract, and studying some of the physical and chemical properties of prepared products, as well as studying the biological activity of hand sanitizers gels products against four kinds of bacteria (Enterococcus faecalis, E. coli, Pseudomonas aeruginosa and Staphylococcus aureus) and two kinds of fungi (Candida albicans and Aspergillus flavus) and comparing them with the standard product (70% hand sanitizer gel). All these previous steps were a summary of the research. The heavy prepared hand sanitizer gels gave high activity or efficacy against types of bacteria and fungi and Similar effectiveness to a commercial standard product.

Acknowledgments
We would like to thank (Pharmacist Tuka S. Al-ameen) for his helpful advice on the topic related to this paper.

References

[1] Jane L J J, Thong P Y, Rajendran J C B, Jason R M, Nagendran T, Thiagarajan M 2020, International Journal of Environmental Research and Public Health, 17 (9): 3326.
[2] Vermeil T, Peters A, Kilpatrick C, Pires D, Allegranzi B, Pittet D 2019, Hand hygiene in hospitals: anatomy of a revolution. Journal of Hospital Infection, 101(4): 383-392.
[3] Prasannaraja C, Kamalanathan A S, Vijayalakshmi M A, Krishnan V 2020, Preparative Biochemistry & Biotechnology, 50 (5): 511-520.
[4] Ascensión M S, María E L C 1, Josefa G M, María J R, Encarna A 2020, Foods, 9 (11): 1542.
[5] Nikola P, Erinda L, Sonila C,Tana S K, Vojislava B, Gorica V, Mirela T S, Aleksandra P, Magdalena C 2020, Sustainability, 12 (8): 3420.
[6] Mohamed A R, Shawkey A E, Rabeh M. A, Abdellatif A O 2020, Journal of Herbal Medicine, 20: 100289.
[7] Silvia S, Nurul I A, Delly C L 2018, International Journal of Applied Pharmaceutics, 10: 216-220.
[8] Mahmood M F, Ezzat H Z, Majed J M 2019, Nano Biomed. Eng, 11 (1): 67-83.
[9] Al Khuzaie M G and Al Majidi A A D 2019, International Journal of Pharmaceutical Research, 11: (4)
[10] Adeel M, Maryam E , Saher P, Huda A A, Amtul B T, Abdullah Y, Kathirvel B ,Arivalagan P 2020, Science of the Total Environment, 742: 140561.
[11] Alberto B, Diego R P, Hamid A M, Lorina B, Iman A B, Giulia B, Marco C and Giovanni F P 2020, International journal of pharmaceutics, 119431.
[12] Sinthia KM, Mahboob M H 2018, American Journal of Microbiological Research, 6 (3): 73-78.
[13] Tsai J-C, Yen- K L, Yen-J H, El W L, Hsiao-Y W, Chia-h W, Yin-T T, Wen-S H, Ka-W T 2016, Infect Control Hosp Epidemiol, 38 (4): 417-422.
[14] Bhoj R S, Prasanna V, Monika B, Vinodh K O, Dharmendra K S and Shiv V S 2016, Pharmaceutica Analytica Acta, 7, 513.
[15] Darioush G, Mohammad K, Zohreh H, Fatemeh B, Seyedeh K B, Ali F F 2015, Journal of Medical Microbiology and Infectious Diseases, 3 (1): 6-10.
[16] Houben E, De Paepe K, Rogiers V 2006, Contact Dermatitis, 54 (5): 261-267.
[17] Yihong Q, Yisheng C, Geoff Z, Lawrence Y, Rao V. M 2016, Developing Solid Oral Dosage Forms (Pharmaceutical Theory and Practice), Academic Press, 2nd Edition, p.(8).
[18] Monice F, Bart H, Wilma F B, Donald V B, Ronald H, Curtis D K, Daniel L, James G M, Ronald C S, Thomas J S, Paul W S, Alan F A 2013, International journal of toxicology, 32, 3: 59-83.

Spectrophotometric determination of atenolol via oxidation and bleaching color reaction for methyl red dye

S A Zakaria[1], R A Zakaria[2] and N S Othman[1,3]

[1]Chemistry Department /College of Science / University of Mosul / Mosul, Iraq
[2]Chemistry Department /College of Education for Pure Science /University of Mosul /Mosul, Iraq
[3]Coreseponding author: Email: nsn20002004@uomosul.edu.iq

Abstract. A selective and sensitive spectrophotometric method has been suggested for the quantitative assay of atenolol (ATNL) as pure and in its manufactural formulation(Tablet). The suggested procedure included oxidation of ATNL with an excess quantity of the oxidant N-bromosuccinimide (NBS), and then the excess of NBS was occupied in bleaching the color of methyl red dye(MRD), then measuring the absorbance of remaining MRD at 518 nm . The absorbance of the unbleached color of MRD corresponds to the ATNL concentration in the sample solution. Beer's law was followed in the range of 0.1-2.0 $\mu g.ml^{-1}$with molar absorptivity value equal to $8.8864x10^4$ $l.mol^{-1}$ $.cm^{-1}$. The suggested method was applied to the assay of ATNL in commercial tablets, with satisfactory results.

1. Introduction

Atenolol (ATNL),chemically identified as 2-[4-[(2RS)-2-hydroxy-3-[(1-methylethyl) amino] propoxy]phenyl] acetamide (Figurl). It is a white powder, it is soluble in ethanol but water is sparingly soluble, {[1] ATNL is a selective β1 receptor antagonist, ATNL accumulated evidence suggests that ATNL is not sufficiently effective as a primary tool to treat hypertension[2] and for reducing morbidity and mortality in patients with heart failure [3].

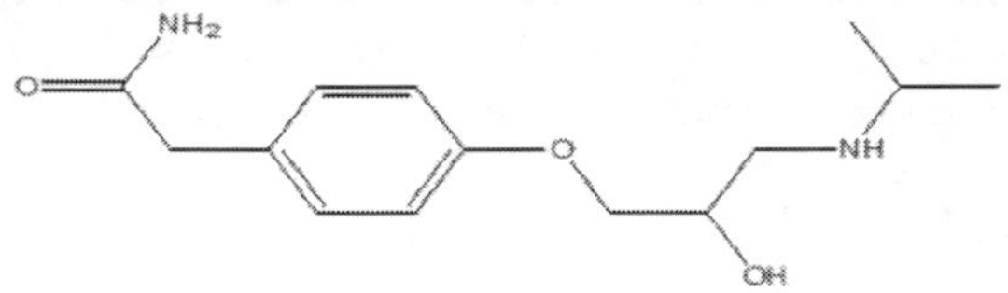

Figure1. Chemical structure of ATNL.

Literature survey exposes that several techniques have been reported for the assay of ATNL, or in combination with other drugs in formulations using flow injection analysis[4-7], HPLC[8-11], spectrofluorometric method[12-14], derivative spectroscopy[15].Various spectrophotometric methods have been reported for determination of ATNL such as: complexation with Fe(III) and Cr(III) ions [16],formation a colored product with bromothymol blue [17] or phenol red[18], condensation reaction with vanillin[19], Schiff's base formation by coupling ATNL with salicylaldehyde[20], ion-pair formation via reaction with bromocresol green in 1, 2-dichloroethane medium[21],oxidative derivatization with $KMnO_4$ in alkaline medium[22], or in acidic medium[23], gas chromatography-mass spectrometry [24], electrochemical[25] and voltammetric determination[26].

Journal of Physics: Conference Series **2063** (2021) 012008 doi:10.1088/1742-6596/2063/1/012008

The objective of the investigation reported in this new work is to evaluate a sensitive and accurate method for the determination of ATNL in an aqueous medium. The proposed method relied on the use of spectroscopy measurements, due to the simplicity and accuracy of spectrophotometric methods, and the spectrometer device is cheap and available in most laboratories, in addition to the fact that the proposed method was distinguished by its simplicity and availability of the required chemical materials in addition to that it has high sensitivity compared to some of the spectral methods mentioned in the literature[18-20,22]. The method based on the oxidation of ATNL in an acidic medium by N-bromosuccinimide, then bleaching the color of methyl red dye by the unreacted N-bromosuccinimide which its intensity proportional to the amount of ATNL.

2. Experimental

2.1. Instruments

The absorbance and the spectrums were achieved by using Jasco UV Spectrophotometric (JascoV-630,Japan) using two silica cells, the pH of solutions estimated by using Philips PW 9420 pH meter with a combined glass electrode.

2.2. Reagents

All chemicals used are pure as possible for analytical studies.

2.3. Preparation of solutions

Working ATNL solution, 20 μg/ml

A 0.0100 g of ATNL (provided from general establishment for medical appliance and drugs / SDI – Samaraa / Iraq) was dissolved in 5 ml of absolute ethanol, then diluted to 100 ml with distilled water in the calibrated flask, 20 ml of the above solution was diluted to 100 ml with distilled water in a calibrated flask.

Methyl red dye solution, $3\times10^{-4}M$

A 0.0080 g of MRD (Fluka) was dissolved in 50 ml of absolute ethanol and diluted to 100 ml with D.W. in a calibrated flask.

N-Bromosuccinimide solution, $4\times10^{-4}M$

A 0. 0071g of N-bromosuccinimide (Fluka) was dissolved in 100 ml distilled water in a calibrated flask.

The hydrochloric acid solution, 1M

An 8.3 ml of concentrated hydrochloric acid was diluted to 100 ml with distilled water in a calibrated flask.

The standard solutions of pharmaceutical formulation

Vascoten tablets solution, 100 μg/ml

A weight equivalent to one tablet(100 mg/Tablet) from five of powder (Atenolol tablets) was dissolved in a 100 ml calibrated flask with distilled water. A solution of 100 μg/ml concentration was prepared by taking 10 ml of the above solution and diluted with distilled water in a 100 ml volumetric flask, 20 and 30μg/ml standard solutions were prepared by appropriate dilution. The same procedure was applied for the preparation of the standard solution of vascoten tablet.

3.General procedure and calibration graph

To a set of 20 ml volumetric flasks, 0.1 to 2.0 ml of ATNL solution (20 μg /ml) were transferred. A 0.7 ml of HCl (1M) and 1.5 ml of N-bromosuccinimide (0.0004M) were added, the solutions were raised at room temperature (25°C) for 25 min., lastly 1.6 ml of MRD solution was added, after 5 min. all solutions were diluted to the mark with distilled water. The absorbance was measured at518 nm versus the blank. Figure 2 indicated the linearity of the method over the range of concentration from 0.1 to 2.0 μg/ml). The value of molar absorptivity is equal to 8.8864×10^{4} $l.mol^{-1}cm^{-1}$ is expresses that the method is sensitive.

Journal of Physics: Conference Series **2063** (2021) 012008 doi:10.1088/1742-6596/2063/1/012008

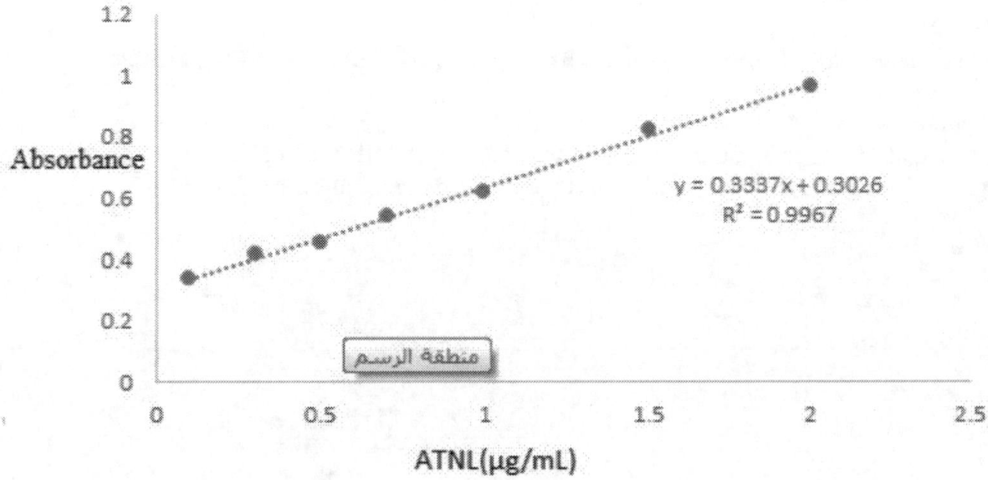

Figure2. The calibration curve of ATNL under the suggested method.

4. **Results and Discussion**

4.1.Principle of method

The method included oxidation of ATNL by adding an excess of NBS.

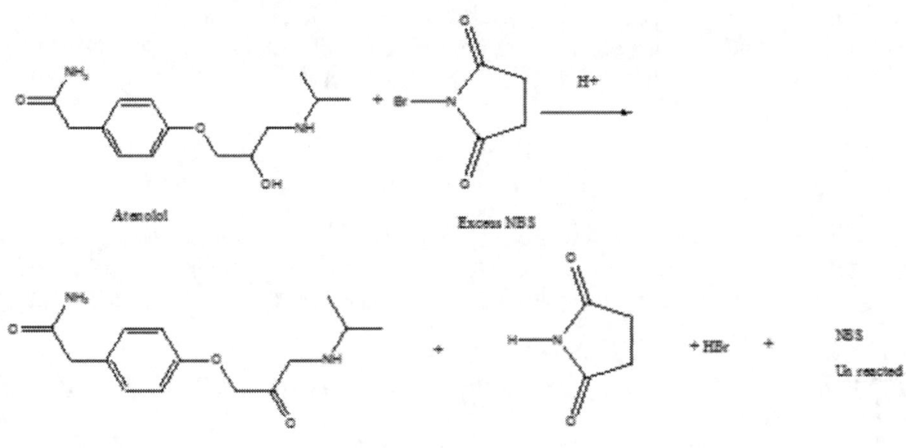

The unreacted NBS was used in bleaching the colour of MRD [27].

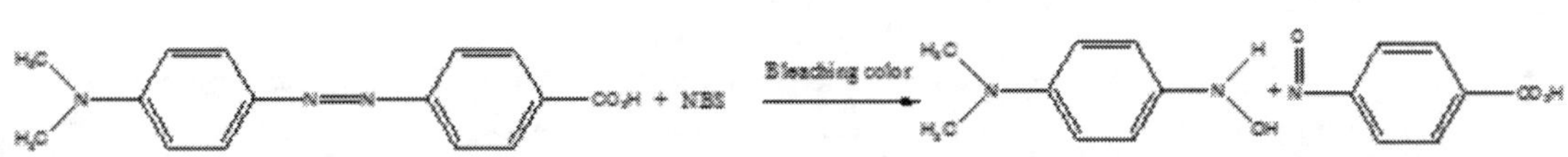

MRD N-4-dimethylamino-phenyl hydroxyl amine 4-nitroso-benzoic acid

The residual MRD measured spectrophotometric, and the absorbance is measured at 518 nm, which is increased linearly with an increasing amount of ATNL.

4.2.Optimunal conditions

The experiments to achieve optimum conditions for the determination of ATNL included:

4.2.1.Amount of MRD

The experimental results in Figure 3 included that 1.6 ml of MRD is considered an optimum amount according to the high intensity and excellent value of the determination coefficient.

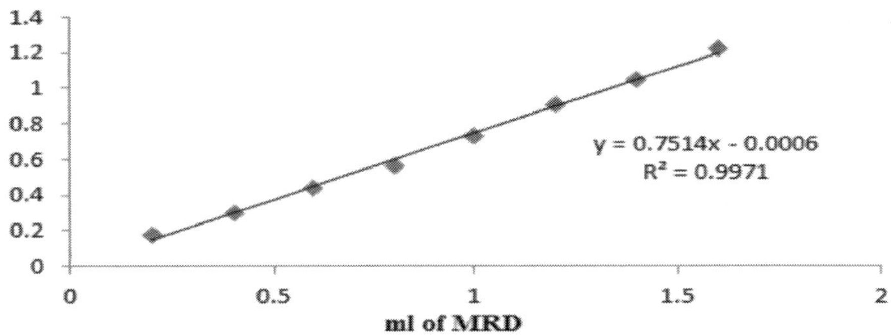

Figure 3. The effect of MRD amount on absorbance.

4.2.2.Type of optimum oxidant

NBS from experimental data was found to be a useful agent, compared with other oxidizing agents such as (NCS and Ce^{+4}) have been also tested, but none offered real advantages over NBS (Figure 4).

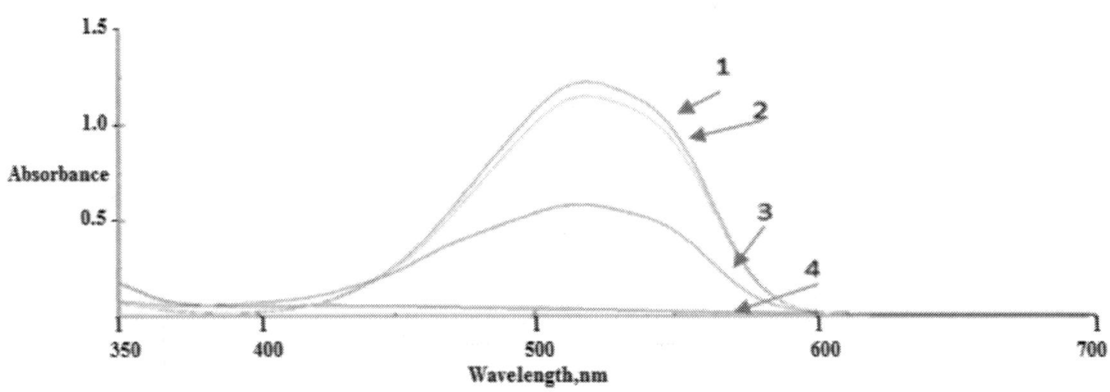

Figure 4. Bleaching the color of MRD :1- MRD only, 2 - MRD with NCS, 3-MRD with ceric ammonium sulphate, and 4- MRD with NBS.

The NBS was chosen as an oxidizing agent for the subsequent experiments because it gave the largest dye bleaching compared to that of ceric ammonium sulphate and NCS.

4.2.3. Effect of oxidant amount

Various volumes from 0.5 to 2.5 ml of 0.0004M of NBS were studied without ATEL to bleach the color of MRD. Figure 5 shows that 1.5 ml of NBS solution was the optimum volume and it's enough to get maximum bleaching of MRD color's, therefore it was recommended in the subsequent experiments.

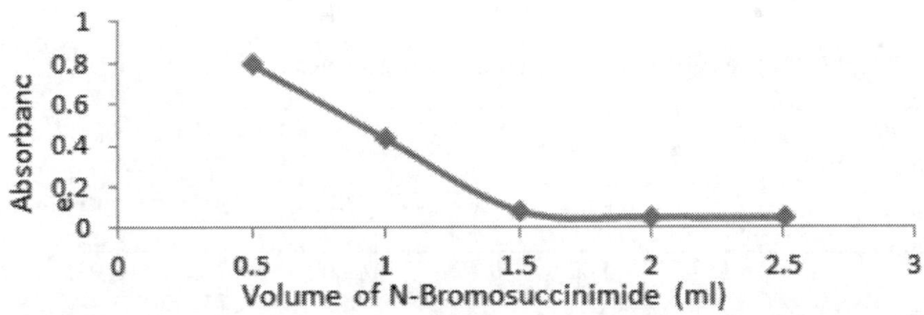

Figure5. The effect of oxidant amount on bleaching color of MRD.

4.2.4. Choice of acid and its amount

The primary experiments have shown that ATNL could oxidize in an acidic medium so that different types of acids were examined (table 1).

Table 1. Selection of acid.

1ml of 1M acid	Absorbance	pH
H_2SO_4	0.204	3.24
HCl	0.394	3.34
HNO_3	0.202	3.32
CH_3COOH	0.189	4.08

The results above in table 1 shows that HCl gave the highest intensity of the residual MRD and is consider as an acid for the present reaction with an optimum amount of 0.7 ml (table 2).

Table2. Effect of acid amount on absorbance.

1M HCl solution(ml)	Absorbance /µg of ATNL				
	15	20	25	30	R^2
0.3	0.292	0.322	0.333	0.404	0.9442
0.5	0.368	0.398	0.420	0.453	0.9973
0.7	0.412	0.420	0.436	0.477	0.9412
1.0	0.388	0.391	0.425	0.437	0.9542

The presence of the acid is essential for the completion of the oxidation process of ATNL, and a volume of 0.7 ml of 1M HCl solution was chosen to give it the highest absorbance of the remaining MRD, compared with volumes 0.5 and 1.0 ml , which indicates a complete oxidation of ATNL.

4.2.5. The effect of time on oxidation and bleaching the color of MRD

The effect of the time on oxidation of ATNL by NBS and the time needed for optimum bleaching of MRD color were studied (table3).

Table 3. The effect of time on oxidation and bleaching of the MRD.

After addition of oxidant, min.	Standing time before dilution, min					
	5	10	15	20	30	40
5	0.415	0.421	0.421	0.426	0.422	0.425
10	0.447	0.442	0.434	0.438	0.436	0.446
15	0.521	0.522	0.527	0.529	0.529	0.522
20	0.590	0.585	0.585	0.571	0.567	0.565
25	0.616	0.616	0.620	0.616	0.622	0.620
30	0.609	0.608	0.607	0.604	0.599	0.606

The results in table 3 show that the standing time of 25 minutes was necessary for the complete oxidation of ATNL and 5 minutes was necessary for bleaching of the color of MRD.

4.2.6. Order addition of components

The order followed as given under general procedure gave high absorbance so that it fixed in the subsequence experiments (table 4).

Table 4. Effect of the order of addition.

Order number	Order of addition	Absorbance
I	ATNL+HCl+NBS+MRD	0.626
II	ATNL+NBS+HCl+MRD	0.293
III	ATNL+NBS+MRD+HCl	0.146
IV	ATNL+HCl+MRD+NBS	0.034

5. Absorption spectrum

When ATN with 20 µg of ATNL was treated according to the recommended procedure, a decrease in the absorbance compared with an absorbance of MRD alone at the maximum absorption of 518 nm (Figure 6).

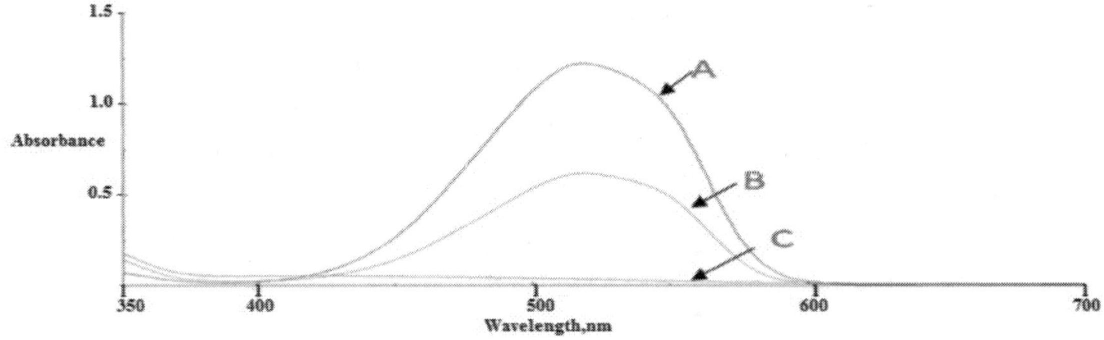

Figure 6. The absorption spectrum of (A) MRD, (B) residual MRD after oxidation 20 µg of ATNL, and (C)blank against distilled water.

6. Application of the method

To test the applicability of the present method, it has been applied to the determination of ATNL in pharmaceutical formulation(tablet). In table 5 the values of RE% and RSD% make sure that good accuracy and precision were obtained.

Table 5. Application of the method.

Drug	Amount taken µg/20ml	Recovery, %	RE*,%	RSD*,%
Atenolol 100 mg /tablet Actavis, Jorden	20	100.11	+0.11	0.40
	30	99.59	-0.41	0.14
Vascoten 100 mg/tablet Limassol-Cyprus	20	99.75	-0.25	0.49
	30	99.59	-0.41	0.11

* The average of four determinations.

The results listed in Table 5 demonstrate the success of the proposed method for estimating ATNL in pharmaceutical formulation (tablet) for two different companies.

7. Estimation of t-test and F-distribution

The suggested method is positively applied for the determination of ATNL in its pharmaceutical formulation(Vascoten tablet). The performance of the suggested method is judged by the application of a t-test at the 95% confidence level for 4 degrees of freedom and F-distribution in comparison with the standard method[1]. The results have been showing that the calculated values for t-exp. and F are less than the tabular values and equal to 1.199 and 7.39 respectively indicating that there is no significant difference between the suggested and standard method for the determination of ATNL.

8. Conclusion

The suggested method described a simple, sensitive and accurate spectrophotometric method for the estimation of ATNL using N-bromosuccinimide as an oxidant agent and the unreacted NBS bleached the MRD color. The ATNL in pharmaceutical preparation has been determined with satisfactory results.

References

[1] British Pharmacopoeia Commission, Great Britain. Medicines Commission, General Medical Council (Great Britain) British Pharmacopoeia (2000).
[2] H, Tomiyama and A. Yamashina. Beta-blockers in the management of hypertension and/or chronic kidney disease. Inter. J. Hyper. 2014,1-7 (2014).
[3] J. Nicolantonio, H.Fares, A. Niazi, S.Chatterjee, F. D'Ascenzo, E.Cerrato, G. Biondi-Zoccai, C. Lavie, D. Bell and J. O'Keefe. β-Blockers in hypertension, diabetes, heart failure, and acute myocardial infarction: a review of the literature. Open Heart 2(1), e000230 (2015).
[4] N. Al-Awadie and A. Khudhair. Determination of atenolol in pharmaceutical formulations by continuous flow injection analysis via turbidimetric (T180o) and scattered light effect at two

opposite position (2N90o) using Ayah 4SW-3D-T180-2N90-Solar-CFI Anal. Iraqi J. Sci. 55(1) ,12-26 (2014).

[5] A. Khataee, R. Lotfi, A. Hasanzadeh, M. Iranifam and S. Joo. Flow-injection chemiluminescence analysis for sensitive determination of atenolol using cadmium sulfide quantum dots. Spectrochim. Acta Part A: Mol. and Biomol. Spectr. 157, 88-95 (2016).

[6] N. Turkey and E. Mezaal. Assessment of long-distance chasing photometer (NAG-ADF-300-2) by estimating the drug atenolol with ammonium molybdate via continuous flow injection analysis. Bagh. Sci. J. 17(1), 78-92 (2020).

[7] E. Mezaal and N. Turkey. Assessment of long distance chasing photometer (NAG-ADF-300-2) by estimating the drug atenolol with povidone-iodine via CFIA. Ibn AL-Haitham Journal For Pure and Appl. Sci. 33(1), 65-83 (2020).

[8] H. Hashem, I. Ehab and E. Magda. A novel stability indicating HPLC-method for simultaneous determination of atenolol and nifedipine in presence of atenolol pharmacopeoial impurities. J. Appl. Pharma. Sci. 5(08), 017-25 (2015).

[9] A. Raoufi, M. Ebrahimi and M. Bozorgmehr. Application of response surface modelling and chemometrics methods for the determination of atenolol, metoprolol and propranolol in blood sample using dispersive liquid–liquid microextraction combined with HPLC-DAD. J.Chroma. B 1132, 121823 (2019).

[10] W. El-Alfy, O. Ismaiel, M. El-Mammli and A. Shalaby. Determination of atenolol and trimetazidine in pharmaceutical tablets and human urine using a high performance liquid chromatography-photo diode array detection method. Inter. J. Anal. Chem. 20191-8 (2019).

[11] Y. Bilal and A. Sakir. Determination of atenolol in human urine by using HPLC. Sep. Sci. Plus. 1(1), 4-10 (2018).

[12] A. Tabrizi, F. Bahrami and H. Badrouj. A very simple and sensitive spectrofluorometric method based on the oxidation with cerium (IV) for the determination of four different drugs in their pharmaceutical formulations. Pharma. Sci. 23(1),50-58 (2017).

[13] N. Safwat, M. Abdel-Ghany and M. Ayad. Sensitive derivative synchronous and micellar enhanced ecofriendly spectrofluorimetric methods for the determination of atenolol, diclofenac and triclosan in drinking tap water. J.AOAC Inter. 104(1), 103–112 (2020).

[14] E. Bakir, M. Gouda, A. Alnajjar and W. Boraie. Spectrofluorometric method for atenolol determination based on gold nanoparticles. Acta Pharma. 68(2), 243-50 (2018).

[15] M. Mohammad, M. Abdullah and S. Sabir. Simultaneous determination of atenolol and amlodipine using second derivative spectroscopy. Polytechnic J. 9(2), 25-9 (2019).

[16] K. Divya and B. Narayana. New visible spectrophotometric methods for the determination of atenolol in pure and dosage forms via complex formation. Indo Amer. J. of Pharma. Res. (IAJ PR) 4(1), 194-203 (2014).

[17] Y. Zhuk and S. Vasyuk. Quantitative determination of atenolol in tablets. Inter. J. of Advances in Pharma. Biol. and Chem. 5(3),350-354 (2017).

[18] I. Elgailani and T. Alghamdi. Development of spectrophotometric method for the determination of atenolol in Normoten drug. Int. J. Chem.9 58-64, (2017).

[19] M. Mohamed, E.Mohamed, M. Samy and A. Mai. Spectrophotometric determination of irbesartan, iosartan, atenolol and hydrochlorothiazide in bulk and dosage forms. Asian J. of Pharma. Anal. and Med. Chem. 4(2), 88-106 (2016).

[20] A. Raghad. Spectrophotometric determination atenolol in pharmaceutical preparation by Schiff's base formation. Kirkuk Univ. J. /Sci. Stud. (KUJSS) 13(2),253-265 (2018).

[21] S. Antakli, L. Nejem and M. Joumaa. Determination of atenolol in tablet formulation by analytical spectrophotometry. Res. J. Pharma. and Technol. 13(2,) 609-14 (2020).

[22] E. Vaikosen, J. Bioghele, R. Worlu and B. Ebeshi. Spectroscopic determination of two beta-blockers–atenolol and propanolol by oxidative derivatization using potassium permanganate in alkaline medium. Rev. Anal. Chem. 39(1),56-64 (2020).

[23] H. Sayyed, R. Katapalle, S. Salim and F. Mazhar. Permanganate study of oxidation of atenolol in acidic medium. World J. Pharma. and Pharma. Sci. 6(2),1281-1289 (2017).
[24] Y. Bilal, and A. Sakir. Determination of atenolol in human urine by gas chromatography - mass spectrometry methods. J. of Chromatogr. Sci. 49,365-369 (2011).
[25] M. Shaterian, A. Aghaei, M. Koohi, M. Teymouri and A. Mohammadi-Ganjgah. Synthesis characterization and electrochemical sensing application of $CoFe_2O_4$/graphene magnetic nanocomposite for analysis of atenolol. Polyhedron. 182, 114479 (2020).
[26] R. Zilberg, V. Maistrenko, L. Kabirova and D. Dubrovsky. Selective voltammetric sensors based on composites of chitosan polyelectrolyte complexes with cyclodextrins for the recognition and determination of atenolol enantiomers. Anal. Meth. 10(16), 1886-94 (2018).
[27] A. Manjunatha, S. Anu and S. Puttaswamy. Oxidative decolorization of methyl red dye with chloramine-T — kinetic and mechanistic chemistry. Indian J. Chem. Tech. 20, 416-422 (2013).

The influence in difference of compatibilizers on the mechanical and rheological properties of LDPE/PLST blends

Tahseen A Saki

Chemistry Department, College of Science, University of Basrah, Basrah, Iraq
Corresponding author E-mail address: Tahseen.saki@uobasrah.edu.iq

Abstract. In this present study, low density polyethylene/plasticizer starch (LDPE/PLST) blends were prepared as a product to be used in disposable packaging (film applications), reducing the negative polymeric environmental effect. Because of their different molecular structures, LDPE blends with starch are fully immiscible; therefore, a compatibility agent is required. Three different polymer and/or copolymer: poly (vinyl alcohol) hydrolyzed 75% (PVOH), styrene-allyl alcohol copolymer (SAA) and polyethylene glycol (PEG) were selected as compatibilizers containing –OH groups. The effects of compatibilizer on the mechanical and rheological properties of LDPE/PLST blends were investigated and compared to LDPE/PLST without compatibilzer. The blends are also characterized by FTIR, which strongly indicates the existence of compatibilizers that can enhance phase interaction and promote compatibility in the blends of LDPE/PLST. Comparing to the blend without a compatibilizer, the tensile strengths of the blends containing PVOH and SAA increased significantly. The elongation at break results shows similar observation. The rheological measurement results suggested that the addition of a compatibilizer exhibited an increase in the shear stress and apparent viscosity comparing to the uncompatibilized blend except the blend with PEG which exhibited phase separation.

1. Introduction

Polyethylene is the highly and commercially used polymer in the world at present. The industrial demand for this polymer is still increasing, a result of the variety of applications based on its superior properties, such as high chemical and mechanical resistance, easy processability and low specific gravity, in addition to low manufacturing costs [1]. Several types of polyethylene, including different levels of density have been produced so far. One of the important types is low-density polyethylene (LDPE). It is a polymer containing long and short chain branches and is well known for its excellent melt characteristics and efficient processability [1]. These properties have contributed to the widespread use of LDPE in many industries, including films for agricultural and food packaging, creating many environmental issues associated with the disposal of plastic after using it. In the soil environment, LDPE is not degraded by microorganisms, which leads to their long-term survival [2].

This subject led researchers to be interested in improving LDPE's biodegradability or partial biodegradability by blending them with an inexpensive and abundant natural polymer like starch [3]. In the case of food packaging films (partially biodegradable) made from LDPE/starch blend, the presence of starch with content greater than 30% by weight has a negative influence on the mechanical properties of the blend [4]. As comparison, LDPE/starch film with starch content less than 20 %, the LDPE matrix dominates plastic deformation, resulting in less crazes and approximately equal to mechanical properties of LDPE films [4, 5]. In general, if these blends are deposited after use it in the environment, the addition of starch to the LDPE blend provides food for the soil microorganisms. The starch is consumed by soil microorganisms, the blend (film) loses its structural stability [6]. This approach can contribute to deterioration in the mechanical properties, promoting the deterioration of other mechanisms and thus allowing the polymeric film blends to invade the bacteria and fungus that reside in the soil [7].

Several compatibilizers containing reactive groups able to format hydrogen bonds (H-bonds) with hydroxyl groups(–OH groups) of starch and interact with non-polar chain of LDPE, thus improve the comaptibilty of LDPE/starch blends [8-11]. Most of these compatibilizers have decreased the biodegradability of LPDE/starch blends and limited their use in humid environments. For this study, selected polymer and/or copolymer having -OH groups to study their compatibility LDPE with starch and its influence on the mechanical and rheological properties. Some of these compatibilizers has been widely used to prepare biodegradable blend with starch such as poly(vinyl alcohol) PVOH [12-14] and polyethylene glycol PEG [15].

2. Experimental part

2.1 Materials

Low density polyethylene LDPE (463) density=0.921 – 0.924, Melt Index= 0.28 – 0.38 was supplied by Petrochemical Company (SCPI), Basrah, Iraq. Starch $(C_6H_{10}O_5)_n$ was supplied by Merck Company. Glycerol supplied by Fluka Company which was added as a plasticizer for starch in the amount of approximately 10% of the total weight of these blends to form plasticizer starch PLST.

2.2 Compatibilizers

The poly (vinyl alcohol) PVOH was prepared by the hydrolysis of poly (vinyl acetate), the degree of hydrolysis was 75%, molecular weight =2000. Styrene-allyl alcohol copolymer SAA, crystalline, allyl alcohol=33 mol % with molecular weight (Mw) average~ 1,600. Both copolymers were supplied by Aldrich Chemical Company. Polyethylene glycol PEG was supplied by BDH Company, waxy with average molecular weight = 1430-1570.

2.3 Blends preparation and characterization

Blends were prepared by mixing at 140°C, 40 rpm and mixing time of 10 min in a Brabender Plasticorder blender. The LDPE/PLST blends were prepared containing 70 wt% LDPE, 20 wt% PLST and 10 wt% compatibilizer. The blends were compressed and molded into sheets about 2 mm thick using 20×20 window frame molds in a hydraulic press. The window was placed between two steel sheets covered by aluminum foil. Then, the assembly was put into the hydraulic press and initially heated to 120 °C at pressure 5 tons for 3 min and for 10 min at a pressure 15 ton. Remove sheets from the press after cooling by water to room temperature. Dumbbell shaped specimens were cut from each sheet to mechanical properties tests.

All FTIR spectra was performed by using a Fourier transform infrared (FTIR)-8400S (Shimadzu, Japan). The mechanical properties measurement of tensile properties, such as tensile strength at yield and at break besides the elongation at break and break energy of the blends was performed using a tensile tester, according to ASTM D882. The evaluated rheological characteristics of all the blends were measured using a capillary rheometer (Instron model 3211). A wide range of cross-head speeds (0.06, 0.2, 0.6, 2.0 and 6.0 cm/min^{-1}) is used for these measurements. The rheological data were recorded for all blends over shear rates of 5.43, 18.1, 54.3, 181, 543, and 1810 s^{-1} at temperatures 150, 160, 170 and 180 °C. The data of shear stress, shear rate and the viscosity were calculated using manual capillary rheometer for 3211 Instron machine (10-364-1A).

3. Results and discussion

3.1. FTIR spectroscopy of blends

A significant and important research topic is the H-bond in polymer blends, it typically increases the miscibility of the blend. Compatibility and also miscibility can be promoted by the inter H-bonds among the components in the blend and has important effects on its properties. The introduction of H-bonds is a routine and successful way to achieve compatibility and changing the properties of the blends. In systems of polymer blends, a variety of experimental methods are used to characterize the H-bonds. FTIR spectroscopy is the most effective and commonly used technique among these methods for investigating the H-bond formation in polymer blends. So, the compatibility of LDPE/PLST blends with various compatibilizers was investigated by FTIR. Figure1 shows the FTIR spectra of PLST and pure compatibilizers: PVOH, SAA and PEG. The complete assignment of the bands can be found in the other references [15-18]. In particular, FTIR showed that the H-bonds in LDPE/comptibilizer/PLST blends were essentially formed between the -OH groups of the sugar units in the starch and the -OH groups of compatibilizers. The interaction between starch and compatibilizer was possible via the -OH groups with a probability of interaction in the glucose ring with the ring oxygen O-5. Scheme 1 shows the H-bond interactions that presumably result among –OH groups of compatibilzer with functional groups of PLST [13].

In all blending spectra, the bands in the 3500-3600 cm^{-1} area were essential for evaluating the extent of the formation of H-bonds due to stretching of the --OH group. The reduction of the -OH peak frequency from 3600 cm^{-1} to lower wavelength and the broadening of this peak after blending provides details on the formation of H-bonding. The peak of the FTIR spectrum (figure 1) of PVOH at 1741 cm^{-1} was due to the carbonyl functional group belongs the residual acetate groups after poly(vinyl acetate) hydrolysis. H-bonding may also be strengthened by these groups, thereby increasing the compatibilization ability of the blend. In the blend, these carbonyl groups shifted from 1741 to 1734-1718 cm^{-1}, suggesting new interactions between PVOH and PLST due to H-bond formation.

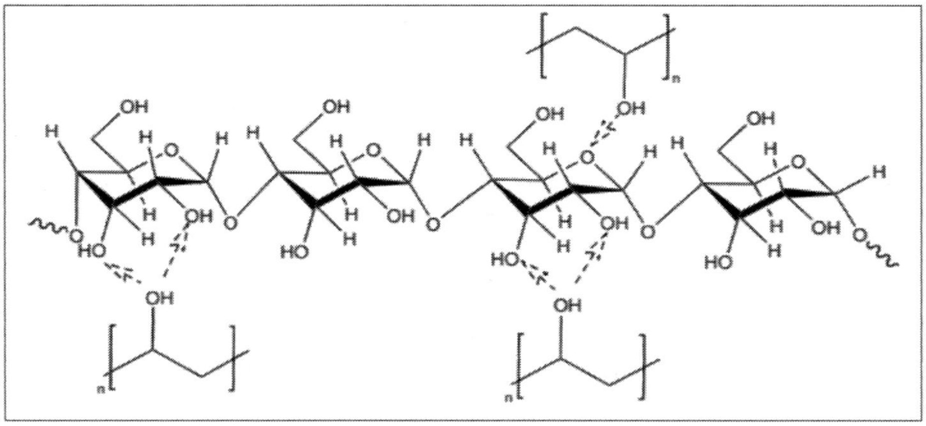

Scheme 1 The possible interaction between polymers containing -OH groups and functional groups in starch through H-bonding

Journal of Physics: Conference Series **2063** (2021) 012009 doi:10.1088/1742-6596/2063/1/012009

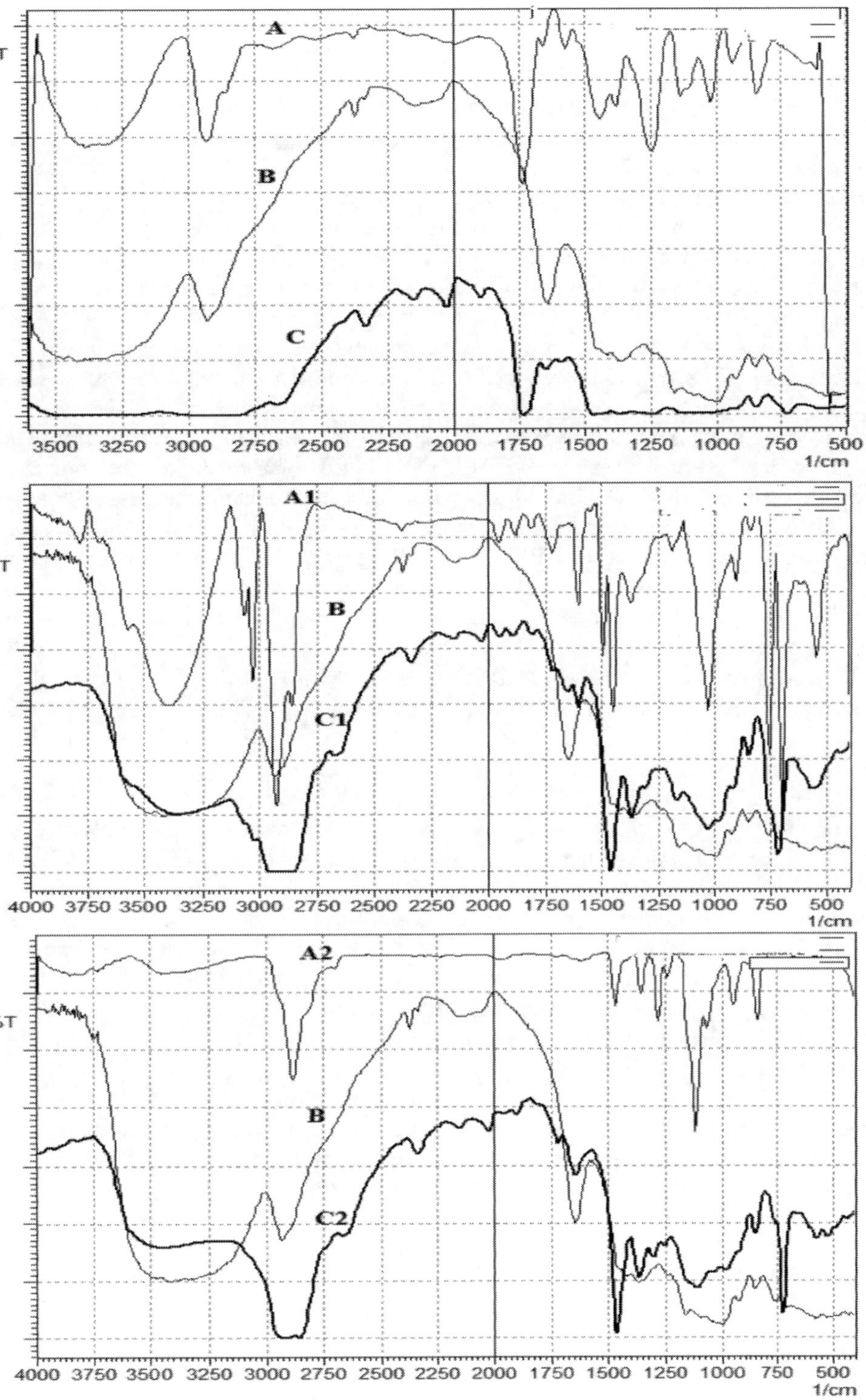

Figure 1 FTIR spectra of (A) PVOH, (A1) SAA, (A2) PEG, (B) PLST, (C) LDPE/PVOH/PLST, (C1) LDPE/SAA/PLST, (C2) LDPE/PEG/PLST

Journal of Physics: Conference Series **2063** (2021) 012009 doi:10.1088/1742-6596/2063/1/012009

3.2 Mechanical properties of the blends

It is well known, comparing to pure LDPE, most of the LDPE/starch blend appears lower values of tensile strength and elongation at break. The tensile strength of LDPE/starch blend has steadily decreased with increasing the percentage of starch in the blend. On the other hand, introducing 10% of starch into LDPE matrix, elongation at break decreasing significantly [19]. When stress force loads were applied to LDPE/starch blend and due to the weak of adhesion between LDPE and starch, it was not able to spread equally, therefore, mechanical properties of blend subsequently deteriorated [20]. Interfacial adhesion must be strong enough to ensure that the transfer of stress between the phases in order to enhance the mechanical properties of blends. Interfacial adhesion is typically improved either by chemical modification of the interfacial region or by the incorporation of third part of interface agents (compatibilizers).

In this study, we have attempted to enhance interfacial adhesion of LDPE/PLST blend with three different compatibilizers which were chosen based on their chemical structure. The influence of these compatibilizers has been assessed on the essential mechanical properties, which was represented to be tensile strength at yield and at break, elongation at break and break energy. The results are presented in figures 2, 3 and 4. In the blends, PVOH is a strong candidate for compatibility. It was containing –OH and unhydrolytic residual groups of poly(vinyl acetate), having H-bonding with the -OH groups of starch ensuring a good compatibility with PLST in the blends. Another important factor is the formation of a graft copolymer PVOH-g-LDPE from macroradical recombination which was generated at melt temperature [21, 22]. This factor probable acts as a reactive compatibilizer promoting increased interfacial adhesion among the blend polymers [23, 24], thus, improves the mechanical properties of the blend, Generally, different from the other compatibilized and uncompatibilized blends, it was observed that the blend with PVOH had a greater influence on the relative tensile strength improvement of the blend. The elongation at break results for blends compatibilized by PVOH was increased in the relative elongation at break of other blends. This is consistency with related findings on LDPE/EVOH/PLST blends [25, 26].

The second compatibilizer tried with LDPE/PLST blend was SAA, it is a random copolymer of polystyrene and poly (allyl alcohol) consist of the vinyl chains in the molecule and it can be compatible with the LDPE phase, while the polarity of the -OH groups and phenyl groups form H-bonding and dipole–dipole interaction, respectively, capable to interact with starch phase. The introduction of SAA copolymer improved the tensile strength of the LDPE/PLST blend, but the overall value, however, remains lower than LDPE/PVOH/PLST blend. Also, compared to the blend with PVOH, it shows a slight decline in tensile strength. At the same time, it reduces the elongation at break and there was significant difference compared to blends with PVOH. A possible explanation for this change, that the phenyl styrene groups presented in the compatibilizer SAA as it supports the toughness and rigidness of the blend.

Polyethylene glycol (PEG) was the third used compatibility agent, which is a low molecular weight linear polymer. It has ethylene groups possible theoretically able to miscible with LDPE hydrophopic chains, and –OH groups in the terminal chains interact with –OH groups in starch. Incorporation PEG in LDPE/PLST blend induces partial miscibility with starch phase while miscibility with LDPE chains led to phase separation due to the ethylene groups in backbone of the PEG immiscible with the LDPE as referred in literatures [25, 27]. Therefore, the results show the deteriorated significantly mechanical properties of blend with PEG as compatibilizer.

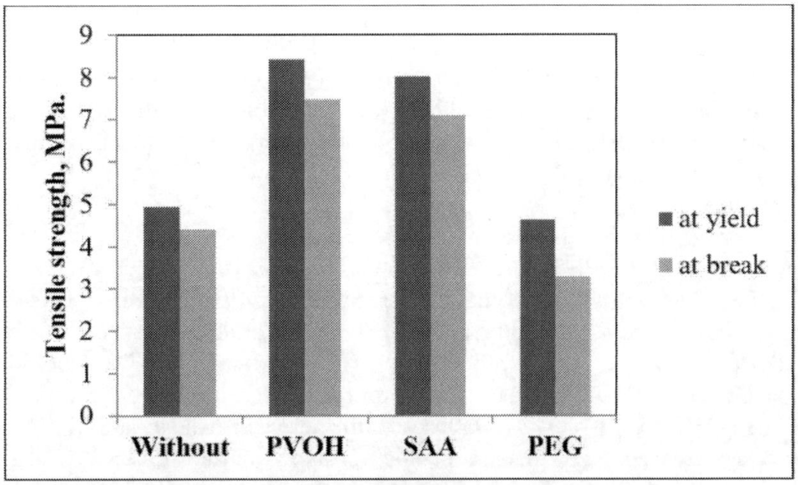

Figure 2 Tensile strength at yield and at break of the blends

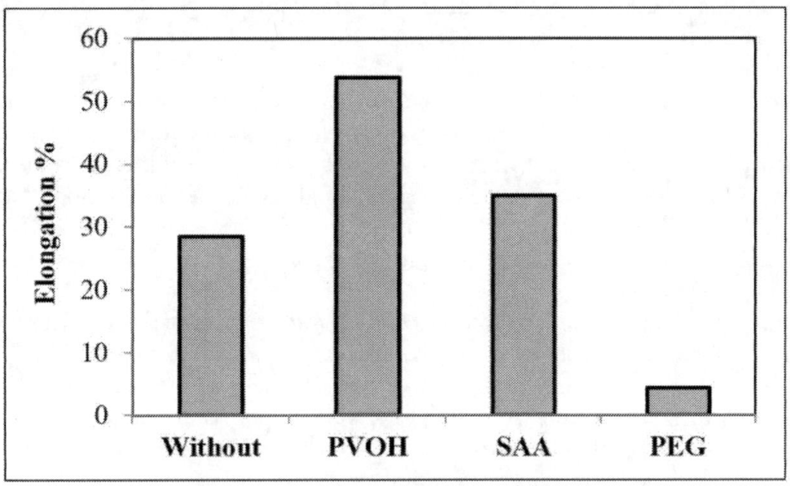

Figure 3 Elongation at break of the blends

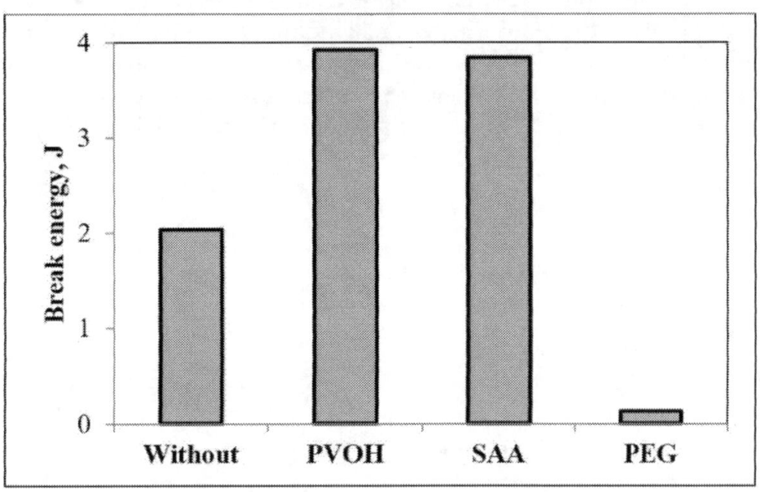

Figure 4 Break energyof the blends

3.3 Rheological properties of the blends

The flow curves of LDPE/PLST blends containing PVOH, SAA and PEG were measured at 150 °C which presented to be typical temperature for standard processing conditions as shown in figure 5. The shear stress increase along with shear rate for all LDPE/PLST blends with and without compatibilizer confirming pseudoplastic non-Newton melt fluids of blends, analogous to most polymeric melts and the relation shear stress with shear rate attached to the power law [28]. In addition, the non-Newtonian index (n) values of LDPE/PLST blends which was calculated from the slope of the fitted lines which were less than one, confirming that the blends were pseudoplastic [29].

The shear stress of LDPE/PVOH/PLST blend is more than sheer stress LDPE/SAA/PLST blend and LDPE/PLST blends without comatibilizer. While LDPE/PEG/PLST blend appeared lower value of shear stress compared to the other blends. However, adding PEG to the LDPE/PLST blend exhibited plasticization effect and enhanced the melt fluidity of blend. Figure 6 shows the result of viscosity versus shear rate of LDPE/PLST blends compatibilized by 10wt% compatibilizer. Noticing that the viscosity of LDPE/PLST blends decreases with increasing shear rate, those results reflecting another indicator of a typical feature of pseudoplastic non-Newton melt fluids.

The viscosity result of blend with PEG is the lowest value while the viscosity result of blend with PVOH is the highest. The highest value of melt viscosity may be explained by the fact that the PVOH was used for this investigation containing –OH and carboxyl groups (vinyl acetate groups) and capable to form H-bonding with starch hydroxyls (as explained above), the intermolecular and interlayer forces of blend fluids were increased thus the viscosity of melt was also increased.

The blend with SAA shows lower values of viscosity comparing to the viscosity of blend with PVOH. However, the viscosity of blend with SAA was higher than the viscosity of blend without compatibilizer and with PEG as a result of the substantial destruction of phenyl rigid groups acting on the PLST and LDPE.PEG had a low molecular weight, so at the processing temperature, the PEG melt had a low viscosity and melt strength. It was expected, due to immiscibly LDPE and PEG and lubricates nature of PEG caused interlayer slippages of the LDPE melt. Therefore, PEG acts only as a plasticizer, reducing the viscosity.

The effect of temperature on melt viscosity for all blends is shown in figure 7. With increasing the temperature, the viscosity of all blends increased except blends with PEG. In the melt state processing, a number of branch points in LDPE with loosely connected tertiary hydrogen's in the polymer molecule can forms free radicals [21]. At the same time, when the temperature reached about 150 °C, the macromolecules of partially hydrolyzed PVOH (which content acetate groups about 25%) degraded by double bonds formation, Furthermore, releasing acetic acid catalyzed the degradation reactions [22, 24]. The macroradicals was generated in the blend melt, causes crosslink between LDPE and PVOH, probably, due to mutual reactions between the tertiary carbon radicals of LDPE and the double bonds of the PVOH activated during melt blending. Therefore, those blends exhibited high viscosity with evaluated temperature. Blend with SAA, exhibits viscosity higher than blend without compatibilizer at a fixed temperature. When the temperature goes up to 180°C, the viscosity of LDPE/SAA/PLST increases sharply. This result attributed to effect polystyrene structure which has high viscosity [30].

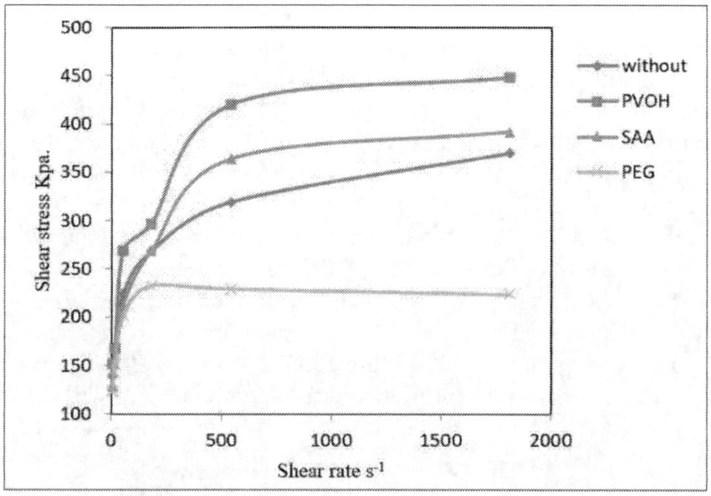

Figure 5 Flow curves of blends at 150 °C

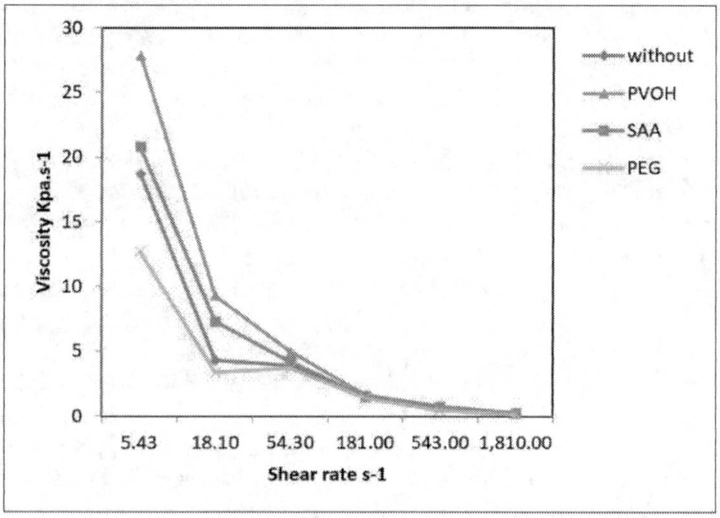

Figure 6 Viscosity versus shear rate of blends150 °C

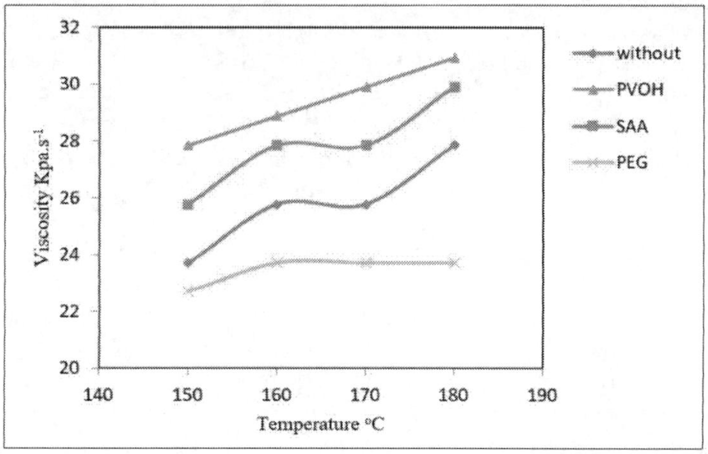

Figure 7 Effect temperatures on viscosity of blends

4. Conclusion

The influences of the three kinds of compatibilizers on the properties of LDPE/PLST blends was a study. Blends with and without compatibilizer was prepared in the melt state using Brabender Plasticorder. The compatibility of blends was characterized by FTIR, mechanical and rheological analyses. FTIR measurement shows all compatibilizers which are able to form H-bonds with –OH groups of starch. Regarding to the results of mechanical properties, PVOH demonstrated better tensile strength results as a compatibilizer among others. Also, using PVOH and SAA as a compatibilizer, the elongation at break values of the LDPE/PLST blends was improved compared to the blend without compatibilizer. Rheological results of the various LDPE/PLST blends showed increase the shear stress and viscosity along with shear rate. The viscosity of the blends was highly influenced by the temperature. Viscosity was increased along with the temperature when increased from 150 to 180 °C. From the results, it was also concluded that adding PEG to the LDPE/PLST blend shows a negative effect on the mechanical and rheological properties due to immiscibility with LDPE caused the phase separation of blend.

References

[1] Nwabunma D and Kyu T 2008 *Polyolefin Blends* (New Jersey: Wiley–Interscience)
[2] Kahar A W M, Ismail H and Othman N 2012 *J. Vinyl Addit. Techn.* **18** 65-70
[3] Tabasum S, Younas M, Zaeem M A, Majeed I, Majeed M, Noreen A, Iqbal M N and Zia K M 2019 *Int. J. Biol. Macromol.* 122 969-96
[4] Psomiadou E, Arvanitoyannis I, Biliaderis G, Ogawa H and Kawasaki N 1997 *Carbohyd. Polym.* **33** 227-42.
[5] Jaserg B, Swanson C, Nelsen T and Doane W 1992 *J. Polym. Mater.* **9** 153-62
[6] Dai L, Zhang J and Cheng F 2019 *Int. J. Biol. Macromol.* **132** 897-905.
[7] Mortazavi S, Ghasemi I and Oromiehie A 2013 *Polym. Test.* **32** 282-91
[8] Oner B, Gokkurt T and Aytac A 2019 *Open Chem.* **17** 557-63
[9] Mazerolles T, Heuzey M C, Soliman M, Martens H, Kleppinger R and Huneault M A 2019 *Carbohyd. Polym.* **206** 757-66
[10] Liu W, Wang Y and Sun Z 2003 *J. Appl. Polym. Sci.* **88** 2904-11
[11] Prinos J, Bikiaris D, Theologidis S and Panayiotou C 1998 *Polym. Eng. Sci.* **38** 954-64
[12] Sreekumar P A, Al-Harthi M A and De S K 2012 *Polym. Eng. Sci.***52** 2167-72
[13] Dean K M, Do M D and Petinakis E and Yu L 2008 *Compos. Sci. Technol.* **68** 1453-62
[14] Siddaramaiah, Raj B and Somashekar R 2004 *J. Appl. Polym. Sci.* **91** 630-35
[15] Kim C H, Kim D W and Cho K Y 2009 *Polym. Bull.***63** 91-9
[16] Pazo C, Rodriguez-Llamazares S, Bouza R, Barral L, Castano J, Muller N and Restrepo I 2018 *J. Polym. Res.* **25** 266
[17] Kahvand F and Fasihi M 2019 *Int. J. Biol. Macromol.* **140** 775-81
[18] Oh S J, Do J S and Ok J H 2003 *Appl. Spectrosc.* **57** 1058-62
[19] Briassoulis D. 2004 *J. Polym. Environ.* **12** 65-81.
[20] Thakore T M, Desai S, Sarawade B D and Devi S 2001 *Eur. Polym. J.* **37** 151-60
[21] Andersson T, Stalbom B and Wesslen B 2004 *J. Appl. Polym. Sci.* **91** 1525-37
[22] Alexy P, Lacik I, Simkova B, Bakos D, Pronayova N, Liptaj T, Hanzelova S and Varosova M 2004 *Polym. Degrad. Stabil.* **85** 823-30
[23] Van den Einde R M, Akkermans C, Van der Goot A J and Boom R M 2004 *Carbohyd. Polym.* **56** 415-22.
[24] Dimonie D, Petrache M, Damian C, Anton L, Musat M, Dima S O, Jinescu C and Maria R 2016 *Int. J. Polym. Sci.* 1-10
[25] YuL, Dean K and Li L 2006 *Prog. Polym. Sci.* **31** 576-602
[26] Sailaja R R N and Chanda M 2002 *J. Appl. Polym. Sci.* **86** 3126-34
[27] Tjong S C and Bei J Z 1998 *Polym. Eng. Sci.* **38** 392-402
[28] Gonzalez F J, Ramsay B A and Favis B D 2003 *Polymer* **44** 1517-26
[29] Saki T A 2015 *Arab. J. Chem.* **8** 191-99
[30] Xu S A, Zhu L, Xie J W and Jiang M 1999 *Polym Int.* **48** 1113-20

Journal of Physics: Conference Series **2063** (2021) 012010 doi:10.1088/1742-6596/2063/1/012010

Synthesis and characterization of azetidin-2-one and 1,3-oxazepine derivatives using Schiff bases derived from 1,1'-biphenyl-4,4'-diamine

H D Hanoon[1,*], H A Abd Al Hussain[2], S K Abass[1]

[1]Department of Chemistry/ College of Science/ University of Kerbala, Iraq
[2]Department of Chemistry/ College of Education for Pure Sciences/ University of Kerbala, Iraq

[*]Corresponding: haitham.alshebly@uokerbala.edu.iq

Abstract. The present study included the synthesis of two series of heterocyclic compounds, azetidin-2-one and 1,3-oxazepine derivatives. All synthesized compounds were characterized using FT-IR and ^{1}H NMR spectra. The study was divided into two parts. The first synthesis Schiff bases derivatives (**1-4**) via the condensation reaction of 1,1'-biphenyl-4,4'-diamine (**A**) with benzaldehyde derivatives (2,5-dimethoxybenzaldehyde, 4-hydroxy-3-methoxybenzaldehyde, furan-2-carbaldehyde and 3-hydroxybenzaldehyde). The second synthesis azetidine-2-one derivatives (**5-8**) from the reaction of Schiff bases with chloroacetyl chloride. Schiff bases also reacted with maleic anhydride to yield 1,3-oxazepine derivatives (**9-12**).

1. Introduction

Schiff base is a functional group that containing the general formula $R_1R_2C{=}N{-}R_3$, where R_3 is phenyl or alkyl groups, which make the Schiff bases stable imines [1]. Schiff base can be prepared by condensation reaction between an primary amines and carbonyl compounds (an aldehyde or ketone) [2]. Schiff bases are important compounds, which have industrial application and used as a material in the synthesis of great bioactive compounds [3], like antiinflammatory [4], antifungal [5] and antibiotic [6].

Azetidines are known important, pharmacological and biological activates [7], like anticancer [8], antioxidant [9], antimicrobial [10] and antitubercular [11].

1,3-Oxazepines are a type of seventh-membered heterocyclic compounds containing two heteroatoms (oxygen and nitrogen) [12]. Oxazepines possesses a wide range of biological activities and pharmaceutical properties like antifungal [13], anitbacterial [14], hypnotic muscle relaxant [15] and antagonistic [16].

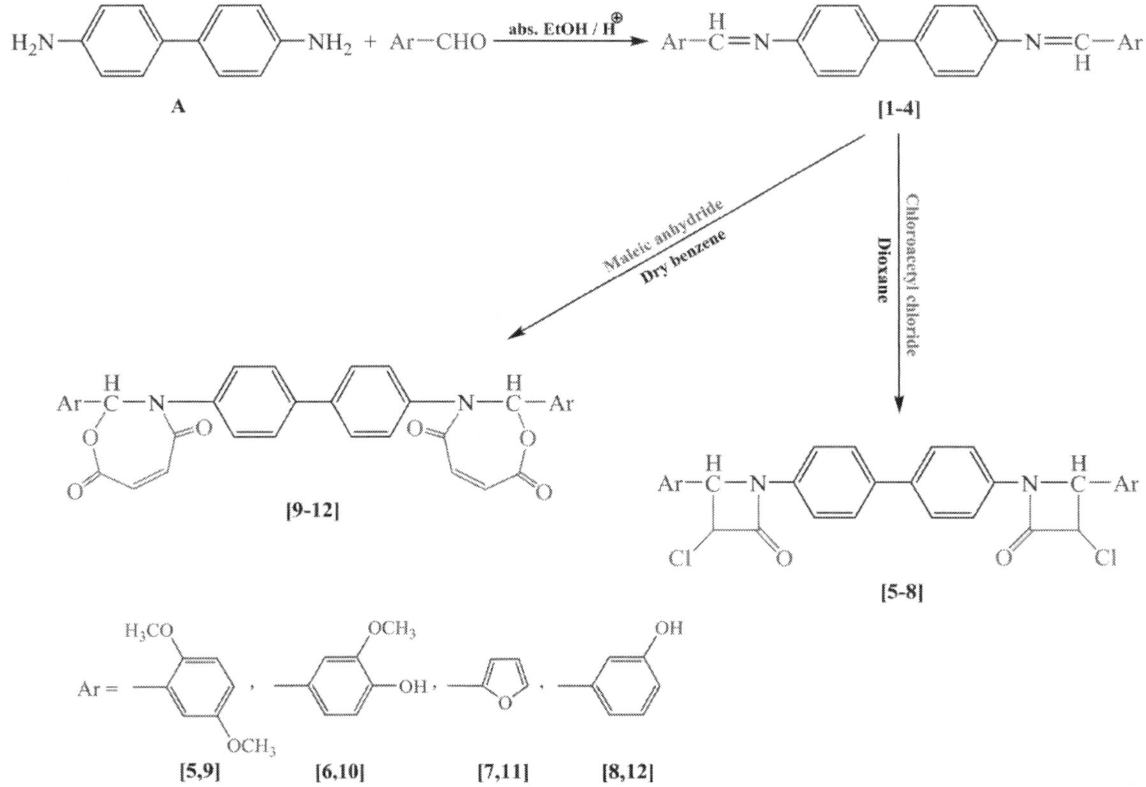

Scheme 1. Synthesis of Schiff base and heterocyclic derivatives.

2. Materials and methods

2.1. Materials

Solvents and reagents were obtained from commercial sources and utilized as they were received. Melting points (m.p) were determined by Electro thermal Stuart melting point apparatus, GOWLLANDS. FT-IR spectra were recorded by Fouriet transform infrared SHIMADZU FT-IR-8400S infrared spectrophotometer. ^{1}H NMR spectra were collected in DMSO-d_6 using INOVA spectrometer operating at 500 MHz.

2.2. Synthesis of compounds

*2.2.1. General method of synthesis of Schiff bases derivatives (**1-4**).* In a round bottomed flask, 1,1'-biphenyl-4,4'-diamine (A) (2 g, 0.01 mol), (30 mL) ethanol, (0.02 mol) aromatic aldehydes and two drops glacial acetic acid were added the reaction was refluxed with stirring for 3 hrs. After the completion of reaction, the mixture allowed to cool. The colored precipitate was filtered and dried. Then recrystallized from ethanol.

*2.2.2. General method of synthesis of Azetidinone derivatives (**5-8**).* Schiff bases (2 g) were dissolved in dioxane (10 mL) and (2 mL) of triethylamine was added, and then chloroacetyl chloride was added slowly at 5-10°C. Ice water to the reaction mixture was added. The separated solid was dried and recrystallized from ethanol

*2.2.3. General method of synthesis of oxazepine derivatives (**9-12**).* The mixture of Schiff bases (0.0012 mol) and maleic anhydride (0.0025 mol) were dissolved in (25 mL) in dry benzene. The mixture was refluxed with stirring for 8 hrs., in water bath at 80°C and excess solvent was removed . The precipitate was filtered and recyrstallized from ethanol.

2.3. Spectral data for the products

N,N'-[(1,1'-biphenyl)-4,4'-diyl]bis[1-(2,5-dimethoxyphenyl) methanimine] (**1**). Yield 90%, m.p. 186–188°C, as a greenish yellow solid. FT-IR spectrum, υ, cm^{-1}: 3026 (Ar-H), 2939, 2835 (C-H), 1624 (C=N), 1215, 1041 (C-O-C).

4,4'-{[(1,1'-biphenyl)-4,4'-diylbis(azaneylylidene)]bis(methaneylylidene)}bis(2-methoxyphenol) (**2**). Yield 83%, mp 181–184°C, as a yellow solid. FT-IR spectrum, υ, cm^{-1}: 3313 (O-H), 3036 (Ar-H), 2872 (C-H), 1626 (C=N), 1269, 1082 (C-O-C), 1192 (C-O).

N,N'-[(1,1'-biphenyl)-4,4'-diyl]bis[1-(furan-2-yl)methanimine] (**3**). Yield 72%, mp 195–197°C, as a dark green solid. FT-IR spectrum, υ, cm^{-1}: 3070 (Ar-H), 1585 (C=N), 1514 (C=C furan), 1280 (C-O).

3,3'-{[(1,1'-biphenyl)-4,4'-diylbis(azaneylylidene)]bis(methaneylylidene)}diphenol (**4**). Yield 76%, mp 198–201°C, as a pale green solid. FT-IR spectrum, υ, cm^{-1}: 3452 (O-H), 3080 (Ar-H), 1627 (C=N), 1172 (C-O).

1,1'-[(1,1'-biphenyl)-4,4'-diyl]bis[3-chloro-4-(2,5-dimethoxyphenyl)azetidin-2-one] (**5**). Yield 78%, mp 200–203°C, as a dark brown solid. FT-IR spectrum, υ, cm^{-1}: 3030 (Ar-H), 2955, 2840 (C-H), 1660 (C=O azetidinone ring), 1398 (C-N),1259, 1031 (C-O-C), 503 (C-Cl). ^{1}H NMR spectrum, δ, ppm: 7.83–7.62 m (6H, Ar-H), 7.55–7.46 m (8H, Ar-H), 4.34 d (2H, HC-Cl), 3.90 d (2H, N-CH), 3.80 s (6H, OCH_3), 2.73 s (6H, OCH_3).

1,1'-[(1,1'-biphenyl)-4,4'-diyl]bis[3-chloro-4-(4-hydroxy-3-methoxyphenyl)azetidin-2-one] (**6**). Yield 70%, mp 170–172°C, as an orange solid. FT-IR spectrum, υ, cm^{-1}: 3323 (O-H), 3030 (Ar-H), 2945 (C-H), 1654 (C=O azetidinone ring), 1315 (C-N), 1215, 1026 (C-O-C), 1180 (C-O), 509 (C-Cl).^{1}H NMR spectrum, δ, ppm: 9.76 s (2H, OH), 8.30–7.62 m (6H, Ar-H), 7.42–6.72 m (8H, Ar-H), 4.28 s (2H, HC-Cl), 3.83 d (2H, N-CH), 3.58 s (6H, OCH_3).

1,1'-[(1,1'-biphenyl)-4,4'-diyl]bis[3-chloro-4-(furan-2-yl)azetidin-2-one] (**7**). Yield 62%, mp 206–208°C, as a dark red solid. FT-IR spectrum, υ, cm^{-1}: 1656 (C=O azetidinone ring), 1581 (C=C furan ring), 1340 (C-N), 1276 (C-O), 553 (C-Cl).^{1}H NMR spectrum, δ, ppm: 8.51–7.13 m (8H, Ar-H), 6.77–6.72 m (6H, =CH, furan ring), 6.71 d (2H, HC-Cl), 6.64 d (2H, N-CH).

1,1'-[(1,1'-biphenyl)-4,4'-diyl]bis[3-chloro-4-(3-hydroxyphenyl)azetidin-2-one] (**8**). Yield 67%, mp 188–190°C, as a brown solid. FT-IR spectrum, υ, cm^{-1}: 3313 (O-H), 3045 (Ar-H), 1660 (C=O azetidinone ring), 1388 (C-N), 1284 (C-O), 509 (C-Cl). ^{1}H NMR spectrum, δ, ppm: 9.91 s (2H, OH), 7.83 d (2H, Ar-H), 7.79–7.48 m (6H, Ar-H), 7.42–7.25 m (8H, Ar-H), 7.14–7.11 m (2H, HC-Cl), 7.05–7.03 m (2H, N-CH).

3,3'-[(1,1'-biphenyl)-4,4'-diyl]bis[2-(2,5-dimethoxyphenyl)-2,3-dihydro-1,3-oxazepine-4,7-dione] (**9**). Yield 77%, mp 162–164°C, as a pale red solid. FT-IR spectrum, υ, cm^{-1}: 3041 (Ar-H), 2937, 2837 (C-H), 1722 (C=O lactone), 1626 (C=O lactam), 1288 (C-N), 1215, 1039 (C-O-C). ^{1}H NMR spectrum, δ, ppm: 8.87 d (2H, Ar–H), 7.75–7.62 m (2H, Ar–H), 7.55 s (2H, Ar–H), 7.35–7.12 m (8H, Ar–H), 6.49 dd (2H, N–CH, oxazepine ring), 6.32 d (2H, =CH, oxazepine ring), 6.24 s (2H, =CH, oxazepine ring), 3.85 s (6H, OCH_3), 3.78 s (6H, OCH_3).

3,3'-[(1,1'-biphenyl)-4,4'-diyl]bis[2-(4-hydroxy-3-methoxyphenyl)-2,3-dihydro-1,3-oxazepine-4,7-dione] (**10**). Yield 71%, mp 215–217°C, as a dark yellow solid. FT-IR spectrum, υ, cm^{-1}: 3294 (O-H), 3070 (Ar-H), 2847 (C-H), 1722 (C=O lactone), 1626 (C=O lactam), 1290 (C-N), 1213, 1028 (C-O-C), 1180 (C-O). ^{1}H NMR spectrum, δ, ppm: 9.77 s (2H, OH), 8.56 d (2H, Ar–H), 7.76–7.64 m (2H, Ar–H), 7.60–7.52 m (2H, Ar–H), 7.45–7.28 m (8H, Ar–H), 6.97–6.91 m (2H, N–CH, oxazepine ring), 6.51–6.32 m (2H, =CH, oxazepine ring), 6.24 s (2H, =CH, oxazepine ring), 3.87–3.84 m (6H, OCH_3).

3,3'-[(1,1'-biphenyl)-4,4'-diyl]bis[2-(furan-2-yl)-2,3-dihydro-1,3-oxazepine-4,7-dione] (**11**). Yield 65%, mp 191-193°C, as a brown solid. FT-IR spectrum, υ, cm^{-1}: 3055 (Ar-H), 1699 (C=O lactone), 1626 (C=O lactam), 1572 (C=C furan), 1321 (C-N), 1257 (C-O). ^{1}H NMR spectrum, δ, ppm: 7.76–7.51 m (8H, Ar-H), 7.36 d (2H, N–CH, oxazepine ring), 6.52–6.47 m (2H, =CH, oxazepine ring), 6.33 dd (2H, =CH, oxazepine ring), 6.27–6.16 m (6H, =CH, furan ring).

Journal of Physics: Conference Series **2063** (2021) 012010 doi:10.1088/1742-6596/2063/1/012010

3,3'-[(1,1'-biphenyl)-4,4'-diyl]bis[2-(3-hydroxyphenyl)-2,3-dihydro-1,3-oxazepine-4,7-dione] (**12**). Yield 61%, mp 186 dec.°C, as a gray solid. FT-IR spectrum, υ, cm^{-1}: 3300 (O-H), 3043 (Ar-H), 1722 (C=O lactone), 1622 (C=O lactam), 1288 (C-N), 1213 (C-O). ^{1}H NMR spectrum, δ, ppm: 9.72 s (2H, OH), 8.59 d (2H, Ar–H), 7.76–7.58 m (6H, Ar–H), 7.45–7.28 m (8H, Ar–H), 6.94–6.74 m (2H, N–CH, oxazepine ring), 6.40 dd (2H, =CH, oxazepine ring), 6.23 s (2H, =CH, oxazepine ring).

3. Results and Discussion

In the current work, the substrate 1,1'-biphenyl-4,4'-diamine (**A**) reacted with selected benzaldehyde derivatives (2,5-dimethoxybenzaldehyde, 4-hydroxy-3-methoxybenzaldehyde, furan-2-carbaldehyde and 3-hydroxybenzaldehyde) in absolute ethanol and the few drops of glacial acetic acid to give the Schiff base derivatives (**1-4**). Two series of compounds namely azetidinone derivatives (**5-8**) and 1,3-oxazepine derivatives (**9-12**) were synthesized using experimental methods as shown in Scheme 1. All the target compounds were characterized by FT-IR and ^{1}H NMR spectra (Figure 1 to Figure 20), which were in agreement with the suggested structures. FT-IR spectra of the prepared compounds (**1-4**) shown absorption peaks at (1585-1627) cm^{-1}, were assigned to the C=N group of Schiff bases. Azetidinone ring derivatives (**5-8**) shown FT-IR bands at (1654–1660) cm^{-1} attributed to the C=O group and appearance peaks at (503-593) cm^{-1} corresponding to the C-Cl group. Further, ^{1}H NMR data of compounds (**5-8**) shown the peaks at (4.28–7.14) ppm and (3.83–7.05) ppm for the protons of azetidinone ring. FT-IR spectra of 1,3-oxazepine derivatives (**9-12**) were observed two important bands at (1699-1722) cm^{-1} assigned to the C=O group of lactone ring and at (1622-1626) cm^{-1} attributed to the C=O group of lactam ring. On the other hand, ^{1}H NMR data of compounds (**9-12**) appeared that (6.49–7.36) ppm, (6.32–6.94) ppm and (6.23–6.33) ppm for the protons of oxazepine ring. These absorption peaks above are evidence for the formation of target compounds (azetidin-2-one and 1,3-oxazepine derivatives).

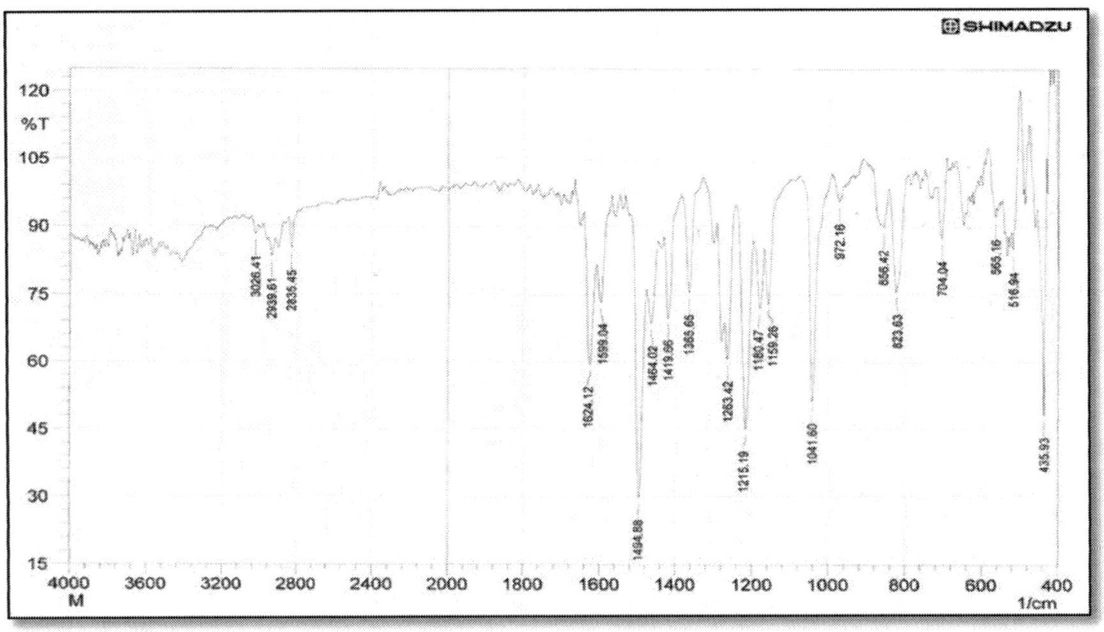

Figure 1. FTIR spectrum of compound **1**.

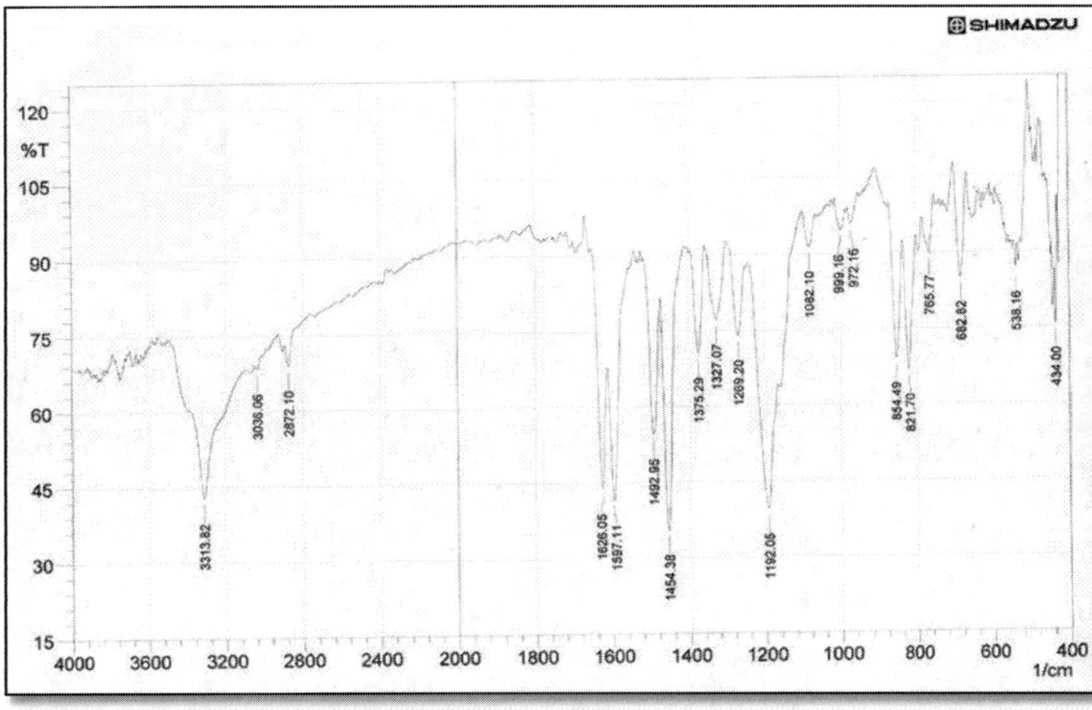

Figure 2. FTIR spectrum of compound **2**.

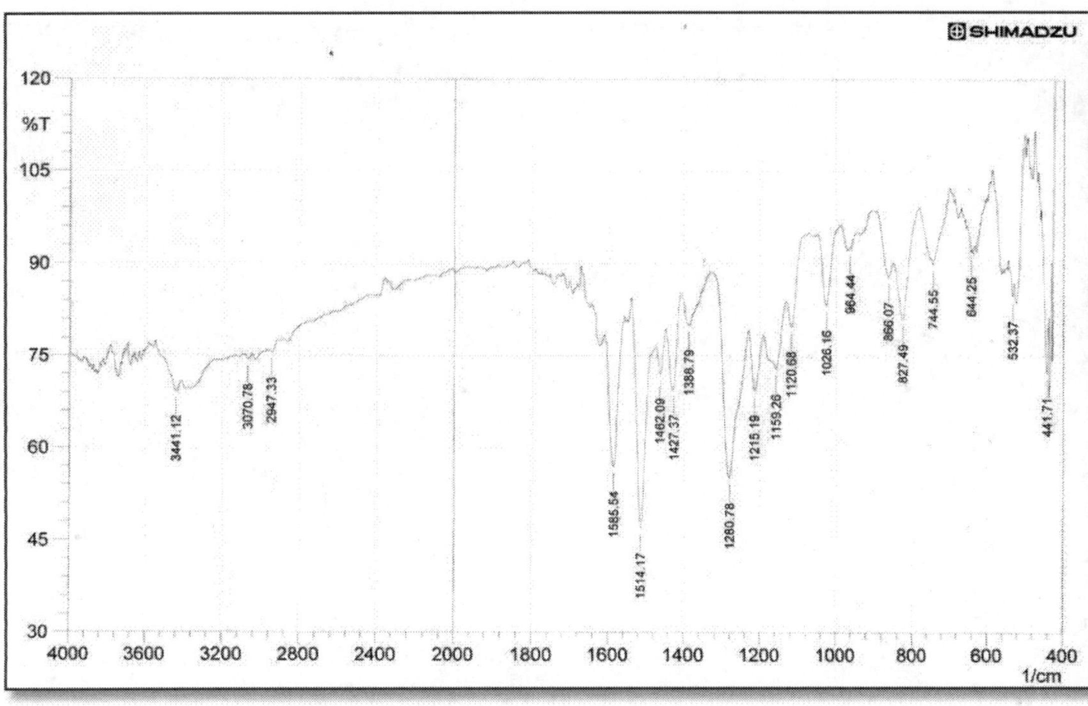

Figure 3. FTIR spectrum of compound **3**.

Journal of Physics: Conference Series **2063** (2021) 012010 doi:10.1088/1742-6596/2063/1/012010

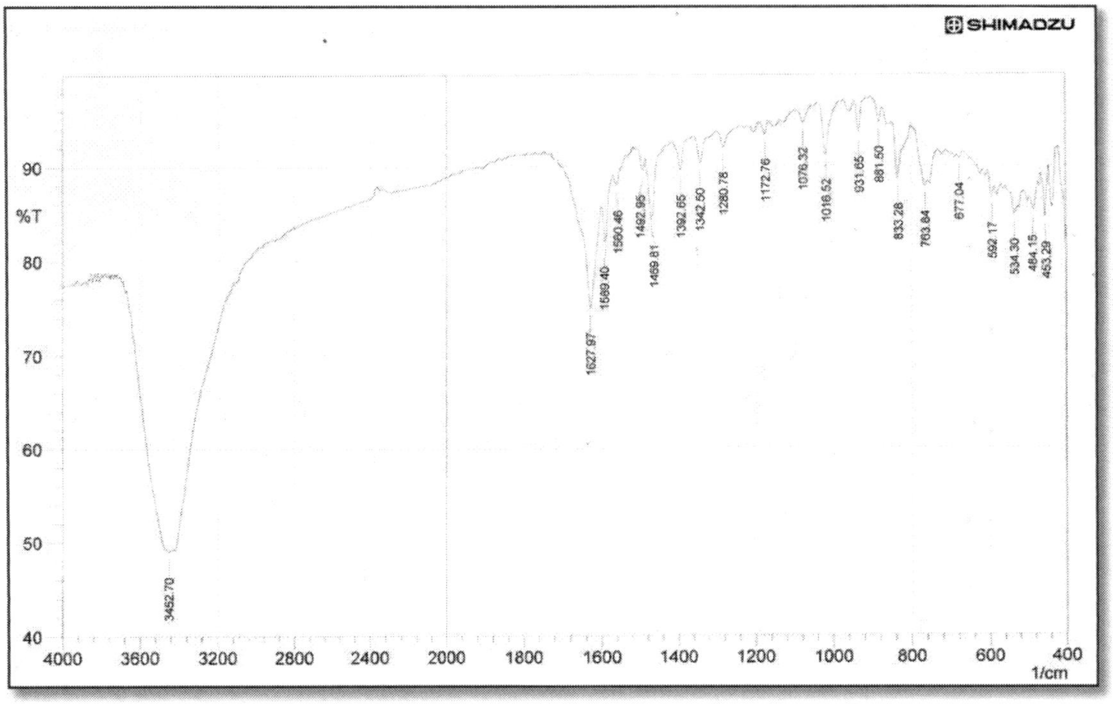

Figure 4. FTIR spectrum of compound **4**.

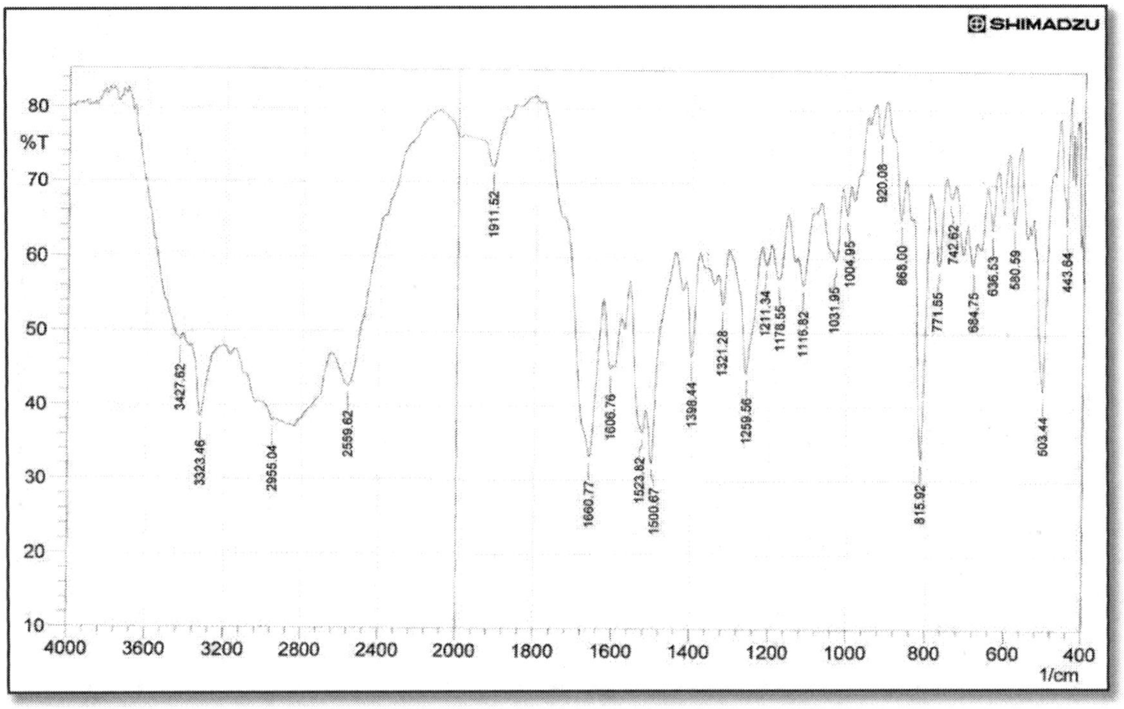

Figure 5. FTIR spectrum of compound **5**.

Journal of Physics: Conference Series **2063** (2021) 012010 doi:10.1088/1742-6596/2063/1/012010

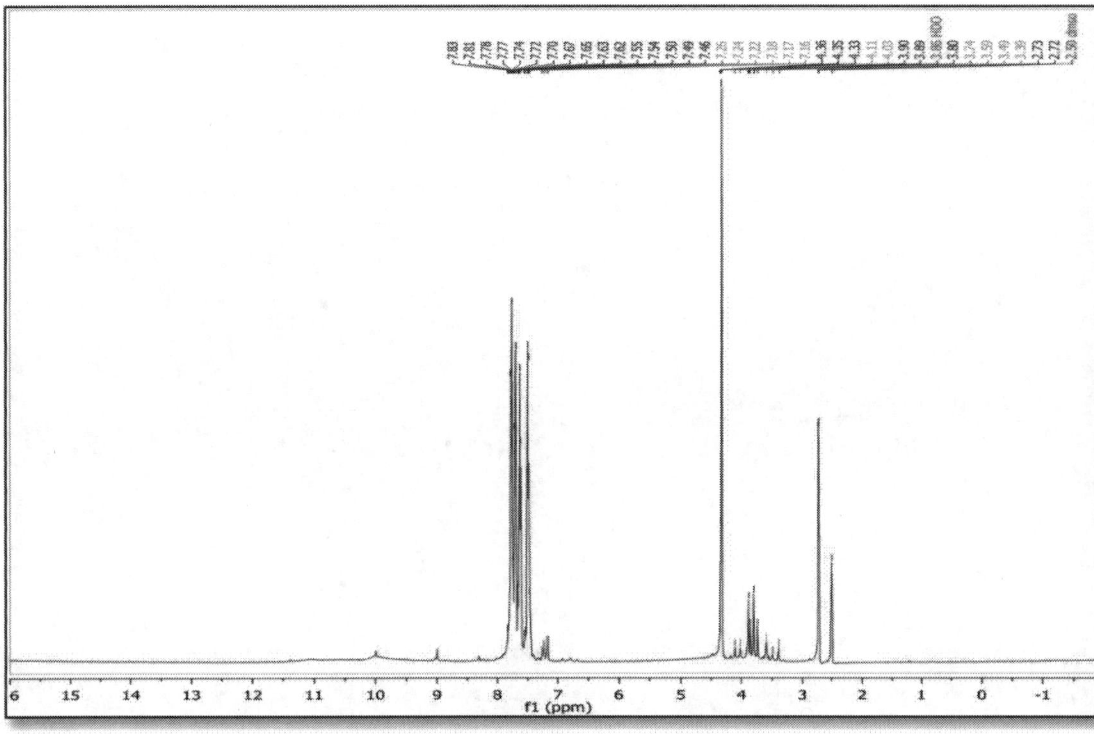

Figure 6. ^{1}H NMR spectrum of compound **5**.

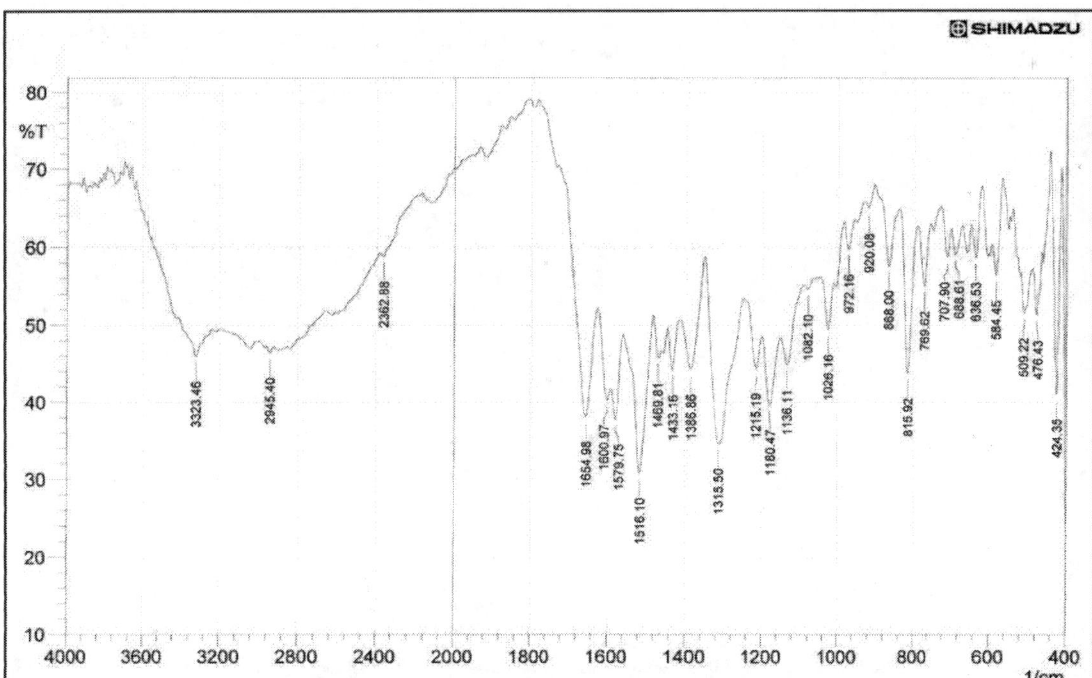

Figure 7. FTIR spectrum of compound **6**.

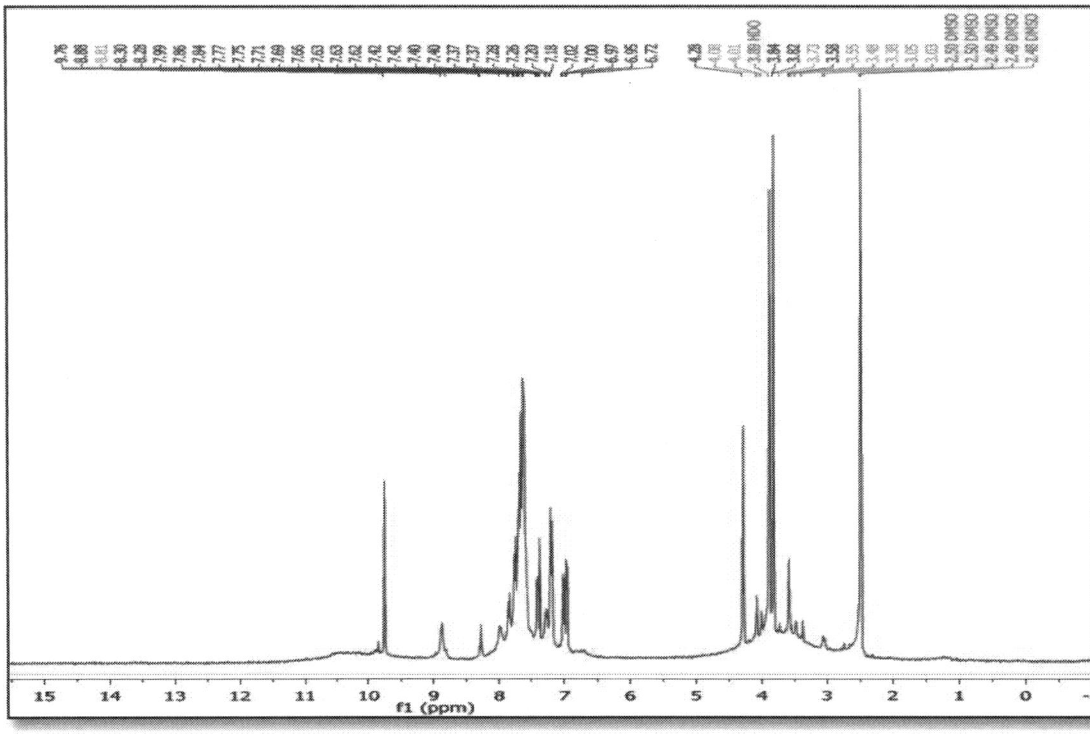

Figure 8. ^{1}H NMR spectrum of compound **6**.

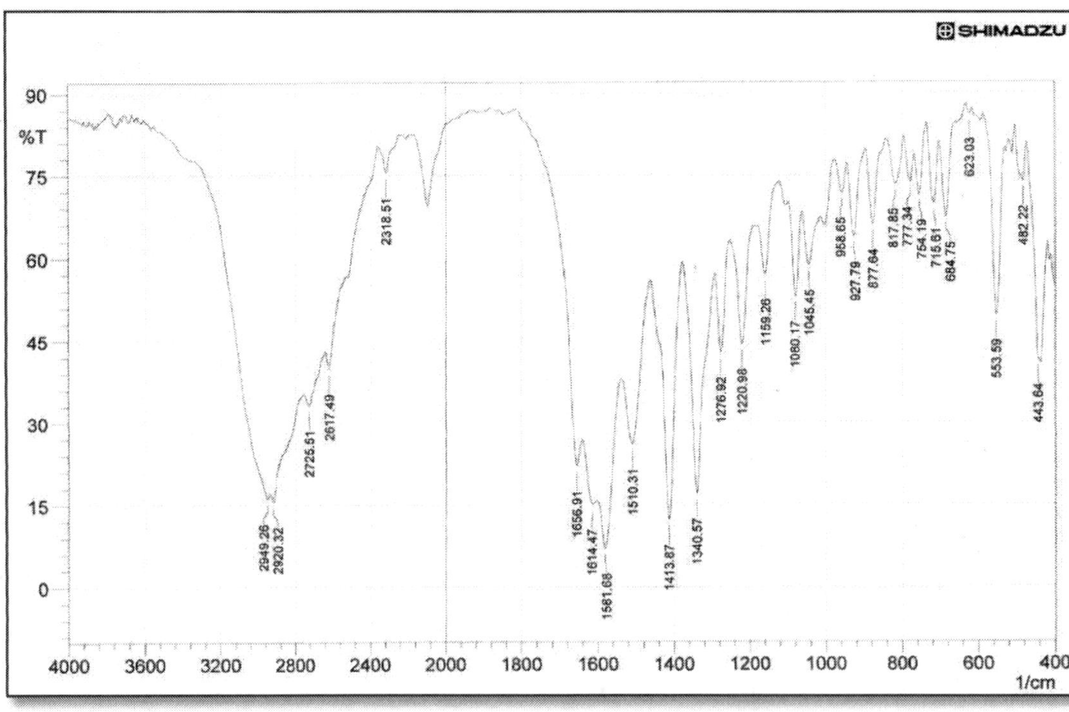

Figure 9. FTIR spectrum of compound **7**.

Journal of Physics: Conference Series **2063** (2021) 012010 doi:10.1088/1742-6596/2063/1/012010

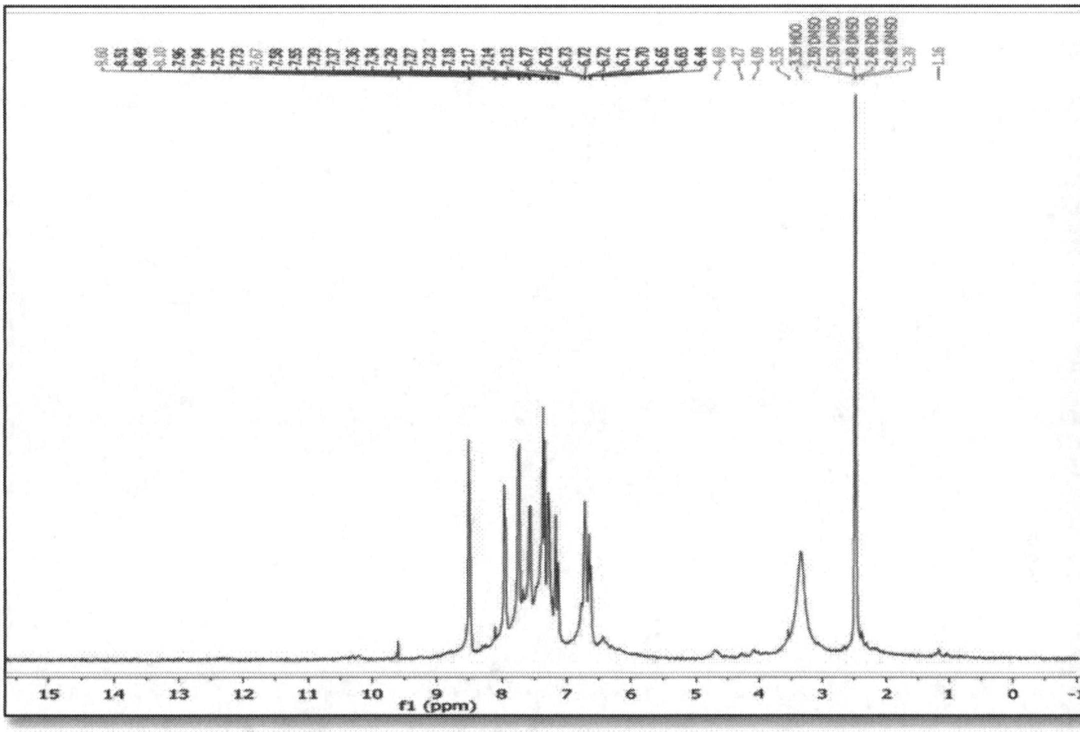

Figure 10. ^{1}H NMR spectrum of compound **7**.

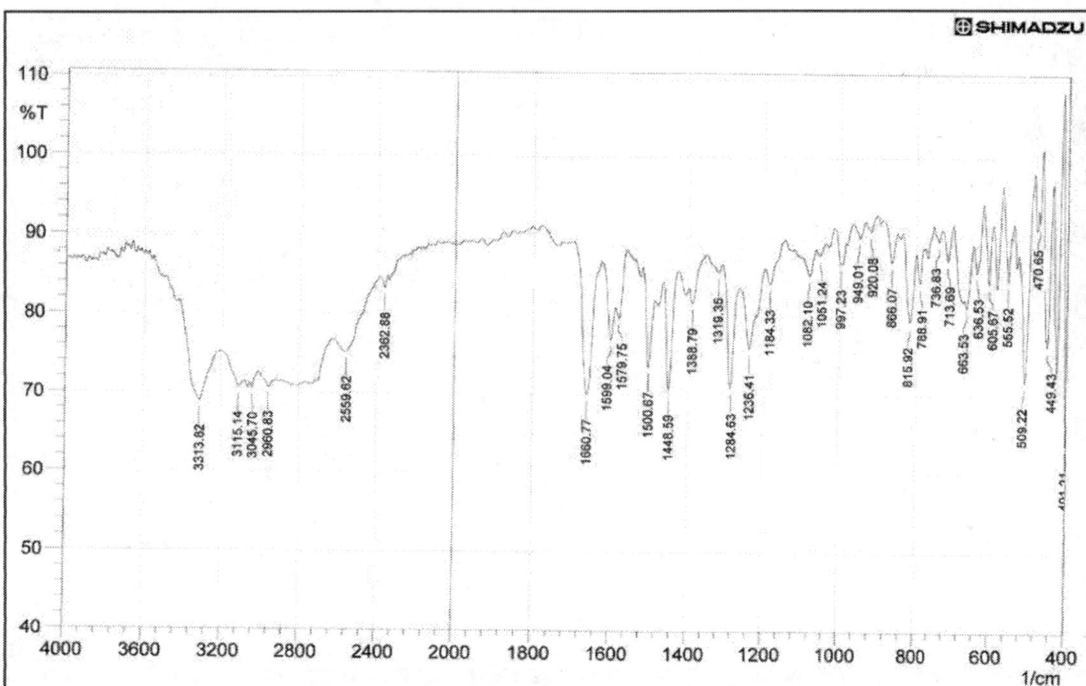

Figure 11. FTIR spectrum of compound **8**.

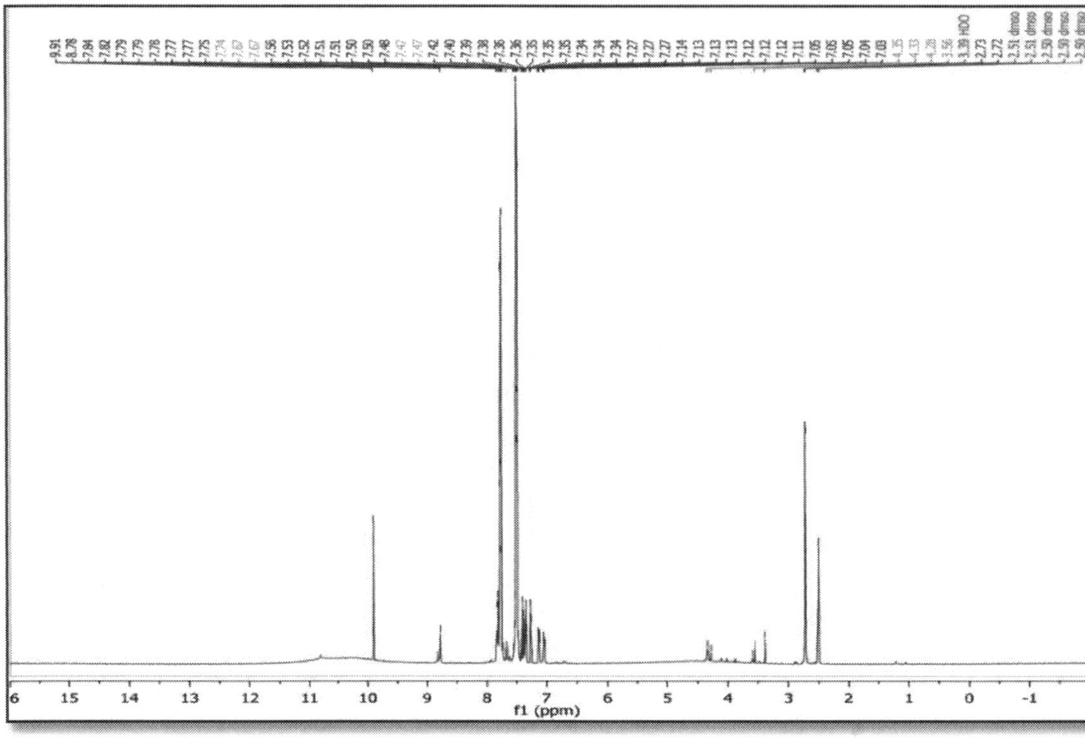

Figure 12. ^{1}H NMR spectrum of compound **8**.

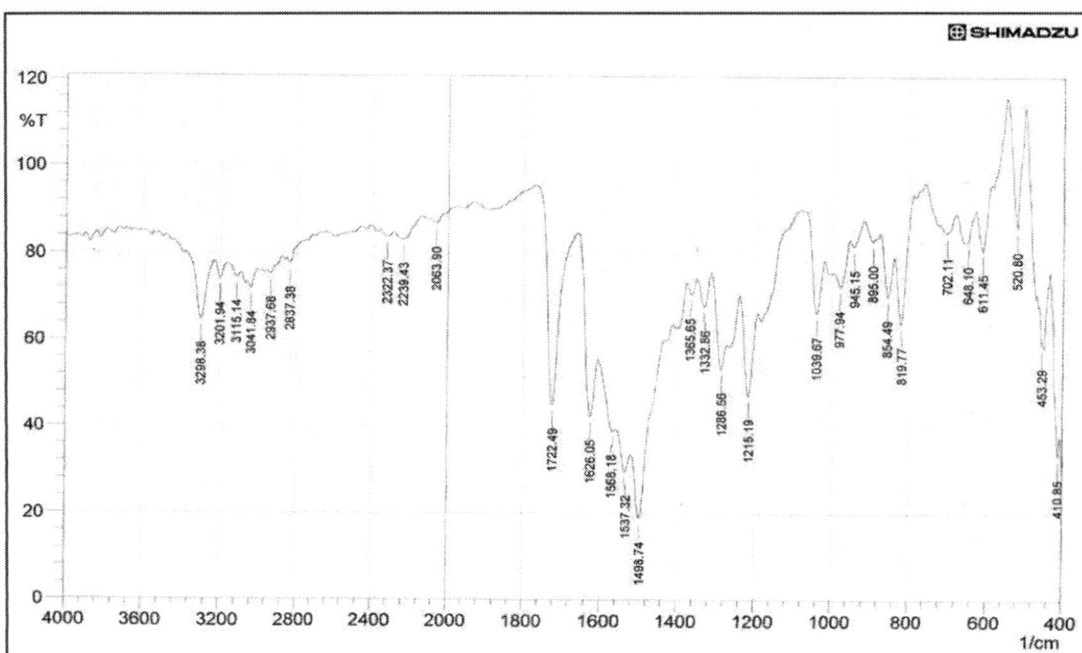

Figure 13. FTIR spectrum of compound **9**.

Journal of Physics: Conference Series **2063** (2021) 012010 doi:10.1088/1742-6596/2063/1/012010

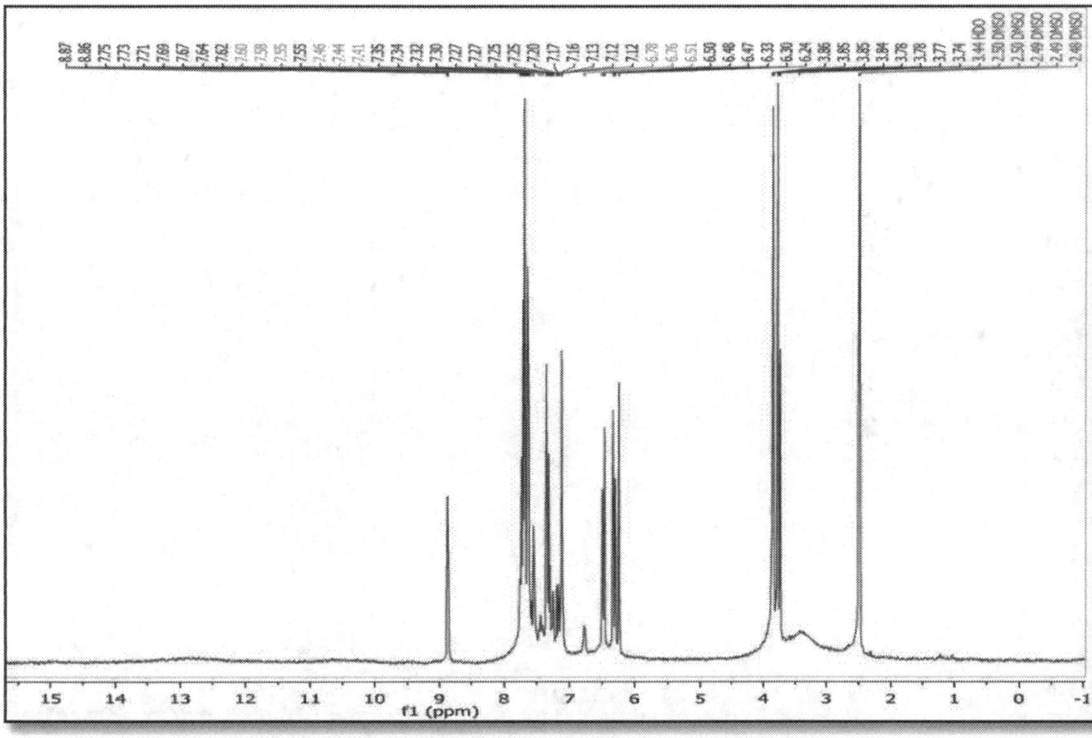

Figure 14. ^{1}H NMR spectrum of compound **9**.

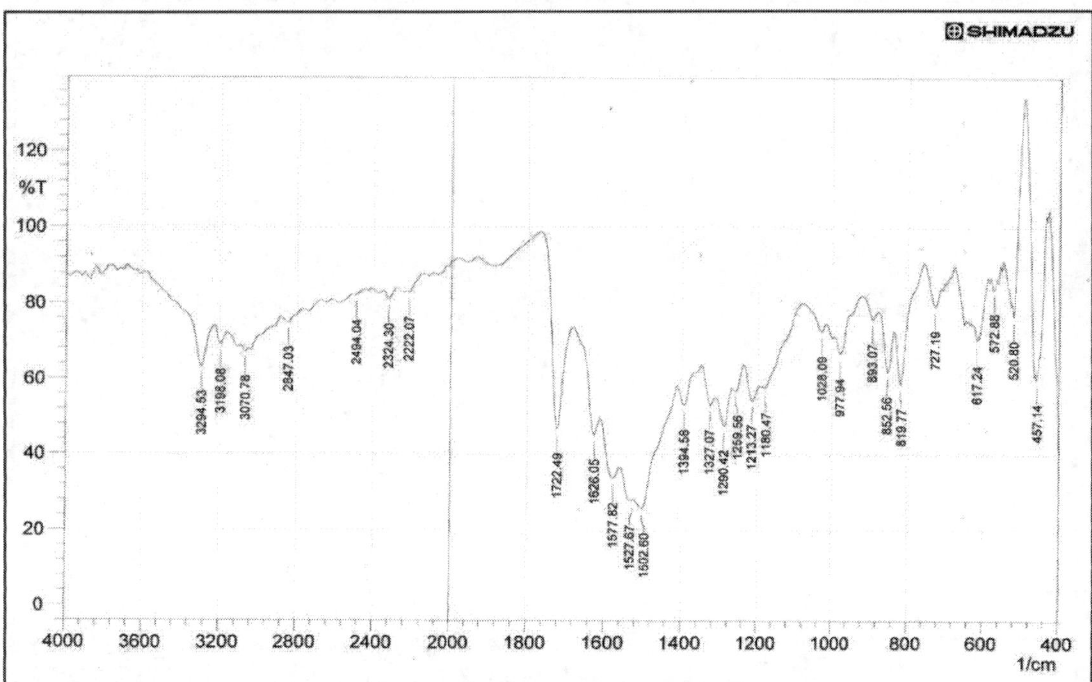

Figure 15. FTIR spectrum of compound **10**.

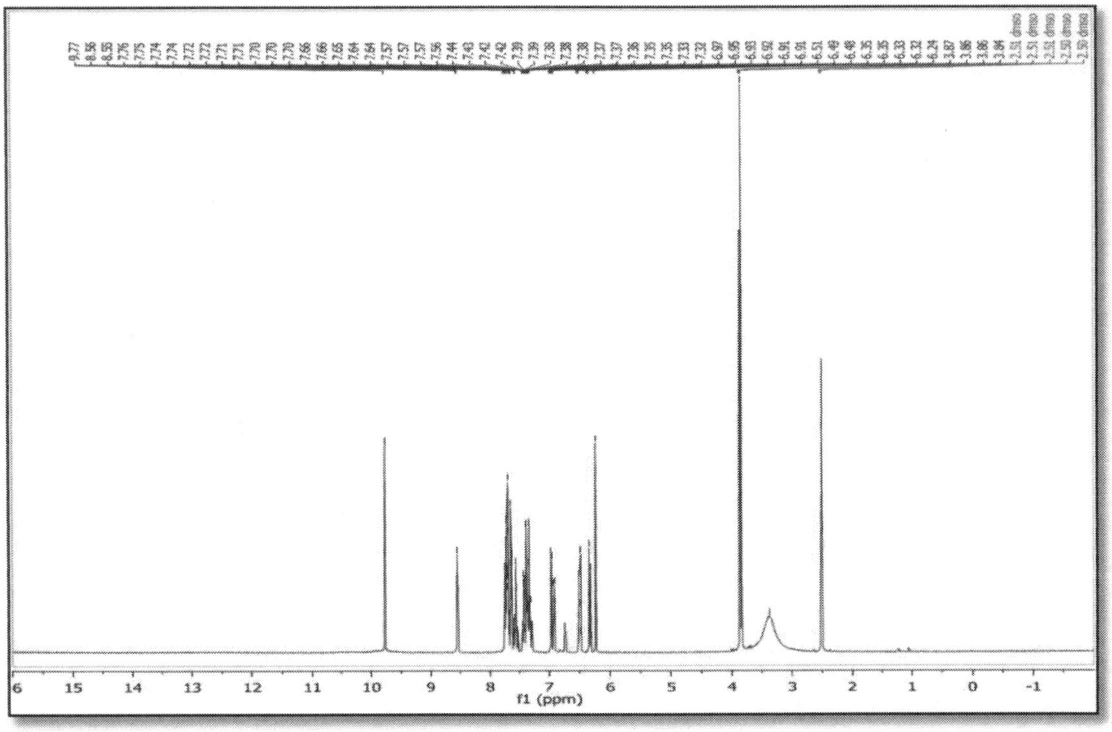

Figure 16. ^1H NMR spectrum of compound **10**.

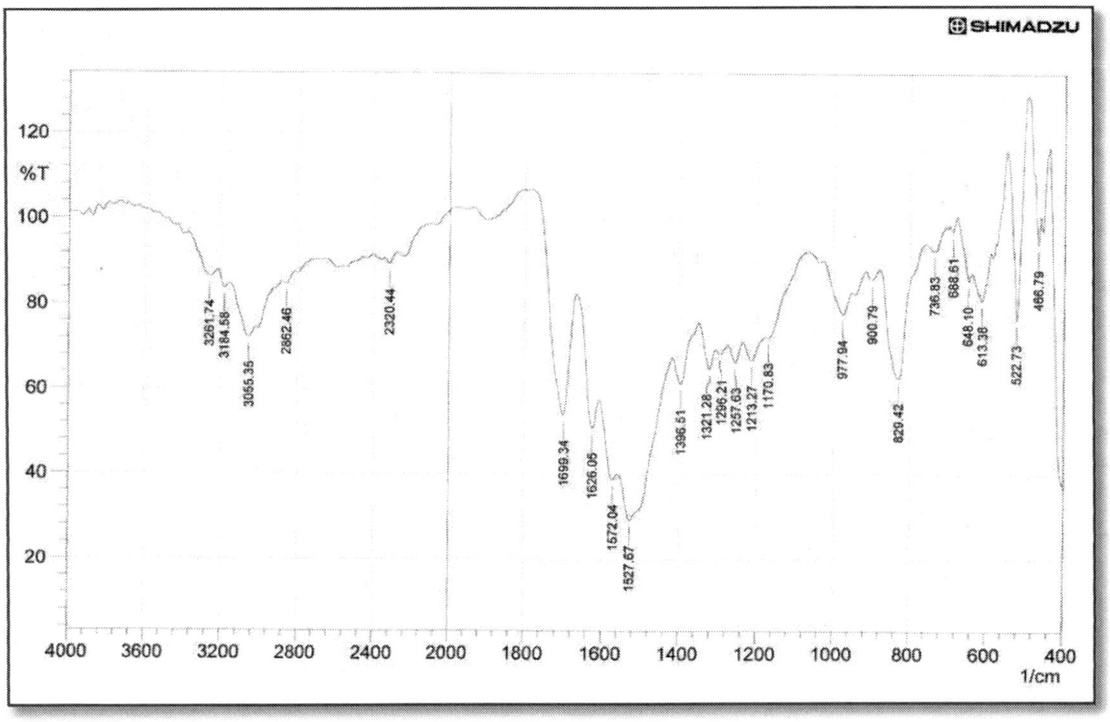

Figure 17. FTIR spectrum of compound **11**.

Journal of Physics: Conference Series **2063** (2021) 012010 doi:10.1088/1742-6596/2063/1/012010

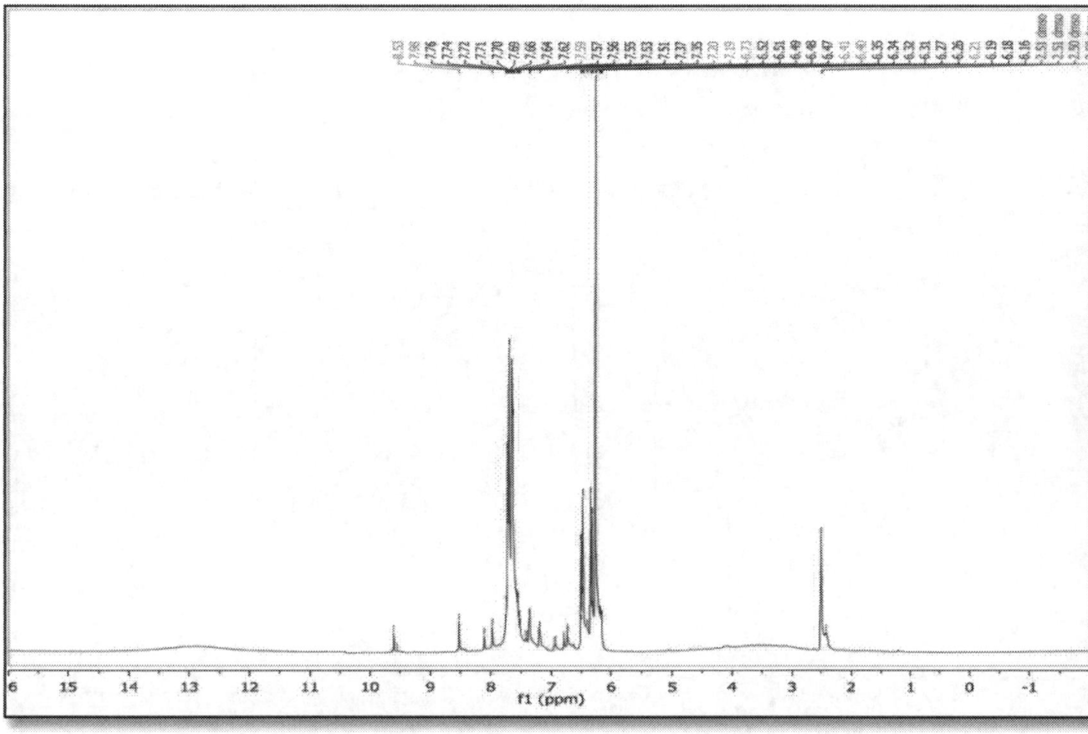

Figure 18. ^{1}H NMR spectrum of compound **11**.

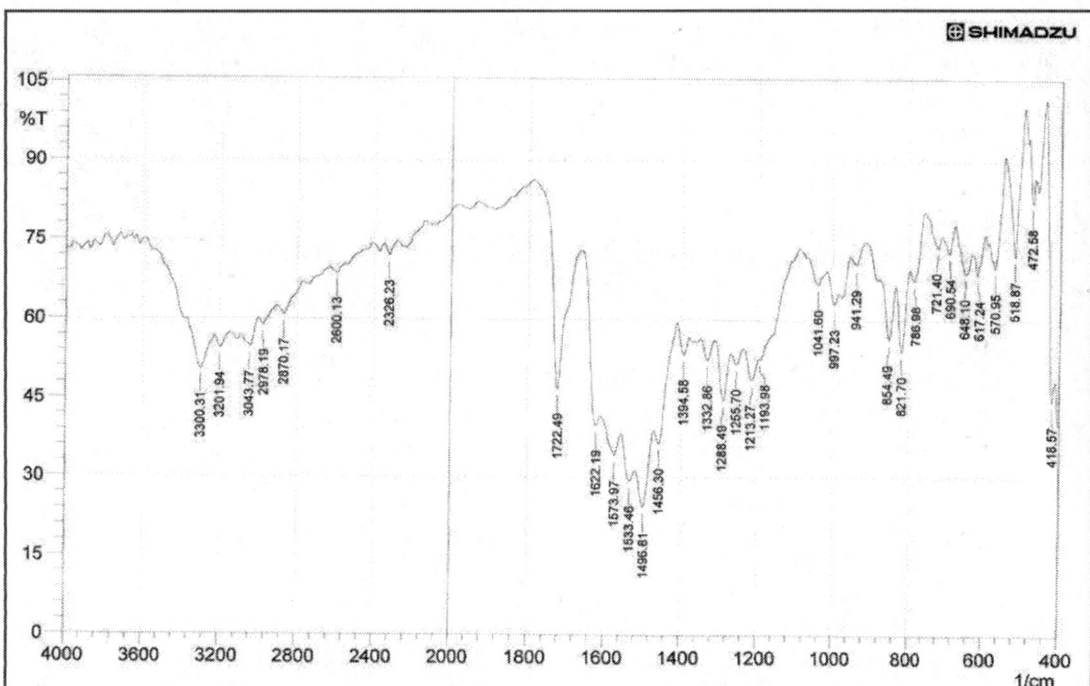

Figure 19. FTIR spectrum of compound **12**.

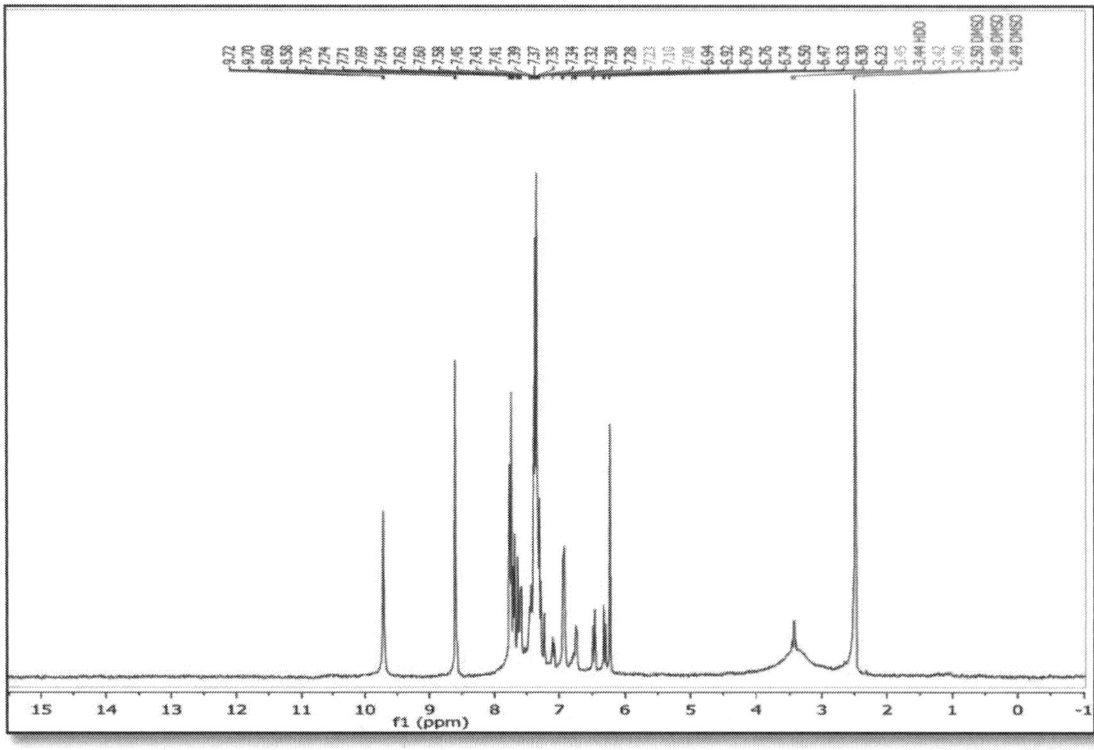

Figure 20. ^{1}H NMR spectrum of compound **12**.

4. Conclusion

In conclusion, the series of azetidin-2-one derivatives and the series of 1,3-oxazepine derivatives were prepared successfully through reaction between Schiff bases with chloroacetyl chloride and maleic anhydride, respectively. The synthesized compounds were determined utilizing FT-IR and ^{1}H NMR spectra. The target compounds may have interesting biological activity.

Acknowledgements

We are thankful to the Department of Chemistry, College of Science, University of Kerbala, for their support.

5. References

[1] Arulmurugan S, Kavitha P H, Venkatraman R P 2010 *Rasayan J. Chem.* **3** 385.
[2] Hussain Z, Yousif E, Ahmed A and Altaie A 2014 *Org. Med. Chem. Lett.* **4** 1.
[3] Przybylski P, Huczynski A W, Pyta K K, Brzezinski B and Bartl F 2009 *Curr. Org. Chem.* **13** 124.
[4] Gupta D, Pathak D P, Kapoor G and Bhutani R 2019 *Int. Res. J. Pharm.* **10** 1.
[5] Chahmana S, Keraghel S, Benghanem F, Ruíz-Rosas R, Ourari A and Morallón E 2018 *Int. J. Electrochem. Sci.* **13** 175.
[6] Cheng K, Zheng Q, Qian Y, Shi L, Zhao J, Zhu H 2009 *Bioorg. Med. Chem.* **17** 7861.
[7] Feula A, Dhillon S S, Byravan R, Sangha M, Ebanks R, Salih M A H, Spencer N, Male L, Magyary I, Deng W, Müller F and Fosse J S 2013 *Org. Biomol. Chem.* **11** 5083.
[8] Rajulu G G, Naik H S B, Kumar G C, Ramaraj S, Sambasivam G and Koppolu K P 2014 **23** 2856.
[9] Nagavolu V R, Velusamy S, Chechugari S, Yalavarthi P R and Karani M 2017 *Indian J. Pharm. Educ. Res.* **51** S707.

Journal of Physics: Conference Series **2063** (2021) 012010 doi:10.1088/1742-6596/2063/1/012010

[10] Patel R and Bhandari A 2014 *Am. J. Adv. Drug Delivery* **2** 104.
[11] Samadhiya P, Sharma R, Srivastava S K and Srivastava S D 2012 *J. Serb. Chem. Soc.* **77** 599.
[12] Abbas S K, Hanoon H D, Abbas Z F, Hussein K A and Radhi S M 2020 *Russ. J. Org. Chem.* **56** 327.
[13] Muslim R F and Saleh S E 2019 *Orient. J. Chem.***35** 1360.
[14] Kais R and Adnan S 2019 *J. Phys.: Conf. Ser.* **1234** 1.
[15] Abedel-Hahez A A and Abdel-Wahab B A 2008 *Bioorg. Med. Chem.* **16** 17.
[16] Hallinan E A, Hagen T J, Tsymbalov S, Husa R K, Lee A C, Stapelfeld A and Savage M A 1996 *J. Med. Chem.* **39** 609.

Removal of Carcinogenic Dyes Congo red (CR) and Bismarck brown Y (BBY) by Adsorption onto Reusable Hydrogels Derived from Acrylamide

Huda S Al-Niaeem[1], *, Ali A Abdulwahid[1] and Whidad S Hanoosh[1]

[1]Chemistry Department, College of Science, the University of Basrah, Garmat Ali, 61004 Basrah, Iraq

*hudaalniaeem@gmail.com

Abstract. Hydrogels of acrylamide (AM), acrylamide\ 2-acrylamido-2-methyl-1-propane sulphonic acid (AMS), and acrylamide\ 2-acrylamido-2-methyl-1-propane sulphonic acid\graphene oxide (AMSGO) were prepared as adsorbents to remove carcinogenic dyes Congo red (CR) and Bismarck brown Y (BBY) from aqueous solutions. Hydrogels were characterized using FSEM and XRD analyses. For both dyes, the synthesized hydrogels demonstrated high adsorption capability at near-neutral pH. Experimental adsorption data were analyzed using the Langmuir and Freundlich isotherm models. It was found that the Langmuir model was more suitable for the experimental data. Kinetic studies found that the pseudo-second-order model demonstrated the best fitting to the experimental data. In addition, thermodynamic studies suggest that the adsorption process was spontaneous and endothermic. The prepared hydrogels were regenerated and reused in four consecutive cycles and it could be applied to remove anionic dyes from aqueous solutions as an effective adsorbent.

1. Introduction

Water is a vital component of both the natural ecosystem and human life. If there is any disturbance in the physicochemical characteristics of water would adversely affect both [1]. Water pollution is the contamination of water by an excess amount of a substance that can cause harm to human beings and/or the ecosystem. It's level subject to pollutant's abundance, pollutant's ecological impact, and water use [2][3]. If these wastewaters are discharged without treatment, they will cause serious damage to the environment [4].

Among water pollutants, dyes (even at very low concentrations) are one of the most dangerous contaminants due to the rapid development of industries. Synthetic materials like plastics, printing, textile, paper, pharmaceuticals, leather, rubber, and paints industries are the main sources of synthetic dye pollution [5]. Azo dyes are the most used dyes, due to their diversity and chemical stability and accounting for more than 60-70% of all dyes [6]. Out of several dyes, Congo red (CR) is a benzidine-based anionic diazo dye [7], possesses mutagenic and carcinogenic properties, and is toxic to human health [8] and harmful to aquatic life [9][10]. Likewise, Bismarck Brown Y (BBY) a cationic diazo dye [11]. Which produced many toxic aromatic amines in water [12].

It is necessary to eliminate the dyes from the water system, but azo dyes are not easily removed from the wastewater by traditional physical, chemical, and biological treatment processes, because of their extreme stability [13]. Hence, the adsorption technique becomes one of the preferable choices to purify the wastewater containing dyes [14], due to its high efficiency, economic viability with a low operational cost, simplicity [15], and easy availability of various adsorbents [16].

Hydrogels are one form of effective adsorbent material that is promising [17], because of their three-dimensional network of polymeric chains and hydrophilic nature [18] resulting from the presences of function such as hydroxyl (–OH), amine (–NH_2), sulphonic (–SO_3H), carboxyl (–COOH), and amide (–$CONH_2$) groups [19]. The chemical structures, crosslinking densities, and compositions of the gels and their sizes have been important factors in their adsorption efficiency [20]. Also, due to their effortless handling, biocompatibility, environmentally friendly nature, and their associated reusability for the development of an effective desorption process [21], Polymeric hydrogels are very effective in dye adsorption. Porous hydrogel structure allows the dye solution to diffuse rapidly into the polymeric network and interact with its functional groups [22]. In past years, hydrogels derived from acrylamide (AM) have gained substantial attention for use as support carriers in biomedical engineering and as specific adsorbents. The introduction of the 2-acrylamido-2-methyl-1-propane sulphonic acid (AMPS) as a co-monomer along with AM can convey the pH responsiveness behavior in AM which can lead to and effectively remove the dye from the gel [23]. AMPS is an anionic monomer susceptible to pH, consisting of a highly ionizable sulfonate group where the acidic groups enhanced the polymeric network's hydrophilicity, which further increases the swelling behavior and may be used for the adsorption of dyes, and other organic contaminants [24]. Graphene oxide (GO) has been used in the hydrogel structure to improve the efficiency of adsorption and to enhance its physical properties. Dye adsorption can be improved by using GO due to its high specific surface area [25] and the presence of various oxygen-rich functionalities that allow either be covalent, non-covalent linkages, or a combination of both [26].

The present study reports an adsorption study of CR and BBY dyes by a hydrogel derived from acrylamide (AM), AM, and AMPS-based hydrogel, and its composite with GO as possible polymeric adsorbent was prepared via free radical polymerization. Here, our interest was to increase the adsorption capacity of AM hydrogels with highly hydrophilic functional groups containing chemical reagents such as AMPS with GO. This hydrogel is used as an adsorbent for the removal of CR and BBY dye from its aqueous solution. Batch adsorption studies are conducted.

2. Materials and Methods

2.1. Materials and measurements

All chemicals used were analytical reagent grades. They were used without further purification. Congo red (CR) and Bismarck brown Y (BBY) days were purchased from Sigma-Aldrich Company. The dye solutions were prepared in distilled water by dissolving accurately weighed dyes. Some characteristics of dyes were given in table 1.

Table 1. Some characteristics of dyes.

Dyes	Molecular Structure	Molar mass ($g.mol^{-1}$)	C.I. Nr.
Congo Red (CR)	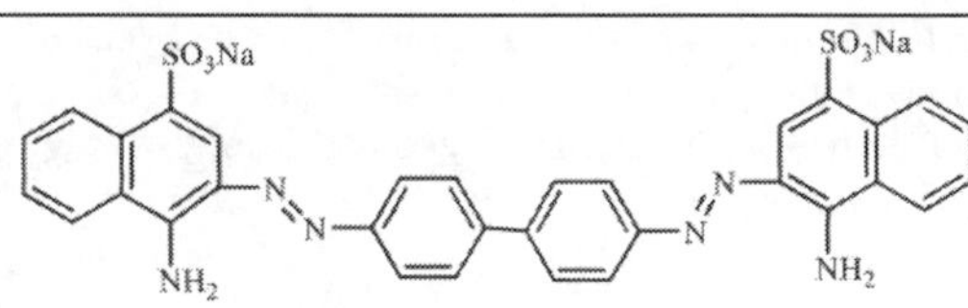	696.7	22120
Bismarck Brown Y (BBY)	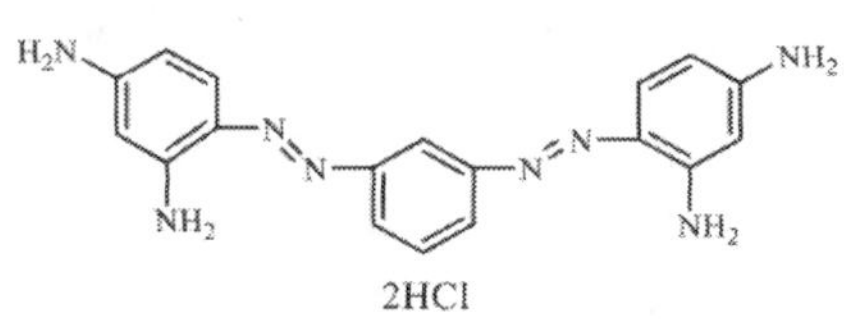	419.31	21000

2.2. Methods

2.2.1. Preparation of graphene oxide (GO)

Modified Hummers method for synthesis graphene oxide was carried out [27]. This was included dissolving 2.0 g of graphite and 1.0 g of $NaNO_3$ in 46.0 ml of concentrated H_2SO_4 under an ice bath.

Journal of Physics: Conference Series **2063** (2021) 012011 doi:10.1088/1742-6596/2063/1/012011

After about 15 min of stirring, 6.0 g of $KMnO_4$ was gradually added to the suspension by stirring as slowly as possible to control the reaction temperature below 20 °C. The suspension was stirred for two hours and then kept for 30 minutes at 35 ° C. 100.0 mL of deionized water was slowly poured into the suspension, allowing the temperature to rise steadily and maintaining the temperature below 98 °C. After 15 min., the suspension was further diluted with warm deionized water to approximately 280.0 ml. 20.0 mL H_2O_2 (30%) was added to remove the residual $KMnO_4$ and MnO_2 to change the color into luminous yellow. Then, the suspension was filtered and washed until no sulfates were found and the pH of the filtrate was changed to 7 percent aqueous HCl aqueous solution and deionized water, respectively. Graphene oxide GO was dried under vacuum at 50 °C.

2.2.2. Preparation of Acrylamide Hydrogel (AM)

A 1.0 g of Acrylamide (Am), and N, N\-methylene bisacrylamide as crosslinker agent (0.2 g, 20% compared to monomer) were dissolved in 5 mL distilled water with stirring. To this homogeneous solution was then added 1 mL ammonium persulfate (10% W/V) was added as an initiator of the polymerization reaction followed by the addition of 5 drops from N, N, N\, N\-tetramethyl ethylene diamine (TMEDA) as an accelerator to the decomposition of initiator with stirring at room temperature for three minutes to complete the polymerization reaction. The resulting hydrogel was dried and washed three times with water to remove any residue of monomer and initiator, filtered, and dried under vacuum at 30 °C.

2.2.3. Preparation of Acrylamide\ 2-Acrylamido-2-methyl-1-propane sulphonic acid (AMS)

A 1.0 g of AMPS was dissolved in 5 mL of distilled water at 35 °C with stirring then 1.0 g of Am and (0.4 g, 20% compared to monomers) N, N\-methylene bisacrylamide as crosslinker agent in 5 mL distilled water was added with stirring until the solution became homogenous. Then added 1 mL ammonium persulfate (10% W/V) was added as an initiator of the polymerization reaction followed by the addition of 5 drops from N, N, N\, N\-tetramethyl ethylene diamine (TMEDA) as an accelerator to the decomposition of initiator with stirring at room temperature for three minutes to complete the polymerization reaction. The resulting hydrogel (Scheme 2.5) was dried and washed three times with water to remove any residue of monomer and initiator, filtered, and dried under vacuum at 30 °C.

2.2.4. Preparation of Acrylamide\ 2-Acrylamido-2-methyl-1-propane sulphonic acid\ Graphene Oxide Composite Hydrogels (AMSGO)

A 1.0 g of AMPS was dissolved in 15 mL of distilled water at 45 °C with stirring then 1.0 g of Am and (0.4 g, 20% compared to monomers) of N, N\-methylene bisacrylamide as crosslinker agent in 5 mL distilled water were added with stirring until the solution becomes homogenous then 0.1 g GO was added to the monomer's solution. Then added 1 mL ammonium persulfate (10% W/V) was added as an initiator of the polymerization reaction followed by the addition of 5 drops from N, N, N\, N\-tetramethyl ethylene diamine (TMEDA) as an accelerator to the decomposition of initiator with stirring at room temperature for three minutes to complete the polymerization reaction. The resulting hydrogel (Figure 1) was dried and washed three times with water to remove any residue of monomer and initiator, filtered, and dried under vacuum at 30 °C.

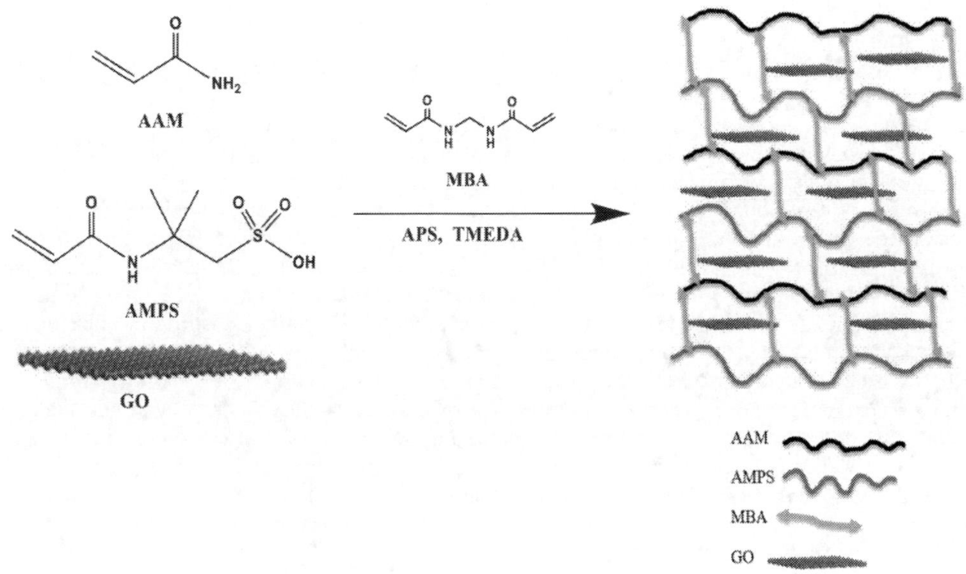

Figure 1. Synthetic routes of preparing AMSGO hydrogel composite.

2.3. Characterization
The morphology of the composite and hydrogels was measured by field-emission scanning electron microscopy (FE-SEM, Nova NanoSEM 450, FEI, USA). The patterns of X-ray diffraction (XRD) were obtained using a Rigaku X-ray Powder Diffractometer (Japan) at a scanning speed of 2 min^{-1} from 5 to 80° (Cu Kα radiation with a wavelength of 1.5406 Å). Absorption studies were determined by UV-Vis spectrophotometer (T80+, PG InL., UK) between (200- 800) nm using quartz cuvettes with a path length of 1 cm.

2.4. Preparation of Aqueous Dyes Solutions

For the adsorption experiments, a stock solution containing 2000 mg L^{-1} of CR and BBY dyes was used. A dilution using deionized water was provided with the required concentrations.

2.5. Adsorption experiments
Equilibrium adsorption isotherms were determined using the batch studies. Batch adsorption experiments were conducted in stoppered flasks containing 25.0 mg adsorbents and 100 mL of dye solutions with different concentrations (100-2000 mg L^{-1}). The adsorption experiments were performed on a thermostat shaker for a night at room temperature (27 °C) at 200 rpm. Then, the hydrogels were filtered off, and the concentration of the remaining dye was determined using a UV–Vis spectroscopy at 494 and 457 nm for CR and BBY, respectively. The effect of adsorption isotherms was studied by carrying out a series of isotherms at different temperatures (27, 45, and 65 °C) and pH= (2-12). After adsorption, the adsorbent was separated from the solution by filtration and the concentration of remaining dyes was measured using UV-Visible Spectrophotometer. Adsorption processes were conducted at optimum conditions applying agitation time (60, 45, and 45 min), with a pH of 5.0 for AM, 7 for AMS, and AMSGO of the CR dye, respectively. While agitation time (60, 30, and 30 min), with a pH of 5.0 for AM, 6 for AMS, and AMSGO of the BBY dye, respectively.

The amount of the CR and BBY dyes on the prepared adsorbents in the equilibrium (q_e) were calculated from the following equation [28]:

$$q_e = \frac{(C_\circ - C_e)}{W} V \qquad (1)$$

where Co and Ce (mg L^{-1}) are the initial and equilibrium concentrations, respectively of the dye in solution, V (L) is the volume of dyes solution and W (g) is the weight of a hydrogel, and q_e (mg/g) is the amount of adsorbed dyes per gram of adsorbents (adsorption capacity).

In the study of adsorption kinetics and the calculation of kinetic parameters, enthalpy (ΔH°), entropy (ΔS°) and free energy (ΔG°), experiments were conducted at (25, 45 and 65 °C) with 25.0 mg of hydrogels at 0.1 L of (400.0, 1200.0 and 2000 mg L^{-1}) CR solution for AM, AMS, and AMSGO, respectively. While 25.0 mg of hydrogels in 0.1 L of (300.0, 1200 mg L^{-1}) and 0.2 L of 1000 mg L^{-1}, BBY solution for AM, AMS, and AMSGO, respectively. The maximum adsorption capacity was calculated at the optimum pH by changing the pH of the dye solution from 2.0 to 12.0; the pH of the dye solution was adjusted using 0.10 M HCl or NaOH solution.

Four adsorption/ desorption experiments were performed to desorb the dye from hydrogels under the following criteria: using the same adsorbents, maximum adsorption of dye conducted the optimum agitation time and the pH-value for each adsorbent in this work. The desorption experiments were carried out by immersing the dye-loaded adsorbent in 100.0 mL of distilled water and the mixture was continuously stirred for 60 min at room temperature, and the desorbed dyes were separated by centrifugation and filtration, then spectrophotometrically determined the concentration of dyes [29]. The efficiency of dye desorption removal (S) was determined by using the following equation[30]:

$$S\,\% = \frac{C_d . V_d}{q_e . m} .100\,\% \qquad (2)$$

3. Results and discussion

3.1. Characterization of GO and prepared hydrogels

3.1.1. Field Emission Scanning Electron Microscopy (FESEM)

The external surface morphology of prepared hydrogels was analyzed by FESEM (Figure 2a-d). The hydrogel micrographs indicate that they have porous structures. From figure 2a, it can be seen that GO shows a layered and wrinkle-like structure which is due to graphite deformation upon the exfoliation process. Also, the surface appears somewhat bristly with edge planes which can enhance the interaction with dye [31]. On the other hand, as demonstrated in figure 2b, The AM hydrogel shown a smooth and neat surface morphology, the porosity was further enhanced with the incorporation of AMPS in AMS hydrogel (Figure 2c). Also, after the addition of GO into the AMS hydrogel, the surface morphology of the AMSGO composite hydrogel becomes rougher and the AMSGO shows an irregular, plat-like structure (Figure 2d). Furthermore, it can be seen that the GO platelets are well distributed as individual platelets in the hydrogel network and no apparent aggregation has been observed.

Figure 2. FESEM images of (a) GO, (b) AM, (c) AMS, and (d) AMSGO.

3.1.2. X-ray diffraction (XRD)

The XRD was used as an important method for studying the crystal structure of surface materials. XRD patterns of the GO, AM, AMS, and AMSGO are depicted in figure 3(a-d). As figure 3a for the GO, an intense peak was seen at 2θ= 10.04° was attributed to the diffraction of (001) crystal plane, which was the characteristic peak of GO. The corresponds to an interlayer distance (d) is 0.876 nm, which was attributed to the presence of oxygenated functional groups on GO sheets. After 12°, The curve progressively tends to be flat, suggesting that the successful oxidation of natural graphite [32]. However, both AM, AMS hydrogels, and AMSGO hydrogel composite exhibit amorphous characters, as shown in figure 2b and c, it also shows a peak at the angle range of (15-30°), which indicates the amorphous nature of the cross-linked chemical structure of both the hydrogels and its composite [33]. As it was observed in figure 2b and c, a broad non crystalline diffraction peak at (2θ= 22.677°), (2θ= 20.783°), corresponding to an interlayer spacing of (d= 3.9180A0), (d= 4.2706A0), respectively [34]. When comparing the XRD pattern of the GO with the composite hydrogel, clearly noticed the disappearance of the peak at 2θ = 10.04° in the XRD pattern of the AMSGO, and this implied that the GO sheets are completely exfoliated and dispersed evenly in the hydrogel network [32][33].

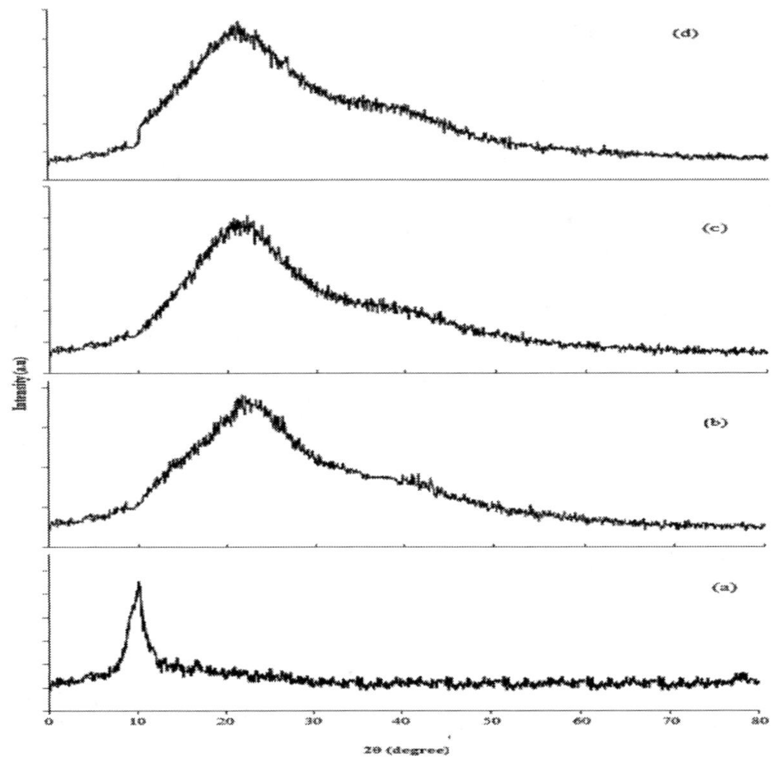

Figure 3. XRD patterns of (a) GO, (b) AM, (c) AMS, and (d) AMSGO.

3.2. Dye Adsorption by the Hydrogel

The CR and BBY dyes adsorption capacity of the prepared hydrogels were examined. The effects of various variables on dye removal were investigated, such as initial dye concentration, pH, time, and temperature, and also other kinetic, isothermal, and thermodynamic parameters.

3.2.1. Effect of pH on Adsorption

The solution's pH value plays an important factor in the adsorption capacity of the dyes onto the hydrogels. The effect of pH on dye's adsorption capacity by prepared hydrogels was investigated at an optimum initial concentration of both CR & BB dyes. The pH was adjusted at the range from 2.0 to 12.0 by 0.1 M NaOH and 0.1 M HCl solution. Figures 3 (a and b) show the pH effect of the adsorption capacities for CR and BBY dyes onto AM, AMS, and AMSGO at 27 °C, respectively. It is apparent that with the increase of pH values, the adsorption capacity of both dyes first increase and then decrease. The maximum adsorption capacity was reached as the pH value approaches 7. With the increasing pH values from 3 to 7, the sulfonic and carboxyl acid groups onto the hydrogels exist as -COOH and $-SO_3H$, which was best suited to the formation of hydrogen bonds between dyes and hydrogels. when pH exceeds 9, adsorption levels of both CR and BBY dyes decrease due to the repulsion of anionic groups on the hydrogel [32].

The results suggest that the optimal pH value was based upon the type of adsorbate used for the adsorption process. Notably, the surface of all hydrogels includes various functional groups, such as sulfonic, carboxylic, and amine groups, these functional groups can ionization by changing the pH values of dye solution.

Journal of Physics: Conference Series **2063** (2021) 012011 doi:10.1088/1742-6596/2063/1/012011

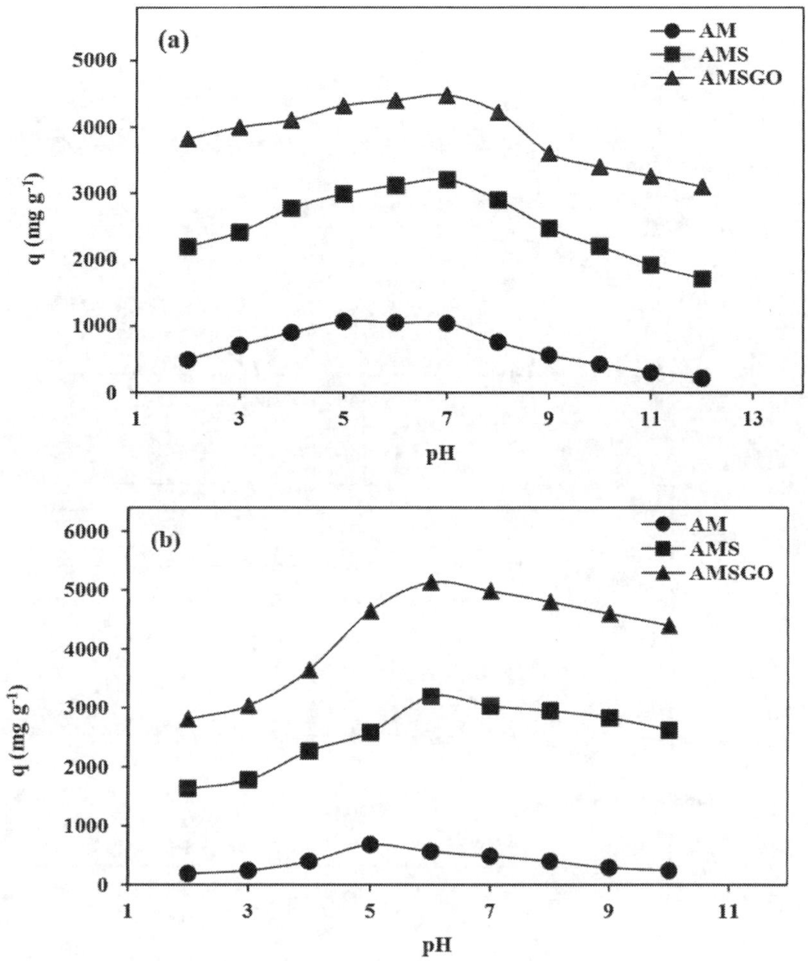

Figure 4. Effect of pH on the adsorption of (a) CR and (b) BBY dyes onto AM, AMG, and AMGGO at 27 °C

3.2.2. Effect of Time and Temperature on Adsorption

The effect of contact time on adsorption of CR and BB dyes onto the prepared hydrogels, at the initial concertation of each adsorbent and optimum pH at different temperatures was investigated. The results are shown in figure 6 and 7, reveals that the adsorption of CR and BBY has rapidly increased from 1 to 60 min, and then the equilibrium is attained within (60-120 min) for AM hydrogel. Also, the behavior of CR dye adsorption by adsorbents AMS and AMSGO as shown in figure 4 has increased rapidly from (1-45 min), and then the equilibrium is achieved within (45-90 min). While figure 5 shows a rapid increase for adsorbents AMS and AMSGO from 1 to 30 min, then the equilibrium is reached within (30-75 min). The trend for both dyes is more or less the same. Initially, the adsorption rate of both dyes was higher because of the availability of a wide number of active sites for adsorption in the hydrogel structure. The adsorption rate gradually decreased due to a reduction in the number of adsorption active sites until equilibrium was achieved. Also, the temperature increases from 25 to 65 ° C, the adsorption capacity for all hydrogels are raised. Since the adsorption of CR and BBY were endothermic, this pattern is predicted for the adsorption capability of the adsorbents. Therefore, for all further experiments, the optimal contact times were selected as 60 min for adsorption of CR dye by adsorbent AM and 45 min for both adsorbents AMS and AMSGO. Whereas 60 min for adsorption of BBY dye by AM adsorbents, and 30 min for adsorption of BBY dye by AMS and AMSGO.

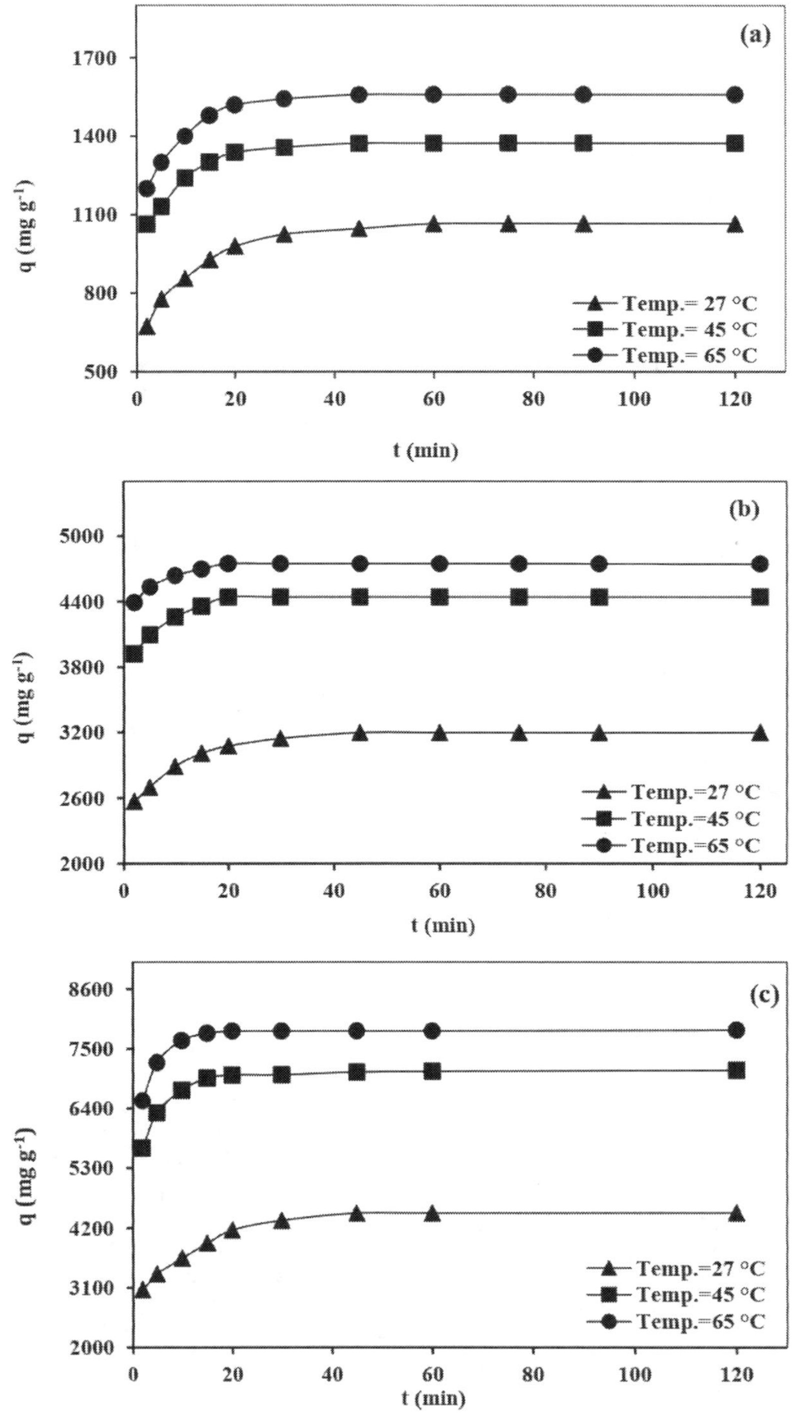

Figure 4. Effect of time on the adsorption of CR dye onto (a) AM, (b) AMG and (c) AMGGO at 27, 45, and 65 °C.

Journal of Physics: Conference Series **2063** (2021) 012011 doi:10.1088/1742-6596/2063/1/012011

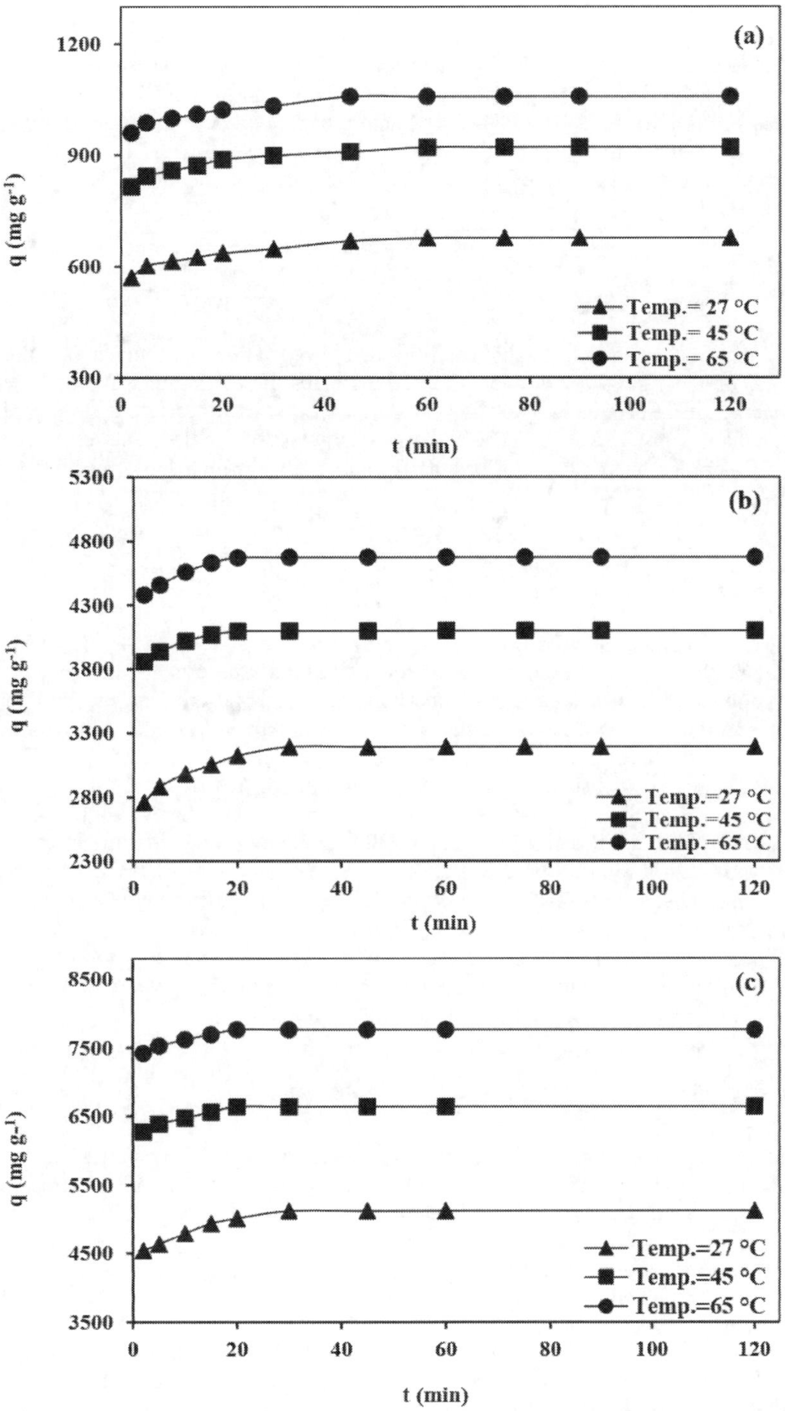

Figure 5. Effect of time on the adsorption of BBY dye onto (a) AM, (b) AMG and (c) AMGGO at 27, 45, and 65 °C.

Journal of Physics: Conference Series **2063** (2021) 012011 doi:10.1088/1742-6596/2063/1/012011

3.2.3. Adsorption Isotherms

The adsorption isotherms indicate the interaction and equilibrium relationship between adsorbate and adsorbent at a constant temperature [35]. It is necessary to find best-fitting isotherms to describe the adsorption process [36][37]. Adsorption data such as Langmuir and Freundlich isotherms are usually described by adsorption models [38][39]. The Langmuir model describes the mechanism of the adsorption of substances on the surface uniformly [19]. It is based on the homogenous monolayer formation of dyes on the surface of hydrogels [24]. The Langmuir isotherm is represented by the following equation [40]:

$$\frac{C_e}{q_e} = \frac{1}{(q_{max} \cdot K_L)} + \frac{C_e}{q_{\max}} \qquad (3)$$

where q_e is the amount adsorbed, C_e the equilibrium concentration, q_{max} the maximum adsorption capacity reflected on a complete monolayer, and K_L represents the Langmuir constant that is related to the apparent energy of adsorption. A plot of C_e/q_e versus Ce should indicate a straight line of slope (1/qmax) and an intercept of $1/(K_L\ q_{max})$. The essential characteristics of the Langmuir isotherms can be described by a separation factor or equilibrium parameter (R_L), which is defined by the following equation [83]:

$$R_L = \frac{1}{1 + (K_L \cdot C_e)} \qquad (4)$$

where R_L values indicate the type of adsorption to be irreversible (R_L= 0), favorable (0 < R_L< 1), liner (R_L= 1) or unfavorable (R_L> 1). This model is the most commonly used isothermal adsorption and has demonstrated good agreement with various experimental data [41][42]. The plots of the Langmuir adsorption isotherms of CR and BBY adsorbed on hydrogels AM, AMS, and AMSGO are shown in figure 6 (a and b), respectively, and table 2 presents the isothermal constants, q_{max}, K_L, R_L, and the correlation coefficient (R^2). As listed in table 2, the K_L offers information on the strength of adsorbate-absorbent interaction. High the K_L value, the interaction is stronger and vice versa. In the present study, the K_L value for all prepared hydrogels was very small, this demonstrates that the interaction between these materials and dyes is weak [24]. Besides, Since the R_L value was found to be (0 < R_L <1), the favorable isotherm has been confirmed and the R^2 was closed to unity, this confirmed that the Langmuir isotherm is applicable for the adsorption of dyes onto the prepared hydrogels [16] under the applied optimum conditions. While in the Freundlich model, which is based on adsorption on the surface of a heterogeneous multilayer, and assumes that the adsorption takes place at sites with varying energy of adsorption [43]. The Freundlich isotherm is represented as follow [44]:

$$\ln q_e = \ln K_F + \frac{1}{n} \ln C_e \qquad (5)$$

where, K_F and $1/n$ are Freundlich constants related to adsorption capacity and adsorption intensity, respectively. K_F indicates the adsorption capacity of the adsorbent toward the adsorbate and the $1/n$ is an indicator for the degree of the surface heterogeneity which describes the distribution of the adsorbed molecules on the adsorbent surface. Overall, a value of n higher than 1 indicates that the adsorbate is favorably adsorbed on an adsorbent, and the higher the n value the stronger the adsorption intensity [43]. Illustrated in figure 7 (a and b), the fitting of Freundlich adsorption isotherm of CR, and BBY adsorbed onto prepared hydrogels and the intercept and slope of the plots of ln q_e against $ln\ C_e$ were used to calculate the Freundlich isotherm constants K_F, $1/n$, and R^2 (Table 2). The value of ($1/n$) indicates if the isotherm type is favorable for adsorption [16]. The adsorption process is favorable when the $1/n$ values are within the 0.1<$1/n$<1 range [45]. When the value of n is less than1, the adsorption involves a chemical process. Otherwise, the adsorption is a physical mechanism [45][46]. In all of these studies, the values of $1/n$ for dyes were between 0.1 and 1 suggesting that the adsorption of both dyes favored chemisorption[45]. Typically, if the value of the correlation coefficient (R^2) approaches unity, the adsorption mechanism is said to obey a certain isotherm model i.e., the Langmuir or Freundlich isotherm model [47]. Therefore, the dye adsorption by the synthesized hydrogels was consistent with the Langmuir isotherm model compare with that of the Freundlich isotherm, as indicated by the results shown in table 2.

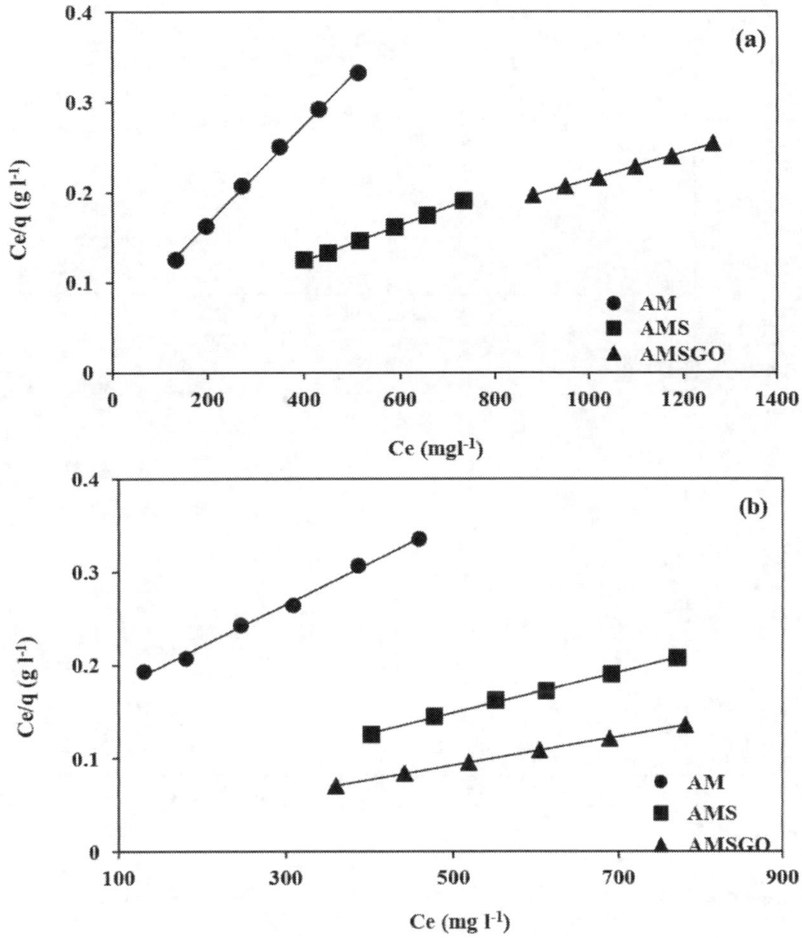

Figure 6. Langmuir adsorption of (a) CR and (b) BBY dyes onto AM, AMG, and AMGGO at 27 °C.

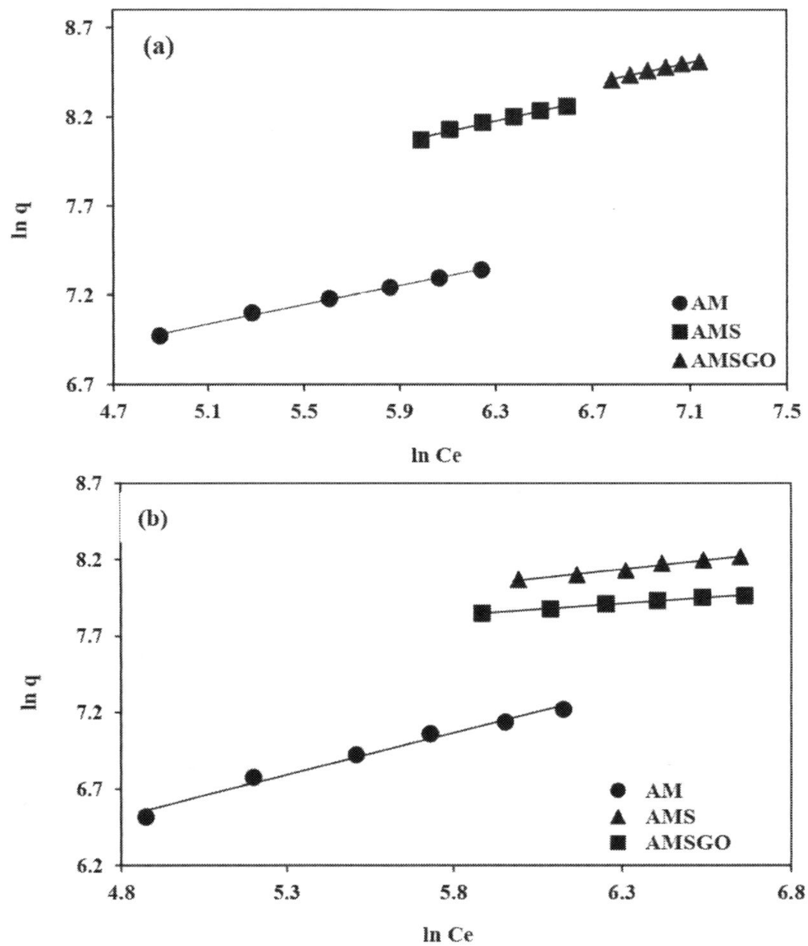

Figure 7. Freundlich adsorption of (a) CR and (b) BBY dyes onto AM, AMG, and AMGGO at 27 °C.

Table 2. Isotherm parameters for adsorption of dyes by hydrogels.

Samples	CR			BBY		
	AM	AMS	AMSGO	AM	AM	AMSGO
Langmuir						
q_{max} (mg g^{-1})	1830.16	5065.856	6600.66	2233.639	4587.156	6468.30
k_L	0.0098	0.00439	0.0024	0.0034	0.0055	0.0103
R_L	0.2027	0.1594	0.1712	0.4933	0.1322	0.0883
R^2	0.9984	0.9986	0.9986	0.9953	0.9977	0.9998
Freundlich						
K_F	286.179	538.56	650.956	49.088	741.400	1032.892
1/n	0.2708	0.2996	0.2851	0.5470	0.2427	0.1544
R^2	0.9967	0.98197	0.9841	0.9843	0.98499	0.9925

3.2.3. Kinetics of Dyes Adsorption

For a better understanding of the adsorption mechanism, the adsorption kinetic curves of the CR and BBY dyes onto prepared hydrogels were studied by kinetic models. In three kinetic models, including pseudo-first-order, pseudo-second-order, and intraparticle diffusion, experimental values were introduced, using the following equations [48]:

$$\ln(q_1 - q_t) = \ln q_1 - K_1 t \qquad (6)$$

$$\frac{t}{q_t} = \frac{1}{K_2 q_2^2} + \left(\frac{1}{q_2}\right) t \qquad (7)$$

$$q_t = k_p t^{1/2} + I \qquad (8)$$

Where q_t, q_1, and q_2 are the amounts of dye adsorbed by the prepared hydrogels at various time t, at equilibrium for the pseudo-first-order and pseudo-second-order, respectively. k_1, k_2, and k_p are the adsorption rate constants of the pseudo-first-order, pseudo-second-order, and intraparticle diffusion models. C is the intraparticle diffusion constant, which was related to the boundary layer thickness.

The experimental kinetic plots for the adsorption of CR and BBY dyes on prepared hydrogel are presented in figure 8-10 and figure 11-13, respectively. The kinetic parameters of kinetic models are summed up in tables 3 and 4: pseudo-first-order, pseudo-second-order, and intraparticle diffusion models. Good matching of the experimental data to the kinetic models was assessed by a high correlation coefficient (R^2) and close accord of the calculated and experimental (q_1 and q_2) values [30]. For all the prepared hydrogels, for CR and BBY dyes, the strong and pronounced disparity between the experimental absorption capacity and the calculated absorption capacity shows the deviation from pseudo-first-order kinetics. While the theoretical equilibrium adsorption capacities obtained from the pseudo-second-order model correlated well with the experimental results, with better correlation coefficients (R2 > 0.999) values, thus indicating a well-fitting of the second-order model, compared to the pseudo-first-order kinetic model [49][50] Also, the adsorption rate k values were not always proportional to temperatures directly. The kinetic model of pseudo-second-order also notes that chemical adsorption is the rate-determining step. According to Eq. 8, if the key rate control step is intraparticle diffusion, the plot should be linear and pass through the origin. As illustrated in figures 10 and 13, the plots for CR and BBY dyes did not pass through the origin that suggests no intercept was found to be zero, which also implied that the diffusion intraparticle was not the only rate-determining step. From table 3 and 4, it is observed that values of k_p did not always directly proportional with temperatures. Values of constants C were nonzero and they increased in increasing temperature.

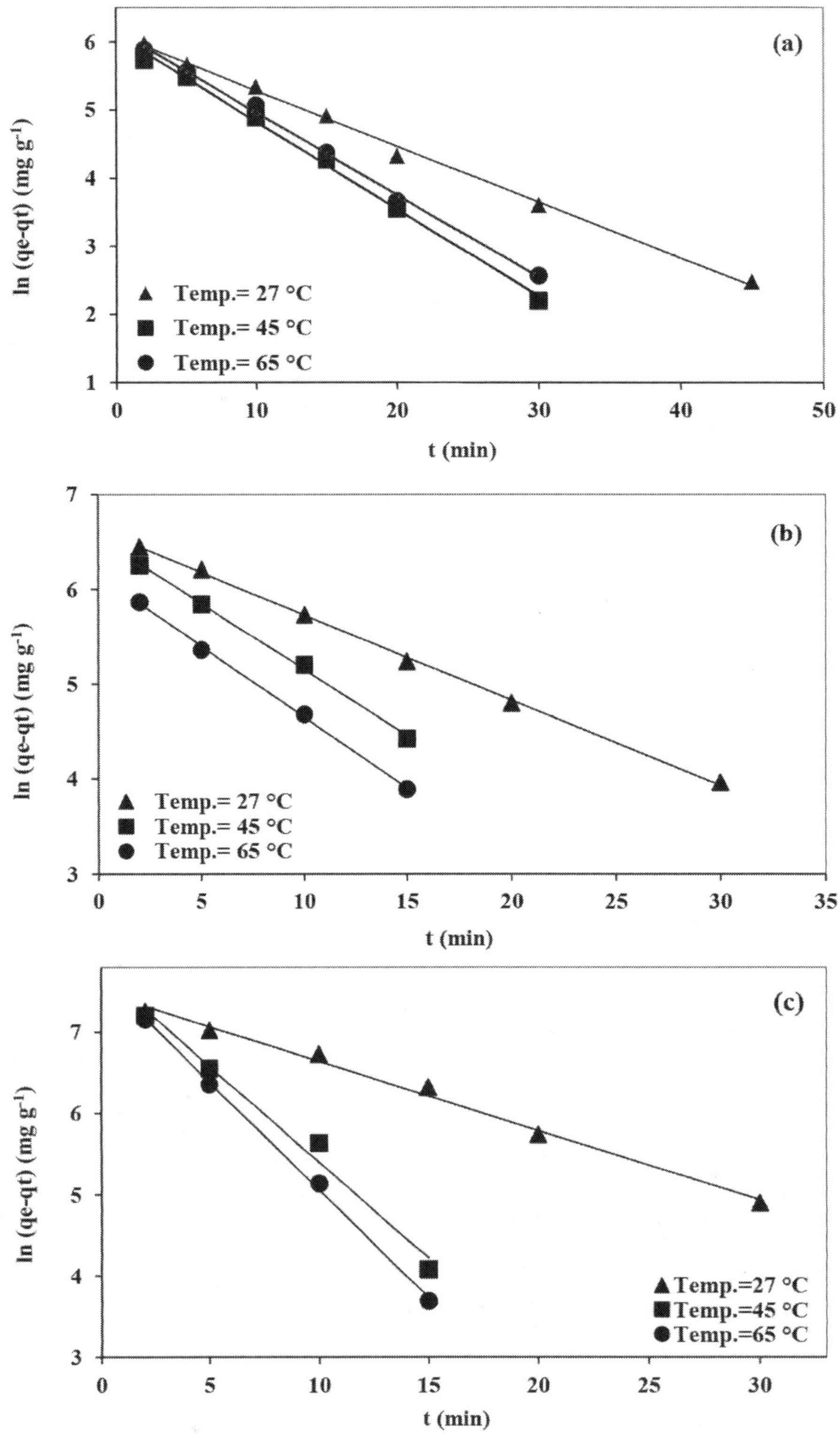

Figure 8. Pseudo-first order plot for the adsorption of CR dye onto (a) AM, (b) AMG, and (c) AMGGO at 27, 45, and 65 °C.

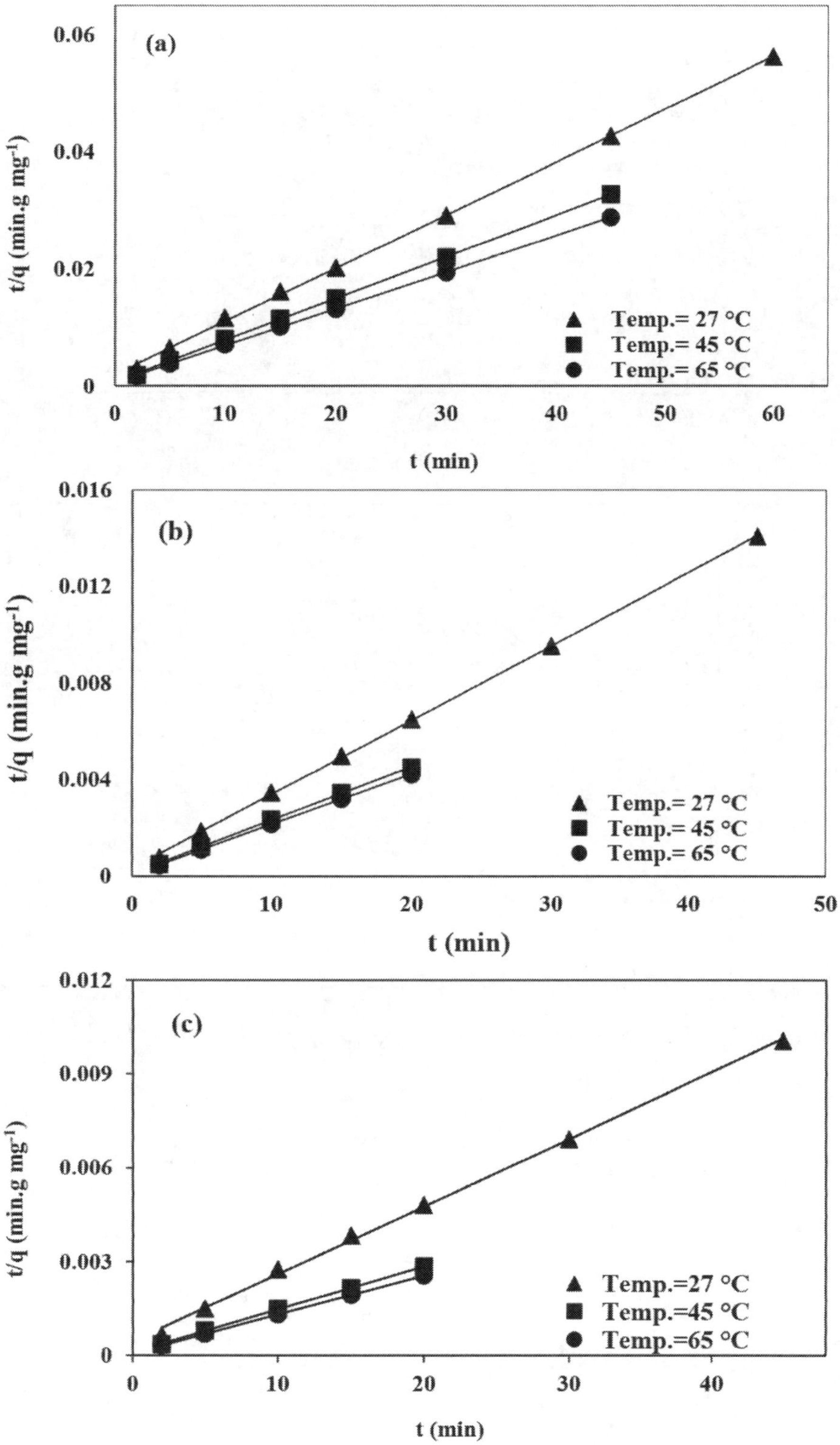

Figure 9. Pseudo-second order plot for the adsorption of CR dye onto (a) AM, (b) AMG, and (c) AMGGO at 27, 45, and 65 °C.

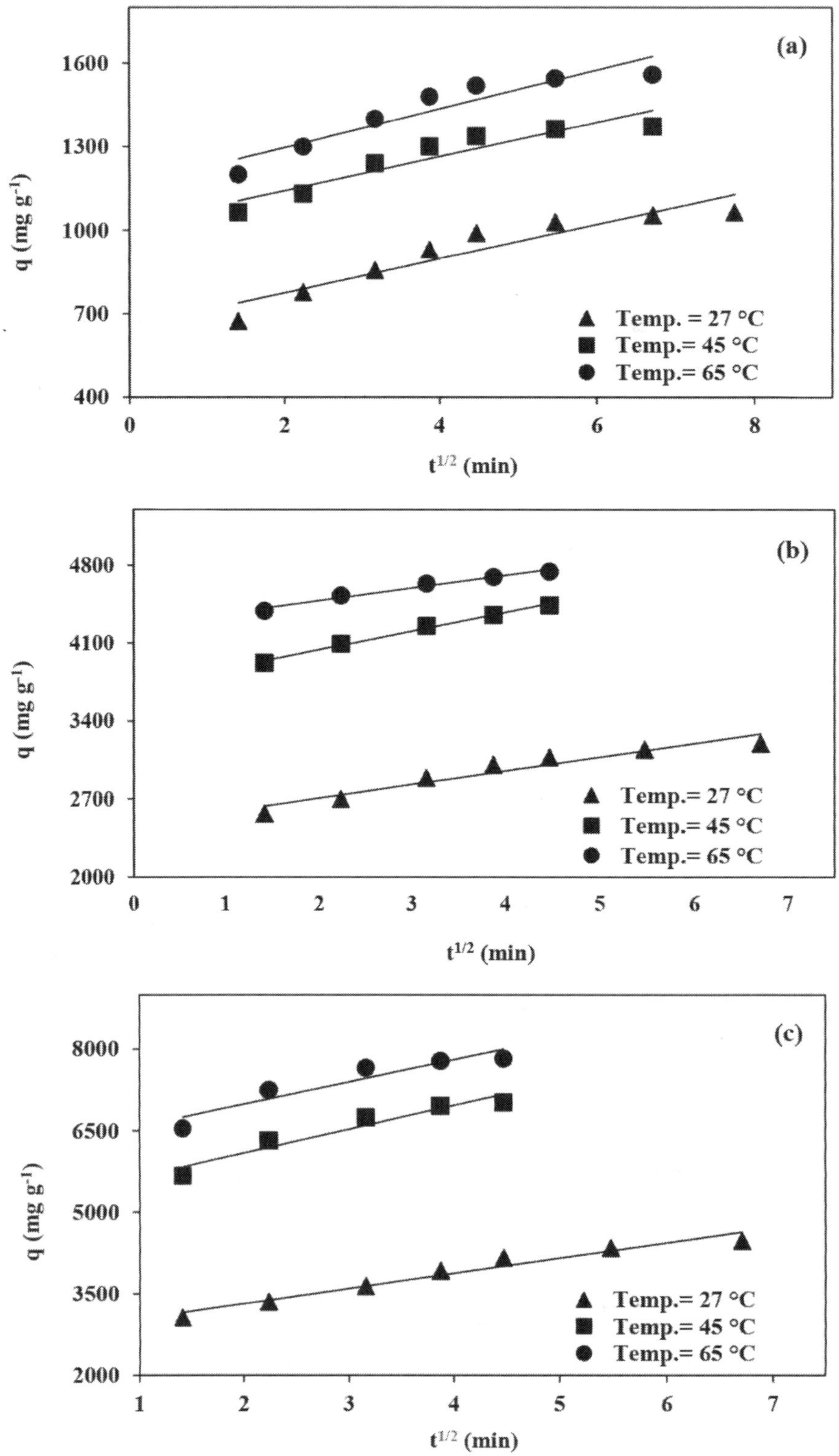

Figure 10. Intra-particle diffusion plot for the adsorption of CR dye onto (a) AM, (b) AMG, and (c) AMGGO at 27, 45, and 65 °C.

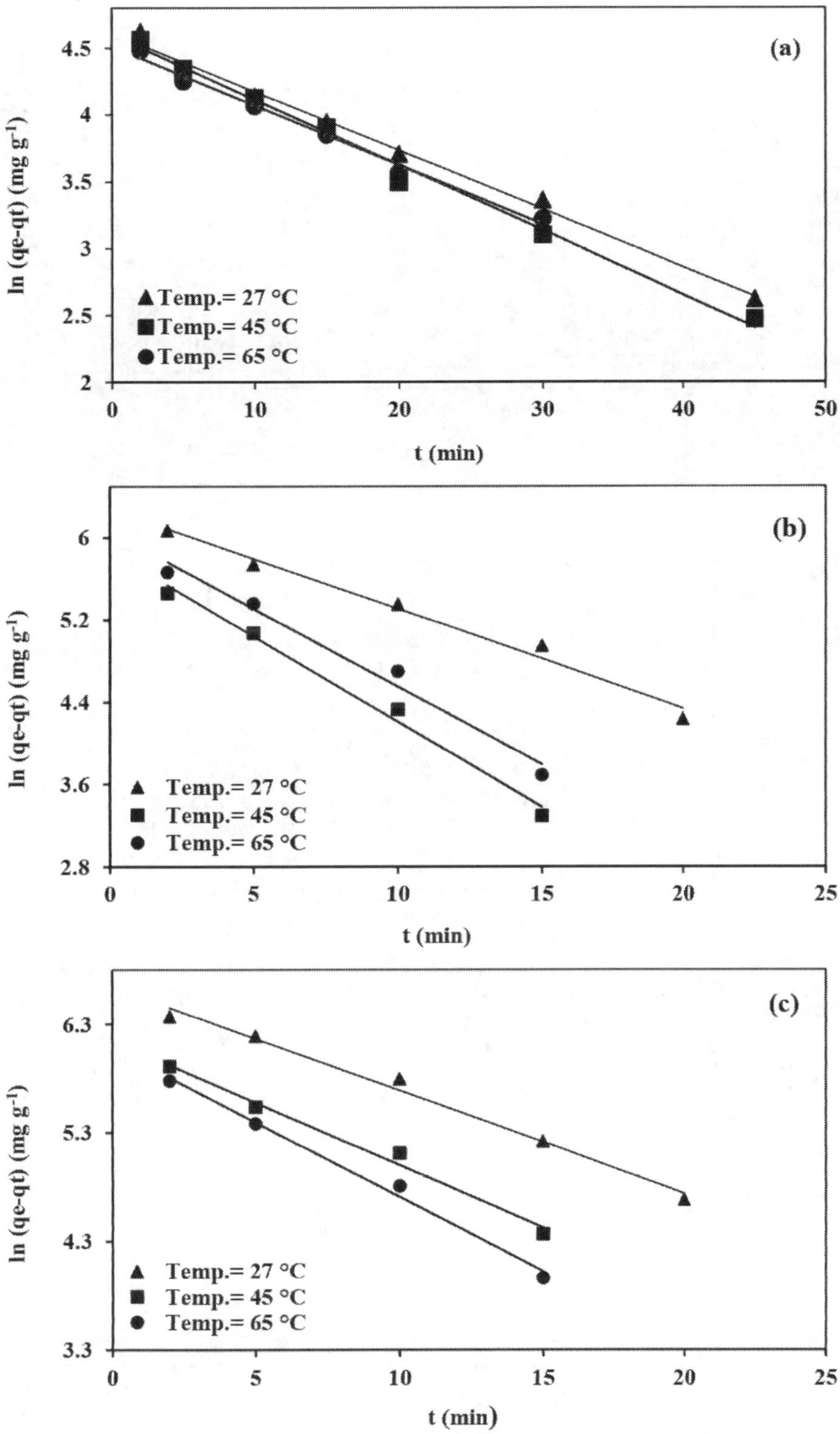

Figure 11. Pseudo-first order plot for the adsorption of BBY dye onto (a) AM, (b) AMG, and (c) AMGGO at 27, 45, and 65 °C.

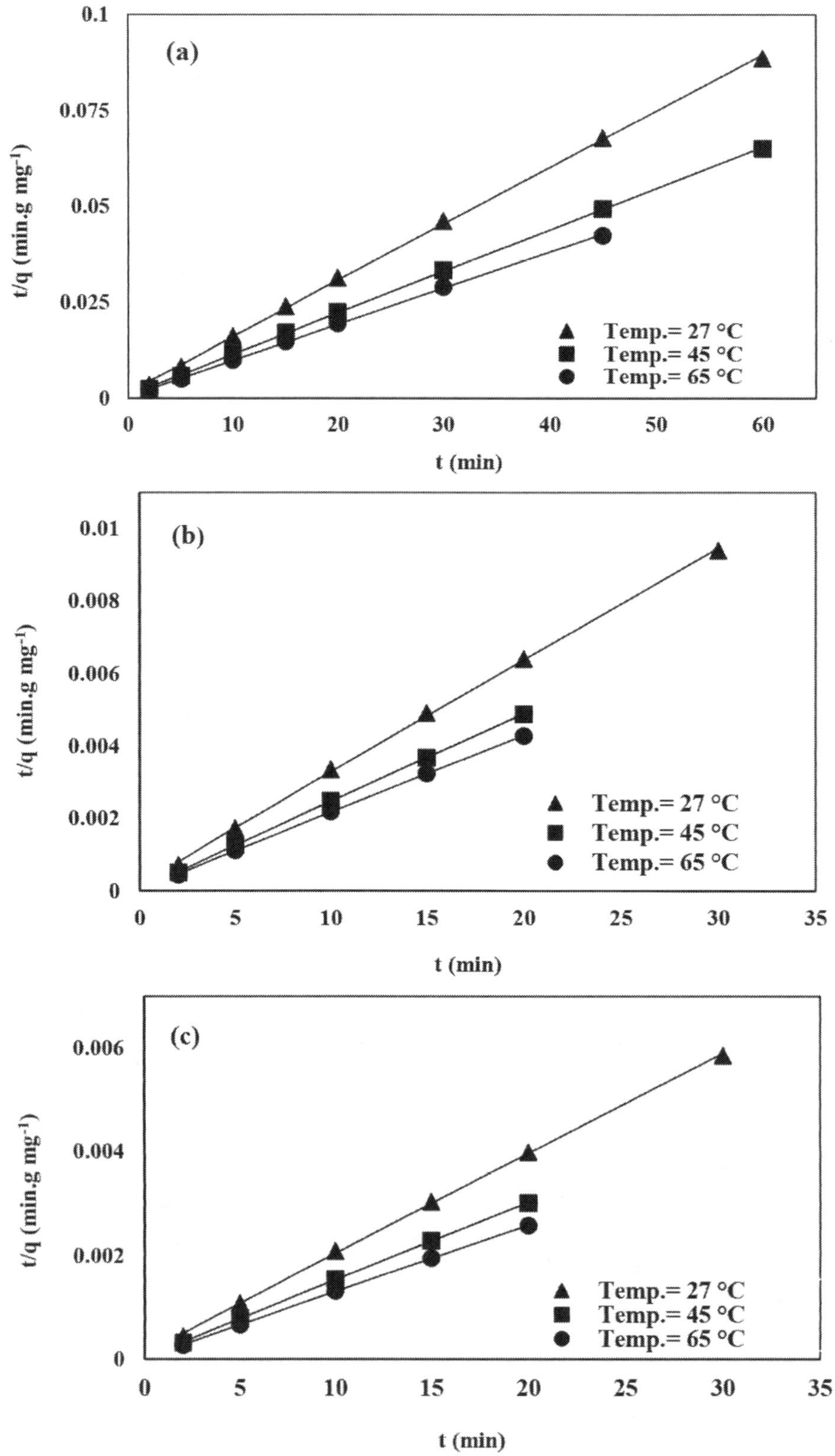

Figure 12. Pseudo-second order plot for the adsorption of BBY dye onto (a) AM, (b) AMG, and (c) AMGGO at 27, 45, and 65 °C.

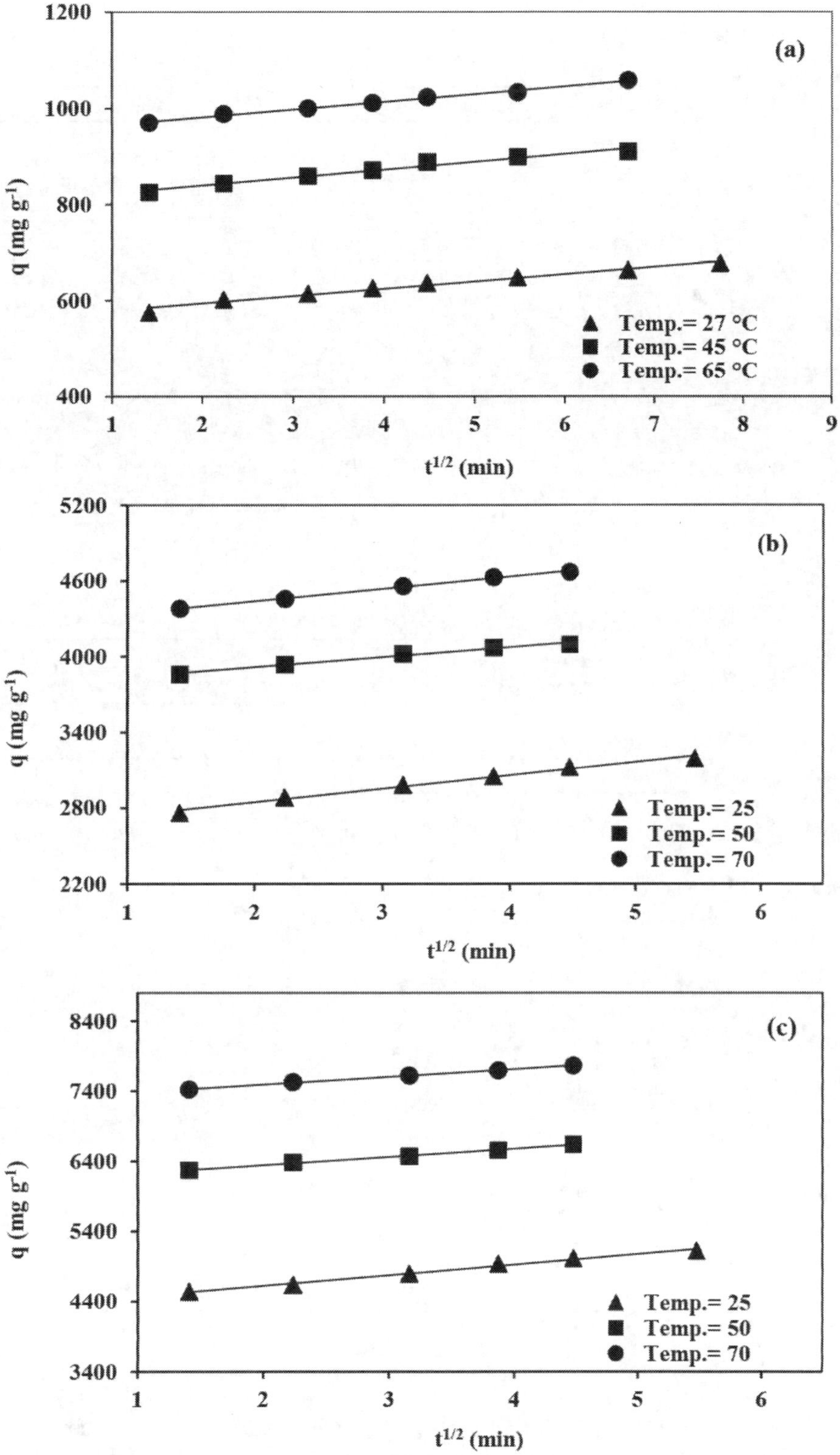

Figure 13. Intra-particle diffusion plot for the adsorption of BBY dye onto (a) AM, (b) AMG, and (c) AMGGO at 27, 45, and 65 °C.

Table 3. Kinetic parameters for adsorption of CR dye by the prepared hydrogels

	AM			AMS			AMSGO		
T (°C)	27	45	65	27	45	65	27	45	65
q_{max} (Exp.) (mg g-1)	1065.7	1372.9	1559	3198.64	4440.13	4744.1	5123	6638.6	7762.5
Pseudo-first order									
q_1 (Cal.) (mg g-1)	446.35	447.85	478.02	753.51	695.95	468.22	764.45	469.64	437.33
K_1 (min-1)	0.0819	0.1280	0.1209	0.0897	0.1392	0.1498	0.0951	0.1153	0.1374
R_1^2	0.9965	0.99658	0.99715	0.9989	0.9979	0.9986	0.9891	0.9869	0.9921
Pseudo-second order									
q_2 (Cal.) (mg g-1)	1102.66	1404.89	1594.39	3256.27	4510.6	4789.27	5186.73	6680.03	7800.31
K_2 (g mg-1 min-1)	1.66E-09	3.78E-10	2.57E-10	2.93E-11	4.91E-12	2.34E-12	4.2E-12	7.28E-13	3.53E-13
R_2^2	0.9994	0.9998	0.9998	0.9997	0.9997	0.9999	0.9996	0.9999	0.9999
Intra-particle diffusion									
K_P	61.651	61.413	69.605	122.89	168.4	113.27	150.99	117.39	109.64
C	651.27	1018.9	1157.4	2461.8	3703.7	4256.1	4322.0	6109.4	7271.1
R_p^2	0.8889	0.8791	0.8783	0.9258	0.9922	0.9922	0.9897	0.9965	0.9979

Table 4. Kinetic parameters for adsorption of BBY dye by the prepared hydrogels

	AM			AMS			AMSGO		
T (°C)	27	45	65	27	45	65	27	45	65
q_{max} (Exp.) (mg g-1)	676.76	921.02	1058.4	3193.1	4096	4669	4476	7020	7823
Pseudo-first order									
q_1 (Cal.) (mg g-1)	107.45	210.51	161.52	533.62	353.83	431.03	1792.65	2324.96	2202.75
K_1 (min-1)	0.0509	0.0470	0.0524	0.0971	0.1660	0.1516	0.0851	0.2355	0.2645
R_1^2	0.9864	0.9868	0.9641	0.9846	0.9897	0.9803	0.9925	0.98499	0.9987
Pseudo-second order									
q_2 (Cal.) (mg g-1)	681.06	925.84	1062.02	3237.29	4130.52	4710.32	4639.94	7243.23	8017.32
K_2 (g mg-1 min-1)	3.17E-09	8.41E-10	4.39E-10	1.9E-11	2.98E-12	2.33E-12	2.07E-11	1.73E-12	9.34E-13
R_2^2	0.9995	0.9998	0.9996	0.9995	0.9999	0.9999	0.9997	0.9997	0.9999
Intra-particle diffusion									
K_P	15.106	15.055	16.004	106.69	78.69	97.28	150.99	117.39	109.64
C	563.69	810.97	949.50	2631.9	3,757.7	4256.1	4322	6109.4	7271.1
R_p^2	0.9768	0.9701	0.9935	0.9872	0.9869	0.9956	0.9897	0.9965	0.9979

3.2.4. Thermodynamic of Dyes Adsorption

To specify whether prepared hydrogels adsorption is an endothermic or exothermic process, the thermodynamic investigation was carried out at various temperatures by calculating thermodynamic parameters namely, standard free energy ($\Delta G°$), standard entropy change ($\Delta H°$) and standard entropy change ($\Delta S°$) using the following equations [51]:

$$K_L = \frac{C_a}{C_e} \quad (9)$$

$$\Delta G^{\circ} = -RT \ln K_L \quad (10)$$

$$\Delta G^{\circ} = \Delta H - T\Delta S^{\circ} \quad (11)$$

$$\ln K_L = \frac{\Delta S^{\circ}}{R} - \frac{\Delta H^{\circ}}{R}\frac{1}{T} \quad (12)$$

where K_L is the thermodynamic equilibrium constant, C_a and C_e are the equilibrium concentrations of CR, and BBY dyes on the hydrogel's adsorbents and in the solution, respectively. T is the absolute temperature, and R is the gas constant (8.314 J/molK). Thus, figures 14 (a and b) show the graphical representation of the obtained data by potting $ln\ K_L$ versus $1/T$. The graph's slope and intercept were used to calculate the ΔS° and ΔH° values and were tabled in table 5. The ΔG° could be calculated at different temperatures by equation (12) or (13).

Also, From the Arrhenius equation, the activation energy (E_a) that can be defined as the minimum amount of energy needed to proceed with the adsorption process was determined by using the following equation [51]:

$$\ln K = \ln A - \frac{E_a}{RT} \quad (13)$$

where K is the rate constant of the pseudo-second-order kinetic model for adsorption system of CR and BBY dyes since the adsorption analyzes are based on the constant obtained from linearized plots R^2 closest to unity, A is the Arrhenius factor, a straight line with a slope of $-E_a/R$ is obtained when $ln\ K$ is graphed versus $1/T$, as shown in figure 15.

The calculated parameters for thermodynamics are shown in table 5. The positive value of ΔH° for the absorption process of CR and BBY dyes showed that the process was endothermic, which was supported by an increase in the absorption capacity of the prepared hydrogels. Similarly, the positive value of ΔS° for the adsorption of both dyes onto the prepared hydrogels suggests the high degree of randomness of the adsorption and affinity of the prepared hydrogels towards the dyes [11][52]. Also, the ΔG° values at all temperatures studied were negative, which underpins adsorption's spontaneous nature and indicates that the adsorption was thermodynamically favorable [53]. As seen in Table 6, with an increasing temperature of (300.15 - 338.15) K, the removal of CR and BBY dyes was less spontaneous. Nevertheless, the spontaneity of sample AM is lower compared to AMS and AMSGO [24]. In addition, the E_a data for the adsorption of CR and BBY dyes were obtained and indicated in table 5., which confirmed the dominant reaction is the chemisorption phenomenon. Also, it can be observed that the adsorption of both dyes onto AM have lower E_a values (less than 40 kJ/mol), this result suggested that the adsorption mechanism was physical adsorption, but the modification of AM could have alternated the adsorption mechanism to chemisorption with a higher E_a value [54].

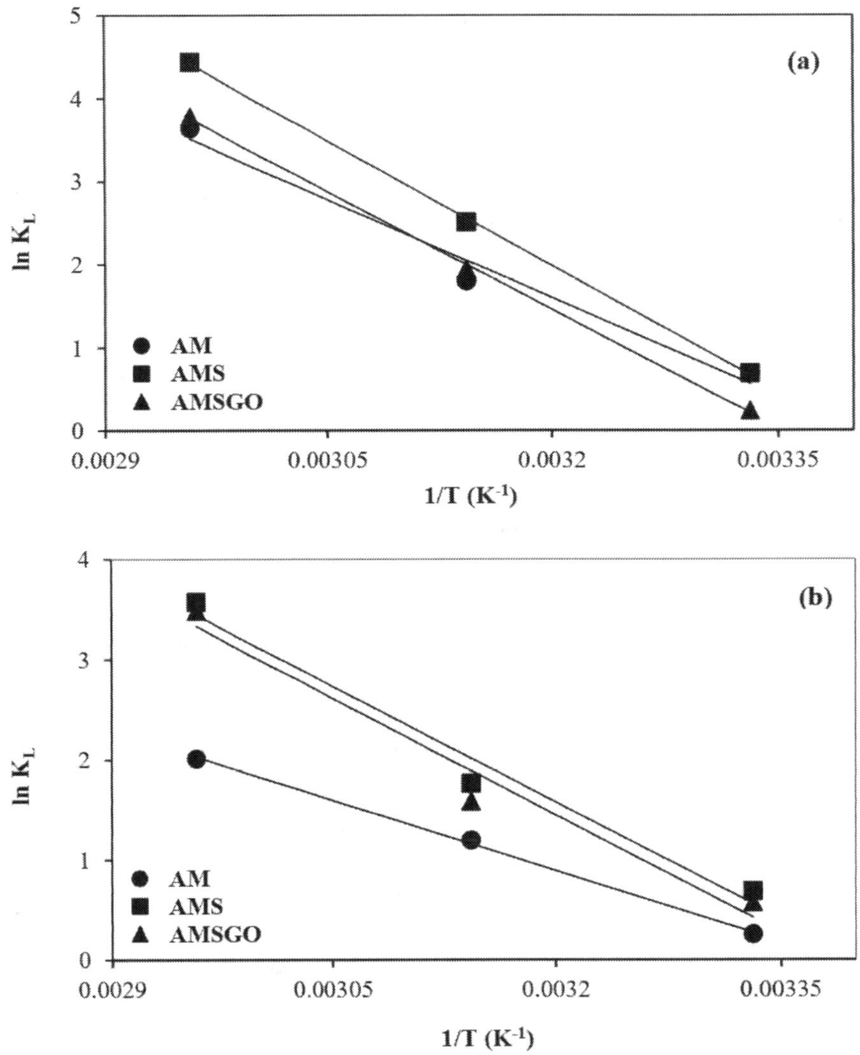

Figure 14. Plots of lnK_L vs. $1/T$ for estimation of thermodynamic parameters for the adsorption of (a) CR and (b) BBY dyes onto AM, AMG, and AMGGO

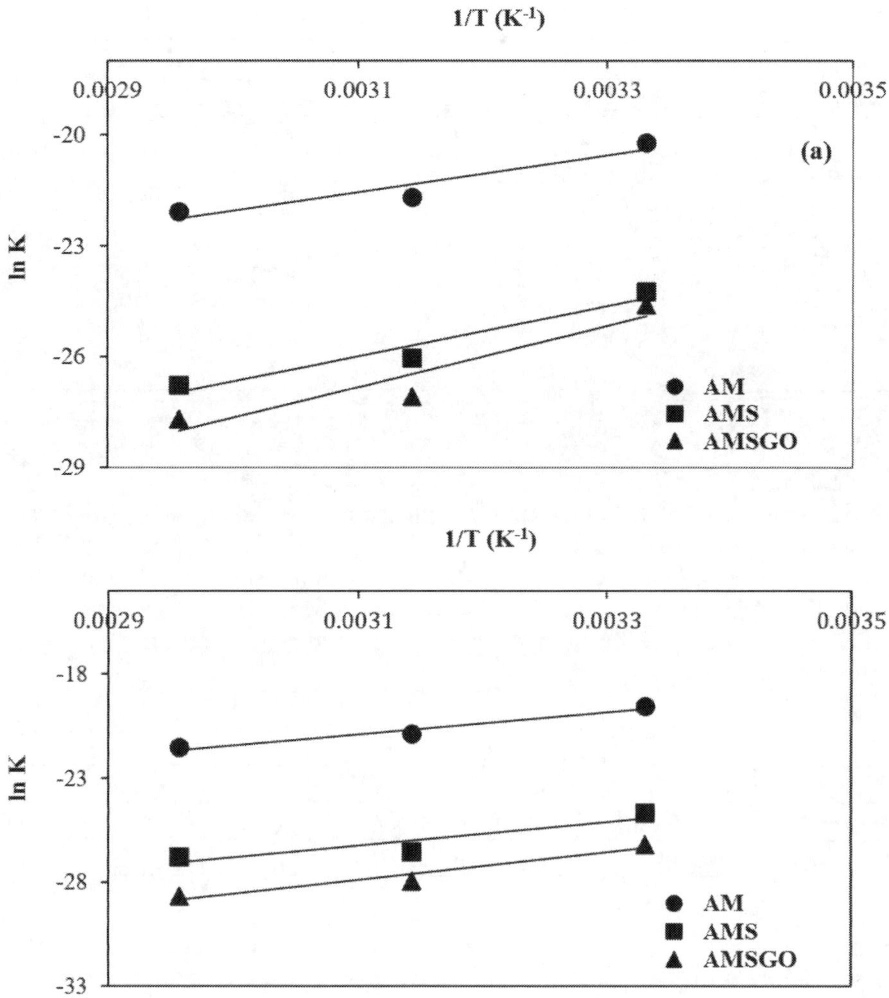

Figure 15. Plots of ln*K* vs. 1/*T* for estimation of activation energy for the adsorption of (a) CR and (b) BBY dyes onto AM, AMS, and AMSGO.

Journal of Physics: Conference Series **2063** (2021) 012011 doi:10.1088/1742-6596/2063/1/012011

Table 5. Thermodynamic parameters for adsorption of CR and BBY dyes by the prepared hydrogels.

Temperature (K)	- ΔG° (KJ.mol−1)					
	CR			BBY		
	AM	AMS	AMSGO	AM	AMS	AMSGO
300.15	1.41	1.68	0.53	0.68	1.40	1.06
318.15	5.42	6.77	5.28	3.07	5.33	4.99
338.15	9.87	12.43	10.57	5.71	9.69	9.37
ΔH° (kJ.mol-1)	65.41	83.24	78.74	38.96	64.08	64.57
ΔS° (J.mol-1.K-1)	222.63	282.91	264.09	132.09	218.18	218.66
Ea (KJ.mol-1)	39.90	56.19	68.87	37.75	46.67	55.00

3.3. Desorption Studies

Desorption studies are essential to understand the mechanism of adsorption [30]. Also, Adsorbent regeneration might improve the adsorption process more economically. Thus, successive desorption processes on the dyes used to evaluate the reusability of the prepared hydrogels over four consecutive cycles were carried out [55] and results are given in table 6, indicate that the percentage desorption was good for the four cycles. Therefore, the prepared hydrogels could be used repeatedly with retaining their good adsorption capacity.

Table 6: Adsorption-desorption for CR and BBY dye onto prepared hydrogels.

	CR						BBY					
	AM		AMS		AMSGO		AM		AMS		AMSGO	
	qe	%S	qe	%S	qe	%S	qe	%S	qe	%S	qe	%S
Cycles												
1	1080.2	95.5	3216.2	87.06	4482.7	80.210	679.6	94.15	3200.4	88.79	5136.6	81.64
2	960.3	90.8	3156.9	79.47	4407.2	71.34	663.5	90.43	3116.7	82.33	4918.9	74.45
3	823.5	82.2	3001.4	68.04	4200.3	60.47	612.5	84.68	2944.3	74.62	4566.3	65.93
4	608.1	75.6	2828.1	58.76	4012.4	48.93	559.6	76.49	2747.6	66.95	4232.6	54.89

References

[1] H. Qadri, R. Ahmad, B. Mohammad, A. Mehmood, and G. Hamid, *Fresh Water Pollution Dynamics and Remediation*. 2020.

[2] A. Inyinbor Adejumoke, O. Adebesin Babatunde, P. Oluyori Abimbola, A. Adelani Akande Tabitha, O. Dada Adewumi, and A. Oreofe Toyin, 'Water pollution: effects, prevention, and climatic impact', in *Water Challenges of an Urbanizing World*, vol. 33, IntechOpen: Rijeka, Croatia, 2018.

[3] S. Ahmed and S. Ismail, 'Water pollution and its sources, effects and managment: a case study of delhi', *Int. J. Curr. Adv. Res.*, 2018, 1, 2, 444.

[4] Z. Li *et al.*, 'Adsorption of congo red and methylene blue dyes on an ashitaba waste and a walnut shell-based activated carbon from aqueous solutions: Experiments, characterization and physical interpretations', *Chem. Eng. J.*, 2020, 388, 124263.

[5] H. Liu and H. Qiu, 'Recent advances of 3D graphene-based adsorbents for sample preparation of water pollutants: A review', *Chem. Eng. J.*, 2020, 393, 124691.

[6] S. K. Ghosh, P. Das Saha, and M. F. Di, *Recent Trends in Waste Water Treatment and Water Resource Management*. Springer, 2020.

[7] K. Z. Elwakeel, A. M. Elgarahy, G. A. Elshoubaky, and S. H. Mohammad, 'Microwave assist sorption of crystal violet and Congo red dyes onto amphoteric sorbent based on upcycled Sepia shells', *J. Environ. Heal. Sci. Eng.*, 2020, 18, 1, 35.

[8] R. L. Kadam, Y. Kim, S. Gailkwad, M. Chang, N. H. Tarte, and S. Han, 'Catalytic decolorization of rhodamine B, Congo red, and crystal violet dyes, with a novel niobium oxide anchored molybdenum (Nb–O–Mo)', *Catalysts*, 2020, 10, 5.

[9] V. K. Gupta, S. Agarwal, R. Ahmad, A. Mirza, and J. Mittal, 'Sequestration of toxic congo red dye from aqueous solution using ecofriendly guar gum/ activated carbon nanocomposite.', *Int. J. Biol. Macromol.*, 2020, 158, 1310.

[10] A. Leudjo Taka, E. Fosso-Kankeu, K. Pillay, and X. Yangkou Mbianda, 'Metal nanoparticles decorated phosphorylated carbon nanotube/cyclodextrin nanosponge for trichloroethylene and Congo red dye adsorption from wastewater', *J. Environ. Chem. Eng.*, 2020, 8, 3, 103602, 2020.

[11] A. A. Mizhir, A. A. Abdulwahid, and H. S. Al-Lami, 'Chemical functionalization graphene oxide for the adsorption behavior of bismarck brown dye from aqueous solutions', *Egyptian Journal of Chemistry*, 2020, 63, 5. 1679.

[12] E. J. Mohammad, M. M. Kareem, and A. J. Atiyah, 'Removal of dye bismarck brown G by photocatalytic reaction over prepared co-oxide Cr2O3-NiO: A kinetic study', *Asian J. Chem.*, 2018, 30, 11, 2527.

[13] J. L. Sousa and A. Aguiar, 'Influence of aromatic additives on Bismarck Brown Y dye color removal treatment by Fenton processes', *Environ. Sci. Pollut. Res.*, 2017, 24, 34, pp. 26734.

[14] B. D. C. Ventura-camargo and M. A. Marin-morales, 'Azo Dyes: Characterization and Toxicity– A Review', *Text. Light Ind. Sci. Technol.*, 2013, 2, 2, 85.

[15] A. Waheed, M. Mansha, I. W. Kazi, and N. Ullah, 'Synthesis of a novel 3,5-diacrylamidobenzoic acid based hyper-cross-linked resin for the efficient adsorption of Congo Red and Rhodamine B', *J. Hazard. Mater.*, 2019, 369, 528.

[16] K. Kaur, R. Jindal, and D. Saini, 'Synthesis, optimization and characterization of PVA-co-poly(methacrylic acid) green adsorbents and applications in environmental remediation', *Polymer Bulletin*, 2020, 77, 6. 3079.

[17] W. Wang *et al.*, 'High-performance two-dimensional montmorillonite supported-poly(acrylamide-co-acrylic acid) hydrogel for dye removal', *Environmental Pollution*, 2020, 257.

[18] A. Verma *et al.*, 'Graphite modified sodium alginate hydrogel composite for efficient removal of malachite green dye', *Int. J. Biol. Macromol.*, 2020, 148, 1130.

[19] T. U. Rehman, L. A. Shah, M. Khan, M. Irfan, and N. S. Khattak, 'Zwitterionic superabsorbent polymer hydrogels for efficient and selective removal of organic dyes', *RSC Adv.*, 2019, 9, 32, 18565.

[20] A. M. Atta, A. K. Gafer, H. A. Al-Lohedan, M. Abdullah, A. M. Tawfeek, and A. O. Ezzat, 'Hybrid Ionic Silver and Magnetite Microgels Nanocomposites for Efficient Removal of Methylene Blue', *Molecules*, 2019, 24, 21, 3867.

[21] F. Bibi, M. Ajmal, F. Naseer, Z. H. Farooqi, and M. Siddiq, 'Preparation of magnetic microgels for catalytic reduction of 4-nitrophenol and removal of methylene blue from aqueous medium', *Int. J. Environ. Sci. Technol.*, 2018, 15, 4, 863.

[22] A. Pourjavadi, M. Nazari, B. Kabiri, S. H. Hosseini, and C. Bennett, 'Preparation of porous graphene oxide/hydrogel nanocomposites and their ability for efficient adsorption of methylene blue', *RSC Adv.*, 2016, 6, 13, 10430.

[23] R. Bera, A. Dey, and D. Chakrabarty, 'Tuning of the swelling and dye removal efficacy of poly(acrylamide-AMPS)-based smart hydrogel', *Sep. Sci. Technol.*, 2016, 52, 4, 743.

[24] T. U. Rehman *et al.*, 'Fabrication of stable superabsorbent hydrogels for successful removal of crystal violet from waste water', *RSC Adv.*, 2019, 9, 68, 40051.

[25] S. Scalese *et al.*, 'Cationic and anionic azo-dye removal from water by sulfonated graphene oxide nanosheets in Nafion nanocomposite membranes', 2016.

[26] J. Y. C. Lim, S. S. Goh, S. S. Liow, K. Xue, and X. J. Loh, 'Molecular gel sorbent materials for environmental remediation and wastewater treatment', *J. Mater. Chem. A*, 2019, 7, 32, 18759.
[27] A. O. da Silva *et al.*, 'Effect of graphene oxide coating on the ballistic performance of aramid fabric', *J. Mater. Res. Technol.*, 2020, v9, 2, 2267.
[28] E. Yilmaz, G. Guzel Kaya, and H. Deveci, 'Removal of methylene blue dye from aqueous solution by semi-interpenetrating polymer network hybrid hydrogel: Optimization through Taguchi method', *J. Polym. Sci. Part A Polym. Chem.*, 2019, 57, 10, 1070.
[29] S. Mohebali, D. Bastani, and H. Shayesteh, 'Equilibrium, kinetic and thermodynamic studies of a low-cost biosorbent for the removal of Congo red dye: acid and CTAB-acid modified celery (Apium graveolens)', *J. Mol. Struct.*, 2019, 1176, 181.
[30] L. Soldatkina and M. Zavrichko, 'Equilibrium, kinetic, and thermodynamic studies of anionic dyes adsorption on corn stalks modified by cetylpyridinium bromide', *Colloids and Interfaces*, 2019, 3, 1, 4.
[31] J. Li, X. Zeng, T. Ren, and E. van der Heide, 'The preparation of graphene oxide and its derivatives and their application in bio-tribological systems', *Lubricants*, 2014, 2, 3, 137.
[32] L. Zhu, Y. Liu, F. Wang, T. He, Y. Tang, and J. Yang, 'Preparation and the swelling properties of sodium alginate graft poly (acrylic acid-co-2-acrylamide-2-methyl propane sulfonic acid)/graphene oxide hydrogel composite', *Adv. Polym. Technol.*, 2018, 37, 8, 2885.
[33] M. Zhong, Y.-T. Liu, and X.-M. Xie, 'Self-healable, super tough graphene oxide–poly (acrylic acid) nanocomposite hydrogels facilitated by dual cross-linking effects through dynamic ionic interactions', *J. Mater. Chem. B*, 2015, 3, 19, 4001.
[34] M. A. Awad and L. S. Jasim Al-Hayder, 'Removel of a Bsoprolol drug from Aqueous Solutions onto Graphene Oxide/Carboymethyl cellulose sodium / Acryl acid polymer Composite by Adsorption', *IOP Conf. Ser. Mater. Sci. Eng.*, 2020, 928, 5.
[35] M. A. Al-Ghouti and R. S. Al-Absi, 'Mechanistic understanding of the adsorption and thermodynamic aspects of cationic methylene blue dye onto cellulosic olive stones biomass from wastewater', *Sci. Rep.*, 2020, 10, 1, 1–18.
[36] N. Ayawei, A. N. Ebelegi, and D. Wankasi, 'Modelling and interpretation of adsorption isotherms', *J. Chem.*, 2017, 2017.
[37] C. Shih, J. Park, D. S. Sholl, M. J. Realff, T. Yajima, and Y. Kawajiri, 'Hierarchical Bayesian estimation for adsorption isotherm parameter determination', *Chem. Eng. Sci.*, 2020, 214, 115435.
[38] B. Dash, B. Dash, and S. S. Rath, 'A thorough understanding of the adsorption of Ni (II), Cd (II) and Zn (II) on goethite using experiments and molecular dynamics simulation', *Sep. Purif. Technol.*, 2020, 240, 116649.
[39] J. Wu, A. Xia, C. Chen, L. Feng, X. Su, and X. Wang, 'Adsorption thermodynamics and dynamics of three typical dyes onto bio-adsorbent spent substrate of Pleurotus eryngii', *Int. J. Environ. Res. Public Health*, 2019, 16, 5, 679.
[40] I. Langmuir, 'The adsorption of gases on plane surfaces of glass, mica and platinum.', *J. Am. Chem. Soc.*, 1918, 40, 9, 1361.
[41] K. Y. Foo and B. H. Hameed, 'Insights into the modeling of adsorption isotherm systems', *Chem. Eng. J.*, 2010, 156, 1, 2–10.
[42] A. H. Sulaymon, T. J. Mohammed, and J. Al-Najar, 'Equilibrium and kinetics studies of adsorption of heavy metals onto activated carbon', *Can. J. Chem. Eng. Technol*, 2012, 3, 4,. 86.
[43] C. M. Hussain, *Nanomaterials in chromatography: current trends in chromatographic research technology and techniques*. Elsevier, 2018.
[44] H. Freundlich, 'The uptake of substances on solid surfaces', *Physicse Chem. Soc.*, 1906, 40, pp. 1361.
[45] S. Saber-Samandari, H. O. Gulcan, S. Saber-Samandari, and M. Gazi, 'Efficient removal of anionic and cationic dyes from an aqueous solution using pullulan-graft-polyacrylamide porous hydrogel', *Water. Air. Soil Pollut.*, 2014, 225, 11.

[46] A. O. Ifelebuegu, 'Removal of steriod hormones by activated carbon adsorption—kinetic and thermodynamic studies', *J. Environ. Prot. (Irvine,. Calif).*, 2012, 3, 6, 469.
[47] K. Bello, B. K. Sarojini, and B. Narayana, 'Design and fabrication of environmentally benign cellulose based hydrogel matrix for selective adsorption of toxic dyes from industrial effluvia', *J. Polym. Res.*, 2019, 26, 3, 1.
[48] F. A. Ugbe and N. Abdus-Salam, 'Kinetics and thermodynamic modelling of natural and synthetic goethite for dyes scavenging from aqueous systems', *Arab. J. Chem. Environ. Res.*, 2020, 7, 1, 12.
[49] J. T. Utsev, R. T. Iwar, and K. J. Ifyalem, 'Adsorption of methylene blue from aqueous solution onto delonix regia pod activated carbon: batch equilibrium isotherm, kinetic and thermodynamic studies', *Agric. wastes*, 2020, 4, 5, 18.
[50] S. Ahmadi and F. K. Mostafapour, 'Adsorptive removal of aniline from aqueous solutions by Pistacia atlantica (Baneh) shells: isotherm and kinetic studies', *J Sci Technol Env. Inf.*, 2017, 5, 1, 327.
[51] M. Erdem, E. Yüksel, T. Tay, Y. Çimen, and H. Türk, 'Synthesis of novel methacrylate based adsorbents and their sorptive properties towards p-nitrophenol from aqueous solutions', *J. Colloid Interface Sci.*, 2009, 333, 1, 40.
[52] A. A. Abdulwahid *et al.*, 'An efficient reusable perylene hydrogel for removing some toxic dyes from contaminated water', *Polym. Int.*
[53] A. Murcia-Salvador *et al.*, 'Adsorption of Direct Blue 78 using chitosan and cyclodextrins as adsorbents', *Polymers (Basel).*, 2019, 11 6, 1003.
[54] S. Chakraborty, A. Mukherjee, S. Das, N. R. Maddela, S. Iram, and P. Das, 'Study on isotherm, kinetics, and thermodynamics of adsorption of crystal violet dye by calcium oxide modified fly ash', *Environ. Eng. Res.*, 2021, 26, 1, 1.
[55] W. T. Al-Rubayee, O. F. Abdul-Rasheed, and N. M. Ali, 'Preparation of a Modified Nanoalumina Sorbent for the Removal of Alizarin Yellow R and Methylene Blue Dyes from Aqueous Solutions', *J. Chem.*, 2016, 2016.

Journal of Physics: Conference Series **2063** (2021) 012012 doi:10.1088/1742-6596/2063/1/012012

Preparation and characterization of graphene nanosheet doped with silver nanoparticles

A H Mohammed and A N Naje*

Department of Physics, College of Science, University of Baghdad, Baghdad, Iraq.

*Corresponding author. email: naje.as75@gmail.com

Abstract. Simple process (exploding wire technique) was used to Prepared sliver nanoparticles (AgNPs). The graphene sheet was added to AgNPs with different concentrations (0.002g/ml and 0.01g/ml). well dispersion of AgNPs are achieved by simple chemistry process. The samples were characterized by ultraviolet-visible spectroscopy (UV-Vis),x-ray diffraction (XRD) , atomic force microscopy (AFM) and Field emission scanning electron microscope (FESEM).The results showed a wide band absorption of AgNPs-graphene (AgNPs-GN) extended from VU to IR region, surface plasmon resonance (SPR) absorption peak position for the AgNPs at (350-600) nm,XRD confirmed the clear distribution of the peaks attributed to polycrystalline for AgNPs appeared at 2θ=38.14°, 44.27°, 64.33,and 77.37° respectively and AgNPs-GN at 2θ=26.51° and 54.65°.The AFM showed that AgNPs have uniformly distribution on the surface of graphene sheet. The average size of AgNPs was confirmed by around (50-80) nm by FESEM and the AgNPs-GN have average particle size (20-40) nm. The AgNPs-GN could become prominent candidate for optoelectronic applications.

1. Introduction

Because of the large specific surface area, optical transparency, high carrier mobility at room temperature, and electrical, mechanical, and thermal properties of graphene, graphene-based technologies have attracted a lot of attention. [1]. Graphene is a planar sheet of carbon content that is one atom thick (mono layer) and tightly packed in a honeycomb crystal lattice structure. Large surface to volume ratio and highly conductive nature allows it to show amusing characteristics [2]. The single or few-layer graphene sheet shows exceptional optical, thermal, mechanical properties and high electrical conductivity [3–6]. Single or few-layer graphene sheets incorporated nanocomposites were extensively studied and analysed for diverse applications such as field-effect transistors (FETs), fuel cells, photovoltaics supercapacitors, resonators and various energy applications [7–9].High-quality graphene can be prepared by various methods including micro mechanical exfoliation of graphite, epitaxial growth and chemical vapor deposition (CVD) [10,11].Because of their unique electrical, optical, physical, chemical, and magnetic properties, nanoparticles have piqued people's interest. Catalysis, optoelectronic materials, magnetic fluids, composite materials, fuel cells, pigments, and sensors are only a few of the applications they have. [12,13].

Silver nanoparticles (AgNPs) can be utilized in a great application; such as, antimicrobial medical materials, surface Plasmon resonance (SPR), sensors, energy storage systems and conducting dough [12,14]. As compared to their bulk principle materials, silver nanoparticles, which are usually smaller than 100 nm and contain 20–15,000 silver atoms, have different physical, chemical, and biological properties. The size and shape of silver nanoparticles have a significant effect on their optical, thermal, and catalytic properties. Silver nanoparticles have also become the most commonly

Journal of Physics: Conference Series **2063** (2021) 012012 doi:10.1088/1742-6596/2063/1/012012

utilized sterilizing nanomaterials in consumption and medical items, such as textiles, food storage bags, fridge surfaces, and personal attention products, due to their broad-spectrum antimicrobial capacity. A laser beam [15], an electron beam, mechanical milling, electrical exploding wire [16,17], chemical vapor deposition, and plasma enhanced chemical vapor deposition [18,19] have all been used to create metal nanoparticles.

The nanoparticles in this study were created using the predominant spark explosion mechanism, which is a modification of the phenomenon known as electro-explosion of wires (EEW), in which both electrodes export particles when surrounded by a liquid medium. When a high current density passes through the electrodes during their physical contact, the two electrodes melt, vaporize, and transform into plasma. The metal plasma bursts at supersonic speeds, causing a shock wave to propagate through the surrounding medium. Finally, the contact with the liquid resulted in the formation of nanoparticles. In comparison to ambient air, the evaporated particles dissolve more quickly in a liquid. In comparison to EEW in gas, this method has only recently been developed to synthesize metallic nanoparticles in a solution; EEW in liquid has received low attention. Because of its simplicity, efficiency, and low cost, it has become one of the most promising methods for synthesis metallic nanoparticles. The synthesis of nanoparticles in liquid does not necessitate the use of a vacuum system. Furthermore, nanoparticles can be made without impurities in water or any other solution [20]. The properties of nanoparticles synthesized by EEW are influenced by a number of factors, including wire dimensions (diameter and length) and content, electrical circuit characteristics, and ambient medium [21,22]. Graphene and nanometals, such as Ag, Au, and Cu, are now commonly used in chemical, biological sensors, nanomedicine, and solar energy generation [23–26]. Because of its high surface free energy, graphene easily forms agglomeration clusters, which limits its use in bulk materials. As metal nanoparticles are incorporated into GNs, they can prevent agglomeration through increased hydrophilicity, they also have novel physical and chemical properties, making them ideal for making high-performance bulk composite materials. [27–30].

At present works the effect of adding different percentage of graphene to silver NPs was studied. the optical, structural and morphology properties were carried out.

2. Materials Methods

2.1 Preparation of AgNPs

As mentioned earlier, this technique involves passing very high currents through thin metal wires in a very short period of time. Here are the most important factors in controlling EEW:

(1) For the explosion phase, the current I must be very large, which means that very high current densities are required, which leads to some nonlinearity in the volt-ampere properties.

(2) The first stage of EEW a sudden reduction in the diameter of the wire at the explosion point, which effectively reducing the cross-sectional area of contact to 1/100th of the rest of the conductor, causing the explosion, which greatly increases the required current density as the contact is made and broken.

(3) The medium in which the explosion would occur.

The wire is guided through a wire guide and blown onto a board of similar material. To conduct a current, a low DC voltage was used before a marked explosion. The explosion occurred in a medium dense like water. A silver wire of 0.3 mm in diameter and 1 meter in length was used.

2.2 Preparation of Graphene-silver nanocomposites

Graphene was prepared by adding GN powder (0.05 and 0.01) g to 5 mL of distilled water. After 1h of sonication, the mixed solution was stirred for 3-5 days to form a homogeneous GN suspension. Then, 2 mL of AgNPs were quickly added to the above solution while stirring. The same steps for preparing graphene were repeated after addition of the AgNPs solution.

3. Results and discussion

3.1 *X-ray diffraction (XRD) studies*

Figure 1 Illustrates the XRD patterns of (a) AgNPs, (b) GN and GN–AgNPs composites. For AgNPs, the crystalline peaks appear at 2θ = 38.14°, 44.27°, 64.33, and 77.37° with d-spacing values of (2.36, 2.04, 1.44 and 1.23) A° correspond to the (111), (200), (220) and (311) planes, respectively. As seen in Figure 1b, the diffraction peaks of graphene appear at 2 θ =26.76° and 54.86° with d-spacing values of 3.327 and 1.671 A° correspond to the (002) and (110) planes, respectively. The diffraction peaks of GN–AgNPs appear at 2 θ = 26.51° and 54.65° with d-spacing values of 3.36 and 1.67 A° correspond to the (002) and (110) planes, respectively. However, after the addition process, the peak of diffraction shifted to 26.51° with 3.36 A° d-spacing. This is seen in the increasingly wide d-spacing of 2.36 to 3.36 Å.

The AgNPs peak has a very low intensity, decreasing compared to GN and GN-AgNPs indicating that the grains tend to have an amorphous crystal structure.

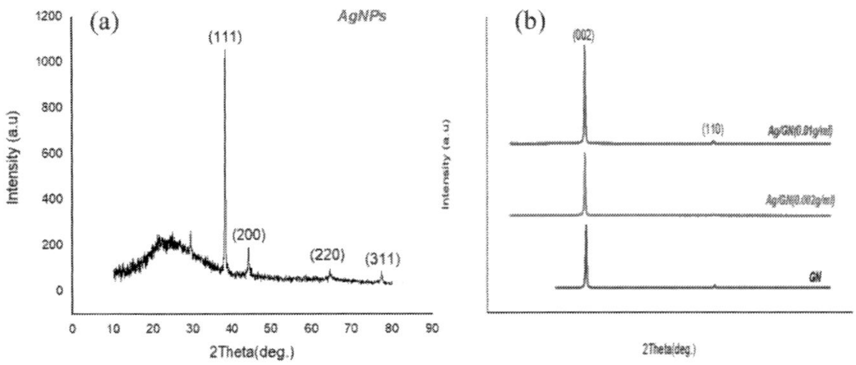

Figure 1. XRD patterns of (a)AgNPs (b)GN and GN-AgNPs with different concentrations of graphene.

3.2 *Field emission scanning electron microscope (FESEM)*

Surface morphology and size of AgNPs were analysis by the Field emission Scanning electron microscopy (FESEM), before and after adding GN. The suspension consisted of nearly spherical particles for the silver with average particle size (50-80) nm. Figure 2 (a, b and c) shows FESEM image for silver nanoparticles before and after adding GN. The distribution of AgNPs on the GN sheets depended on the mixing of AgNPs and GN. Figure (2a) shows FESEM image for silver nanoparticles before adding GN. Figure 2 (b and c) show FESEM images for silver nanoparticles after adding GN for different concentrations of GN (0.002g/ml and 0.01g/ml) and the AgNPs-GN have average particle size (20-40) nm.

The graphene is very thin and wrapped around silver particles, as shown in Figure 2(b and c). Silver particles, on the other hand, prevented graphene agglomeration, suggesting that the graphene was well dispersed during the ultrasonic crushing process.

(a) AgNPs (b) Ag/GN (0.002g/ml) (c) Ag/GN (0.01g/ml)

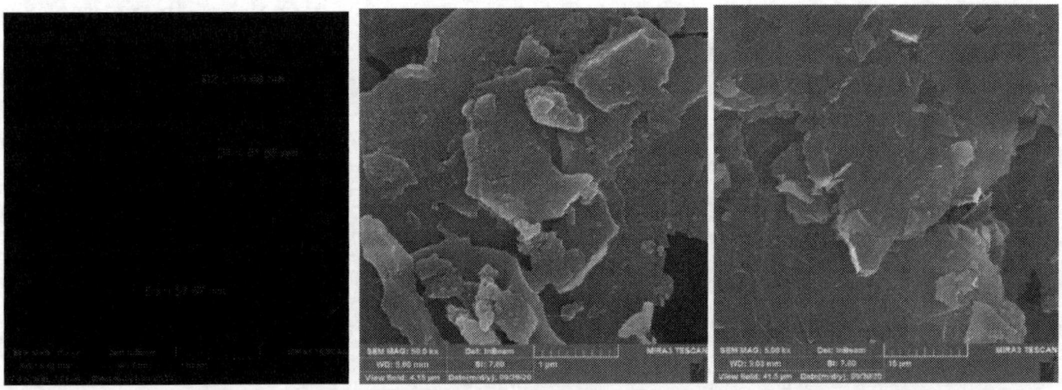

Figure 2. FESEM images of AgNPs (a) before adding GN and (b, c) after adding GN.

3.3 Atomic force microscopy (AFM)

Figure 3 (a, b and c) shows AFM image for silver nanoparticles before and after adding GN. Figure (3a) shows AFM image for AgNPs before adding GN.Figure 3 (b and c) show AFM images for after AgNPs adding GN for different concentrations of GN (0.002g/ml and 0.01g/ml). The existence of uniformly distributed AgNPs on the GN nanosheets was revealed by AFM analysis. The rippled like structure of the GN is further confirmed by the AFM study.

(a) AgNPs (b) Ag/GN (0.002g/ml) (c) Ag/GN (0.01g/ml)

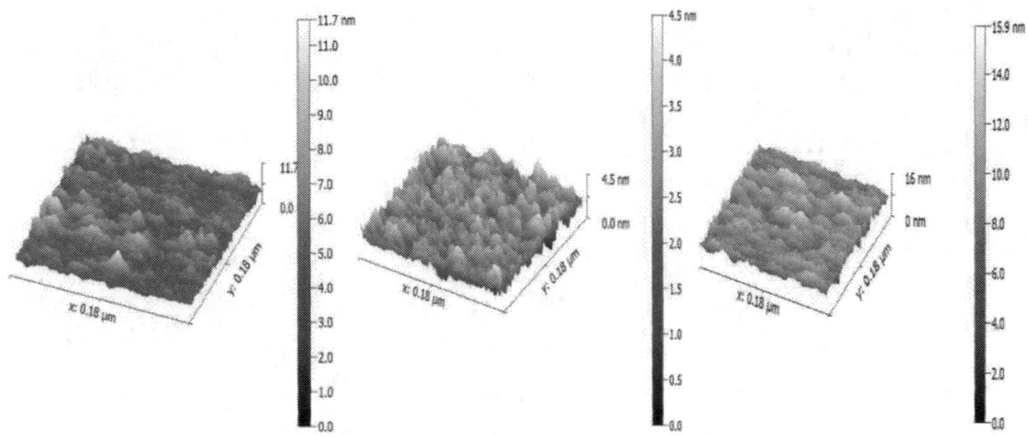

Figure 3. Three-dimensional AFM image of Ag NPs (a) before adding GN and (b, c) after adding GN.

3.4 UV-vis Spectroscopy

Figure 4 shows UV–visible spectra of (a) AgNPs, (b) GN and (c) Ag/GN with different concentrations of graphene. We have observed characteristic surface plasmon band at range (350-600) nm, which indicates the Ag nanoparticles formed as shown in fig.4(a). Figure 4(b) shows UV-visible spectra of GN.

The UV-visible spectra of GN contains two distinct absorption bands. At 193 nm, the absorption band was based. Figure 4 (c) shows the UV-vis spectra of Ag/GN composites made with various graphene concentrations. At 386 nm, an SPR band of silver nanoparticles appears in Ag/GN nanocomposite samples prepared at 0.002gm/ml graphene The SPR band maxima is 377 nm when graphene concentration is increased from 0.002gm/ml to 0.01gm/ml, which can be due to a decrease in silver nanoparticle scale. The lack of AgNPs and graphene absorption peaks in this case indicates that the percentage of silver nanoparticles is very low, which is not picked up in optical absorption measurements.

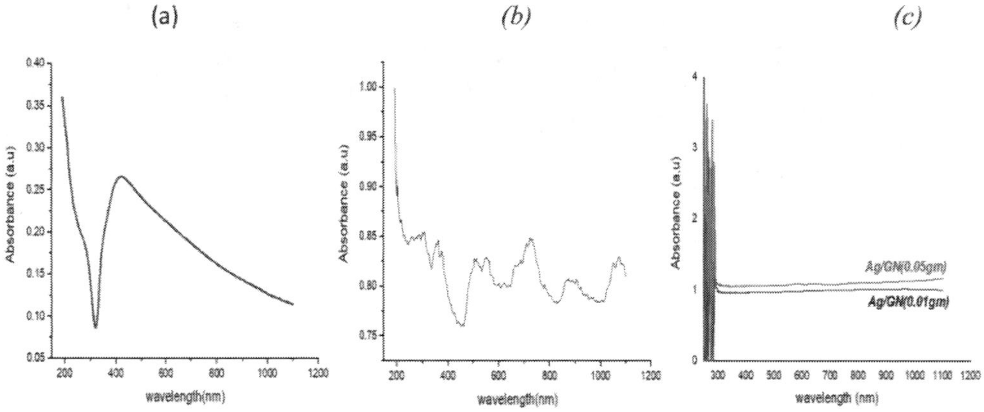

Figure 4. UV-Vis spectra of (a)AgNPs (b)GN and (c) GN-AgNPs with different concentrations of graphene.

4. Conclusions

In this study, well dispersion of AgNPs is achieved by a simple chemistry process. The XRD confirmed the crystalline structure for Ag NPs and graphene nanosheet. The absorption spectrum (especially the SPR region) indicates the formation of Ag NPs. Morphology test indicated the uniform distribution for graphene and Ag NPs.

Acknowledgment

This work was supported by department of physics, College of science, University of Baghdad, Baghdad, Iraq.

References

[1] Novoselov, K. S., Geim, A. K., Morozov, S. V., Jiang, D., Zhang, Y., Dubonos, S. V.,and Firsov, A. A. (2004) Electric field effect in atomically thin carbon films. Science **306**: 666-669.
[2] Su, Q., Pang, S., Alijani, V., Li, C., Feng, X., & Müllen, K. (2009) Composites of graphene with large aromatic molecules. Advanced materials.21:3191-3195.
[3] Bai, H., Sheng, K., Zhang, P., Li, C., & Shi, G. (2011) Graphene oxide/conducting polymer composite hydrogels. Journal of Materials Chemistry.21: 18653-18658.
[4] Liao, Q., Zhang, H. J., Wu, K., Li, H. Y., Bao, S. N., & He, P. (2010) Oxidation of graphene on Ru (0 0 0 1) studied by scanning tunneling microscopy. Applied surface science. 257: 82-86.

[5] Singh M. K., Titus, E., Gonçalves, G., Marques, P. A., Bdikin, I., Kholkin, A. L., & Gracio, J. J. (2010) Atomic-scale observation of rotational misorientation in suspended few-layer graphene sheets. Nanoscale.**2**: 700-708.

[6] He, H., Klinowski, J., Forster, M., & Lerf, A. (1998) A new structural model for graphite oxide. Chemical physics letters.287 : 53-56.

[7] Pumera, M., Ambrosi, A., Bonanni, A., Chng, E. L. K., & Poh, H. L. (2010) Graphene for electrochemical sensing and biosensing. TrAC Trends in Analytical Chemistry.29: 954-965.

[8] Kang, X., Wang, J., Wu, H., Aksay, I. A., Liu, J., & Lin, Y. (2009) Glucose oxidase–graphene–chitosan modified electrode for direct electrochemistry and glucose sensing. Biosensors and Bioelectronics.25: 901-905.

[9] Zhou, K., Zhu, Y., Yang, X., Luo, J., Li, C., & Luan, S. (2010) A novel hydrogen peroxide biosensor based on Au–graphene–HRP–chitosan biocomposites. Electrochimica Acta.55: 3055-3060.

[10] Biswas, Chandan, and Young Hee Lee. (2011) Graphene versus carbon nanotubes in electronic devices.Advanced Functional Materials. 21: 3806-3826.

[11] Cancan Huang, Chun Li and Gaoquan Shi. (2012) Graphene Based Catalysts. Energy & Environmental Science.5: 8848-8868.

[12] D.B. Sanchez, (2007) The Surface Plasmon Resonance of Supported Noble Metal Nanoparticles: Characterization, Laser Tailoring, and SERS Application. Ph.D Thesis, Madrid University.

[13] M Raffi, F. Hussain, T. M. Bhatti, J. I. Akhter, A. Hameed and M. M. Hasan, J. (2008) Antibacterial characterization of silver nanoparticles against E. Coli ATCC-15224. Journal of Materials Science and Technology.24 :192-196.

[14] Park, E. J., Lee, S. W., Bang, I. C., & Park, H. W. (2011) Optimal synthesis and characterization of Ag nanofluids by electrical explosion of wires in liquids. Nanoscale research letters.6 :1-10.

[15] Ullmann, Marc, Sheldon K. Friedlander, and Andreas Schmidt-Ott. (2002) Nanoparticle formation by laser ablation. Journal of Nanoparticle Research.4: 499-509.

[16] Araujo, L., R. Löbenberg, and J. Kreuter. (1999) Influence of the surfactant concentration on the body distribution of nanoparticles. Journal of drug targeting.6 :373-385.

[17] Wonbaek Kim, Je-shin Park, Chang-yul Suh, Jong-GwanAhn and Jae-chun Lee. C.(2008) Cu–Ni–P alloy nanoparticles prepared by electrical wire explosion. Journal of alloys and compounds. 465: L4-L6.

[18] Singh,S., Ashfaq, M., Singh, R. K., Joshi, H. C., Srivastava, A., Sharma, A., & Verma, N. (2013) Preparation of surfactant-mediated silver and copper nanoparticles dispersed in hierarchical carbon micro-nanofibers for antibacterial applications. New biotechnology.30: 656-665.

[19] Sarkar, Anirban, Hao Wang, and Theda Daniels-Race. (2014) Surface enhanced Raman spectroscopy on silver-nanoparticle-coated carbon-nanotube networks fabricated by electrophoretic deposition. Electronic Materials Letters.10 :325-335.

[20] Kotov, Yu A. (2003) Electric explosion of wires as a method for preparation of nanopowders. Journal of Nanoparticle Research.5 : 539-550.

[21] Vandana and P. Sen, J. (2005) Nanometre scale surface modification in a needle–plate exploding system. Journal of Physics: Condensed Matter.17 :5326.

[22] Cho, C. H., Park, S. H., Choi, Y. W., & Kim, B. G (2007) Production of nanopowders by wire explosion in liquid media. Surface and Coatings Technology.201: 4847-4849.

[23] Yang, W., Xu, X., Gao, Y., Li, Z., Li, C., Wang, W. and Li, Y. (2016) High-surface-area nanomesh graphene with enriched edge sites as efficient metal-free cathodes for dye-sensitized solar cells. Nanoscale.8:13059-13066.

[24] Lin, D., Motlag, M., Saei, M., Jin, S., Rahimi, R. M., Bahr, D., and Cheng, G. J. (2018) Shock engineering the additive manufactured graphene-metal nanocomposite with high density nanotwins and dislocations for ultra-stable mechanical properties. Acta Materialia.150 :360-372.

[25] Murphy, Sean, Libai Huang, and Prashant V. Kamat.(2013) Reduced graphene oxide–silver nanoparticle composite as an active SERS material. The Journal of Physical Chemistry C.117: 4740-4747.

[26] Deng, Z., Zhou, J., Miao, L., Liu, C., Peng, Y., Sun, L., and Tanemura, S.(2017) The emergence of solar thermal utilization: solar-driven steam generation. Journal of Materials Chemistry A. 5 :7691-7709.

[27] Wang, J., Li, J., Gao, M., and Zhang, X. (2017) Self-assembling covalent organic framework functionalized magnetic graphene hydrophilic biocomposites as an ultrasensitive matrix for N-linked glycopeptide recognition. Nanoscale. 9 :10750-10756.

[28] Wang, P., Wang, L., Sun, Q., Qiu, S., Liu, Y., Zhang, X., and Zheng, L (2016) Preparation and performance of Fe3O4@ hydrophilic graphene composites with excellent Photo-Fenton activity for photocatalysis. Materials Letters.183 :61-64.

[29] Fonseca, A. F., Liang, T., Zhang, D., Choudhary, K., and Sinnott, S. B (2016) Probing the accuracy of reactive and non-reactive force fields to describe physical and chemical properties of graphene-oxide. Computational Materials Science.114 :236-243.

[30] Li, M., Che, H., Liu, X., Liang, S.,and Xie, H (2014) Highly enhanced mechanical properties in Cu matrix composites reinforced with graphene decorated metallic nanoparticles. Journal of materials science.49 :3725-3731.

Journal of Physics: Conference Series **2063** (2021) 012013 doi:10.1088/1742-6596/2063/1/012013

Neural Network Model for Synthesis of Thermally Sprayed (Al/Al_2O_3) Composite Protective Coatings

Thamir Abdul-Jabbar Jumah[1], Saad Ali Ahmed[2]

[1]Al-Nahrain University, College ,of Sciences , [2]Baghdad University, College of Education Ibn-Haytham of Pure Science

*Corresponding author Email: drthamir05@yahoo.com

Abstract. Al_2O_3 and Al_2O_3–Al composite coatings were deposited on steel specimens using Oxy-acetylene gas thermal spray gun. Alumina was mixed with Aluminum in six groups of concentrations (0, 5, 10,12,15 and 20%) Al2O3, Specimens were tested for corrosion using Potentiodynamic polarization technique. Further tests were conducted for the effect of temperature on polarization curve and the hardness tests for the coated specimens. At first, Modelling was carried out using MINITAB-19, least square method, as a 2nd degree nonlinear model, bad results were achieved because of the high nonlinearity. Better result was achieved using neural network fitting tool. The network was designed using five neurons in the hidden layer and the input was I input with two layers, the electrical potential and alumina concentration.

1. Introduction

Aluminium composites are widely used as anticorrosion material for marine applications. (Berenika et al,2020) reviewed and summarized corrosion mitigation methods applied in the field of marine thermal spray coatings and the control parameters acting on the protection performance for aluminum thermally spray coatings used for steel [1].

The wetting behaviour between poly- crystalline alumina substrate and molten aluminium doped with magnesium and pure aluminium was studied. (Sangghaleh et. l.2008) [2]. To improve wettability between Al_2O_3 and Al in the fabricating of composite by employing the infiltration method (Szafran et. al. 2006) [3]. The structural energetic and adhesion strength for both solid and liquid phases of Al/Al_2O_3 interfaces by using the reactive force field was studied by (Zhang et. al. 2006) [4]. Silicon based alloy and Al_2O_3 particles with volume fraction of (5, 10, 15) wt.% and size of 44, 85 and 125 μm, were Characterized by (Sevak et. al. 2006) [5]. Effect of polarization on stress corrosion cracking of aluminium alloy type7075 was studied by (Liu et. al. 2004) [6]. The corrosion behaviour of Al- Zn/ Al2O3 in NaCl solution was investigated by (T.G.Durai et. al. 2008) [7]. They conclude that the corrosion depends on the weight fraction of the reinforcing particles.

(Edoziuno et al,2020) reported that the optimization and development of predictive models for the corrosion inhibition of mild steel in Sulfuric acid by Methyl-5-benzoyl-2-benzimidazole Carbamate (Mebendazole). Design methodology, weight loss measurement, open-circuit potential analysis, Tafel polarization, etc. were used for the evaluation of inhibition efficiency of mebendazole for mild steel in H_2SO_4. The corrosion inhibition process parameters were optimized and predictive mathematical models developed using the Response Surface Methodology using the central composite design tool of Design Expert software version 11 [8].

Uang et al, reported the Cross-sectional FESEM views of the Al and Al-Al_2O_3 coatings deposited by flame spray, XRD patterns of the starting Al powder, and the Al-20 wt.%Al_2O_3 powder the flame spray Aluminium and the flame spray of Al-Al_2O_3 coatings. They concluded that aluminium-alumina composite coatings fabricated by thermal flame spraying have a hard skeleton structure, it is attributed to the distribution of alumina splats in coatings, the cost effective for this protective coating coatings is promising for potential applications as the barrier and sacrificial protective layers in the marine environment [9]. The aim of this project is improving the corrosion resistance of steel due to composite and thermally treatment.

2 Materials and methodology

2.1 Specimens Preparation

Corrosion Specimens were firstly slicing with dimension of about (2.5mm) thickness and (25mm) diameter. The specimens were cleaned carefully with distill water and then by alcohol solution. Grinding and polishing by metallographic papers of grits 1200 and 2000 were achieved. Finally, the specimens should have cleaned by rinsed in distill water and acetone to remove any traces of the cleaner and organic which it adherent, and then kept in desiccator according to the ASTM [10].

2-2 Composite

Spray powders of (Al /Al_2O_3) mixed were prepared for (Al_2O_3) concentration (0 ,5 ,10, 12,15 & 20 wt.%) the rest is Al- powder. The percentage of cumulative particle size of Al- powder lied between (10-74) μm, while the particle size of Al_2O_3 lied between (5-45) μm. The powders were mixed carefully for 8 hr. before spraying process.

2.3 Thermal spray

Thermal spraying coating technique of (Al/Al_2O_3) composite for specimens surface of steel. Thermal spray gun with an Oxy-acetylene gas of pressure 5 bar was used. The operation conditions as follows: Substrate Temp 50 ^{o}C. Distance between the nozzle and substrate equal 15 cm. with two flame passes and 90° vertical spraying angle.

2.4 Potentiodynamic tests

Potentiodynamic polarization technique was carried out using WENKING Mlab multi channels potentiostat and SCI-Mlab corrosion measuring system from Bank Electroniks-Intelligent control GmbH, Germany 2007. The coating specimens achieved by the thermal spraying technique.

Corrosion tests were conducted electrochemically by Potentiodynamic method. It is a polarization technique in which the potential of the electrode is varied over a relatively large potential domain at a selected rate by the application of a current through the electrolyte. The research is conducted on potential domain of (-600~ -1300 mV).

Calculation of polarization resistance is performed by illustrate Tafel curves. the anodic and cathodic slopes and the corrosion current (α, β and i_{cr}) respectively were predicted from graphs. by substitution of these slopes in the equation (1) we got the polarization resistance. The prediction of these values is elucidated in figure 1'

$$R = \frac{\alpha \cdot \beta}{2.3(\alpha+\beta)i_c} \ldots.(1)$$

Journal of Physics: Conference Series **2063** (2021) 012013 doi:10.1088/1742-6596/2063/1/012013

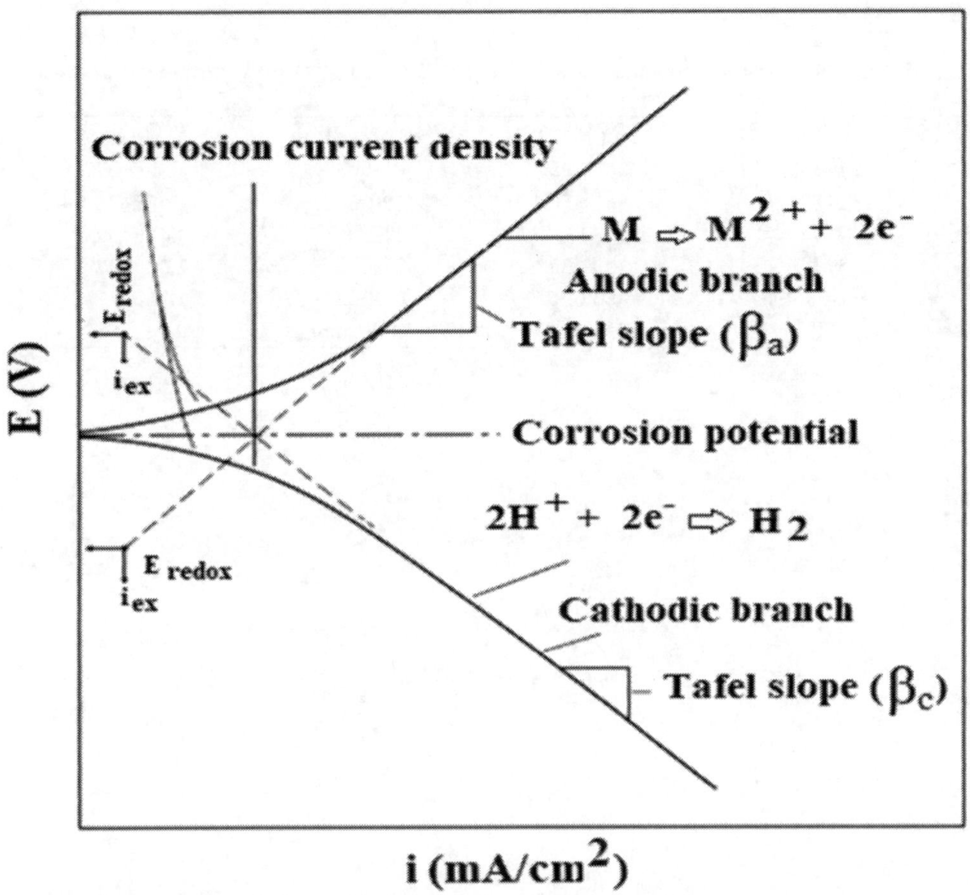

Figure 1. Prediction of Tafel equation parameters from polarity curve.

2-5 Hardness

The specimens are tested according to Vickers technique (HV) for hardness to conform the durability of the protective coating.

3 Results and discussion

3.1Polarization analysis

Results of Aluminium / Alumina concentration effect are elucidated in figure 2. The figure. shows The Potentiodynamic polarization curves of the coating samples acquired in (3.5M) NaCl solution. Tafel curves were illustrated from experimental data. The slopes of anodic and cathodic lines as well as the corrosion current is predicted. The polarization resistance was calculated using equation (1), the data is shown in 'table 1'.

At the corrosion potential, the cathodic potential and anodic potentials are equal (see 'table 1'). light implies that the higher electrical resistance exists at 10wt.% alumina, when spread in the aluminium it made splats and a hard Skeleton that improves the corrosion and erosion resistance. The corrosion potentials of the (Al-Al_2O_3) composite coating and the (Al) coating show increased in the electrical potential and increase in the current density, (-1275mV) for Al only surface coating to (-1063mV) for (20% Al_2O_3), the current density of Al-Al_2O_3 composite protective coating, (2.95 $\mu A/cm^2$) for Al only and (9.9 $\mu A/cm^2$) for 20% Al_2O_3, this meant that adding alumina powder enhances the corrosion resistance.

Table 1. Corrosion resistance and Tafel line slops for various alumina weight %.

Al_2O_3%	α (Ω.cm2)	B (Ω.cm2)	i_c (A)	R(Ω.cm2)
0	1086.957	1086.957	8E-6	2.95E+07
5	1123.596	561.5	1.10E-05	1.48E+7
10	2778	1111.11	1E-5	3.45E+07
12	3111	1333.333	2E-5	2.03E+07
15	-1086.96	766.6667	1.00E-06	1.95E+7
20	782	252	9.90E-6	9.21E+6

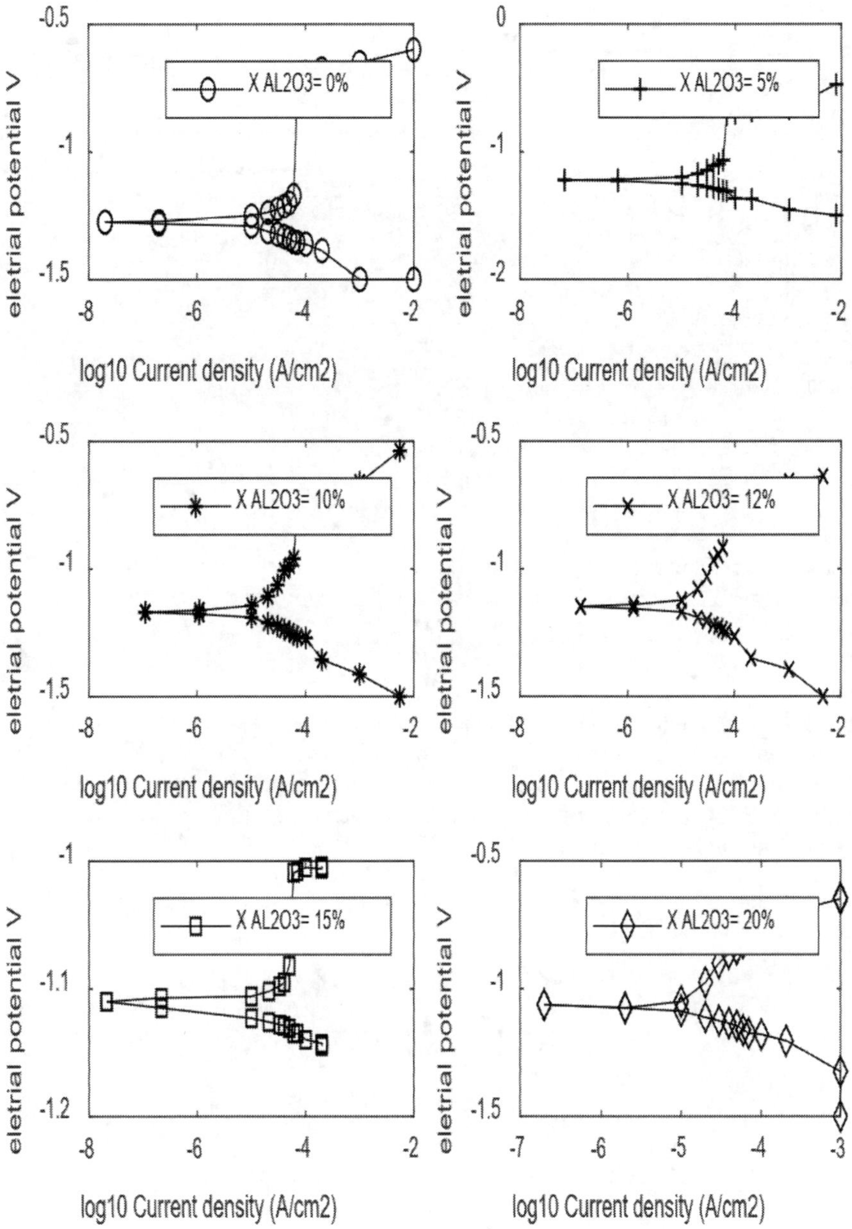

Figure 2. Polarization curves

Journal of Physics: Conference Series **2063** (2021) 012013 doi:10.1088/1742-6596/2063/1/012013

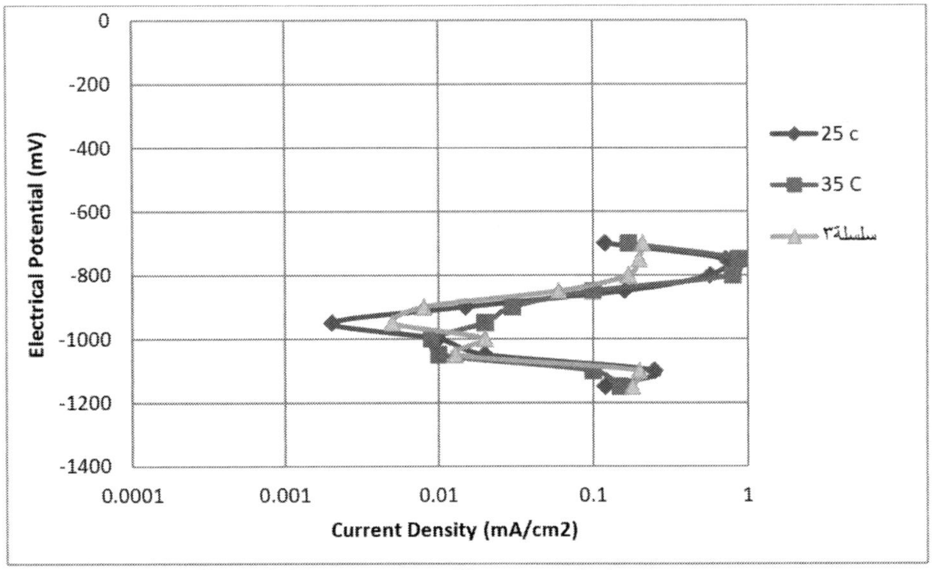

Figure 3. Temperature effects on polarization curve

It is reasonably, increasing the temperature increase the corrosion as shown in figure 2. Temperature effects on polarization are shown in figure 3, increasing temperature speeds corrosion.

3.2 Hardness

Hardness of the tested specimens is enhanced according to the results tabulated in 'table 2'. It was done by that the Vickers method for micro hardness analysis. The method is based on an optical measurement system. A square base pyramid shaped diamond is used for testing in the Vickers scale. A test force of (50 gf) load was used to perform the test. A schematic drawing for the test indenter is shown in figure 4.

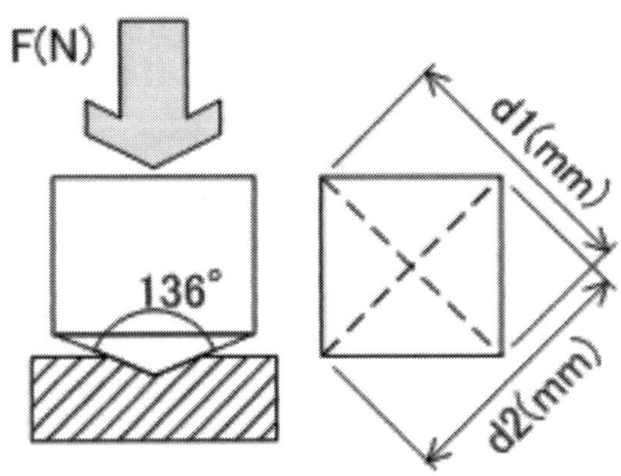

$$HV = 0.102\frac{2F\sin\frac{136^\circ}{2}}{d^2}$$

Figure 4. Vickers test indenter

Table 2. Hardness of Aluminum thermal spray coating

Al_2O_3 (wt.%)	Hardness (HV)
0	85
5	90
10	95
12	100
15	105
20	95

One can conclude that the coated specimens by alumina particles act as obstacles against any mechanical deformation such as indentation and dislocation when the additives increase up to 15.%. The resultant is when the specimens coated by ceramic such as Al/Al_2O_3. Increasing alumina which is less in particle size will decrease the porosity and increase the adhesion strength and that is reflected increase in the hardness. Adding more alumina increase the brittleness of the surface and decrease the hardness.

3.3 Microstructure analysis:

The microscopic photographs revealed the implies grains of different concentrations of alumina on the substrate. The aluminium metal matrix composites reinforce with Al_2O_3 of different weight fraction, will be reinforced the protection layer anti-corrosion as shown in 'figure 5', the thermally sprayed Al-Al_2O_3 coating, respectively. Compared with the thermally sprayed Al coating, the composite coating is much denser. The larger grain size of alumina on surface could be attributed to the accumulation of the molten droplets onto substrate surface, and this state may be is due to the flight length of droplet [1,3]. For the Al-Al_2O_3 composite coatings, it is clear that alumina barriers are equally distributed in the coating [7,8] 'figure 6', showing undetectable defects at Al/Al_2O_3 interfaces. While in 'figure 6 ', the Al/Al_2O_3 does not cover the substrate evenly and different spots shown [2,9]. 'Figure 7' Shows the microscopic image of surface after corrosion occurrence. The present experiment shows that the corrosion resistance is reduced with increasing volume portions of alumina particles. This state is attributed to the Galvanic coupling in between alumina and matrix [8,9].

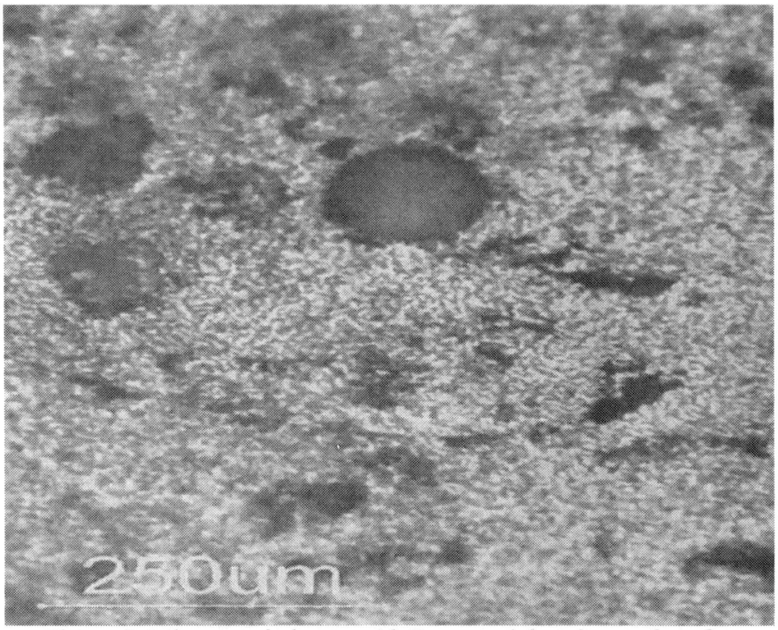

Figure 5. Microscopic image: Al_2O_3 material at the steel surface.

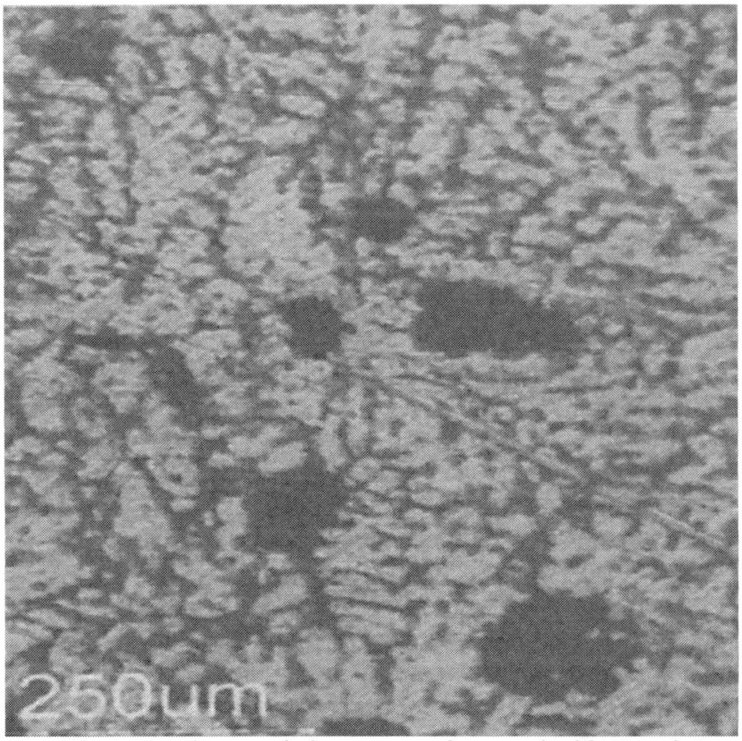

Figure 6. Microscopic image: sample spryed with 7% Al_2O_3

Journal of Physics: Conference Series **2063** (2021) 012013 doi:10.1088/1742-6596/2063/1/012013

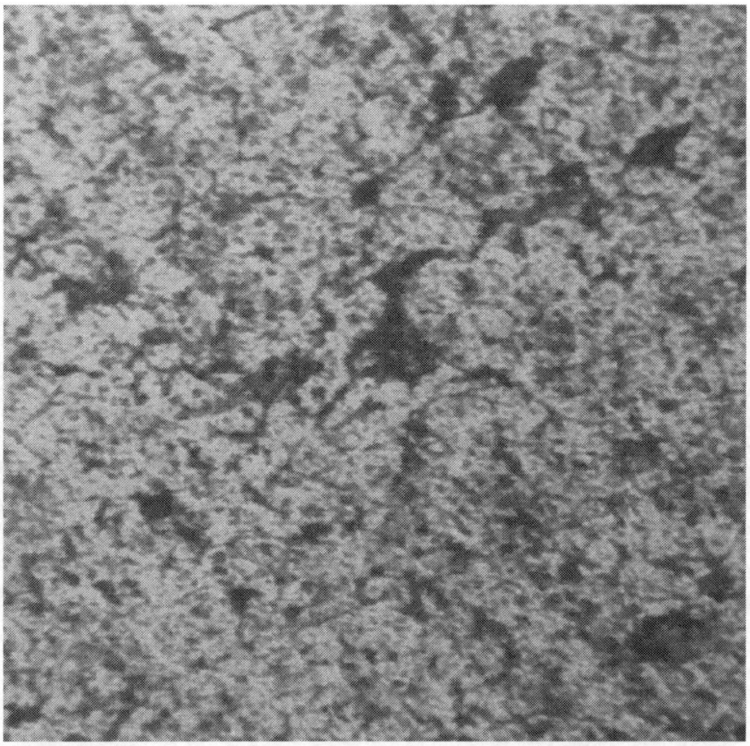

Figure 7. Microscopic image: protective coating layer after corrosion

3.4 Statistical analysis

Treatment of results are carried out through Minitab and MATLAB as mentioned below.

3.4.1 Minitab:

Response surface design is created using statistical software (Minitab- 19) : model is created for electrical potential as a function of aluminium concentration and electrical anode and cathode current density. It is appeared that the square of weight percent, the interaction between weight percent and cathodic current, and the interaction between anodic and cathodic current has significant values.

'Table. 2' Response surface analysis X: Alumina %, IA: Anodic current density, IC, cathodic current density, E: electrical potential [11].

Journal of Physics: Conference Series **2063** (2021) 012013 doi:10.1088/1742-6596/2063/1/012013

Response Surface Regression: E versus X, IA, IC

Analysis of Variance

Source	DF	Adj SS	Adj MS	F-Value	P-Value
Model	9	0.000142	0.000016	17.43	0.000
Linear	3	0.000057	0.000019	20.99	0.000
X	1	0.000000	0.000000	0.51	0.478
IA	1	0.000001	0.000001	0.67	0.415
IC	1	0.000002	0.000002	2.36	0.129
Square	3	0.000008	0.000003	2.77	0.048
X*X	1	0.000005	0.000005	5.84	0.018
IA*IA	1	0.000001	0.000001	0.87	0.354
IC*IC	1	0.000000	0.000000	0.03	0.858
2-Way Interaction	3	0.000008	0.000003	2.82	0.045
X*IA	1	0.000000	0.000000	0.11	0.741
X*IC	1	0.000005	0.000005	5.09	0.027
IA*IC	1	0.000000	0.000000	0.26	0.611
Error	68	0.000062	0.000001		
Total	77	0.000203			

The model summary values are shown in 'Table. 3', the fitting of model gives R^2=.69 but the predicted values for R^2 is less than (0.5), the fitting is not satisfied.

Table3. Model Summary

S	R-sq	R-sq(adj)	R-sq(pred)
0.0009511	69.76%	65.75%	48.79%

The regression equation is:

E = -0.031 + 0.00195 X - 0.0032 IA - 0.023 IC - 0.000011 X*X + 0.0136 IA*IA + 0.0059 IC*IC + 0.000181 X*IA + 0.001171 X*IC - 0.0209 IA*IC

3.5 Neural Networks

3.5.1 Temperature effects by neural networks:

Three temperature sets were treated 298 ^{O}K, 308 ^{O}K and 318 ^{O}K. They were treated separately to ensure the accuracy and monitoring the results using MATLAB, 2017b neural networks Fitting tools. The artificial neural network which was used for nonlinear modelling of electrical potential with the current density is shown in figure 8. The architecture is used for all temperatures [11].

Journal of Physics: Conference Series **2063** (2021) 012013 doi:10.1088/1742-6596/2063/1/012013

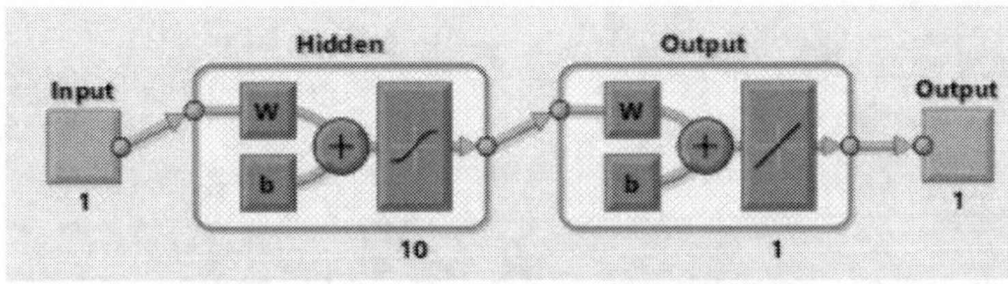

Figure 8. Neural network fitting function.

The RMS and R values for the three temperature ranges are listed in 'Table. 4'.

Table 4. MSE and R values for the neural network model of temperature changes

T (^{O}K)		Samples	MSE	R
298	Training	6	4.94706 e-3	0.989944
	Validation	2	0.189061	1.000000
	Testing	2	6.32511 e-2	1.000000
308	Training	6	1.45586 e-21	1.000000
	Validation	2	3.18962 e-1	.9999999
	Testing	2	0.158381	-1.000000
318	Training	6	3.27960 e-4	0.99261
	Validation	2	1.98587 e-3	1.000000
	Testing	2	0.701652	-1.000000

It was shown in 'Table. 4 ', at validation a reverse fitting sign. That is mean a change from cathode protection to anodic region. In figure 9, the three temperature zones are shown for the experimental and neural network output for the current density versus potential.

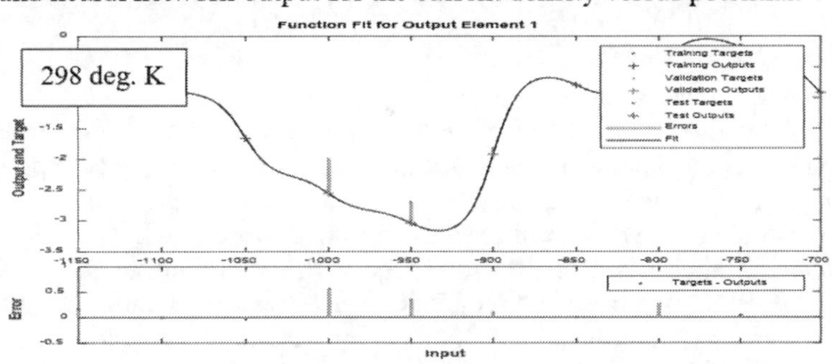

Figure 9.a Errors at 298°K

Journal of Physics: Conference Series **2063** (2021) 012013 doi:10.1088/1742-6596/2063/1/012013

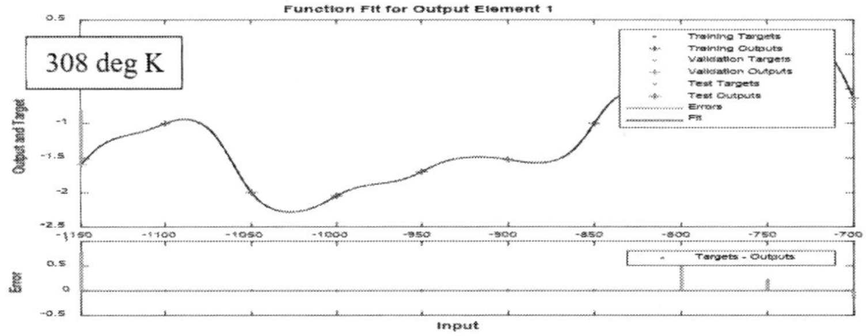

Figure 9b. Errors at 308°K

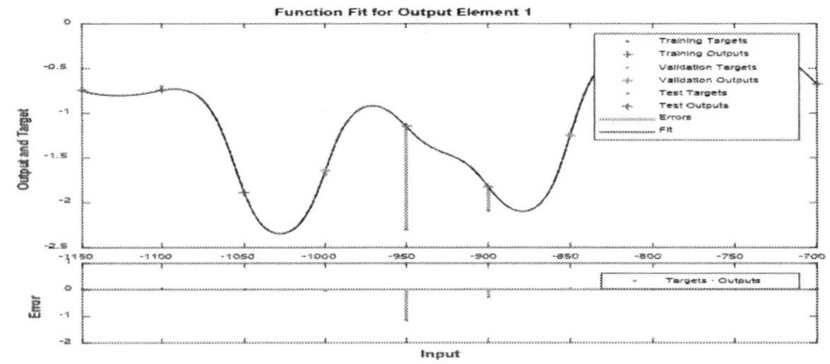

Figure. 9c .errors at 318°K

Figure 9. Fitting current density for 298, 308 and 318 °K with error diagram

3.5.2 Corrosion resistance by Neural Networks

A new ANN is designed to model the current density as a function of the electrical potential, and the weight fraction. The neural network is designed to accept three inputs, alumina wet percent, current density and electrical potential, the output of the network is the electrical resistance to corrosion. The data used for training the network is highly nonlinear, MINITAB statistical software failed to model it, so we tried with neural networks. Multi hidden layers succeeded in achieving this target. Three hidden layers was good choice for this duty, the 1st hidden layer contains 12 neurons, the 2nd hidden layer includes 10 neurons and the 3rd has 12 neurons. All of them are (tansig) type activation function. The output layer has one neuron of linear type. The architecture of the neural network is shown in figure 10. Number of Neurons is was set by trial.

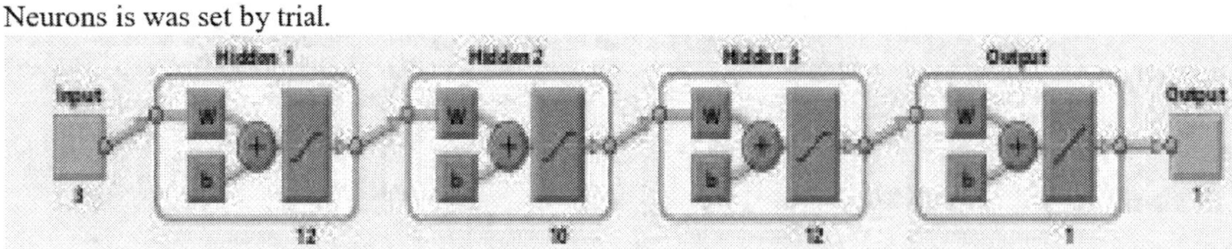

Figure. 10. Architecture of the neural network used in modelling corrosion resistance

Journal of Physics: Conference Series **2063** (2021) 012013 doi:10.1088/1742-6596/2063/1/012013

Inputs for the neural network are Alumina content, current density and the electrical potential. it was considered the electrical resistance is a function of these three inputs. the network gives good results, so it is important to thought of new terms.

$ANN = f(E, i, x_{Al2O3})$, the output of the ANN is R

A neural network is trained with 70% of input data, while the rest is 25% for validation and 15% for test. Good results were achieved for this highly nonlinear behaviour with experimental disturbances, RMS an R values for training, validation and test [12,13]. Comparing the neural network output with the experimental result shows complete alignment as shown in figure 11.

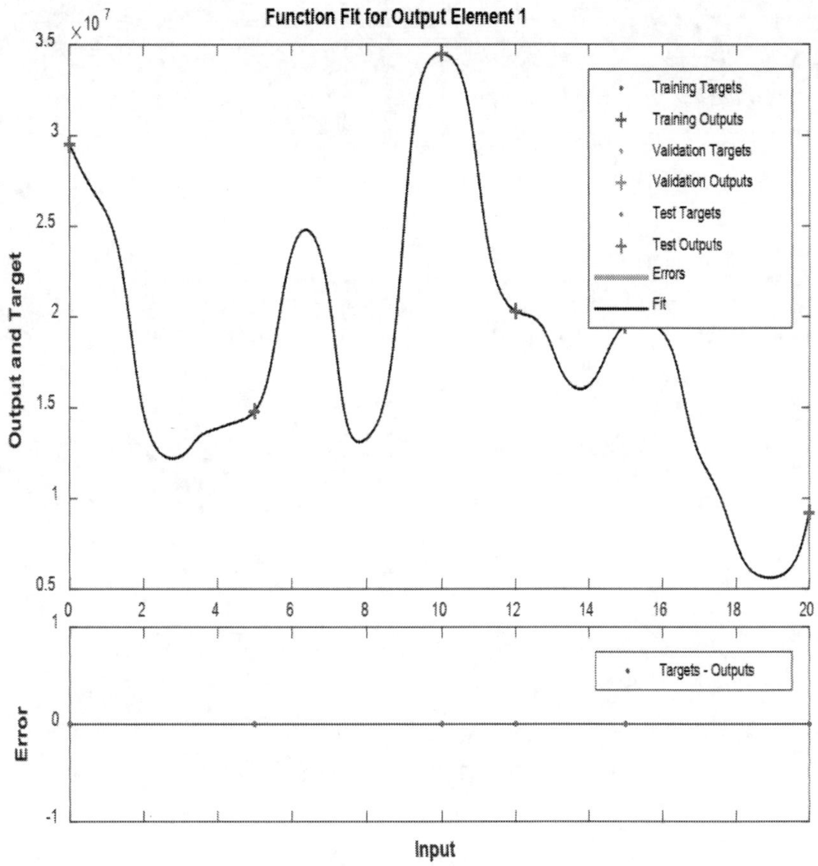

Figure 11. Neural network and experimental values of current density data series

Regression plots are shown in figure. 12.

Journal of Physics: Conference Series **2063** (2021) 012013 doi:10.1088/1742-6596/2063/1/012013

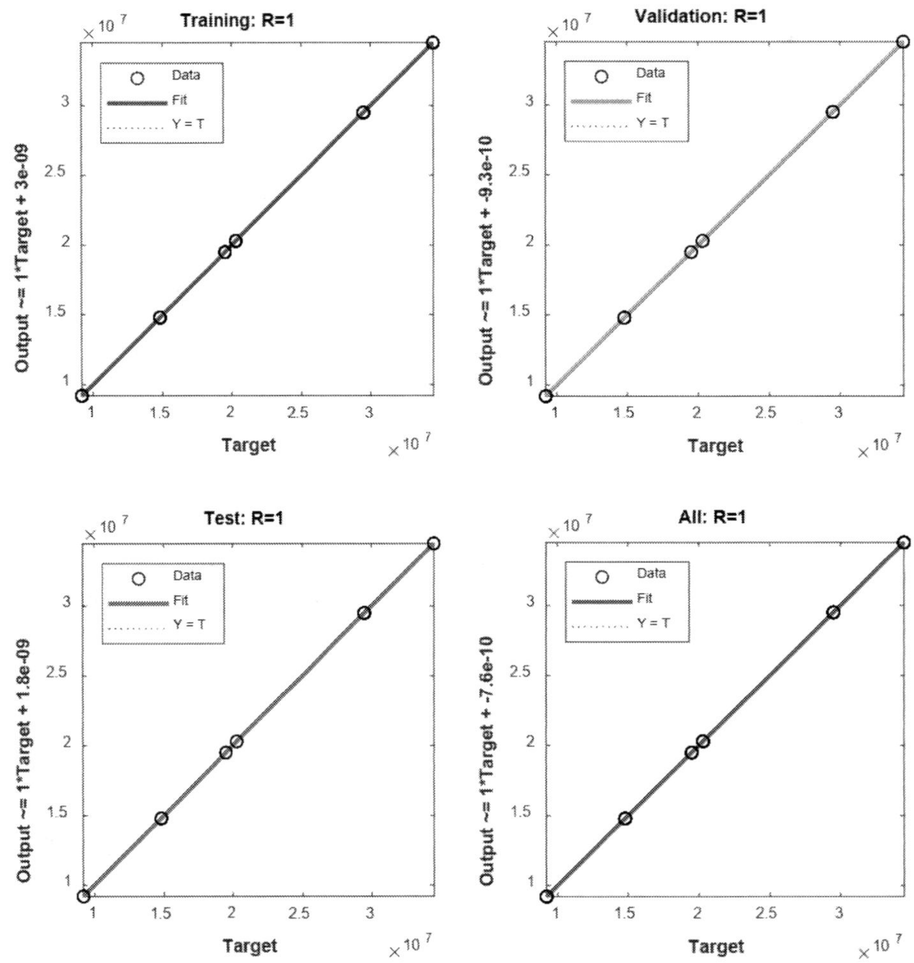

Figure.12 Regression plots for outputs and targets for the neural network

4. Conclusions

1-Corrosion rate is enhanced as the weight percentage of alumina

2- Alumina particles act as obstacles to dislocate movement and in the same time acts to increase the surface hardness.

3- Thermal spraying by gun is efficient technique and low cost for applies the coating.

4- Neural was better method than modeling such nonlinear corrosion case.

References

[1] B. S. Gerstenkorn 1, S. Paul and A. J. Davenport 1,:Sacrificial Thermally Sprayed Aluminium Coatings for Marine Environments: A Review, *Coatings,* Vol. 267, pp. 1-19, 2020.

[2] A. Sangghaleh and M. Halali, : An investigation on the wetting of polycrystalline alumina and aluminum , journal of material and technology, Vol. 197, pp. 156-160, 2008.

[3] K.Konopka and M.Szarfran, : Fabrication of Al2O3 composites by infiltration method and their characteristic, journal of material processing technology , Vol. 175, pp.266- 270, 2006.

[4] Q. Zhang and Cagin: adhesion and nonwetting – wetting transition in the Al/α-Al2O3 interface, physical review 69(4). Art, No. 045423.ISSN O163 1829 , 2006.

[5] H. Sevik and S. Can Kurnaz, : properties of alumina particulate reinforced aluminum alloy produced by pressure die castin, Vol.27, pp. 676 683, 2006.

[6] L. Ji-hua and L. pei-ying,: effect of polarization on SCC of 7075 aluminum alloys, materials forum journal, Vol.28, 2004.

[7] T.G.Durai , Karabi, and S. Das,: corrosion behavior of Al- Zn/ Al2O3, Journal of Alloys and Compounds , VOL. 462, ,PP. 410-415, 2008.

[8] S.K.Chaudhury and S.C.Panigrah, :Role of processing parameter on micro structural evaluation of spray formed Al-2Mg alloy and Al- 2Mg- TiO_2 of composite, Journal. of materials processing technology, Vol. 182, pp. 343-351, 2007.

[9] "Optimization and development of predictive models for the corrosion inhibition of mild steel insulphuric acid by methyl-5-benzoyl-2-benzimidazole carbamate (mebendazole)," *Cogent Engineering,* vol. 7, 2020.

[10] ASTM G1-3 Standard practice for preparing, cleaning and evaluating corrosion test specimens; ASTM International; west conshohocken, PA,USA, 2003.

[11] F. T. M.Noori,: Experimental,kinetic, isotherm and quantum chemical calculation; corrosion inhibitor for mild steel in acidic medium, Journal.WULFENIA, Vol. 23, No. 2, pp. 493-500, 2016.

[12] J. Huang, Y. Liu, J. Yuan, and H. Li, :Al/Al_2O_3 Composite Coating Deposited by Flame Spraying for Marine Applications: Alumina Skeleton Enhances Anti-Corrosion and Wear Performances, *J Therm Spray Tech Journal of Thermal Spray Technology,* Vol. 23, pp. 676-683, 2014.

[13] Edgar, T. F. Optimization of chemical processes / Thomas F. Edgar, David M. Himmelblau, Leon S. Lasdon.-2nd ed. p. cm.--(McGraw-Hill chemical engineering series.) ISBN 0-07-039359, 2012.

Synthesis, Biological Activity of Trimethoprim derivative and the Complexes

Hassan H Albahadly, Nahed H Al-Haidery, and Bushra K AL-Salami

Department of Chemistry ,College of Scinece. Unviersity of Basrah , Iraq

hhralbahadly@yahoo.com

Abstract. This study demonstrates the synthesis of new ligands derived from trimethoprim with their Iron(III) and copper (ll) complexes. At the beginning, preparing new ligands , the first namly N,N'-(5-(3,4,5-trimethoxybenzyl)pyrimidine-2,4-diyl)bis(3,4,5-trihydroxy benza mide)and the Ligand second namly N,N'-(((5-(3,4,5-trimethoxybenzyl)pyrimidine-2,4-diyl) bis(azanediyl))bis(carbonthioyl) bis(3,4,5-trihydroxybenzamide) which prepared by nucleophilic addition of trimethoprim to Gallic chloride and prepared also by nucleophilic addition of trimethoprim to the Solution of Ammonium thiocyanate and Gallic chloride . Ferric Ion (III) and Copperic Ion (ll) complexes have been prepared with molar ratio [1:2]. The synthesis ligands have been characterized by Uv-Visible, FT-IR, ^{1}HNMR, ^{13}CNMR and EI-mass, while the complexes have been characterized by elemental analysis, Uv-Visible, FT-IR, Conductivity measurements and thermogravemtric (TgA) analysis. The ligands acts as multiple sites coordinating with ferric ion and copper ion , Via lone pair of nitrogen atom of NHC=O and phenolic oxygen. In Vitro, the ligands and complexes have been tested for their growth Inhibitory activity against Gram negative bacteria *Salmonella Spp* and Gram Positive *Staphylococcus Spp*. The results of the test indicate that the synthesized compounds possessed a high inhibition effectiveness comparative with trimethoprim.

1. Introduction

Antibiotics are among the medicines that are widely used in the field of health, as there are many about these antibiotic compounds and their derivatives, which have succeeded in eliminating a group of bacteria and fungi that infect living organisms[1].

It is known that most of these compounds are susceptible to being interfered with by hemoglobin in the blood by nitrogen or oxygen and sulfur atoms present within the formulations of these drugs [4].

The binding property leads to a reduction in the proportion of iron it is found in the body of people who take large amounts of these antibiotics, and it is noticed that the diuretic changes of these people during the course of treatment with these medicines.

Therefore, thinking and searching for ways to reduce this effect of these drugs began while preserving or increasing their therapeutic efficacy, as the combinations of these antagonists have been used and contain groups through which to add groups that have the ability to supply the body with iron instead of reducing its percentage in the body and increasing its therapeutic effectiveness. the research dealt with one of these types of antibiotics that are widely used to treat ear infections Central, urinary tract infections and diseases that affect the respiratory system[2], which is Trimethoprim[3] and its abbreviation TMP Trimethoprim[5] as in the Figure1 and Gallic acid[6] derivatives Figure 2 and

Journal of Physics: Conference Series **2063** (2021) 012014 doi:10.1088/1742-6596/2063/1/012014

Figure 3 have the ability to be persistent with many ions of transitional elements, and among these ions that the body of the organism needs are the iron ions[7], and some of its derivatives are able to bind with the antibiotic containing the amine groups that have been used.

In this research, by increasing the efficacy of trimethoprim, preventing its association with hemoglobin, and increasing the second derivative of the new antibiotic and its complexes in water and increasing its biological effectiveness[8].

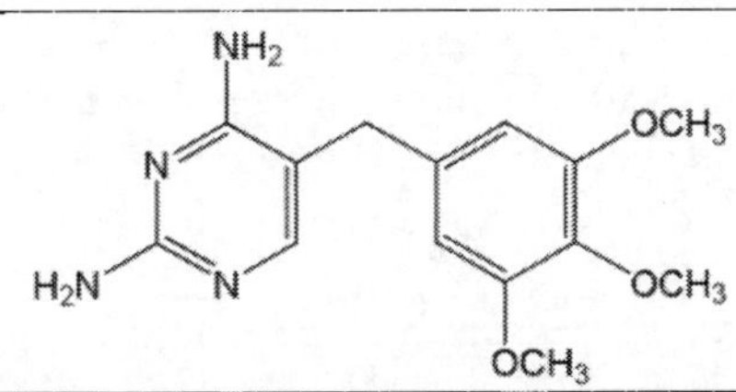

Figure 1. The structure of TMP.

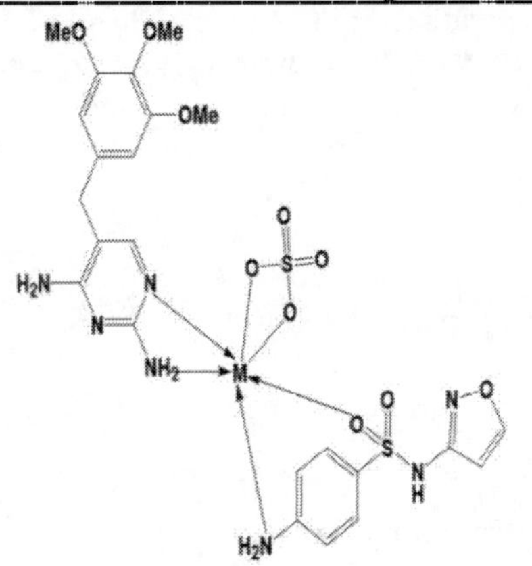

Figure 2. Trimethoprim derivative , their Iron (III) Complex.

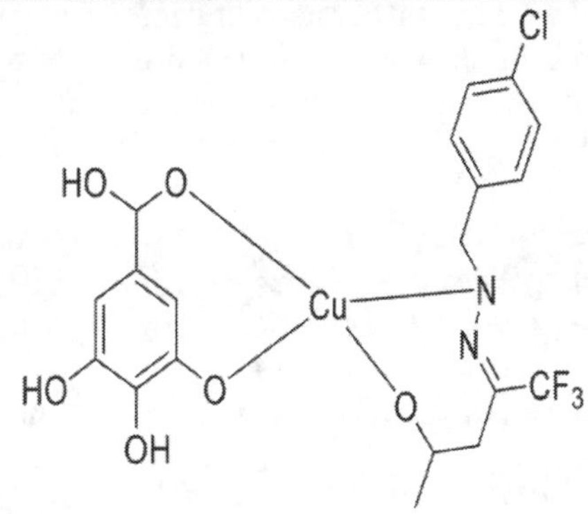

Figure 3. Gallic Acid derivative , their Copper (II) Complex .

2. Experimental

2.1. Materials

Trimethoprim was purchased from Aldrich . Gallic acid , Thionyl chloride , Ammonium thiocyanate, Copper nitrate of dehydrate ($Cu(NO_3)_2.3H_2O$) and Ferrie chloride ($FeCl_3$) were obtained from Fluka, all solvent used were of analytical grade and used without further purification.

2.2. Instrumentation

Infrared (IR) were recorded for KBr pellets using Shimadzu FTIR model 8400s in the range 4000-400 cm^{-1}. Electronic spectra for the synthesized compounds were recorded by using scan 80 D (England) at range 200 – 800 nm using H_2O as solvent and 1 cm pathway quartz cells.The ^{1}H-NMR ,^{13}CNMR were recoded on a Bruker (400 MHz for H-NMR and 500 MHz, for CNMR) using (DMSO-d6) as a solvent, and tetra methyl silane (TMS) as an internal standard.Thermal analyses measurements (Tg and DTG) were recoded and a Rheometric Scientific Inc. 1998. Nitrogen flow rate 10 cm^3/min and heating rate 10 °C min^{-1}. Mass spectrum scanned by EI-technique at 70 ev using Agilent Technologies 5975 C spectrometer.

2.3. Synthesis of Gallic chloride

Gallic Acid (1.7 g, 10 mmol) has been dissolved in anhydrous DCM (20 mL), in thionyl chloride (1.6 mL, 10 mmol) added to the solution with stirring with reflux at 60 °C for 90 min, Then the mixture has been washed with dried toluene (20 mL) to afford Gallic chloride as a pale yellow product.[9]

2.4.Synthesis of N,N'-(5-(3,4,5-trimethoxybenzyl)pyrimidine-2,4-diyl)bis(3,4,5-trihydr oxybenzamide) (L1)

Gallic chloride (1.508g, 4mmol) was suspended in 20 ml of acetone. (0.58g , 2 mmol) of Trimethoprim was added dropwise ,with constant stirring .The reaction mixture was reflux for 2 hrs. The solid product which obtain , filtrated off , washed with ethanol and dried . yield 72% , Rf = 0.87, M.p.188-190 °C .

2.5. Synthesis of N,N-(((5-(3,4,5-trimethoxybenzyl)pyrimidine-2,4-diyl)bis(azanediyl))bis(carbono thioyl)) bis(3,4,5-trihydroxybenzamide)(L2)

The solution of thiocyanate (0.761 g .10 mmol) in 20 ml acetone was added to (1.88 g . 10 mmol) of Gallic chloride ,with constant stirring, the reaction mixture refluxed for 2 hrs.The mixture was filtered .The filtrate solution added dropwise to solution of Trimethoprim (1.6 g, 5 mmol) with continuous stirring. The resulting solution refluxed for 3 hrs. The solid white product which obtain was filtrated off, washed with ethanol and dried. yield 66%, Rf = 0.56, M.p. 190-192 °C .

2.6. Synthesis of Copper (ll) Comlex.$Cu_2L1.4NO_3.H2O$ (C1)

(L1) (10 mmol) dissolved in ethanol and (20 mmol) of aqueous solution of Copper nitrate of dehydrate has been added dropwise with constant stirring at room temperature for 2 hrs with molar ratio 2:1 (M:L).Green precipitate formed which was filtered , washed several time with water and dried ,yield 63% , m.p. 220-222 °C. The obsered physical properties are given in Table 1.

2. 7. Synthesis of Iron(lll) Comlex. $Fe_2L1.(Cl)_6.2H_2O$ (C2)

Trimethoprim derivative (L1) (10 mmol) dissolved in ethanol and (20 mmol) of aqueous solution of ferric chloride has been added dropwise with constant stirring at room temperature for 2 hrs with molar ratio 2:1 (M:L). Black precipitate formed which was filtered , washed several time with water and dried ,yield 86% , m.p. 195-197 °C. The obsered physical properties are given in Table 1.

2.8. Synthesis of Copper (ll) Comlex.$Cu_2L1.4NO_3.H2O$ (C4)

(L2) (10 mmol) dissolved in ethanol and (20 mmol) of aqueous solution of Copper nitrate of dehydrate has been added dropwise with constant stirring at room temperature for 2 hrs with molar ratio 2:1 (M:L).Green precipitate formed which was filtered , washed several time with water and dried ,yield 78% , m.p. 176-178 oC. The obsered physical properties are given in Table 1.

2.9. Synthesis of Iron(lll) Comlex. $Fe_2L1.(Cl)_6.2H_2O$ (C5)

(L2) (10 mmol) dissolved in ethanol and (20 mmol) of aqueous solution of ferric chloride has been added dropwise with constant stirring at room temperature for 2 hrs with molar ratio 2:1 (M:L). Black precipitate formed which was filtered , washed several time with water and dried ,yield 81% , m.p. 245-244 °C. The obsered physical properties are given in Table 1.

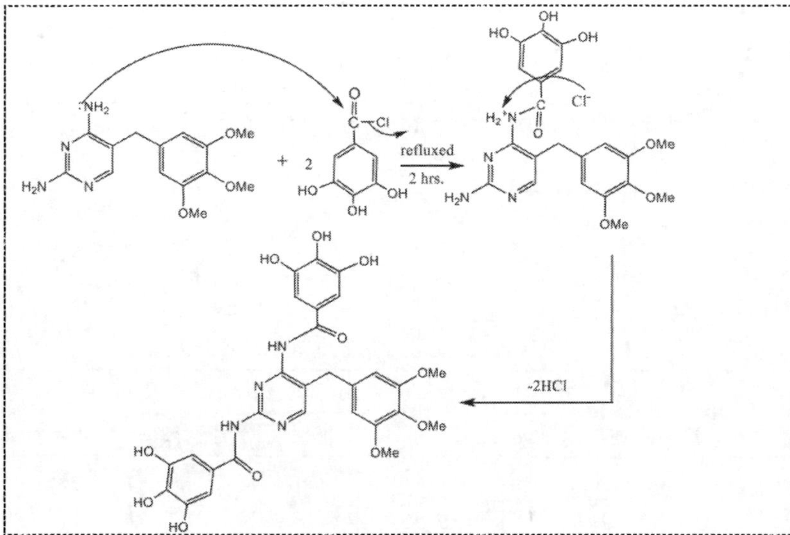

Figure 4. The reaction mechanism of L1.

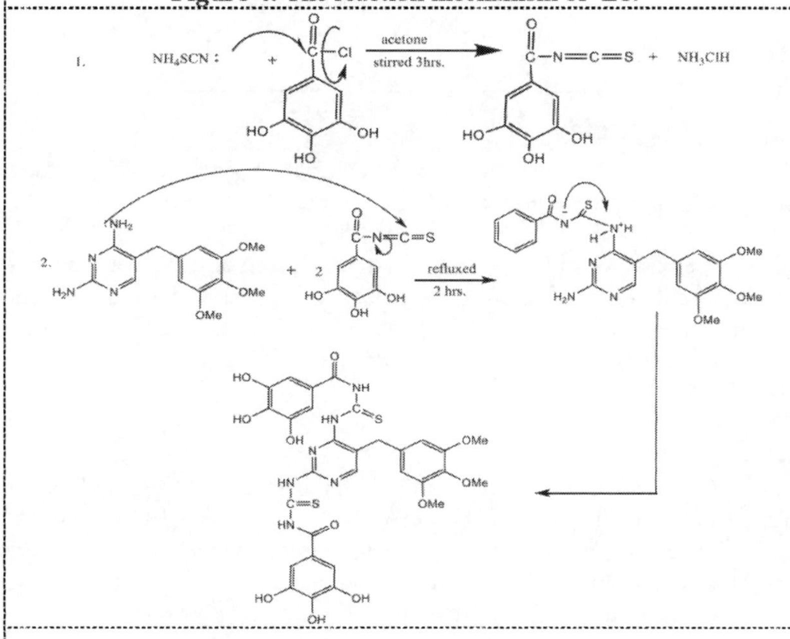

Figure 5. The reaction mechanism of L2.

$$L + Cu(NO_3)_2.3H_2O \xrightarrow[\text{Stir .2h}]{EtOH\ /\ H_2O} Cu_2L.nH_2O \quad (1)$$

$$L + 2FeCl_3.5H_2O \xrightarrow[\text{Stir .2h}]{EtOH\ /\ H_2O} Fe_2L.nH_2O \quad (2)$$

Table 1. Analytical and physical data of ligands and complexes

	Colour	M.p. °C	Ohm^{-1} Cm^2.mol^{-1}	Yield %
L1	White	190-188	-	72
L2	White	190-192	-	66
C1	Green	222-220	180.5	86
C2	Black	197-195	182.7	79
C4	Green	178-176	137.1	81
C5	Black	245-244	202.3	69

Table 2. CHN results for the prepared complexes

		Elemental of	analysis found	(calc.)	
Compound	Formula	% C	% H	%N	% S
C1	$C_{28}H_{30}Cu_2N_6O_{19}$	42.20(42.37)	4.89(4.32)	7.50(7.06)	-
C2	$C_{28}H_{22}Fe_2N_4O_{11}$	39.84(39.66)	3.67(3.09)	6.79(6.61)	-
C4	$C_{30}H_{32}Cu_2N_6O_{13}S_2$	41.79(41.14)	3.11(3.68)	9.85(9.60)	7.68(7.32)
C5	$C_{30}H_{36}Fe_2N_6O_{15}S_2$	34.23(34.71)	3.46(350)	8.55(8.09)	6.62(6.91)

3. In vitro antimicrobial activity

The ligands and complexes were Screend for antibacterial activity in vitro against two kinds of organisms species . Gram(-ve) *Salmonella spp* and gram(+ve) *Staphylococcus spp* , by agar well diffusion method .

These compounds were dissolve in DMSO to prepared solution of different weights (5μg ,4μg, 3μg) in (0.1ml) of DMSO . Sterile discs dipped in this solutions , dried it and place on nutrient agar plate spreaded with the bacteria .

The plate were further incubated for 24 to 48 hours at 37 °C and the diameters of inhibition zone measured in millimeter[10] .

4.Result and Discussion

New oregano Ferric (lll) and Copperic (ll) were synthesized by the reaction of Trimethoprim derivative in molar ratio 2:1 [M:Ligand] . The Ferric (lll) and Copper(ll) complexes are subjected to elemental analysis. The result obtained is in good agreement with those calculated for suggested formula of the complexes. The high value of molar conductance of aqueous solution of complexes confirm the electrolytic nature and indicate that the chlorine ion and nitrate ion out of coordination sphere. Table 1.

The IR spectroscopy is important to diagnose the Prepared complexes, as through it is possible to infer and determine the coordination sites between the ligand and the metal ion. It's can be the deflection the Values of IR due to coordination between metal and ligand as shown in the table 3.The IR spectra of the ligands (L1 and L2) indicate broad bands at (3367, 3437) cm^{-1} respectively which attributed to stretching of OH group . While the bands within ranges (3290, 3163) cm^{-1} characteristic of stretching vibration of NH group[11]. The ligand(L1) showed two strong bands at the rang 1701 and 1666 cm^{-1} that attributed to C=O moiety . Also two strong bands appeared within rang 1539 cm^{-1} which attributed to asymmetrical and symmetrical stretching of aromatic C=C . The Ligand(L2) showed one strong band at the rang 1670 cm^{-1} that attributed to C=O and showed one strong band at 2080 cm^{-1} that attributed to C=S. The IR spectra of the ligands showed weak bands at the range (3073, 3090) cm^{-1} respectively which attributed to aromatic C-H stretching , the aliphatic C-H stretching bands appeared at rang (2847, 2939) cm^{-1} respectively further more the rang at (1246 , 1238) cm^{-1} can

Journal of Physics: Conference Series **2063** (2021) 012014 doi:10.1088/1742-6596/2063/1/012014

be attributed to stretching vibration of C-N group . In other hand the L1 showed aband at 1030 cm^{-1} that attributed to stretching vibration of C-O while the L2 showed the band of C-O at 1130 cm^{-1} [12] [13] .

Table 3. IR data spectrum of ligands and complexes.

	υ(O-H) cm^{-1}	υ(N-H) cm^{-1}	υ(C=O) cm^{-1}	υ(C=S) cm^{-1}	υ(N-H bond) cm^{-1}	υ(C-H Arom.) cm^{-1}	υ(C-H al ph.) cm^{-1}	υ(C-O) cm^{-1}	υ(C-N) cm^{-1}	υ(C-Hbend.) cm^{-1}
L1	3367(br)	3290(br)	1701(s)	-	1620(m)	3070(w)	2847(w)	1450(m) 1342(m)	1246(s)	1030(m)
C1	3452(br)	3390(br)	1705(s)	-	1620(m)	3080(w)	2850(w)	1446(m) 1338(m)	1242(s)	1118(s)
C2	3452(S)	3190(m)	1666(s)	-	1593(m)	3075(w)	2974(w)	1458(w) 1334(w)	1238(s)	1118(s)
L2	3437(m)	3163(m)	1670(s)	2080(s)	1531(m)	3090(w)	2939(w)	1458(w) 1346(w)	1238(s)	1130(m)
C4	3452(S)	3136(br)	1666(s)	2180(m)	1593(m)	3070(w)	2978(w)	1462(w) 1334(w)	1234(s)	1118(s)
C5	3406(br)	3171(br)	1732(m)	2066(m)	1589(m)	3080(w)	2850(w)	1404(w) 1303(w)	1238(s)	1126(s)

s: strong, m: medium, w: weak, br : broad

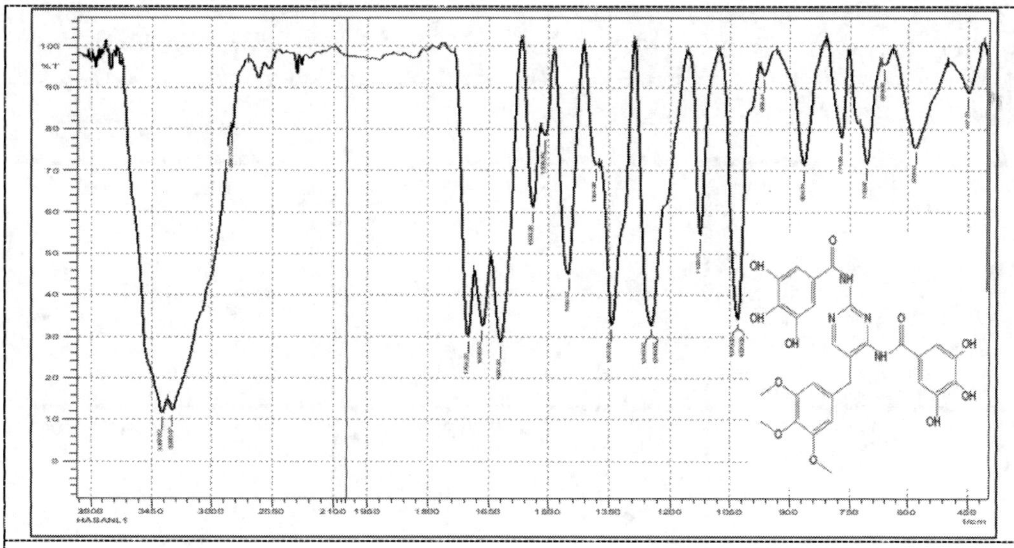

Figure 6. IR spectrum L1.

Journal of Physics: Conference Series **2063** (2021) 012014 doi:10.1088/1742-6596/2063/1/012014

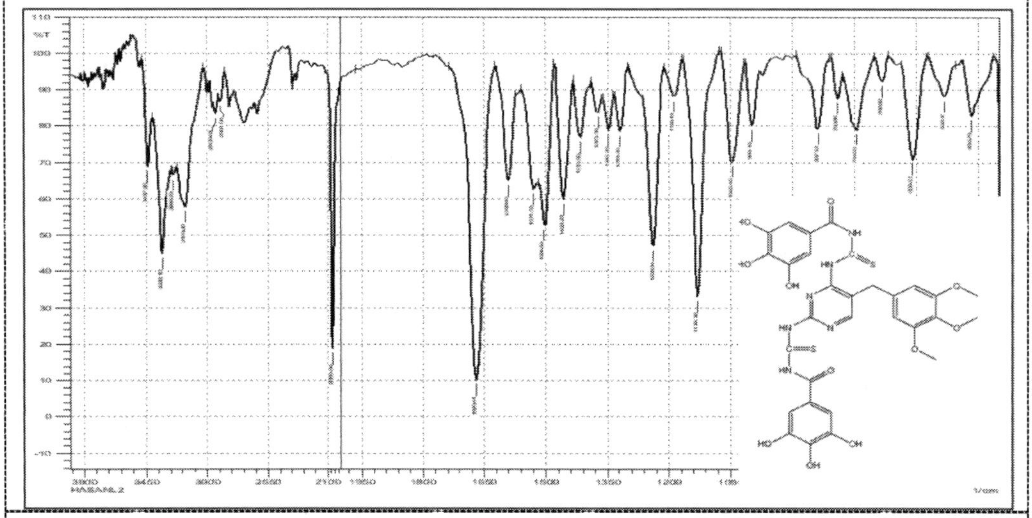

Figure 7. IR spectrum L2.

The UV-visible spectra of the ligands and their complexes have been studied (figure 8) at 1x10-3 M aqueous solution at range 200-800 nm by using quartz cell. The UV-vis. Spectrum of the ligand L1 showed four absorption region at 245 nm, 290 nm, 360 nm and 405 nm while the ligand L2 showed two absorption at the region at 210 nm and 270 nm which may be all these attributed to n→π* transition of the aromatic heterocyclic group of the Trimethoprim molecule .

The Electronic transitions for UV-visible spectra of the Ferric(lll) and Copper(ll) complexes have the spectrum of the electronic range (200-400)nm which attributed to n→π * while the spectrum of the electronic at range (400-800)nm in the Ferric(lll) which attributed to transition type $^{2}T_{2}g \rightarrow {}^{2}Eg$ and the Copper(ll) complexes which attributed to transition type $T_2 \rightarrow E$ [14] [15].

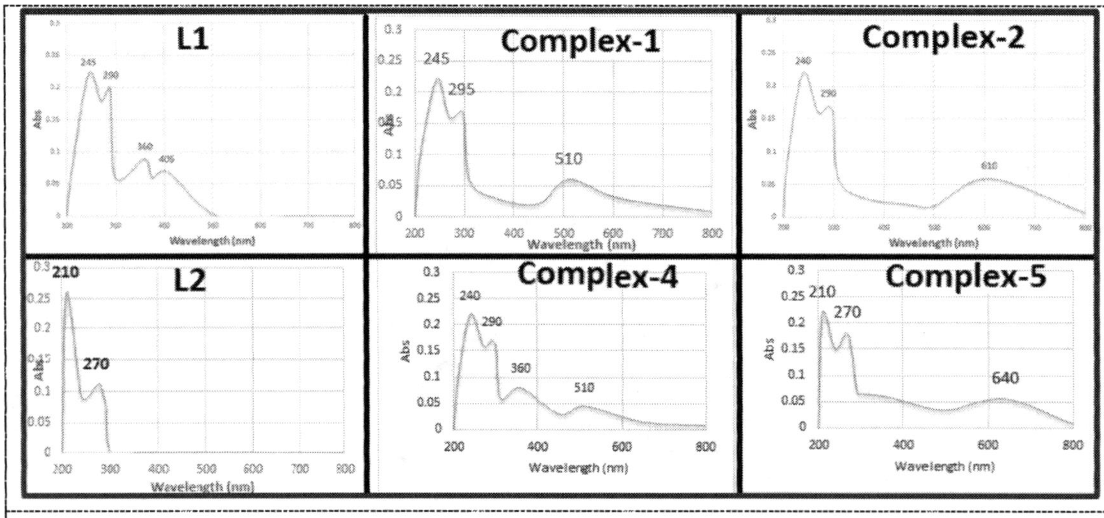

Figure 8. UV-vis. Spectrum of Prepared Compounds.

The ^{1}H-NMR ,^{13}CNMR of the prepared Ligands showed in figure(9,10,11) DMSO-d6 used as solvent . The spectrum of ^{1}H,^{13}C-NMR of the Ligand (L1) the signals showed as expected and can be given as follows:-

Journal of Physics: Conference Series **2063** (2021) 012014 doi:10.1088/1742-6596/2063/1/012014

The ^{1}H-NMR spectra of L2 shows a singlet signal at δ 3.58 ppm which attributed to two protons of methylene group (CH_2). Also appear a singlet signal at δ 3.64-3.75 ppm which attributed to OCH_3 while the signal appeared at δ 6.60 ppm assigned to NH group . Also the Ligand L1 showed a multiple signal which assigned to aromatic range at δ 6.76 – 6.94 ppm. The signal that appeared at the rangs δ 7.21 – 7.53 ppm can be attributed to phenolic group (OH) .The ^{1}H-NMR spectra of L2 shows of a single signal at δ 3.56 ppm which attributed to two protons of methylene group (CH_2).Also the spectrum showed a single signal at the range δ 3.63 – 3.74 ppm that assigned to nine protons of methoxy group (OCH_3) . The compounds is characterized by showing abroad signal at δ 4.93 ppm which can be assigned to amino group (NH). Furthermore the multiple signals that appear at δ 6.94-7.52 ppm can be attributed to aromatic rings of this Ligand. In addition the phenolic groups OH were observed as single signals that appeared at the range δ 6.23 – 6.58 ppm .[11][12][16].

Similarly, the 13C-NMR spectra of the ligands L1 and L2 the Ligand first L1shows the single signal at δ 33.42 ppm which attributed to methylene group and the signal at the range δ 56.28 – 60.44 ppm is due to the methoxy groups (OCH3) and signals at the range δ 138.34,145.88,153.18 and 162.81 ppm which attributed to C-OCH3 , C-OH , C-N and C=N respectively . Also the signal of aromatic carbon of these Ligands observed at the range δ 106.32 -136.22 ppm . Additionally the signal at carbonyl group C=O of these compounds observed at δ168.30 ppm . while the Ligand second L2 which confirms the composition of these Ligands shows a singlet signal at δ 32.60 ppm which attributed of methylene group (CH2) and the signal at the range δ 56.38 – 60.45 ppm is due to the methoxy groups (OCH3) and signals at the range δ 130.09,133.43,136.65 and 140.28 ppm which attributed to C-OH , C-OCH3 , C-N and C=N respectively . Also the signal of aromatic carbon of these Ligands observed at the range δ 106 -136 ppm . Additionally the signal at carbonyl group C=O of these compounds observed at δ153.39 ppm While this ligand differed from the ligand L1shows the single signal at δ 164.52 ppm which attributed to (C=S) group .[17][18][12].

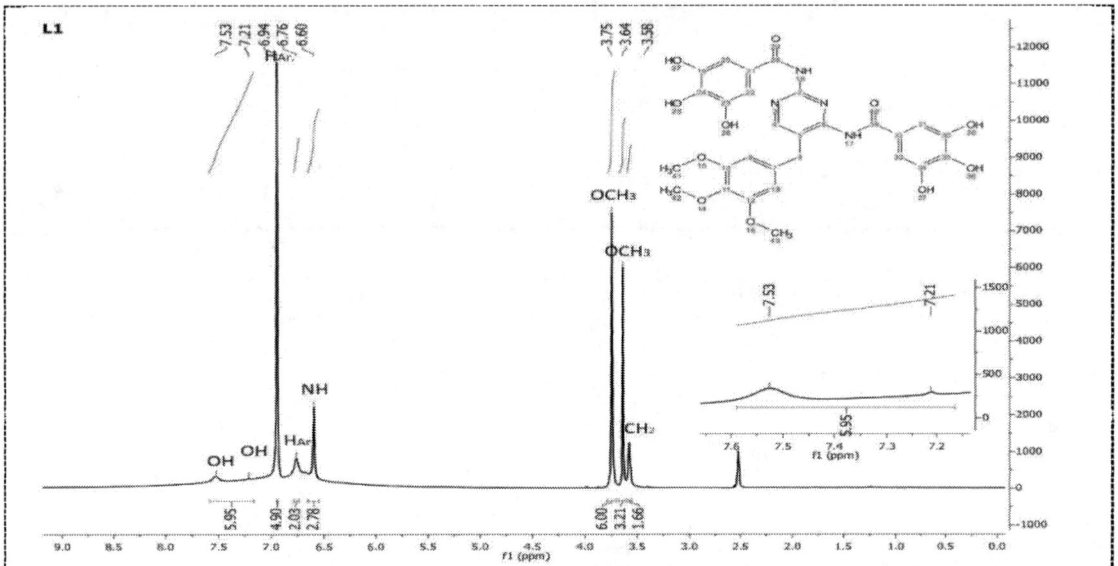

Figure 9. ^{1}HNMR spectrum L1.

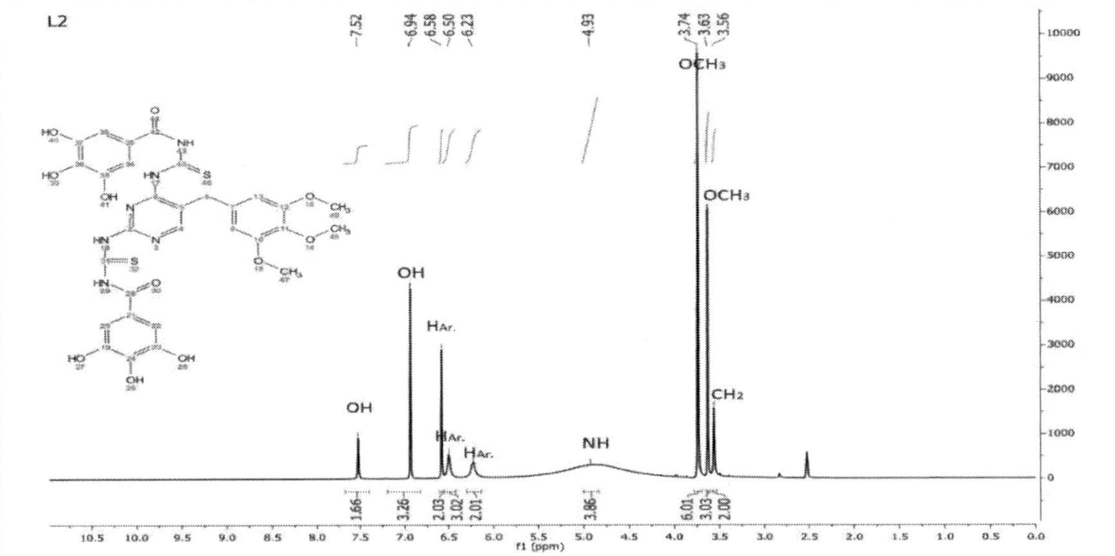

Figure 10. ^{1}HNMR spectrum L2.

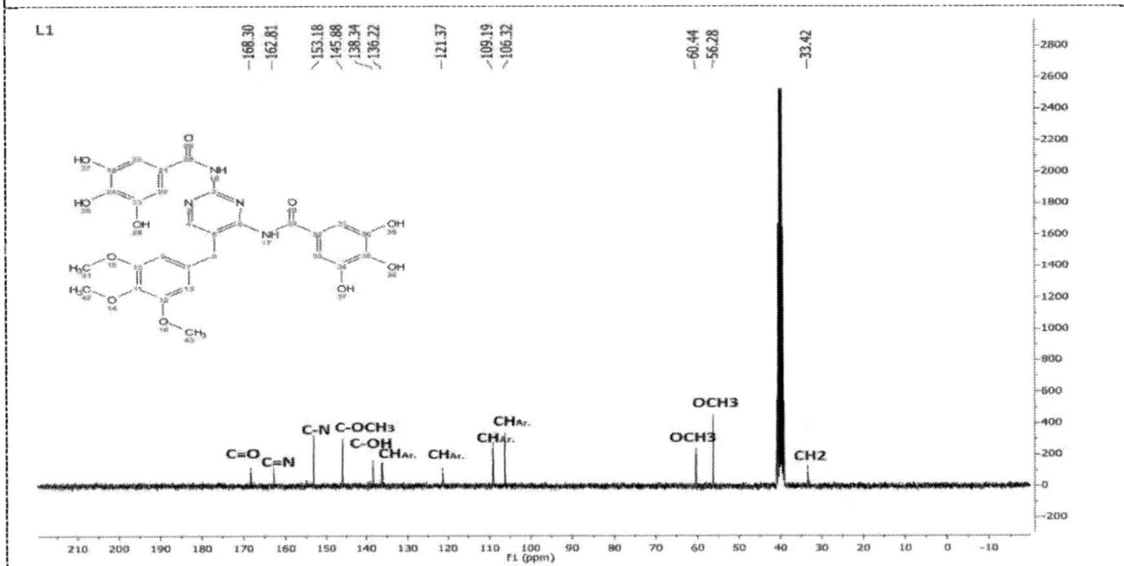

Figure 11. ^{13}C NMR spectrum L1.

El-Mass: The mass spectrum of ligand(L1) Figure 12 . show the exact molecular ion at m/z 594 with relative low abundance which indicate the condensation between Trimethoprim and Gallic acid has taken place resulting into the formation of this derivative with the rate (2:1) which agreement with the formula that suggested. Also the spectrum show peak at m/z 288 attributed to base peak [$C_{14}H_{16}N_4O_3$] $^+$, while the peaks which appear at m/z 275, m/z 259 and m/z 243 with relative abundance 76% , 89% and 51% can be attributed to the ion $[C_{13}H_{15}N_4O_3]^+$, $[C_{12}H_{11}N_4O_3]^+$ and $[C_{11}H_7N_4O_3]^+$ respectively ,while The mass spectrum of ligand(L2) figure (13) shows the exact molecular ion at m/z 712 with relative low abundance which indicate the reaction between Trimethoprim and the product who would be from interacting of gallic chlorid with ammonium thiocyanate resulting into the formation of this derivative with the ratio (2:1) agreement with the formula that suggested. Also the spectrum show peak at m/z 288 that attributed to base peak

[$C_{14}H_{16}N_4O_3$] $^+$, while the peaks which appear at m/z 58, m/z 259 and m/z 275 with relative abundance 20% , 54% and 41% can be attributed to the ion $[CHNS]^+$, $[C_{13}H_{15}N_4O]^+$ and $[C_{13}H_{15}N_4O_3]^+$ respectively .

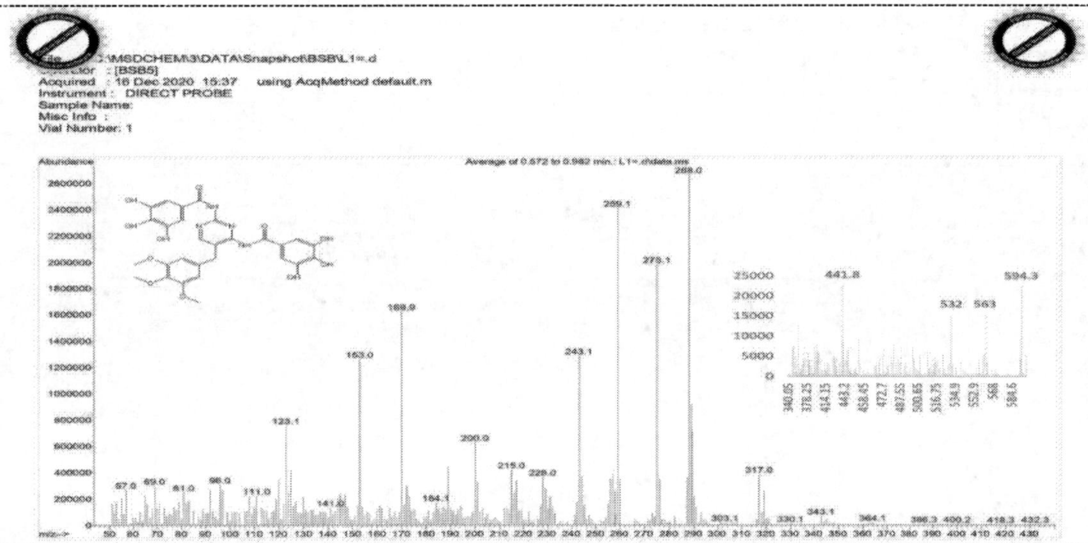

Figure 12. Mass spectrum L1.

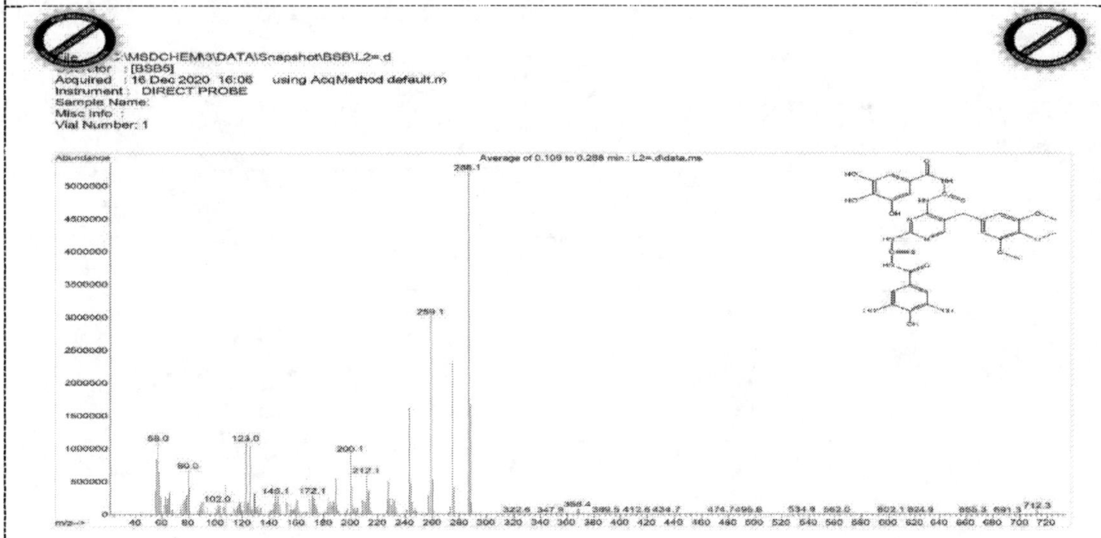

Figure 13. Mass spectrum L2.

Thermal analysis : The thermal investigation was carried out from 80 – 600 °C under nitrogen atmosphere (20 ml/min) with heating rate 10 °C /min . The thermos gram of the Ferric lll complex and Cu (ll) complex[19],[20]]. The TGA curves for complex C1 (figure 14) indicates three stages of weight loss from a temperature of 100 – 360 °C with a loss of approximately 67.19%, the first stage at a temperature of 80 - 150 °C with a loss of 6.068% in practice (theoretically 7.10%) Which is attributed to the loss of four water molecules that are symmetrically bound with the metal [21], the second phase at a temperature of 270- 150 °C with a loss rate of 41.2% in practice (theoretically 41.09%) attributable to the loss of $C_{18}H_{13}N_4O_3$ while the third phase is at a degree Heat 350-270 °C with a loss rate of

19.93% practically (18.50%) which is attributed to the loss 6 molecules of NO_3 [22]. While the TGA curves for complex C2 (figure 15) indicates three stages of weight loss from a temperature of 80-600 °C with a loss of approximately 60.84, the first stage at a temperature of 80-100 °C with a loss of 5.68% in practice (theoretical%) (4.27), which is attributed to the loss of two crystallized water molecules, the second stage at a temperature of 250 - 350 °C with a loss of 34.40% in practice (theoretically 34.19%) that is attributed to the loss of $C_{14}H_{18}N_4O_3$, while the third stage at a temperature of 550-370 °C , a loss of 20.76% in practice (%) 21.34) which is attributed to the loss of $C_{10}H_{13}O_3$.

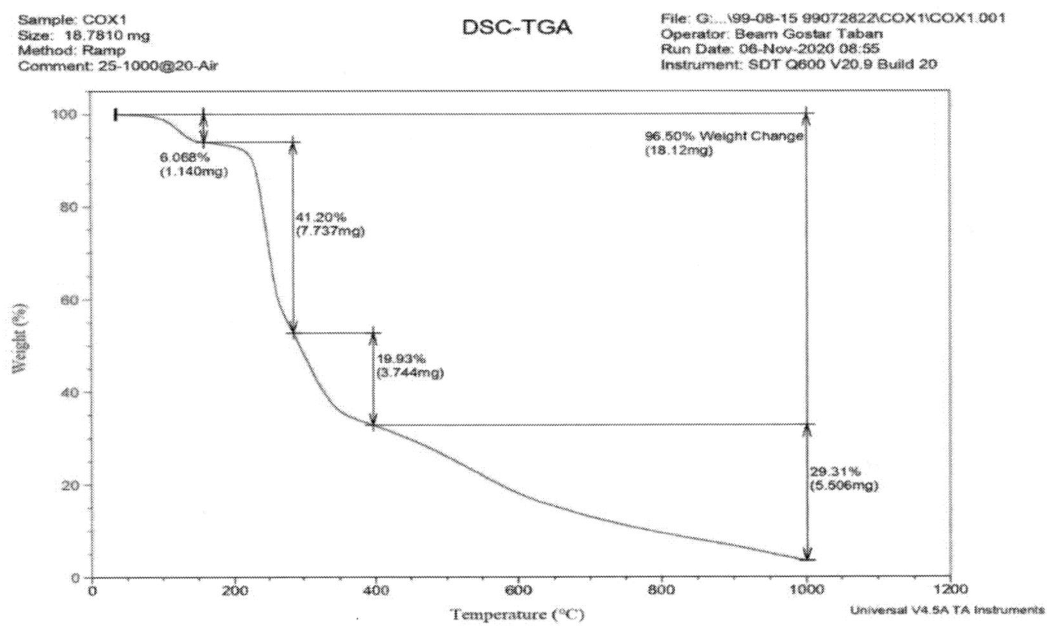

Figure 14. DSC-TGA spectrum C1.

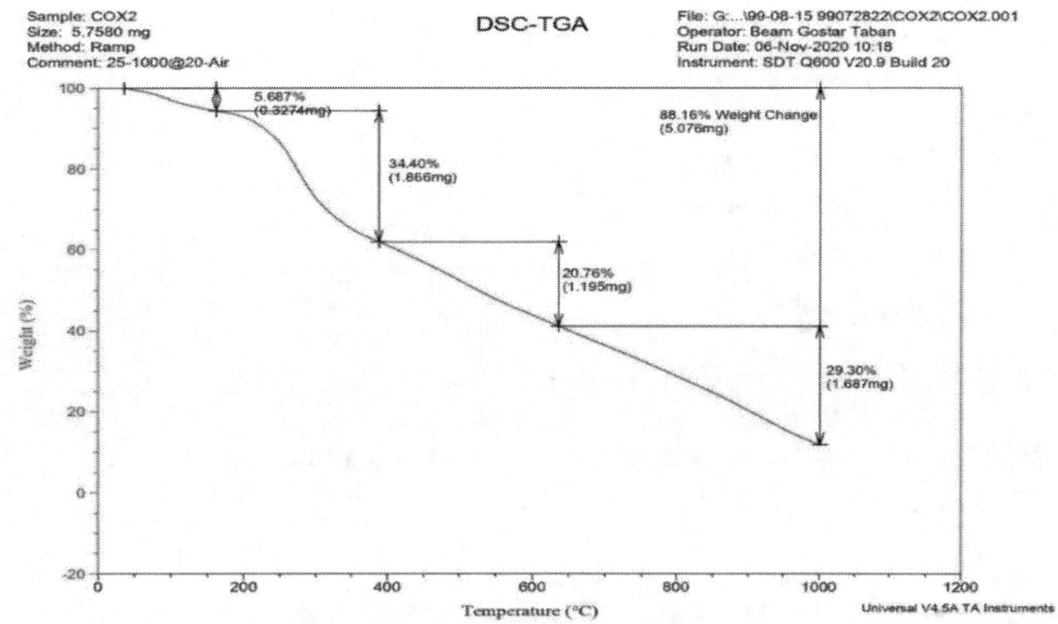

Figure 15. DSC-TGA spectrum C2.

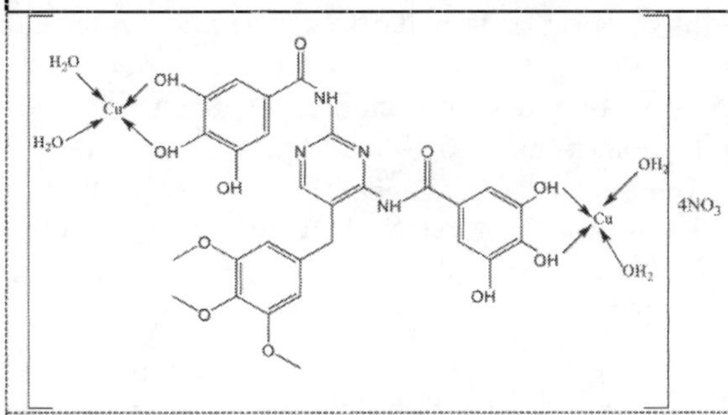

Figure 16. Suggested structure of (C1) $[Cu_2L1(H_2O)_4](NO_3)_4$.

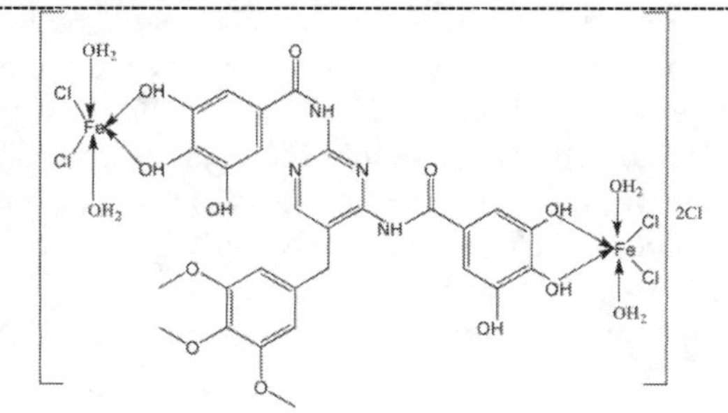

Figure 17. Suggested structure [5] of (C2) $[Fe_2L1(Cl)_4](Cl)_2.2H_2O$.

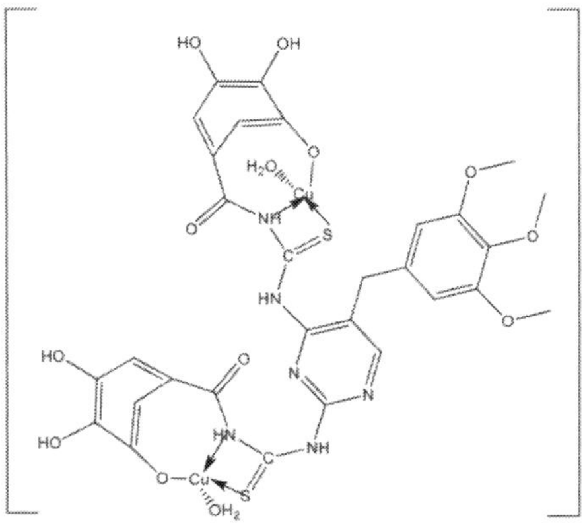

Figure 18 . Suggested structure [6] of (C4) $[Cu_2L2(H_2O)_2](NO_3)_2$.

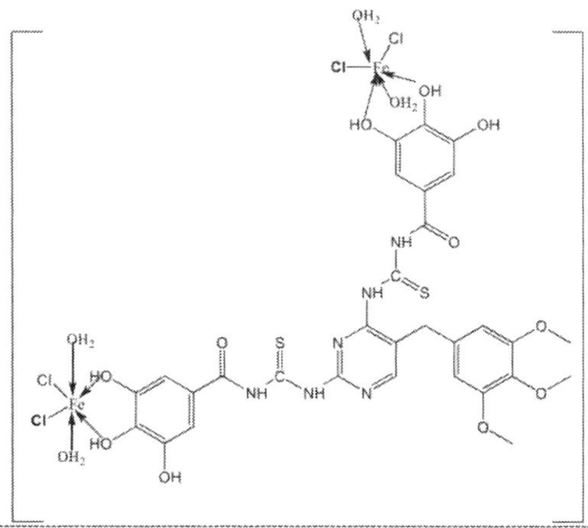

Figure 19. Suggested structure of (C5) $[Fe_2L2(Cl)_4(H_2O)_4](Cl)_2$.

The study also showed the biological effectiveness of the prepared compounds compared to the starting material against two types of bacteria As shown in the Table 4 and Figure20.The results indicate that the solvent has no inhibitory activity towards the studies of bacteria[23],[24], which confirms that the measured inhibition diameters are due to the activity of the prepared compounds. All compounds have biological efficiency for both sexes of bacteria. It is noted from the results that the compounds (L1,L2,C1,C2,C4 and C5) have a greater inhibitory activity compared to the compound trimethrim on both sexes of bacteria because the presence of a double membrane surrounding each bacterial cell. Although all bacteria have a membrane The inner cell, and Gram-negative bacteria, have a unique outer membrane. This outer membrane excludes some Cell-penetrating medicines and antibiotics [25] but the C1 haven't inhibitory activity towards bacteria *Staphylococcus app* for all weights . A table (4) shows the results of the biological effectiveness of the prepared compounds and their comparison with the anti-trimethprim (Contr.) [26],[27].

Table 4. In vitro antimicrobial activity of prepared compounds

	The diameter of the inhibition zone for negative bacteria is in mm *Salmonella* Weight			Contr.In millimeters	The diameter of the inhibition zone for cationic bacteria is in mm *Staphylococcus* Weight			Contr.In millimeters
	5μg	4μg	3μg		5μg	4μg	3μg	
DMSO	0	0	0	0	0	0	0	0
L1	35	26	25	23	30	28	27	25
L2	23	23	21	26	20	16	16	20
C1	25	20	20	26	0	0	0	20
C2	31	30	29	26	28	20	20	20
C4	26	24	22	26	20	20	18	20

C5	20	20	22	26	21	20	20	20

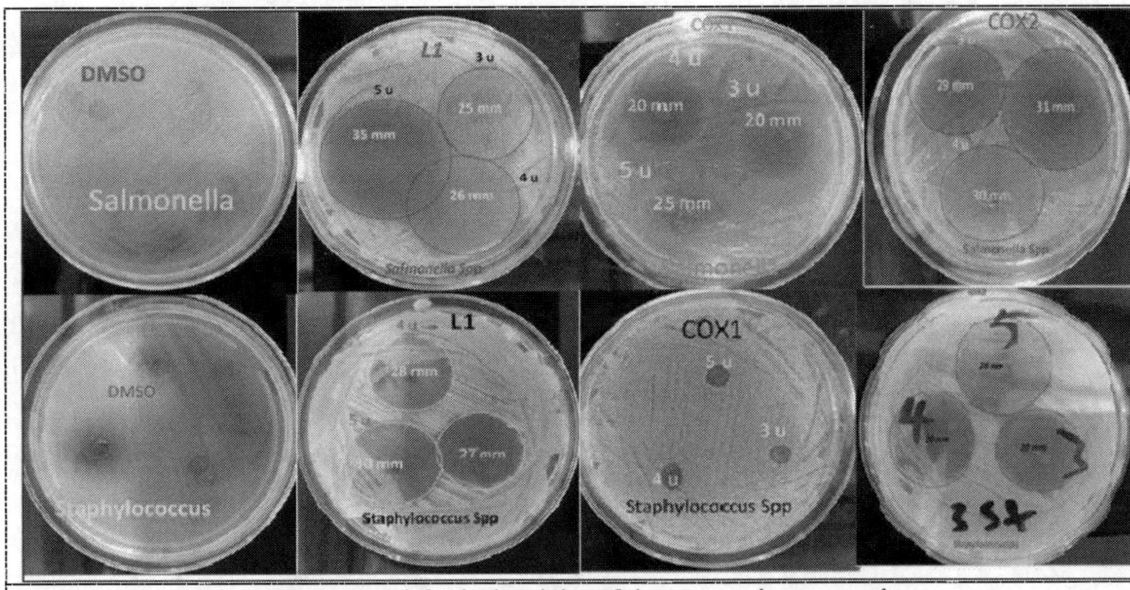

Figure 20. Biological activity of the prepared compounds.

The cytotoxicity of all compounds and complexes prepared using human blood were studied according. Different concentrations (5, 4, 3) µg / ml were prepared in DMSO as a solvent for all the prepared compounds . The blood solution was prepared by mixing (1 ml) of human blood with (20 ml) of normal saline and their cytotoxicity was measured for periods of time (60, 30, 15 minutes), and observing what happens in terms of turbidity or the appearance of sediment, it is evidence of toxicity to the prepared compounds. while no observing any chang of all compounds and concentrations .It isn't toxic compounds .[28]

5.Conclusions

Many measurements were made on the compounds prepared in this research, and through the analysis of the results, this study proved the expected formula for it, as the results of the analyzes of the infrared spectrum, the nuclear magnetic resonance spectrum and the mass spectrum of the prepared compounds were made and compared with the chemistry literature in conformity with what was planned in preparing these Compounds . which are part of a series of compounds prepared in a lengthy study of a group of compounds prepared and will be published later.

The study of thermal analysis of TGA of the prepared complex, which is part of an integrated study of the complexes of the prepared complexes of TMP derivatives, also proved that it is identical with the structural formula of the complex prepared in the study.

References

[1] Nasiri, M. I., Naqvi, S. B. S., Zaidi, A. A., Saeed, R., & Raza, G. (2013). Comparative study on resistant pattern of clinical isolates against Levofloxacin and Cefepime. Pakistan journal of pharmaceutical sciences, 26(2).

[2] Williams, M. F., Doss, E. P., & Montgomery, M. (2017). Possible Trimethoprim–Sulfamethoxazole-Induced Hemolytic Anemia: A Case Report. Journal of pharmacy practice, 30(6), 653-657.

[3]Figueiras, A., Cardoso, O., Veiga, F., de Carvalho, R. B., & Ballaro, G. (2015). Preparation and

characterization of Trimethoprim inclusion complex with Methyl-β-Cyclodextrin and determination of its antimicrobial activity. Pharm Anal Acta, 6, 405-409.

[4] Vilchèze, C., & Jacobs, W. R. (2012). The combination of sulfamethoxazole, trimethoprim, and isoniazid or rifampin is bactericidal and prevents the emergence of drug resistance in Mycobacterium tuberculosis. Antimicrobial agents and chemotherapy, 56(10), 5142-5148.

[5] Osowole, A. A., Wakil, S. M., & Alao, O. K. (2015). Synthesis, characterization and antimicrobial activity of some mixed Trimethoprim-Sulfamethoxazole metal drug complexes. World Applied Sciences Journal, 33(2), 336-342.

[6] Poonam, I., Shreya, M., & Varun, H. (2019). DNA Interaction Studies by UV of Copper Complex based on Gallic Acid. Journal of Innovation in Pharmaceutical Sciences, 3(2), 6-8.

[7] Fazary, A. E., Taha, M., & Ju, Y. H. (2009). Iron complexation studies of gallic acid. Journal of Chemical & Engineering Data, 54(1), 35-42.

[8] Somwanshi, J. L. (2020). Synthesis and characterization of bioactive chef bases derived from Trimethoprim. Journal of Global Biosciences, 9(4), 6963-6973.

[9] Zhang, X., Jin, C., Jiang, Y., Liu, G., Wu, G., & Kong, Z. (2017). A novel gallic acid-grafted-lignin biosorbent for the selective removal of lead ions from aqueous solutions. BioResources, 12(3), 5343-5357.

[10] Al-Diksin, A. A., & Al-Amood, H. K. (2015). Synthesis and Biological Activity Study of Some New Pteridine Derivatives. RESEARCH JOURNAL OF PHARMACEUTICAL BIOLOGICAL AND CHEMICAL SCIENCES, 6(6).

[11] ElShaer, A., Hanson, P., Worthington, T., Lambert, P., & Mohammed, A. R. (2012). Preparation and characterization of amino acids-based trimethoprim salts. Pharmaceutics, 4(1), 179-196.

[12] Pike, E. R. (1960). Introduction to soft x-ray spectroscopy. American Journal of Physics, 28(3), 235-242.

[13] Sridharan, K. (2016). Spectral Methods in Transition Metal Complexes. Elsevier.

[14] Onda, K., Yamochi, H., & Koshihara, S. Y. (2014). Diverse photoinduced dynamics in an organic charge-transfer complex having strong electron–Phonon interactions. Accounts of chemical research, 47(12), 3494-3503.

[15] Housecroft, C. E., & Sharpe, A. G. (2006). Anorganische Chemie. Pearson-Studium.

[16] Cho, Y. S., Kim, S. K., Ahn, C. B., & Je, J. Y. (2011). Preparation, characterization, and antioxidant properties of gallic acid-grafted-chitosans. Carbohydrate Polymers, 83(4), 1617-1622.

[17] Abdalsahib, N. M., Numan, A. T., & Atiyah, E. M. (2009). Preparation, Characterization and Biological Activity Studies of Some Metal Complexes with New Schiff Base for Trimethoprim Ligand.

[18] Romer, F. H., Underwood, A. P., Senekal, N. D., Bonnet, S. L., Duer, M. J., Reid, D. G., & Van der Westhuizen, J. H. (2011). Tannin fingerprinting in vegetable tanned leather by solid state NMR spectroscopy and comparison with leathers tanned by other processes. Molecules, 16(2), 1240-1252.

[19] Guo, Z., Xu, L., Liu, C., Sun, F., Kang, Y., & Liang, S. (2016). Comparison of physicochemical properties of activated carbons derived from biomass wastes by H4P2O7 activation: adsorption of trimethoprim. Desalination and Water Treatment, 57(46), 21957-21967.

[20] Chicot, D., Mendoza, J., Zaoui, A., Louis, G., Lepingle, V., Roudet, F., & Lesage, J. (2011). Mechanical properties of magnetite (Fe3O4), hematite (α-Fe2O3) and goethite (α-FeO· OH) by instrumented indentation and molecular dynamics analysis. Materials Chemistry and Physics, 129(3), 862-870.

[21] Al Zoubi, W., Al-Hamdani, A. A. S., Ahmed, S. D., & Ko, Y. G. (2018). A new azo-Schiff base: Synthesis, characterization, biological activity and theoretical studies of its complexes. Applied Organometallic Chemistry, 32(1), e3895.

[22] Prabhumirashi, L. S., & Khoje, J. K. (2002). TGA and DTA studies on en and tmn complexes of Cu (II) chloride, nitrate, sulphate, acetate and oxalate. Thermochimica acta, 383(1-2), 109-118.

[23] Beier, R. C., Pillai, S. D., Phillips, T. D., & Ziprin, R. L. (Eds.). (2008). Preharvest and postharvest food safety: contemporary issues and future directions (Vol. 16). John Wiley & Sons.
[24] Strauss, M., Zoabi, K., Sagas, D., Reznik-Gitlitz, B., & Colodner, R. (2020). Evaluation of Bio-Rad® discs for antimicrobial susceptibility testing by disc diffusion and the ADAGIO™ system for the automatic reading and interpretation of results. European Journal of Clinical Microbiology & Infectious Diseases, 39(2), 375-384.
[25] Faure, L. M., Fiche, J. B., Espinosa, L., Ducret, A., Anantharaman, V., Luciano, J., ... & Mignot, T. (2016). The mechanism of force transmission at bacterial focal adhesion complexes. Nature, 539(7630), 530-535.
[26] Choubey, S., Varughese, L. R., Kumar, V., & Beniwal, V. (2015). Medicinal importance of gallic acid and its ester derivatives: a patent review. Pharmaceutical patent analyst, 4(4), 305-315.
[27] Nilchan, N., Phetsang, W., Nowwarat, T., Chaturongakul, S., & Jiarpinitnun, C. (2018). Halogenated trimethoprim derivatives as multidrug-resistant Staphylococcus aureus therapeutics. Bioorganic & medicinal chemistry, 26(19), 5343-5348.
[28] Nair, M. G., Putnam, A. R., Mishra, S. K., Mulks, M. H., Taft, W. H., Keller, J. E., ... & Lynn, D. G. (1989). Faeriefungin: a New board-spectrum antibiotic from Streptomyces griseus var. autotrophicus. Journal of natural products, 52(4), 797-809.

Journal of Physics: Conference Series **2063** (2021) 012015 doi:10.1088/1742-6596/2063/1/012015

Synthesis of Composite from Bacterial Cellulose and Gold Nanoparticles

Saad Ali Ahmed, [2]Alaa Abdulnabi Ahmed and [1]Mohamed Hassan Abdul Latif
[1]University of Baghdad, College Ibn Al Haitham for Pure Science, Iraq
[2] Ministry of Education, Iraq
*Corresponding author E-mail:Saad.a.a@ihcoedu.uobaghdad.edu.iq

Abstract. A process of bacterial cellulose gold nanocomposite has been investigated based on experimental work and cited literature. A literature review on the production process is carried out in this study. Bacterial cellulose is a high crystalline fabric material generally used in biomedical applications. A Nanocomposite was made by synthesis from gold and bacterial cellulose. The experimental work includes growing, and isolating bacterial cellulose, preparation of gold Nanoparticles and preparation of Nano composite. Nanoparticle's formation and adsorption on the cellulose tissue have been observed visually, where a colour change was observed. The predicted particle size for the gold nanoparticles was (2-100) nm.

1. Introduction

Bacterial cellulose (BC) was invented by Brown in 1886, he found a creation of material in a glucose solution created by a special kind of bacteria " Acetobacter xylinum" whose structure is similar to that of cellulose [1]. Chemical formula $(C_6H_{10}O_5)_n$. Nnaturally, the bacteria synthesize extracellular polysaccharides. The chemical structure is shown in Figure 1.

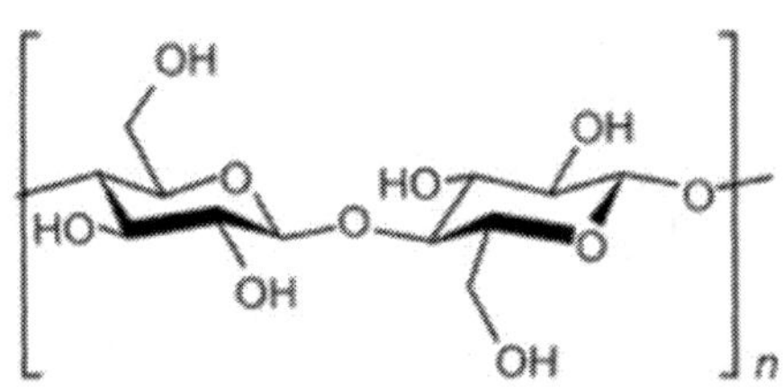

Figure 1. Chemical structure of Bacterial cellulose [2].

BC built from linear glucan molecules attached with hydrogen bonds seems similar to plant cellulose [3]. BC has been characterized by high porosity. BC is a nanomaterial that appears as a gel. Its particle size (10 -100 nm), so it has a capability for water retention [4]. (See Figure 2)
Drug-resistant bacterial infections have serious problems accompanied by wound healing. Bacterial cellulose decorated by organic carrying materials and modified gold nanoparticles were discovered as a dressing nanocomposite for treating bacterially infected wounds. Gold (AuNP) nanoparticles process

unique optical, antibacterial, and catalytic properties, usually in situ synthesis of these NPs in BC for the fabrication of functional materials [5].

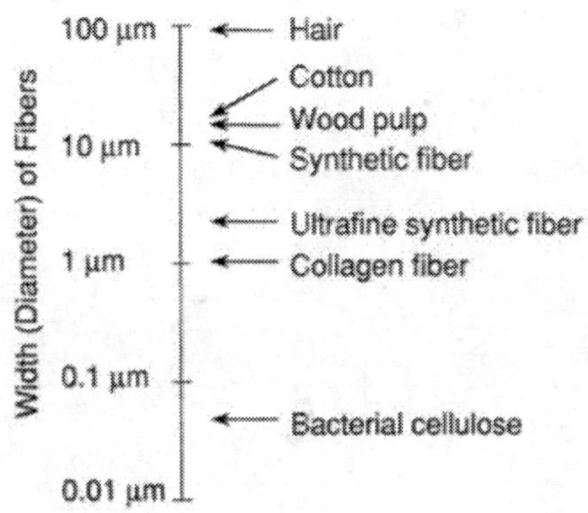

Figure 2. Sizes of synthetic and naturally occurring fibers [6].

1.1 BC Applications

Table 1. Applications of Bacterial Cellulose

Application	Example	Refs.
Biomedical	Wound dressing material, Wound healing, BC-based materials for skin tissue repair dental implants	[7] [8][9]
cosmetology	Emulsion stabilizer.	[10]
Fabrics	Textiles.	[11]
Paper industry	Durable Papers.	[12]
Membranes	Ultrafiltration equipment	[13][14]
Food	Thickening and gelling, stabilizing, water-binding, and as a packing material	[15]
Acoustics	Membrane speaker	[16]
Science	Protein immobilization, culture medium	[17]
Catalysts	Catalyst support	[18] [19].

1.2 *BC Preparation and Production*

The production process faces many important restrictions like low yield, biochemical reaction time, the material of construction –etc. all of these are considered as reasons which limit the construction of industrial processing of BC when evaluated from an economic view. Many studies were conducted to overcome the limitations. Optimization, evaluation of substrate mass transfer on the biochemical reaction [20], and effect of reaction container wall on surface culture [21]. Research and development

Journal of Physics: Conference Series **2063** (2021) 012015 doi:10.1088/1742-6596/2063/1/012015

on reactors and processes continued. Cellulose production depends heavily on several factors such as the growth medium, environmental conditions, and the formation of byproducts. The best carbon sources in the fermentation medium which give the highest yield is glucose, followed by fructose, sucrose, and ethanol [22]. The problem with using glucose is the formation of gluconic acid side product, which lowers the pH of the culture and reduces the cellulose product , this problem was solved by the addition of (1%, w/v) lignosulfonate[23], or (2%) acetic acid [24]. Selection of reactor types should be conducted on pilot results. Several reactor types were reported for the biochemical fermentation reaction (see Table 2)

Table 2. Reactors used for Synthesize of Bacterial Cellulose

Reactor	C g/L	Time	Ref.
Classic reactors	1.69-2.5	12-20 day	[25]
Internal loop air lift reactor	3.8 - 8.7		[26][27]
Rotating disk reactor*1	-----------	84 h	[28][29]
Agar surface modified reactors	5.472	140h	[30]
Biofilm reactor	7.05	120h	[29]
STR +Spin filter*2	5.65	140h, pH 5	[31]

*1 Horizontal fermenter equipped with rotating discs or rollers rotated at 4 rpm.

*2 Fermentations are equipped with a turbine impeller and a spin filter.

It's important to review preparation methods to assess the process design requirements and process flow diagram. For economic requirements, researchers aim is to reduce the cost of products. using Acetobacter xylinum. The challenge for minimizing cost through changing raw materials such as low-quality date syrup in Static batch fermentation method [25], or equipment design as shown in Table 2. It was reported for the process of cellulose formation from sugar. It was found that the production process was controlled by the diffusion of atmospheric oxygen [32].

1.3 *Preparation of Nano Gold*

Gold nanoparticles (AuNPs) have been widely used in biotechnology due to their unique properties and multiple surface functionalities. [33] [34] [35]. One of the synthetic methods for producing AuNPs, mixing ($HAuCl_4$) with citric acid in boiling water, where the citrate acts as both a reducing and stabilizing agent [36]. particle size controlled by gold to citrate ratio [37].

This protocol has been used to synthase dilute solutions of moderately stable spherical AuNPs with diameters of (10 - 20 nm) [38].

Different methods for composites fabrication of gold nanoparticles, bacterial cellulose was studied. They looks similar in the principle, such all of them used reducing agents. They differ in number of steps or the type of reducing and linking agents like the one step poly-ethyleneimine as the reducing and linking agent with the addition of different halides [39]. Two steps amidoxime surface- functionalized bacterial cellulose [18]. Radical polymerization method was carried out using a monomer of 2-hydroxyethyl methacrylate to get a uniform distribution of the AuNPs on the membrane surface during the synthesis stage [18]. In situ polymerization of aniline by ammonium persulphate[40].

Journal of Physics: Conference Series **2063** (2021) 012015 doi:10.1088/1742-6596/2063/1/012015

An invention related to drying of BC, called drying under tension, the method integrates the dehydrating and drying the BC produced in agitated culture followed by homogenization. The aim was to restore the properties of BC after drying [41].

Purification of BC can be done by boiling distilled water followed by soaking in NaOH solution for 24 hr. then rinsing with distilled water till reaching a pH of 7 [42].

2. Materials and Methods

2.1 Materials

- Gold (99.99%): local market
- Hydrochloric acid (37%)
- Nitric acid (68-70%)
- Butanedioate, 2-hydroxy-, sodium salt
- Iraqi mother vinegar was brought up from province of Diyala, east of Baghdad.
- Zahdi date, length of date piece (3.4 - 4 cm), diameter 2.4-2.5 cm, core form about 11% its weight, total sugars about 67%. Total dissolved materials 77.8%.

2.2 Methods

2.2.1 Preparation of Zahdi juice:

The date juice was prepared in a method that is similar to the procedure in the cited reference [43]. A 500gm date of cleaned and pitted date, cut into small pieces and leached with distilled water (2:1) at 70°C. The extracted juice was filtered and then concentrated to the required concentrations by evaporation at 70°C. The prepared date juice concentration of 40%.

2.2.2 Preparation of BC

Bacteria were put in water at (10% ~ 15%) concentrations for 10 days. A (2 - 3) drop of Ammonium acetate were added frequently every two days as an activator. the bacterial cellulose was mixed by magnetic stirrer and distributed into sterilized clean pots.

Vinegar bacteria was added with Two drops of ammonium acetate and the stirring continued for five minutes. Then containers were closed and sealed, incubated at a temperature of (30 C°). the cans were opened and fed by date syrup every 4 days frequently within two weeks period. Homogeneous bacterial cellulose was enlarged to three times its volume.

The bacterial cellulose colony was transferred to an open dish. Washed with distilled water three times then bleached by immersion in sodium hypochlorite solution bacterial, the resulted white colour BC was poured in distilled water.

2.2.3 Gold Nanoparticles:

0.5g Gold was melted in a mixture of (6ml HCl (0.1 M) + 2 ml $HNO3$ (0.1 M)) with boiling then dissolved in 500 ml to distilled water to give concentration (1000 g/L).

2.2.4 Preparation of gold Nanocomposite:

Chloroauric acid is prepared by dissolving 0.5 gm. of gold in aqua regia $(4HCl + HNO_3)$

$$Au + HNO_3 + 4\,HCl \rightarrow HAuCl_4 + NO + 2\,H_2O$$

Journal of Physics: Conference Series **2063** (2021) 012015 doi:10.1088/1742-6596/2063/1/012015

In the next step, 8.0 mL of the Chloroauric acid was added to a solution mixture of (10 ml sodium maleic + 10 ml sodium hydroxide) in a test tube fixed in a water bath near boiling point, after that, the mixture was subjected to further heating without stirring. Red colour was achieved after 5min at this stage a piece of bacterial cellulose was added. Colour changes continued after 10min. a brown hue was shown. After 15min. a dirty gray colour appears, and after 20min. aggregated particles were visible. (See Figure 3).

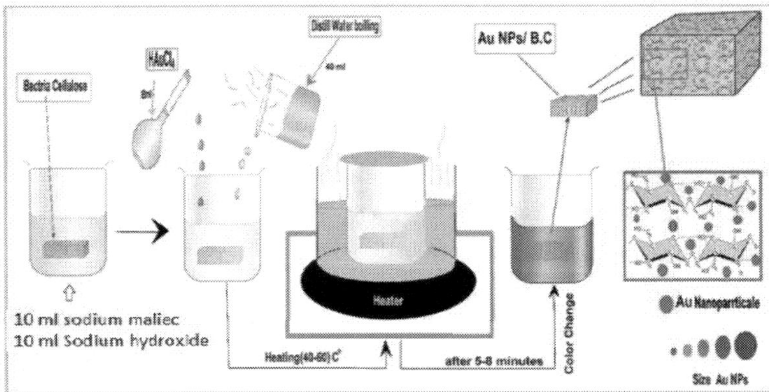

Figure 3. Preparation pieces of gold/bacterial cellulose nanocomposite.

3. Results and Discussion

Bacterial cellulose has high purity due to a lack of lignin and hemicellulose; this property makes bacterial cellulose. Bacterial cellulose is considered as a non-cytotoxic, non-genotoxic, and highly biocompatible material. Bacteria size increased after 10 days. It was observed that the anaerobic fermentation led to the growth of bacterial cellulose is observed as shown in Figure 4.

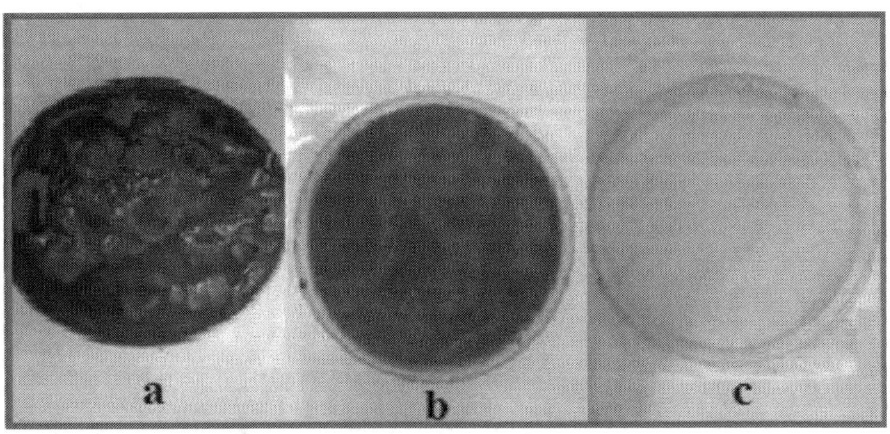

Figure 4. a- Mother of vinegar. b- after 21 days in an incubator at 30 ^{O}C. (c) Bleached bacterial cellulose.

The formation and adsorption of nanoparticles on the surface of cellulose tissue have been observed visually, a colour change was observed. The colour change is attributed to the Plasmon surface effects. Butanedioate, 2-hydroxy-, sodium salt used as reducing agent. (See Figure 5).

Figure 5. Gold/bacterial cellulose nanocomposite.

Particle size for nanomaterials prophesied from absorbance. Moreover, wavelength values. The predicted value for the size of gold nanoparticles, 2-100 nm by maleic It is corresponds to the top absorbance value of 305-623 as shown in (Figure 6).

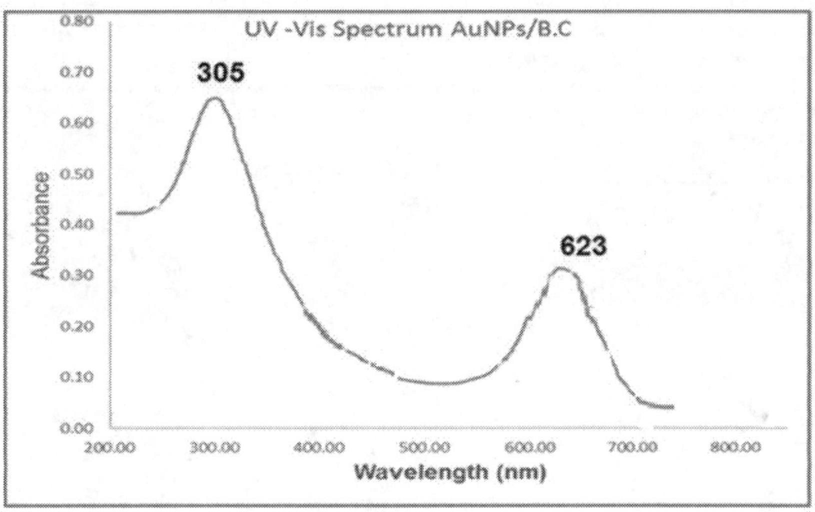

Figure 6. UV-Vis spectrum AuNPs. /B.C.

Morphology analyses illustrated in Figure 7, images show a cellulose tissue composed of porous fibers with a diameter of (0 - 70 nm). A scanning electron microscope explicates the structure of bacterial cellulose, ultrafine fibers are distributed in nano dimensions of infinite precision interrelated in a woven form that helps gold ions to spread in a spongy structure.

Journal of Physics: Conference Series **2063** (2021) 012015 doi:10.1088/1742-6596/2063/1/012015

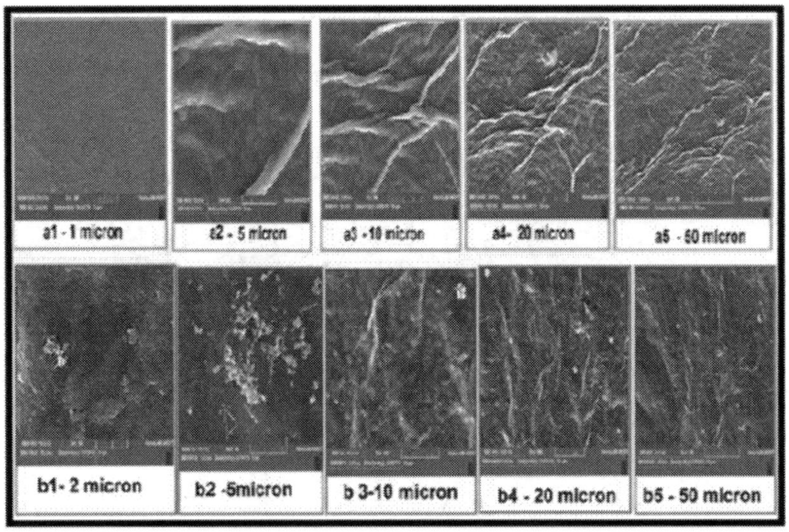

Figure 7. SEM micrographs of the a- pure bacterial cellulose
b- by malic salt (Sodium2- hydroxy Butanedioate).

Atomic force microscope (AFM) Was used to examine the pure bacterial cellulose and the Au NPs/B.C. It used to predict pure cellulose and 3D as shown in Figure 8, it was being adsorbed on the bacterial cellulose tissue that sizes (47nm), (35nm). This indicates successful adsorption of gold nanoparticles on bacterial cellulose tissue.

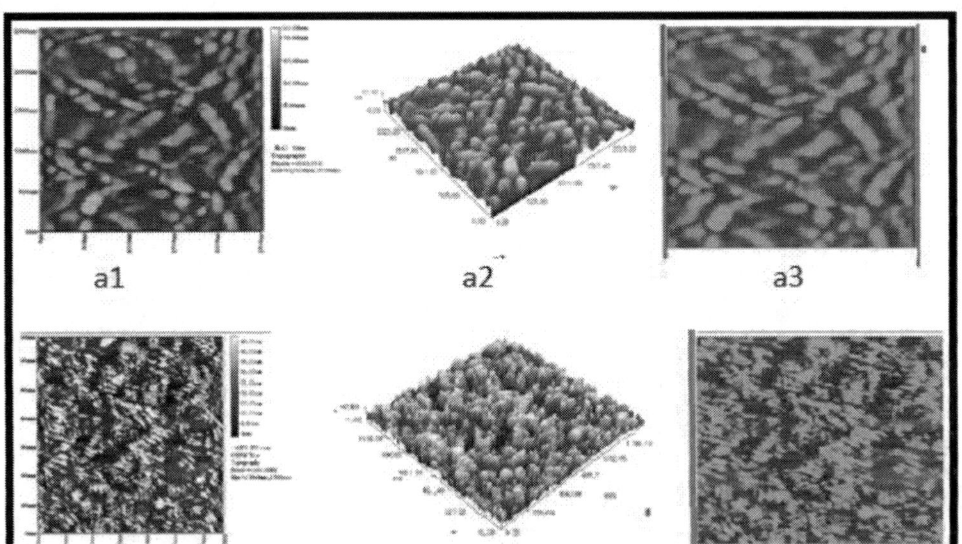

Figure 8. AFM image of bacterial cellulose. 1- 2D 2-3D 3- photo film at 25 °C

3.1 Suggested process

The suggested process is for processing 1ton Zahdi date pilot plant for the production of gold nanoparticles - bacterial nano cellulose.

Journal of Physics: Conference Series **2063** (2021) 012015 doi:10.1088/1742-6596/2063/1/012015

Basis 1 Tons Zahdi date, Sugar 67%.\
Date syrup: for each 1 kg date, 2 L water required to produce syrup 22-25%bx
Recovery of Sugar 60%. A preliminary outline of the suggested process is described by Figure 8-a for the date syrup process and Figure 8-b for bacterial cellulose- Au-NP.

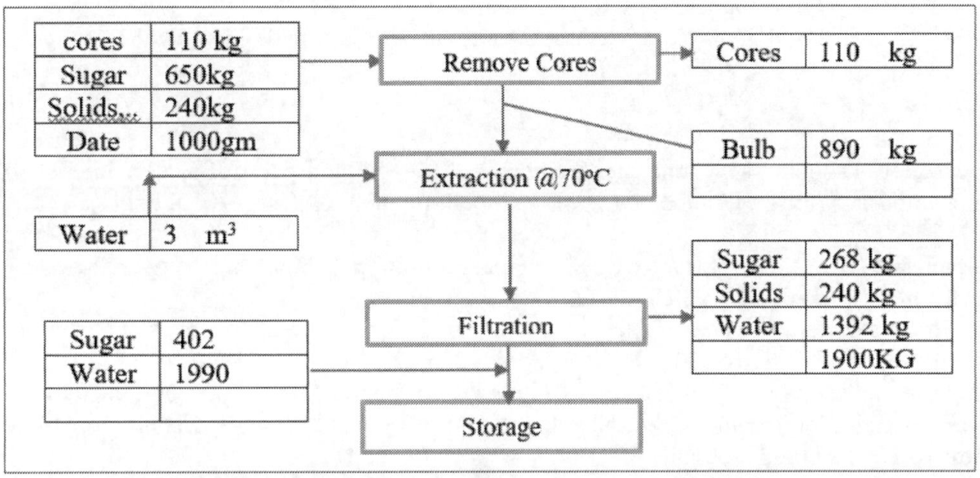

Figure 9(a). Schematic drawing for date syrup block diagram.

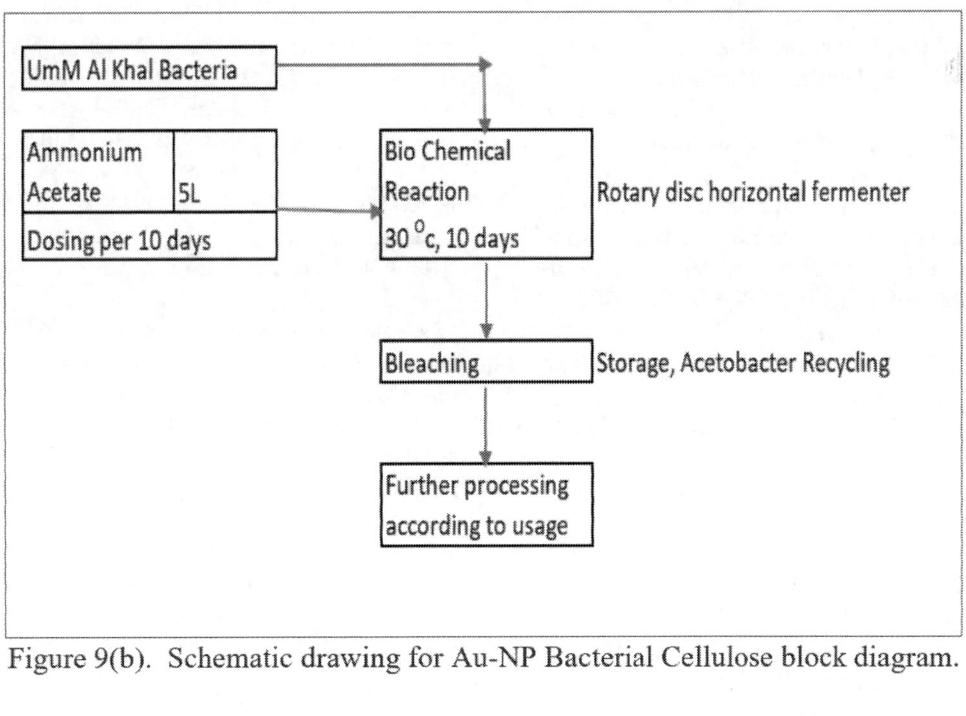

Figure 9(b). Schematic drawing for Au-NP Bacterial Cellulose block diagram.

Journal of Physics: Conference Series **2063** (2021) 012015 doi:10.1088/1742-6596/2063/1/012015

4. Conclusions

- Laboratory results have been proved that the Iraqi mother vinegar is a good bacteria source for synthase of bacterial cellulose.
- Successful adsorption of gold nanoparticles on bacterial cellulose tissue.
- Based on morphology analyses, it is concluded that cellulose tissue is composed of porous fibers with a diameter of 0 to 70 nm size.
- This method gives extra pure nano bacterial cellulose with good mechanical properties

References

[1] P. Jacek, F. Dourado, M. Gama, and S. Bielecki, "Molecular aspects of bacterial nanocellulose biosynthesis," *Microb. Biotechnol.*, vol. 12, no. 4, pp. 633–649, Jul. 2019, doi: 10.1111/1751-7915.13386.

[2] "Bacterial Cellulose." https://www.sciencedirect.com/topics/biochemistry-genetics-and-molecular-biology/bacterial-cellulose.

[3] Y. Huang, C. Zhu, J. Yang, Y. Nie, C. Chen, and D. Sun, "Recent advances in bacterial cellulose," *Cellulose*, vol. 21, Feb. 2014, doi: 10.1007/s10570-013-0088-z.

[4] Y. Li *et al.*, "Composites of Bacterial Cellulose and Small Molecule-Decorated Gold Nanoparticles for Treating Gram-Negative Bacteria-Infected Wounds," *Small*, vol. 13, May 2017, doi: 10.1002/smll.201700130.

[5] S. Li and P. S. Lee, "Development and applications of transparent conductive nanocellulose paper," *Sci. Technol. Adv. Mater.*, vol. 18, no. 1, pp. 620–633, Dec. 2017, doi: 10.1080/14686996.2017.1364976.

[6] F. Yoshinaga, N. Tonouchi, and K. Watanabe, "Research Progress in Production of Bacterial Cellulose by Aeration and Agitation Culture and Its Application as a New Industrial Material," *Biosci. Biotechnol. Biochem.*, vol. 61, no. 2, pp. 219–224, Jan. 1997, doi: 10.1271/bbb.61.219.

[7] L. Fu, J. Zhang, and G. Yang, "Present status and applications of bacterial cellulose-based materials for skin tissue repair," *Carbohydr. Polym.*, vol. 92, pp. 1432–1442, Feb. 2013, doi: 10.1016/j.carbpol.2012.10.071.

[8] T. Carvalho, G. Guedes, F. L. Sousa, C. S. R. Freire, and H. A. Santos, "Latest Advances on Bacterial Cellulose-Based Materials for Wound Healing, Delivery Systems, and Tissue Engineering," *Biotechnol. J.*, vol. 14, no. 12, p. 1900059, Dec. 2019, doi: https://doi.org/10.1002/biot.201900059.

[9] R. Portela, C. R. Leal, P. L. Almeida, and R. G. Sobral, "Bacterial cellulose: a versatile biopolymer for wound dressing applications," *Microb. Biotechnol.*, vol. 12, no. 4, pp. 586–610, Jul. 2019, doi: https://doi.org/10.1111/1751-7915.13392.

[10] A. Stanisławska, "Bacterial Nanocellulose as a Microbiological Derived Nanomaterial," *Adv. Mater. Sci.*, vol. 16, Dec. 2016, doi: 10.1515/adms-2016-0022.

[11] K. Kamiński *et al.*, "Hydrogel bacterial cellulose: a path to improved materials for new eco-friendly textiles," *Cellulose*, vol. 27, Jun. 2020, doi: 10.1007/s10570-020-03128-3.

[12] M. Zhang, X. Wu, Z. Hu, Z. Xiang, T. Song, and F. Lu, "A Highly Efficient and Durable Fluorescent Paper Produced from Bacterial Cellulose/Eu Complex and Cellulosic Fibers," *Nanomaterials (Basel, Switzerland)*, vol. 9, no. 9. State Key Laboratory of Pulp and Paper Engineering, South China University of Technology, Guangzhou 510640, China., 2019, doi: 10.3390/nano9091322.

[13] P. C. S. Faria-Tischer, C. A. R. Costa, I. Tozetti, L. H. Dall'Antonia, and M. Vidotti, "Structure and effects of gold nanoparticles in bacterial cellulose–polyaniline conductive membranes," *RSC Adv.*, vol. 6, no. 12, pp. 9571–9580, 2016.

[14] W. Shao, J. Wu, H. Liu, S. Ye, L. Jiang, and X. Liu, "Novel bioactive surface functionalization of

bacterial cellulose membrane," *Carbohydr. Polym.*, vol. 178, pp. 270–276, 2017.
[15] Z. Shi, Y. Zhang, G. O. Phillips, and G. Yang, "Utilization of bacterial cellulose in food," *Food Hydrocoll.*, vol. 35, pp. 539–545, 2014, doi: https://doi.org/10.1016/j.foodhyd.2013.07.012.
[16] T. Nge, J. Sugiyama, and V. Bulone, "Bacterial Cellulose-based Biomimetic Composites," 2010.
[17] N. Petersen and P. Gatenholm, "Bacterial cellulose-based materials and medical devices: current state and perspectives," *Appl. Microbiol. Biotechnol.*, vol. 91, no. 5, pp. 1277–1286, 2011, doi: 10.1007/s00253-011-3432-y.
[18] M. Chen, H. Kang, Y. Gong, J. Guo, H. Zhang, and R. Liu, "Bacterial Cellulose Supported Gold Nanoparticles with Excellent Catalytic Properties," *ACS Appl. Mater. Interfaces*, Sep. 2015, doi: 10.1021/acsami.5b07150.
[19] M. Abdulwahab, A. Khamkeaw, B. Jongsomjit, and M. Phisalaphong, "Bacterial Cellulose Supported Alumina Catalyst for Ethanol Dehydration," *Catal. Letters*, vol. 147, Sep. 2017, doi: 10.1007/s10562-017-2145-y.
[20] M. Hornung, M. Ludwig, A. Gerrard, and H.-P. Schmauder, "Optimizing the Production of Bacterial Cellulose in Surface Culture: Evaluation of Substrate Mass Transfer Influences on the Bioreaction (Part 1)," *Eng. Life Sci.*, vol. 6, pp. 537–545, Dec. 2006, doi: 10.1002/elsc.200620162.
[21] M. Hornung, M. Ludwig, A. Gerrard, and H.-P. Schmauder, "Optimizing the Production of Bacterial Cellulose in Surface Culture: Evaluation of Product Movement Influences on the Bioreaction (Part 2)," *Eng. Life Sci.*, vol. 6, pp. 546–551, Dec. 2006, doi: 10.1002/elsc.200620163.
[22] E. P. Çoban and H. Biyik, "Effect of various carbon and nitrogen sources on cellulose synthesis by Acetobacter lovaniensis HBB5," *African J. Biotechnol.*, vol. 10, no. 27, pp. 5346–5354, 2011.
[23] S. Keshk and K. Sameshima, "Influence of lignosulfonate on crystal structure and productivity of bacterial cellulose in a static culture," *Enzyme Microb. Technol.*, vol. 40, pp. 4–8, Dec. 2006, doi: 10.1016/j.enzmictec.2006.07.037.
[24] K. Toda, T. Asakura, M. Fukaya, E. Entani, and Y. Kawamura, "Cellulose production by acetic acid-resistant Acetobacter xylinum," *J. Ferment. Bioeng.*, vol. 84, pp. 228–231, Dec. 1997, doi: 10.1016/S0922-338X(97)82059-4.
[25] M. Moosavi-Nasab and A. R. Yousefi, "Biotechnological production of cellulose by Gluconacetobacter Xylinus from agricultural waste," *Iran. J. Biotechnol.*, vol. 9, pp. 94–101, Jan. 2011.
[26] Y. Chao, M. Mitarai, Y. Sugano, and M. Shoda, "Effect of addition of water-soluble polysaccharides on bacterial cellulose production in a 50-L airlift reactor," *Biotechnol. Prog.*, vol. 17, no. 4, pp. 781–785, 2001.
[27] M. Hornung, M. Ludwig, and H. P. Schmauder, "Optimizing the Production of Bacterial Cellulose in Surface Culture: A Novel Aerosol Bioreactor Working on a Fed Batch Principle (Part 3)," *Eng. Life Sci.*, vol. 7, no. 1, pp. 35–41, Feb. 2007, doi: https://doi.org/10.1002/elsc.200620164.
[28] A. Krystynowicz, W. Czaja, A. Wiktorowska-Jezierska, M. Gonçalves-Miśkiewicz, M. Turkiewicz, and S. Bielecki, "Factors affecting the yield and properties of bacterial cellulose," *J. Ind. Microbiol. Biotechnol.*, vol. 29, no. 4, pp. 189–195, Oct. 2002, doi: 10.1038/sj.jim.7000303.
[29] K.-C. Cheng, J. M. Catchmark, and A. Demirci, "Enhanced production of bacterial cellulose by using a biofilm reactor and its material property analysis," *J. Biol. Eng.*, vol. 3, p. 12, Jul. 2009, doi: 10.1186/1754-1611-3-12.
[30] N. Shah, J. H. Ha, and J. K. Park, "Effect of reactor surface on production of bacterial cellulose and water soluble oligosaccharides by Gluconacetobacter hansenii PJK," *Biotechnol. Bioprocess Eng.*, vol. 15, no. 1, pp. 110–118, 2010.
[31] J. Y. Jung, T. Khan, J. K. Park, and H. N. Chang, "Production of bacterial cellulose by Gluconacetobacter hansenii using a novel bioreactor equipped with a spin filter," *Korean J. Chem. Eng.*, vol. 24, no. 2, pp. 265–271, 2007, doi: 10.1007/s11814-007-5058-4.

[32] X. Li, Q. Feng, D. Li, C. Narh, H. Ke, and Q. Wei, "Reusable Surface-Modified Bacterial Cellulose Based on Atom Transfer Radical Polymerization Technology with Excellent Catalytic Properties," *Nanomaterials*, vol. 9, p. 1443, Oct. 2019, doi: 10.3390/nano9101443.
[33] T. Zhang *et al.*, "A new strategy improves assembly efficiency of DNA mono-modified gold nanoparticles," *Chem. Commun. (Camb).*, vol. 47, pp. 5774–5776, May 2011, doi: 10.1039/c1cc11337b.
[34] P. Mukherjee *et al.*, "Potential therapeutic application of gold nanoparticles in B-chronic lymphocytic leukemia (BCLL): enhancing apoptosis," *J. Nanobiotechnology*, vol. 5, p. 4, May 2007, doi: 10.1186/1477-3155-5-4.
[35] L. Calzolai, F. Franchini, D. Gilliland, and F. Rossi, "Protein–Nanoparticle Interaction: Identification of the Ubiquitin–Gold Nanoparticle Interaction Site," *Nano Lett.*, vol. 10, no. 8, pp. 3101–3105, Aug. 2010, doi: 10.1021/nl101746v.
[36] J. Turkevich, P. C. Stevenson, and J. Hillier, "A study of the nucleation and growth processes in the synthesis of colloidal gold," *Discuss. Faraday Soc.*, vol. 11, pp. 55–75, 1951.
[37] G. Frens, "Controlled nucleation for the regulation of the particle size in monodisperse gold suspensions," *Nat. Phys. Sci.*, vol. 241, no. 105, pp. 20–22, 1973.
[38] Y.-C. Yeh, B. Creran, and V. M. Rotello, "Gold nanoparticles: preparation, properties, and applications in bionanotechnology," *Nanoscale*, vol. 4, no. 6, pp. 1871–1880, Mar. 2012, doi: 10.1039/c1nr11188d.
[39] T. Zhang *et al.*, "Biotemplated Synthesis of Gold Nanoparticle–Bacteria Cellulose Nanofiber Nanocomposites and Their Application in Biosensing," *Adv. Funct. Mater. - ADV FUNCT MATER*, vol. 20, Apr. 2010, doi: 10.1002/adfm.200902104.
[40] P. Faria Tischer, C. Costa, I. Tozetti, L. Dall'Antonia, and M. Vidotti, "Structure and effects of gold nanoparticles in bacterial cellulose-polyaniline conductive membranes," *RSC Adv.*, vol. 6, Jan. 2016, doi: 10.1039/C5RA25332B.
[41] A. S. Kunihiko Watanabe, all of K. Hiroshi Ougiya, T. Y. Nobuya Hioki, and all of J. Morinaga, Kawasaki, "method for processing bacterial cellulose," 6,153,413, 2000.
[42] M. Chanthiwong, W. Mongkolthanaruk, S. Eichhorn, and S. Pinitsoontorn, "Controlling the processing of co-precipitated magnetic bacterial cellulose/iron oxide nanocomposites," *Mater. Des.*, vol. 196, p. 109148, Nov. 2020, doi: 10.1016/j.matdes.2020.109148.
[43] M. H. Matloob, "Zahdi Date Vinegar: Production and Characterization," *Am. J. Food Technol.*, vol. 9, no. 5, pp. 231–245, 2014, [Online]. Available: https://scialert.net/abstract/?doi=ajft.2014.231.245.

Journal of Physics: Conference Series **2063** (2021) 012016 doi:10.1088/1742-6596/2063/1/012016

Synthesis, spectroscopic, and thermal studies of azo compounds from luminol and procaine with acetylacetone and their complexes

Noor Al AlBaheley*,T A Fahad and Asaad A Ali
Department of Chemistry, Basrah University's College of education for Pure Sciences.Iraq
corresponding author's e-mail address : noor86hakim@gmail.com

Abstract. This The study entails the synthesis of two newly synthesized azo dyes luminol and procaine with acetylacetone (N_1 and N_2 correspondingly) .Elementalanalysis ,^{1}HNMR, T.G.A, and FTIR. have all been used to characterize dyes. These new dyes were reacted with Cpper and Nikel ionin 1:2 molar ratios to form of complexes of metals (II) with a general stoichiometry; CuL2, and NiL2 in complexes., FT IR, as well as the corresponding metal (II) complex, were used to characterize them. The dye acts as a bidentate ligand, according to elemental analysis and spectral results.The thermal properties of these compounds were investigated using thermogravimetric analysis (TGA).

Thermal decomposition of these compounds is a process that occurs in stages.,

1.Introduction

Azo compounds used in a wide range of practical applications, including fiber coloring [1,2 3], and got a special importance in pharmaceutical industry because play an active role inbiological system[4],Non-linear and photonic creative methods [5], intelligent storage [6], nitrogen fixation, and carcinogenic [7-14].Whereas certain azo compounds are moderately toxic and produce detectable adverse effects over time [15].Azo–diketones exhibit a broad range of biological activity, acting as a major component of bimolecular compounds such as hemoglobin, vitamin B12 [15].Because the –N=N–group and diketone moiety are combined in a single compound, its ability to form complexes increases and their stability increases [16,17].Due to their interesting electronic and geometrical properties, as well as their possible applications in molecular memory storage, Azodyes complexes applied in nonlinear optical elements, and printing systems. [18-19]. Azo-metal complexes have a number of significant advantages. During synthesis, azo compounds have the potential to improve the ability to interact with the substrate to which they are applied, as well as increase the color intensity.

Recent research indicates that forming a complex between an azo dye and a metal such as Co(II) reduces cytotoxicity even at low complex concentrations[20].Many azo dyes and their complexes have been studied and published [21].The purpose of this work was to synthesis and characterization of two azo dyes (ligands N1 and N2) derived from luminol and procaine with acetylacetone .Using a variety of analytical instruments, we report on the coordination activity of these azo compounds with Cu and Ni ions.

2. Experimental.

All chemicals are of a superior quality.On an ashimadzu 8400 FTIR Spectrometer, Complexse and ligands were analyzed using KBr pelts in the infrared (400-4000)cm-1..Euro EA elemental analyzers were used to conduct the elemental analysis (C.H.N). On a BRUKER-500 MHz spectrophotometer, the 1HNMR spectrum of the ligand was determined in CDCl3 (internal standard TMS).),Using an Agilent Technologies 5975C mass spectrometer, the ligand's mass spectrum was estimated at 70e.The electrical conductivity of the chelate complexes was measured (10^{-3}M) in DMF solvent,The metal content of the complexes was determined using an atomic absorption meter Jenway Pcm3Phoenix-986.The melting points of the synthesis compounds were measured using an Electro thermal melting point 9300., A TGA Q500-V6.7 was used to conduct thermal analysis (TG) in a complex under nitrogen gas at 10°C per minute heating rate

2.1. prepare azo ligands

N1 (5-amino-2,3-dihydro-1,4-phthalazinedione azo-2,4-pentadione) N2, 3-[(2-(Diethylamino)ethyl 4-aminobenzoate azo-2,4-pentadione) [3,21]

Azo ligands were synthesized by dissolving 0.004 moles of each(luminol and procaine) in 1.4 ml concentrated HCl and 20 ml distilled water and 20 ml distilled water, then drop by drop, at - 5 oC, a stirring solution of (0.304 g sodium nitrite in 20 ml distilled water) was added to form diazonium salt.The above diazonium salt was added to an alkaline solution of 0.004mole of (Acetyl acetone) in (1.2 percent w/v. NaOH).Then, by adding diluteHCl (to maintain a pH value of 6.5 – 7.5), convert the prepared dye from sodium to hydrogen.To remove any remaining un reacted substances.The product was purified by washing with distilled water and recrystallized by using ethanol as solvent and dried in a vacuum at 50°C for several hours scheme 1

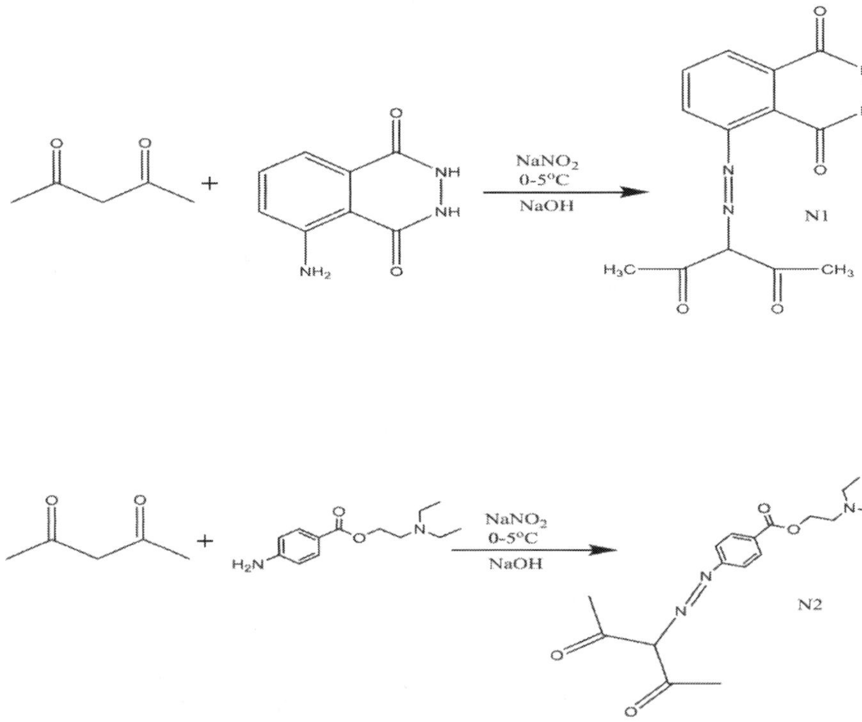

Scheme 1. Synthesis of azo dyes (N1&N2)

2.2 Preparation of azo complexes of nickel and copper [21]

Nickel and Cupper acetate (1 mmol) in ethanol (10 mL) was applied to(2 m mol) the ligand solution in 20 mL of an ethanol-chloroform mixture in a flask with a round bottom and a magnetic stirrer.We noticed a definite change in the color of the garment as well.

While the reflux was happening, the mixture was stirred for 3 hours.The purple mass was removed and washed several times with ethanol; then it was vacuum-dried...

3.Results and Discussion

The new azole and its respective complexes have demonstrated good stability in DMF, methanol, ethanol, and aceton and do not degrade in air at room temperature .Table 1 summarizes the results of the CHN of complexes are strong accordance with the suggested molecular formula that the ligand behaves bidentate in prepared complexes

Table(1): Elements analysis and physical properties for novel azo- ligands) and their metal complexes

Compound	Molecular formula (Mol.Wt)	Yield (%)	Color	m.p °C	Elemental Analysis Found (Calc.) %C	%H	%N	%M	Cond. S cm2 mol−1
N_1	$C_{13}H_{12}N_4O_4$	83	brown	235	54.16 (53.03)	4.19 (4.34)	19.34 (19.11)	-	-
N_1Cu	$C_{26}H_{28}N_8O_{10}Cu$	77	Brown green	>300	46.19 (45.82)	4.17 (4.09)	16.57 (16.38)	9.39 (9.25)	15.34
N_1Ni	$C_{26}H_{30}N_8O_{11}Ni$	73	Brown yellow	>300	45.31 (44.22)	4.38 (4.24)	16.25 (16.13)	8.51 (8.42)	18.25
N_2	$C_{18}H_{25}N_3O_4$	82	black	224	62.22 (61.01)	7.25 (7.18)	12.09 (12.01)	-	-
N_2Cu	$C_{36}H_{54}N_6O_{10}Cu$	79	Dark brown	>300	54.43 (53.24)	6.85 (6.72)	10.57 (10.39)	7.99 (7.82)	20.58
N_2Ni.	$C_{36}H_{56}N_6O_{11}Ni$	75	Light brown	235	53.54 (52.42)	6.89 (6.79)	10.40 (10.31)	7.62 (7.19)	15.45

3.1. *The infrared spectra*

Table 2 shows the details on azoligands and their metal chelat complexes. The infrared spectrum of ligand displays a medium band about 3300-3400 cm^{-1}. , which might be related to stretching vibrations (N-H) and (O-H).As a result, the (N-H) group's remaining hydrogen means that the ligand is not involved in metal ion coordination [22].Due to the fact that the novel azo ligands contain (C-N),, there is a strong strength band in the range of 1300-1270cm-1.In metal complexes, this band changed, suggesting imine nitrogen atom covalent bonding to metal ions[23]. At, one can observe the azo group. 1482 cm^{-1}., and At this time, this category is shifting to a lower frequency. (1450-1435) cm^{-1}, indicating that it is involved in metal ion bonding [20].The (M-O) vibration is responsible for the appearance of new bands at (543-572)cm^{-1} which aren't found in the free ligand.The presence of weakly absorbing bands in IR spectra of complexes (420-480)cm^{-1} region, that can be traced back to (M-N)stretching , also indicated metal oxygen binding to form chelat complexes.The azo acts as a bidentate chelating agent according to IR spectra data.

A new band corresponding to C-O vibration appeared around 1260-1240 in the spectra of the complexes, and strong bands At low frequencies, the symbol v N=N appeared.

We can deduce from IR spectral data that azo act as an N,O-bidentate ligands.[23]

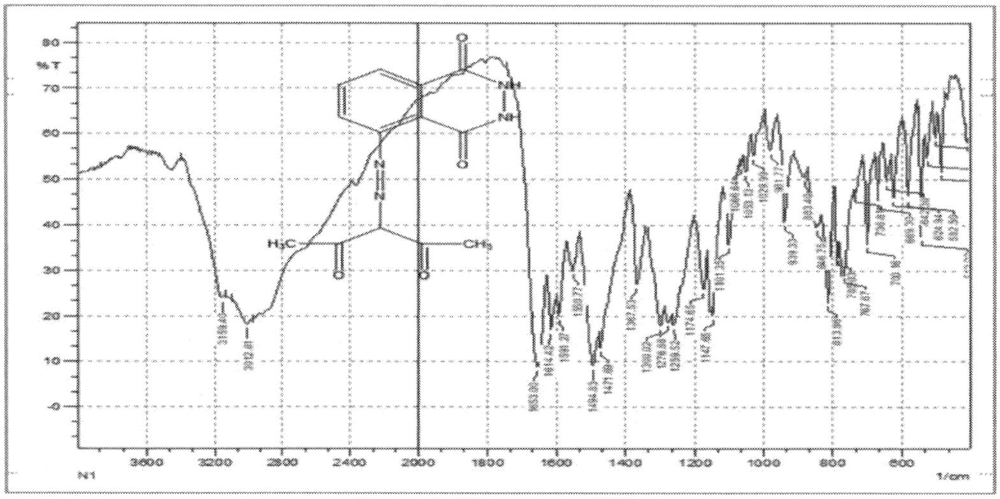

Figure1. IR spectra of N1

TABLE-2 .For azo-ligands and their complexes, the infrared spectral data

Compound	ν(O-H) ν(N-H)	νC-H Arom.	ν N=N	νC=C Ar	ν C-N	νC=O	ν C_O	ν (M_O)	V(M_N)
N_1	3400 m	3159.4 w	1494 .8 s	1591.2 m	1300.0 m	1653.0 S	-	-	-
[N1Cu]	3250 w	3008.9 m	1492 . s	1591.2 m	1367.5 s	1651.0 S	1245 s	543. s	450 w
[N1Ni]	3250 w	290.23 m	1481 s	1571.9 m	1357.8 m	1643.3 S	1255 m	550w	420w
N_2	3300	3300w	1508 .S	1604.7 s	1271.0 m	1689.6 s	-	-	-
[N2Cu]	3300	2918.3 m	1398 .s	1560.4 m	1276.8 m	1606.7 m	1255.8 m	561.2 m	476.w
[N2Ni]	3310	2918. m	1386 .s	1597.0 s	1315.4 m	1666.5 m	1254m	572.8 w	480.w

m = medium, s = strong, w = weak

3.2-TheNMR spectra

The 1H-NMR spectra (N1 and N2) of the ligands synthesized were carried out in DMSO solutions, with TMS serving as an internal assessment. (figure.2and 3).

The DMSO solvent produced a peak at 2.4 ppm in H-NMR spectra, and dissolved water in DMSO produced a sharp peak at 3.4 ppm in some spectra. [23]

TABLE 3.: N1 and N2 compounds 1H-NMR spectra.

Comp	ppm (Chemical shift)
N_1	2.4(S,6H-DMSO) ,3.4(S,H,HOD),2.3 (S,6H,CH3),6.5-8.5(m,3H,Ar-H-H)-11.5(s,2H,NH)
N_2	2.4(S,6H-DMSO) ,3.4(S,H,HOD)-1.23(s,6H,CH_3) 7.6-8.2(m,4H,Ar-H) -2.59(s,6H,CH_3)14, (s, 1H, OH/NH),

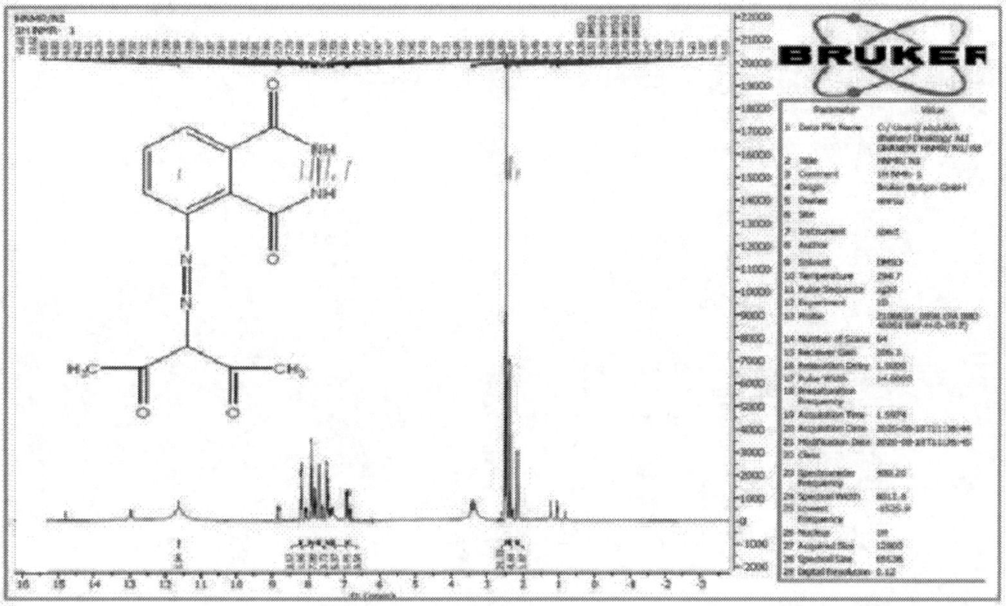

Figure 2. NMR spectra of N1

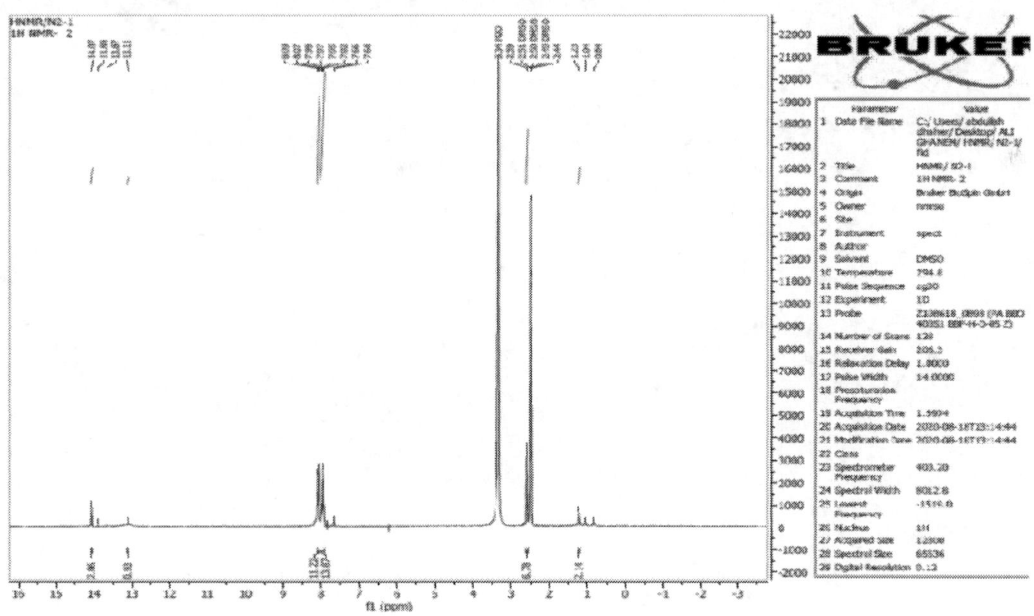

Figure 3. NMR spectra of N2

3.3-Mass spectra of N1and N2

The structure of transition metal(II) complexes can be elucidated using mass spectra of synthesized compounds.The stoichiometry of complexes was compared using mass spectra obtained at room temperature(21) (Table 4). (figure 4and 5).Each synthesized complex matched the molecular ion fragment and supported the complexes' proposed structure. [24]

Journal of Physics: Conference Series **2063** (2021) 012016 doi:10.1088/1742-6596/2063/1/012016

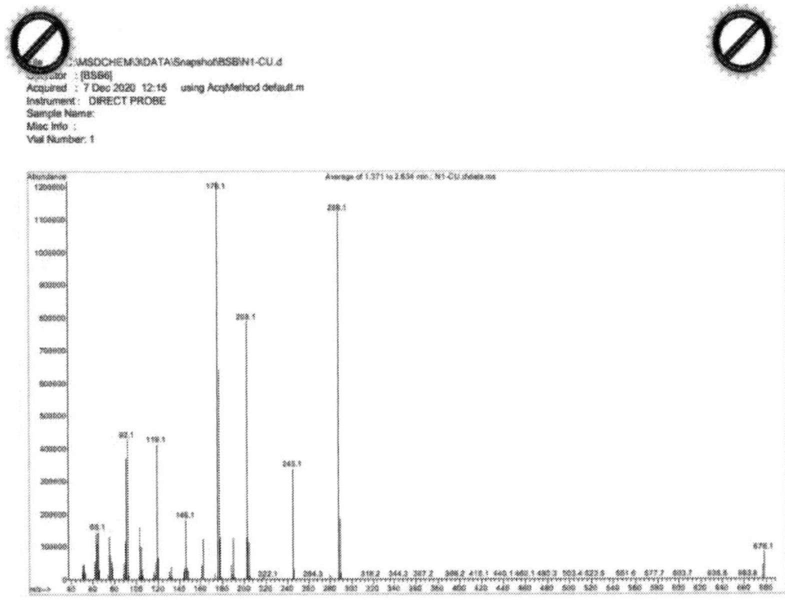

Figure 4. Mass spectrum of Cupper complex of N1

N1cu

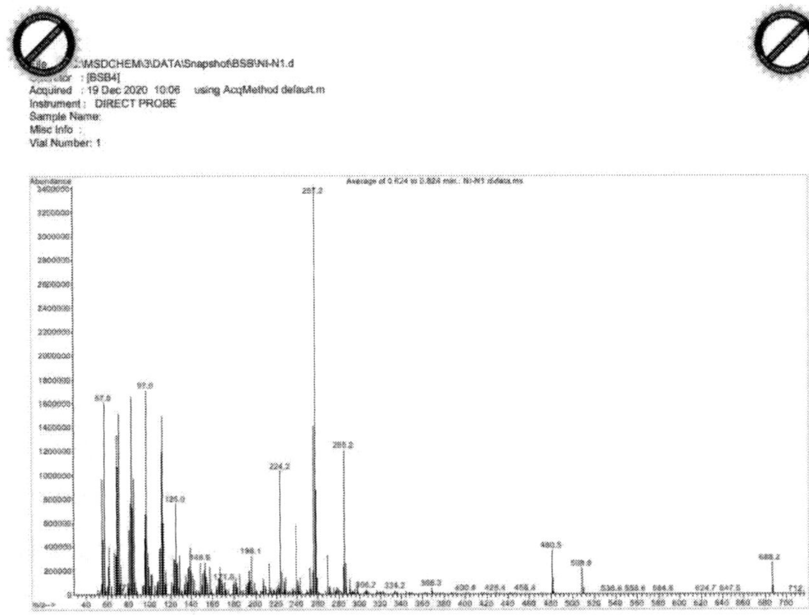

Figure 5. Mass spectrum of Nickel complex of N1

(Table 4)- Mass spectra of synthesized complexes

Compounds	Calculated mass	Obtain mass m/z	Peak assigned
[Cu (N_1)2]2H20	**676**	**676**	**M+**
[Ni ((N_1)$_2$(H2O)2]H2O	**689.22**	**688**	[M-H]+
Cu(N2)$_2$]H2O	**794**	**693**	[M-H]+
[Ni ((N2)2(H2O)2]H2O	**809**	**793**	[M-H_2O]+

Journal of Physics: Conference Series **2063** (2021) 012016 doi:10.1088/1742-6596/2063/1/012016

Electrolytic studies of metal complex solutions shed light on their nature and composition.
These studies provide insight into the number of ions present in a solution that are responsible for current conduction, and thus allow for the extraction of quite significant structural information.[25]
The chelate complexes prepared in this study had molar conductance values ranging from (15-20) S.mole-1.cm2 in DMSO (10-3M) at room temperature, indicating that the Ni(II),Cu(II),complexes are not electrolytes and that no conductive species exist., indicating their non-electrolyte nature tale (1) [26]

3.4.Interpretation of the structural
The structure of all ligand-containing complexes can be interpreted as having a common distribution of similar coordinating sites, as seen in fig(6), which is covalently linked to copper(II) and nickel (II) [27]

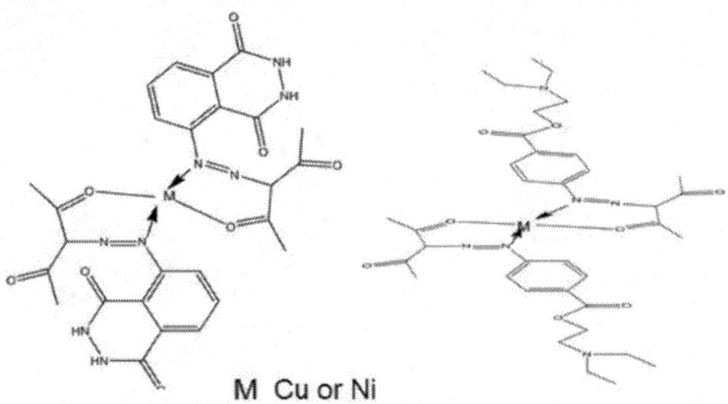

Figure 6.Metal complexes: proposed structural formulae

3.5.Thermal properties
At room temperature, the Cu(II) and Ni(II) complexes are stable and can be stored indefinitely.
The obtained compounds were subjected to thermogravimetric (TG) analysis in N2 atmospheres at temperatures ranging from ambient to 800 degrees. The TG curves are depicted in Figure7,8)

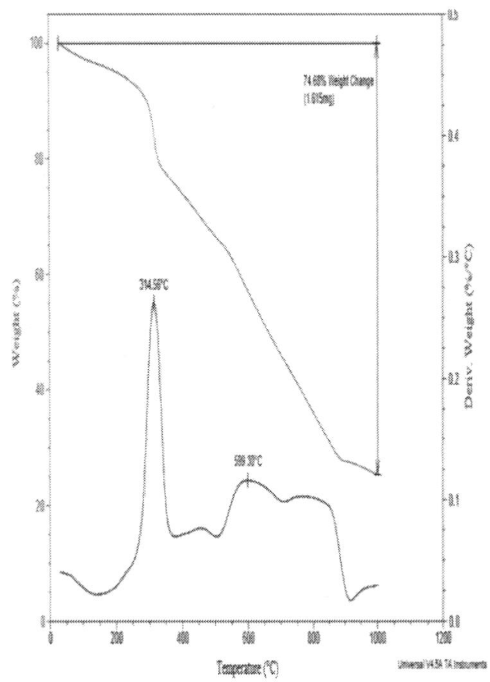

A

Figure 7.A-TG curve of N1 Cupper complexe

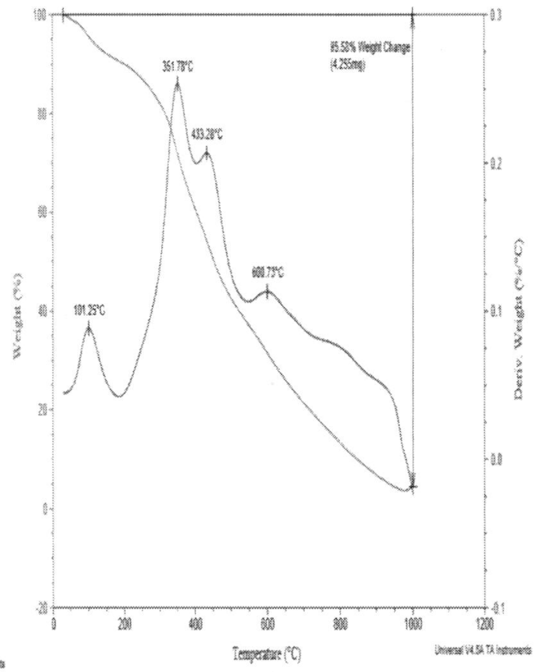

B

Figure 7.B-TG curve of N1 Nikelcomplexe

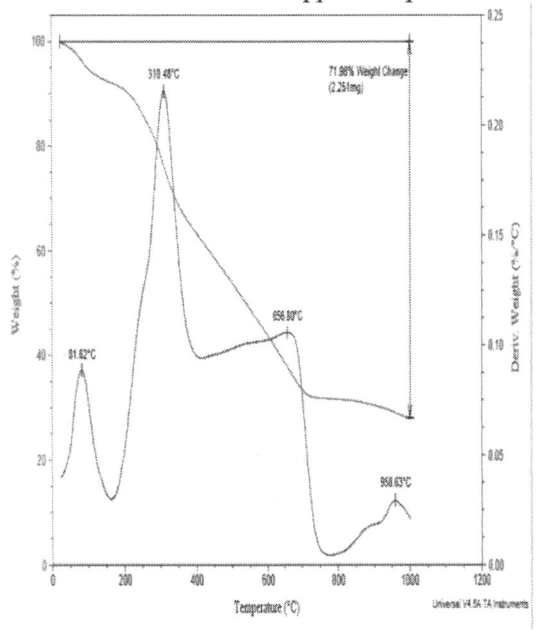

A

Figure 8.A-TG curve of N2 Cupper complexe

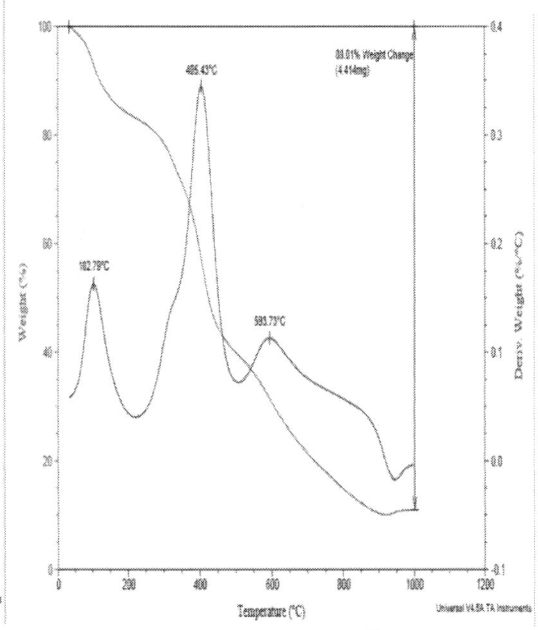

B

Figure 8.B-TG curve of N1 Nikelcomplexe

The data from TGA clearly show that the complexes are decomposed in three or four phases.

Between 25 and 160 0C, crystal water molecules were lost, and between 160 and 250 0C, coordinated water was lost[28].The loss of organic moiety was the second step, and the loss of metal oxide and some organic species was the third and fourth steps..

Journal of Physics: Conference Series **2063** (2021) 012016 doi:10.1088/1742-6596/2063/1/012016

3.6. *Antibacterial study*

The ligands and their complexes were found to have antibacterial activity was determined and tabulated against both gram positive and gram negative bacteria table(5).

TABLE(5)- Antibacterial effect of the ligands their cupper complexes against bacteria

Isolates chemicals	Conc.mg/ml	*Inhibition zone(mm)*	
		S. aureus	*E. coli*
L1	100	23	-
L2Cu	100	14	-
L2	100	25	-
L2Cu	100	11	-

The orbitals of each metal ion is made to overlap with the ligand's atoms. The delocalization of – electrons in the chelate ring increases the lipophilicity of complexes as activity rises. Increased lipophilicity can result In some instances, the cell's permeability barrier is compromised..[29]

4. References

[1] Hajipour, A.R., Zahmatkesh, S., Roosta, P. et al. Synthesis and characterization of new optically active poly(azo-ester-imide)s via interfacial polycondensation. *Amino Acids*, **36**, 511–518 (2009). https://doi.org/10.1007/s00726-008-0113-y

[2] Benkhaya, S. M'rabet, and A. El Harfi, (2020)Classifications, properties, recent synthesis and applications of azo dyes,*Heliyon*,**6(1)**: e03271

[3].Ali A.A., Fahad,T.A and Al-muhsin,A.A.(2020) Preparation and pectroanalytical Studies of Two New Azo Dyes Based on Luminol IOP Conf. Series: Materials Science and Engineering **928** :52007,IOP Publishing,doi:10.1088/1757-899X/928/5/052007

[4]. Aljamali,N.M.,(2015) Review in Azo Compoundsand itsBiological Activity,*Biochem Ana.l Biochem.*,**4:2** .,http://dx.doi.org/10.4172/2161-1009.100016

[5].Mavila,S.S., Schitram,C., H.Jsekar,,(2017) Spectroscopic, DFT and Z-scan supported investigation of dicyanoisophorone based push-pull NLOphoric styryl dyes,*Optical Materials*, **64**: 453-460

[6]. Bae,J.S., Freemanand,H.S., El-Shafei,A. (2003). Metallization of non-genotoxic direct dyes, Dyes and Pigments, **57(2):**121-129

[7]. Badea, M., Olar, R., Cristurean, E., Marinescu, D., Emandi, A.,Budrugeac, P., Segal, E., 2004. Thermal stability study of some azo-derivatives and their complexes, Part 2. New azo-derivative pigmentsand their Cu(II) complexes. *J. Therm. Anal. Calor.* **77 (3)**: 815–824

[8]. Broekaert, J.A.C (1981) Separation of yttrium and rare earth elements from geological materials,*Anal. Chim. Acta*, **124**: 421-425.

[9]. Ggarg,H., and Prakash,C. (1972) Imidazo[1,2-a]pyridine anthelmintic and antifungal agents.*J Med. Chem.*, **15(4)**: 435-436

[10]. VIDYA,V.G., and SADASIVAN,V.,(2018) Synthesis, Characterization and Applications of SomeAzo Dyes Derived from Various Aromatic Amines, *Asian Journal of Chemistry*, **30 :** 2049-2053

[11].Chung,K.T.,(2000) Mutagenicity and carcinogenicity of aromatic amines metabolically produced from Azo Dyes, *J. Environ, Sci.Health,Part C: Environ. Carcin. Ecotoxicol. Rev*, , **18**: 51-74

[12]. Zhang,D., Zhang,M., Liu,Z., Huang,C.,(2006) Highly selective colorimetric sensor for cysteine and homocysteine based on azo derivatives,*Tetrahedron Lett.*, **47:** 7093

[13]. Gupa,R., Giziroglub,E., Kırkan,B.,(2007) Synthesis and spectroscopic properties of new azo-dyes and azo-metalcomplexes derived from barbituric acid and aminoquinoline, *Dyes and Pigments,***73**: 40-46

[14]. Patil,C.J., and Rajput,S.V.,(2019) Coupling reactions involving aryldiazonium salt: part-ix. review on synthesis of azo-phenolic derivatives, their applications and biological activities, *International Journal of Recent Scientific Research*,. **10**: 32144-32156

[15]. Aljamali,N.M., Jawd,S.M., JawadZ.M., Alfatlawi,I.O.,(2017) Inhibition Activity of (Azo – Acetyl acetone) on Bacteria of Mouth. *Research J. Pharm. and Tech.*, 10(6): 1683-1686. doi: 10.5958/0974-360X.2017.00297.9

[16]]. Abdallah,S.M., (2012)Metal complexes of azo compounds derived from 4-acetamidophenol and substituted aniline,*Arabian Journal of Chemistry*, 5: 251–256.

[17]. Gouda, M.A., Eldien, H.F., Girges, M.M., Berghot, M.A.,(2016) Synthesis and antitumor evaluation of thiophene based azo dyes incorporating pyrazolone moiety. *J. Saudi Chem. Soc.* **20 (2)** : 151–157

[18]. Mutlu, H., Geiselhart, C.M. , Barner-Kowollik, C.(2018) Untapped potential for debonding on demand: The wonderful world of azocompounds. *Mater. Horizons*, **5**:162-183,

[19] .Wang, S. Shen,S., Xu,H., (2000) Synthesis, spectroscopic and thermal properties of a series of azo metal chelate dyes, *Dyes Pigments*, **44**:195-198

[20] Farghaly, T.A. ; Abdalla, M.M.,(2009) Synthesis, tautomerism, and antimicrobial, anti-HCV, anti-SSPE, antioxidant, and antitumor activities of arylazobenzosuberones. *Bioorg. Med. Chem.*, **17(23)**: 8012-8019

[21]. Dhumad,A.M., Hassan,Q.M.,C.A., Fahad,T.A., Emshary, Raheem,N.A., Sultan,H.A., (2021),SpectrochimicaActa Part A: Molecular and Biomolecular Spectroscopy 251- 119487

[22]. Saritha,T.J., Metilda,P., (2021) Synthesis, spectroscopic characterization and biological applications of some novel Schiff base transition metal (II) complexes derived from curcumin moiety ,*Journal of Saudi Chemical Society,*25: 101245

[23]. Pesyan, N. N., Gholsanamloo, V, Par, M., Rashidnejad, H, Gharib, A.,Nejati,K(2019) Synthesis, characterization and spectroscopic properties of new azo dyes derived from aniline

derivatives based on acetylacetone and azo-metal (II) complexes and singular value decomposition (SVD) investigation. *Iranian Chemical Communication*, **7**:1-19

[24]. Dhumad,A.M., Hassan,Q.M.,C.A., Fahad,T.A., Emshary, Raheem,N.A., Sultan,H.A.,(2021) Synthesis, structural characterization and optical nonlinear properties of two azo-β-diketones, *Journal of Molecular Structure* 1235:130196

[25]. Ali,I., W AWani, and Saleem,K., (2013),Synthesis and Reactivity in Inorganic, Metal-Organic, and Nano-Metal Chemistry, 43:1162–1170,

[26]. El Wakiel,N.A., RizkHF,S.A.,Ibrahim.A., (2017) Synthesis and characterization of metal complexes of azo dye based on 5-nitro-8-hydroxyquinoline and their applications in dyeing polyester fabrics. *ApplOrganometal Chem.*,e3723. https://doi.org/10.1002/aoc.3723

[27]. Chandra, S.; Gupta, L.K., (2004) Spectroscopic characterization of tetradentate macrocyclic ligand: its transition metal complexes, *Spectrochim. Acta A,* 60:2767–2774.

[28]. Devi, J.M.,Tharmaraj,P., Ramakrishnan, S. K.,Ramachandran,(2008) On the thermal properties of metal (II) complexes of chalcone, *Materials Letters* ,**62**:852–856.

[29]. Vinodkumar, Keshavayya, J. , Pushpavathi, I., Keerthikumar C. T., Maliyappa M. R., Ravi B. N.,(2020) Synthesis, characterization, computational and biological studies of nitrothiazole incorporated heterocyclic azo dyes. *Struct Chem*, **31**:1317–1329

Spectrophotometric assay of Nicotinamide in Pharmaceutical Dosages

Basima A A Saleem*[1,a], Raghad A M Hamoo[2,b], Ghassan Q Ismail[3,c]
[1,2,3] Mosul University, College of Science, Department of Chemistry, Mosul, Iraq

[b]email: raghadham@uomosul.edu.iq, [c]email:ghassan.ir.74@gmail.com
*Corresponding author: [a]email: basmasaleem@uomosul.edu.iq

Abstract: The reaction of nicotinamide and alizarin reagent using charge transfer reaction at a pH of 5.54 lead to produce a red colored compound measured at 527 nm., while the blue colored complex was formed using the oxidation reduction reaction between nicotinamide and chromate at pH 3.49 in the presence of an indigo cochineal dye. Theses tow colored products were measured at 527 and 610 nm respectively using two simple, fast and an accurate spectrophotometric methods. The linearity of the charge transfer method was followed Beer's law 0.4 - 32 µg, while the oxidation reduction method was obeyed Beer's law from 1.6 - 40 µg in depending on the concentration range. Molar absorptivity was 1.95×10^4 and 2.16×10^4 mol^{-1} cm^{-1} for the red and blue colored complex respectively. Finally, the values of Sandal's sensitivity were 0.00626 and 0.00565 μg^{-2} cm^{-1} for the first and second methods respectively. These two methods have been applied to quantify the amount of nicotinamide in pharmaceuticals with good recovery.

Keywords: Nicotinamide, Assay, Alizarin, reaction, Indigo carmine, Dosage form.

1. Introduction

Nicotinamide (NAM), chemically named as 3-pyridinecarboxamide, vitamin PP, and vitamin B_3.(Figure 1). There are two important sources of NAM, the first one is medication, and the second one is food supplements, like meat, vegetables, mushrooms, nuts, and found in trace amounts in fish. The niacin deficiency caused pellagra disease, which is treated with NAM, as well as NAM used for treated acne. The most important compounds NADH and NAD + were contain the important part of NAM [1-6].

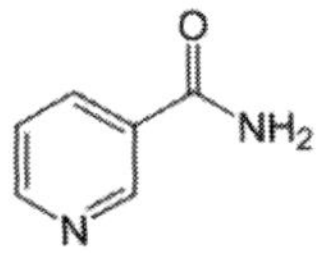

Figure 1. Chemical structure of NAM

There are several methods have been used to estimate NAM in literatures, some of them are spectrophotometric methods for determination of NAM [7], also, chromatography methods were used for estimation NAM such as High-Performance Liquid Chromatography methods (HPLC)[8-9], HPTLC method[10], pre-column derivatization and high-performance liquid chromatography method combined with detection of fluorescence [11], High-Performance Liquid Chromatography method combined with UV detection [12].

In this paper, two an organic reagents were used (Figure 2), the first is alizarin which was chemically named as 1,2-dihydroxyanthraquinone. Alizarin was derived from madder genus' plant roots. The most important uses of alizarin as a prominent dye [13]. The second an organic reagent used in this project is indigo carmin dye, this dye is a oxidation reduction indicator. Indigo carmine chemically named with 5,5′-indigodisulfonic acid. The important uses of indigo carmine were a food colorant in industry, and in the manufacturing of capsule[15].

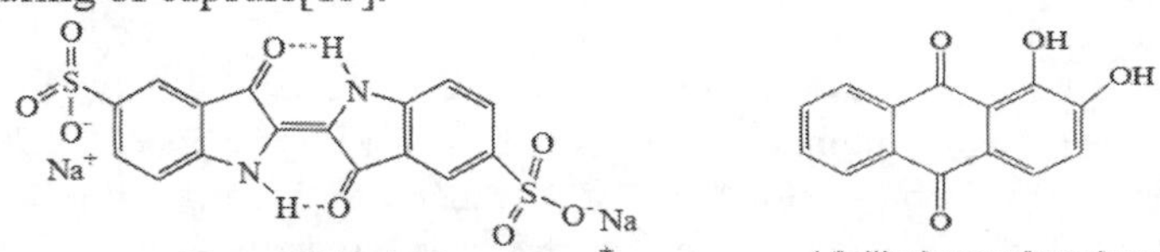

M.wt : 466.36 g/mol
Indigo Carmine dye

1,2-dihydroxyanthraquinone
$C_{14}H_8O_4$
M.wt. = 240.21 g.mol^{-1}

Figure 2. Chemical structure of Indigo carmine and Alizarin

The aim of this project was to suggest spectrophotometric methods for the determination of NAM in pharmaceutical dosages, depending on charge transfer and oxidation reduction reactions.

2. Experimental:

2.1. Equipment and chemical materials:

All the measurements of absorbance and final spectrum were carried out with JASCOV–630UV-Visible spectrophotometer, and 1 cm matched cells. While the measurements of pH were carried out with HANA pH meter.

2.2. Chemical materials:

The chemical solutions were prepared using chemical materials of an analytical reagent grade.

Nicotinamide solution, 1000 μg.mL^{-1}: To prepare a stock solution of NAM, 0.1000 g of NAM was dissolved in distilled water in a final volume of 100 ml using an dry and clean 100 mL volumetric flask, Then prepared from this solution NAM working solution of 100 μg/mL by appropriate dilution of stock solution with distilled water.
Alizarin solution 2.5×10^{-3} M: This solution was prepared when 0.0600 g of Alizarin (BDH) was dissolved with distilled water using a 100 mL volumetric flask.
Chromate solution, 8.6×10^{-4} M solution: this solution was prepared by dissolving 0.0167 g of potassium chromate (Fluka) in distilled water using a 100 mL volumetric flask.
Indigo Carmine, 1×10^{-3} M solution: This solution was prepared by weighing 0.1165 g of indigo carmine (BDH) and dissolved in distilled water using a 250 ml volumetric flask.
Hydrochloric acid solution, 1N: An appropriate dilution of concentrated hydrochloric acid with distilled water in a final volume of 250 mL using volumetric flask to prepare 1N of acid solution.

2.3. Pharmaceutical dosages :

Multivitamins Tablets*: (Multivitamin orange tablets –Germany (T&D Pharma GmbH), 10 tablets of multivitamins have been weighed, crushed, then, weighed one tablet and dissolved in a suitable volume of distilled water and stirred until the material was dissolved completely in solution. Then filtered transferred into a100.0 mL volumetric flask.
Multivitamins Capsules**: (Vitamin B-complex Capsules –India (Brawn laboratory Ltd.), ten capsules of dosage form were emptied quantitatively, mixed and weighed, then, an average weight of one capsule was dissolved with distilled water and treated similar to tablet preparation.

*Multivitamin orange tablets –Germany (T&D Pharma GmbH) AM (18), vitamin C (60), vitamin E (10), vitamin B_5(6), vitamin B_6(2) vitamin B_2(1.6), vitamin B_1(1.4), folic acid (0.2), biotin (0.15),vitamin B_{12}(0.001)
**Vitamin B-complex Capsules –India (Brawn laboratory Ltd.) NAM (10), vitamin B_5(3), vitamin B_6(1) vitamin B_2(2), vitamin B_1(5),vitamin B_{12}(0.010).

3. Results and discussion

3.1. Study of optimum conditions:
Many analytical parameters which affected on the intensity of absorbance on the colored system was studied and optimum conditions have been selected.

3.2. Effect of buffer
To check the effect of pH on the absorbance intensity of colored complex, various amounts (0.1-4.0) of 1N acids (Hydrochloric acid, Sulphuric acid, Acetic acid) were added to an aliquot of solution containing 100 µg of NAM for both methods (charge transfer and oxidation-reduction methods). The addition any amount of acid were omitted for charge transfer method because of giving no useful results. While, 1 ml of 1N hydrochloric acid was selected as an optimum amount for oxidation-reduction method as shown in fig.3.

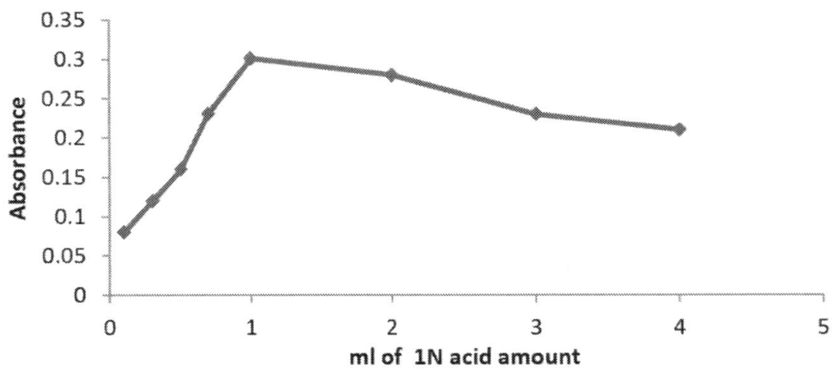

Figure 3. Effect of acid added

After addition of 2 ml of alizarin reagent, The pH was 5.54, therefore different buffers (Purrin and Dempsey, 1974) at pH 5.54 were prepared such as citric acid-NaOH, sodium acetate-acetic acid, Succinic acid-NaOH, KH-Phthalate-NaOH, and Tris-Buffer. From the experimental results found that all kinds of buffers were decrease the absorbance intensity of colored system so that this study was omitted from the subsequent results.

3.3. Effect of chromate ion amount
Chromate ions for oxidation-reduction method have been studied to test an optimum amount of chromate. So, various amount of chromate 0.5-3.0 mL 8.6×10^{-4} M were added to various amounts of NAM (50-500) µg in a final volume 25 mL for all components. 1.5 ml of chromate solution was selected as an optimum amount due to the higher value of determination coefficient (R^2=0.997139) therefore, it was recommended for the subsequent experiments (Table 1).

Table 1. Effect of chromate (VI) ion

ml of chromate solution 8.6×10^{-4} M	Absorbance / µg NAM in 25 ml				R^2
	50	100	250	500	
1.0	0.189	0.271	0.361	0.411	0.9634
1.5	0.211	0.313	0.501	0.761	0.9975
2.0	0.182	0.258	0.441	0.679	0.9912

3.4. Effect of reagents

The amount of reagents have been studied using various amount (0.5-3.0) ml of reagents and (50-500) µg of NAM for both methods, the optimum amount of first method (charge transfer method) was 2 ml of alizarin reagent 2.5×10^{-3} M (R^2 =0.99743), while 1 ml of 1.0×10^{-3} M indigo carmine dye for oxidation-reduction method was an optimum amount (R^2 =99811).

3.5. Effect of order of addition and stability period

Effect of orders was tested experimentally for the second method (oxidation-reduction method). Table 2 showed that the order (I) was the optimum as due to the highest value of absorbance as well as the stability at the selected wavelength 610 nm of the reaction mixture (Fig.4). therefore it was selected for the subsequent experiments. The first method (charge transfer method) have only one order of addition colored system measured at 526 nm.

Table 2. Effect of order of addition

Reaction components	Order number	Absorbance
NAM + Ox + A + I.C	I	0. 309
NAM + Ox + A + I.C	II	0.282
NAM + A + Ox + I.C	III	0.273

NAM =Nicotinamide, Ox=Chromate, A=Hydrochloric acid, I.C=Indigo carmine.

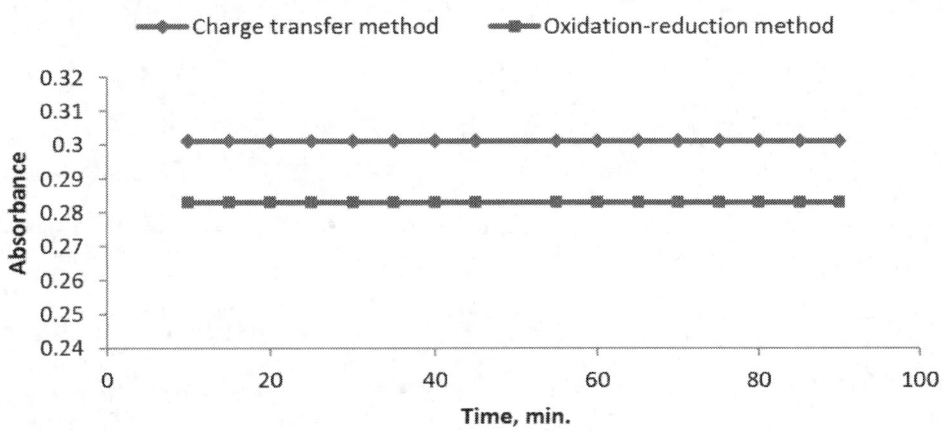

Figure 4: Effect of time and stability period.

3.6. Absorption spectra

NAM was treated as the suggested two spectrophotometric method (charge transfer reaction for first method and oxidation reduction reaction for second method), the spectrum and calibration curve were shown in Fig.5 and 6. Charge transfer and oxidation reduction methods have been obeyed Beer's law at 0.4-32 and 1.6-40 µg respectively in a final volume of 25 mL. Also, molar absorptivity and Sandal sensitivity were 1.95×10^4 and 2.16×10^4 $l.mol^{-1}.cm^{-1}$, 0.00626 and 0.00565 $\mu g^{-2}.cm^{-1}$ for charge transfer and oxidation reduction methods respectively (Figure 7). These two spectrophotometric methods have been applied for determination of NAM in pharmaceutical dosage forms with good recoveries.

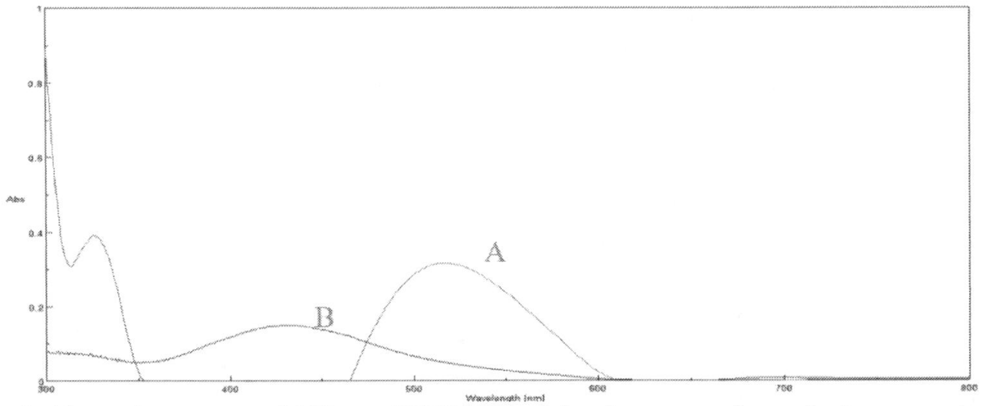

Figure 5: Absorption spectra of 100 μg of NAM /25 ml for charge transfer method measured against (A) reagent blank, (B) blank against distilled water

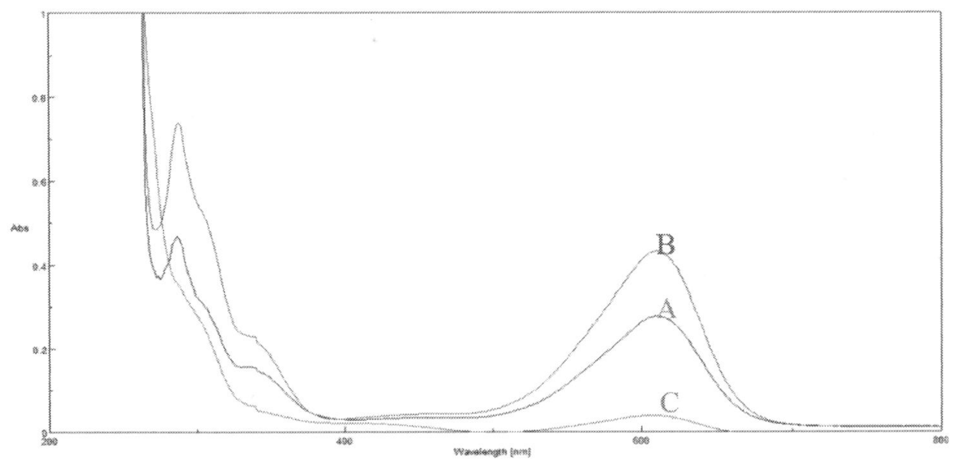

Figure. 6: Absorption spectra of 100 μg of NAM /25 ml for oxidation reduction method measured against (A) reagent blank, (B) distilled water, (C) blank against distilled water.

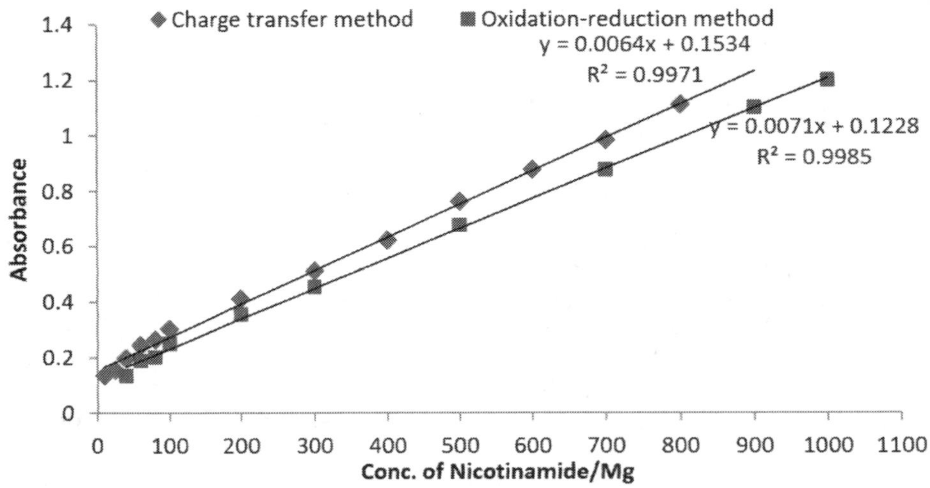

Figure 7: Calibration graph of NAM in both methods.

3.7. Accuracy and precision

The accuracy and precision of present methods for calibration curve were checked for determination of NAM at three concentrations, the results in table 3 show that the accuracy as well as precision were reliable.

Table 3. Accuracy and precision

Amount of NAM taken, µg/25ml	Recovery*, %	RSD*, %
100	99.31	± 0.276
400	99.73	± 0.241
600	100.13	± 0.187

* Average of five determinations.

3.8. Nature of the reactions

The continuous variations method (Job's method) has been used to study the reaction ratio of NAM and alizarin at the first method (charge transfer method), as well as with chromate at second method (oxidation reduction reaction), so that, the obtained results as shown in fig. 8 indicate that the ratios of NAM to alizarin and NAM to chromate was 1:1.

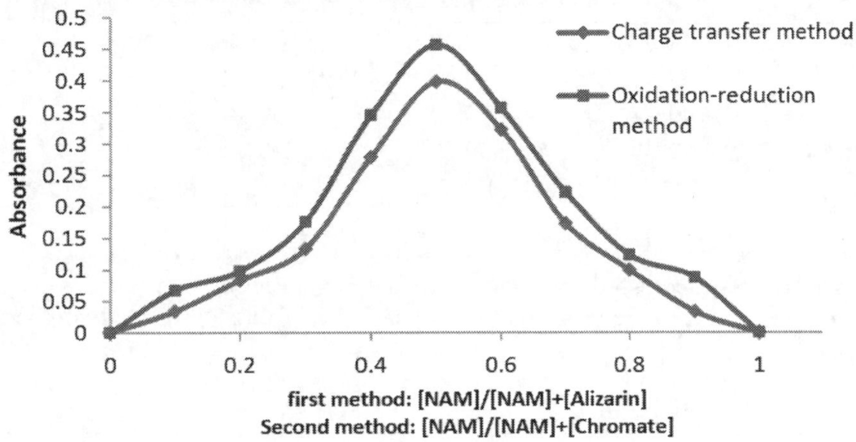

Figure 8: Job's plot for ATN - chromate

As a result the following reaction is suggested for charge transfer method (Figure 9) [13]:

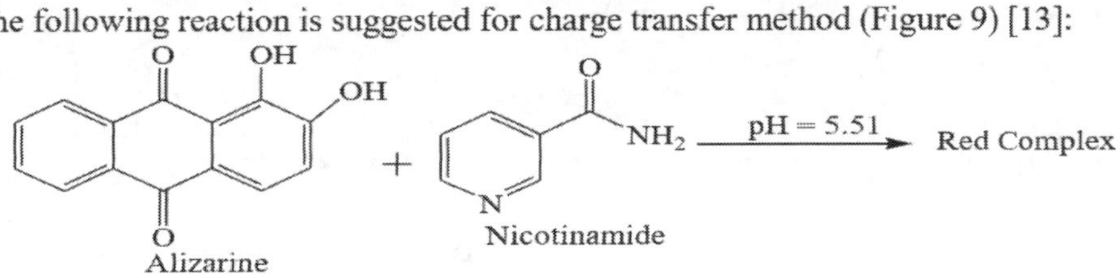

Figure 9: Job's plot for ATN – chromate

So, the reaction for oxidation-reduction method was suggested as below (Figure 10) [14]:

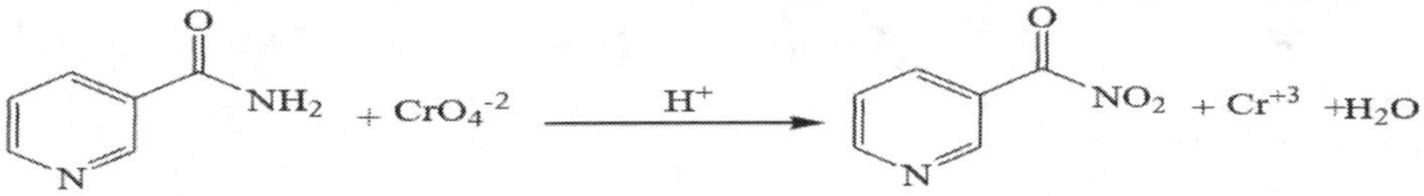

Figure 10: Job's plot for ATN - chromate

Journal of Physics: Conference Series **2063** (2021) 012017 doi:10.1088/1742-6596/2063/1/012017

3.9. Effect of interferences

The selectivity and efficiency of the proposed methods was checked towards some foreign components such as (glucose, lactose, acacia, starch and menthol) that are usually founded in dosage forms. The addition of various foreign substances to 100 µg NAM in a final volume of 25 ml was studied and observed that the studied foreign materials has not interfering in the two suggested methods (Table 4).

Table 4. Effect of interferences

Interferences	**Recovery(%) of 100 µg NAM / µg of interference added**			
	100	**300**	**500**	
Starch	99.64	99.91	99.93	**Charge transfer method**
Lactose	99.89	99.82	99.89	
Glucose	99.71	99.76	99.78	
Menthol	100.39	100.72	100.24	
Acacia	100.36	100.59	100.34	
Interferences	**Recovery(%) of 100 µg NAM / µg of interference added**			
	100	**300**	**500**	
Starch	100.34	100.39	100.48	**Oxidation-reduction method**
Lactose	99.34	99.45	99.67	
Glucose	99.89	99.82	99.79	
Menthol	99.91	99.89	99.92	
Acacia	100.42	100.62	100.45	

3.10. Application of the method

The proposed method was successfully applied to the determination of NAM in its pharmaceutical preparation (tablet and capsule). The results which are shown in Table 5 indicate that good recoveries were obtained.

Table 5. Analytical applications

Amount of NAM, µg	**Recovery(%) of NAM***	
	Multivitamin orange tablets – Germany (T&D Pharma GmbH)	Vitamin B-complex Capsules – India (Brawn laboratory Ltd.)
100	99.75	97.62
300	99.59	98.43
500	99.27	97.69

* Average of five determinations.

To calculate t-test values [15], the proposed method has been compared with literature methods [7] Table 6 showed that t-test did not exceed the theoretical values at the 95% confidence level for five degrees of freedom when it compare with literature methods.

Table 6. The results of t-test analysis

Drug	**Pharmaceutical preparation**	**t-test**	**Tabulated t-test**
Vitamin B-complex Capsules –India (Brawn laboratory Ltd.)	Capsules	1.605	
Multivitamin orange tablets –Germany (T&D Pharma GmbH)	Tablets	1.425	2.571

3.11. Comparison of the methods

The comparison between present methods with other literature spectrophotometric methods was shown in Table 7 which indicate that the proposed methods is sensitive and can be applied successfully to the determination of NAM in pharmaceutical preparation (Tablet and Capsule).

Table 7. Comparison of the methods

Analytical parameters	Present method		Literature method [7]	
	First method	Second method		
Reaction	Charge transfer	Oxidation-reduction	potassium iodide and potassium iodate	potassium iodide and potassium iodate
λ_{max} (nm)	526	610	350	288
Reagent	Alizarin	Indigo-carmine	Tri iodide ions	Tri iodide ions
Beer's law range (µg/ml)	0.4-32	1.6-40	2.5-20.0	20.0-2.5
Molar absorptivity (l.mol^{-1}.cm^{-1})	1.95×10^4	2.16×10^4	0.89×10^4	1.25×10^4
Reaction time (min)	90	90	----	-----
Color of the product	Red	Blue	Yellow	Yellow
Sandell's sensitivity (µg.cm^{-2})	0.00626	0.00565	0.028	0.019
R.S.D. (%)	±0.14	±0.21	±0.16	±0.22
Application of the method	Pharmaceutical preparation	Pharmaceutical preparation	Pharmaceutical preparation	Pharmaceutical preparation

4. Conclusion

Two spectrophotometric methods were proposed for the determination of nicotinamide in its pure form as well as in its pharmaceutical preparations. These methods were rather sensitive, simple, fast, economical, and with good accuracy. The analytical parameters and statistical comparison justify this method for application the proposed method for the determination of nicotinamide in both pure and dose images. Also, the procedure does not include any tedious sample preparation or critical reaction conditions. The recommended methods have been very suitable for the determination of nicotinamide in pharmaceuticals.

5. Acknowledgments

I extend my thanks and gratitude to University of Mosul, College of Science, Department of Chemistry.

6. References

[1] British Pharmacopoeia on CD-ROM (2001) Version 5, 3rd Ed., Vol. 1, Copyright by System Simulation Ltd, The Stationery Office Ltd., London.

[2] MacKay, D., Hathcock, J., Guarneri, E. (2012) Niacin: chemical forms, bioavailability, and health effects. Nutrition Reviews. 70 (6): 357–66.

[3] Niren, N.M., (2006) Pharmacologic doses of nicotinamide in the treatment of inflammatory skin conditions: a review. Cutis. 77 (1): 11–6.

[4] Inocha, R., Damian, DL., Halliday, GM., (2018) Melanoma and nonmelanoma skin cancer chemoprevention: A role for nicotinamide. Photodermatology, Photoimmunology & Photomedicine. 34 (1): 5–12.

[5] Kaanders, JH., Bussink, J., van der Kogel, AJ. (2002) ARCON: a novel biology-based approach in radiotherapy. The Lancet. Oncology. 3 (12), 728–37.

Journal of Physics: Conference Series **2063** (2021) 012017 doi:10.1088/1742-6596/2063/1/012017

[6] Sauve, AA., (2008) NAD^+ and vitamin B_3: from metabolism to therapies. Journal of Pharmacology and Experimental Therapeutics 324, 883–893.

[7] Khateeb, M., Elias, B., AL Rahal, F., (2015) Validated Spectrophotometric Method to Assay of B_6 and B_3 Vitamins in Pharmaceutical Forms Using Potassium Iodide and Potassium Iodate. International Letters of Chemistry, Physics and Astronomy, 60,113-119.

[8] Abd Al-Timimia, F., Abid Al-Ganib, M., and Al-Dulimyia, E., (2013) Determination of some water- soluble vitamins in different species of garlic extracts by using high performance liquid chromatography. Journal of Chemical and Pharmaceutical Research, 5(10):15-20.

[9] Muszalska, I., Kiaszewicz, K., Ksoń, D. and Sobczak, A., (2013) Determination of nicotinamide (vitamin B_3) in cosmetic products using differential spectrophotometry and liquid chromatography (HPLC). Journal of Analytical Chemistry 68,1007–1013.

[10] Kabita, B., Simhachalam, R., Gupta,B K., (2019) Chromatographic Estimation of Niacinamide in Anti-aging Dermatological product. Asian J. Pharm. Ana. 9(2):62-66.

[11] Baolei Fan, Jinmao You, Yourui Suo, (2018) A novel and sensitive method for determining vitamin B_3 and B_7 by pre-column derivatization and high-performance liquid chromatography method with fluorescence detection 13(6).

[12] Antakli, S., Sarkees, N., Sarraf, T., (2015) Determination Of Water-Soluble Vitamins B_1, B_2, B_3, B_6, B_9, B_{12} And C On C_{18} Column With Particle Size 3 Mm In Some Manufactured Food Products By HPLC With Uv-Dad/Fld Detection . Int J Pharm Pharm Sci 7(6), 219-224.

[13] Saleem, B.A.A., (2019) Spectrophotometric determination of sulphite and thiamin hydrochloride using proton transfer reaction-application to water sample and pharmaceutical formation. Tikrit Journal of Pure Science 24(1), 78-81.

[14] Saleem, B. A. A., (2019) Spectrophotometric Determination of Atenolol Using Indigo Carmine Dye. kirkuk Univ. J. Sci. Stu. 14(2):15-39.

[15] Christian, G D., (2004) Analytical chemistry John (USA : Wiley and Sons) 6th ed. p 83-99.

Journal of Physics: Conference Series **2063** (2021) 012018 doi:10.1088/1742-6596/2063/1/012018

Lab build Fully Automated microfluduic flow injection System

Ihsan A Al-Asady, K H Al-Sowdani
Chemistry Department, College of education for Pure Science, University of Basrah
E-mail: -Mrihsan1979@yahoo.com

Abstract. A fully automated microfluidic Flow Injection spectrophotometric system was designed in our laboratory with methylene blue used as an example. The device consists of a double-line microfluidic chip with dimensions (30 µl x 4 cm), each 15 µl volume intended to allow a very small volume of reagent and sample to be used. The microfluidic system also consists of two types of Arduino software. The first was the UNO of a type used to control the two mini peristaltic pumps to drive the water and methylene blue as carrier stream and sample respectively. The Mega type, which was equipped with homemade software, signal-to-peak, was the second Arduino which used as a data logger to record the results in the form of peaks using Microsoft Excel 2016. The peak's height was corresponding to concentration. The linearity was in the range (0.025-0.125 µg / ml) with a five-point regression coefficient (0.9981), RSD% for ten replicates is 0.025 µg / mL. The detection limit was (0.001 µg / ml) and the sample throughput was 300 samples per hour and each sample required 25 µl of chemical reagents. Therefore, 240 samples need only (5.0 ml) from the chemical reagents, which clearly indicates that the consumed reagent and the waste were very low. Therefore, the method for determining methylene blue with a fully automated system is very environmentally friendly. save time, consumption regent, Improve quality, decrease cost-consumption reducing the consume of sample and inherited the reproducibility and accurate from flow injection analysis.

1. Introduction

Automation is the use of technology to perform processes without human aid[1]. In the situation of analytical chemistry, automation may include operations such as measurement of signals and computation of the results, which corresponding to the concentration of the calculated sample[2].

Nowadays, these processes can be easily operated and interfaced with analytical instruments by using a lot of smart devices, which are available in low prices in local markets[3].Recently, many papers were published concern with the design of homemade microfluidic flow injection systems from our lab for phosphate, chloride, and nitrite determination[4].In this study, it was decided to developed these systems from semi-automated to fully automated systems.Two microcontrollers types Arduino were used, one UNO to manipulating the two lab-made mini peristaltic pumps to drive the carrier reagent and to suck the right volume of sample by selects the desired speed.The second one Mega was equipped with a home-made software ,single- to –peak , which was used as a data logger to recorded the signal as smooth peak height to concentration with aid of Microsoft Excel 2016[5,6,7,8,9,10]

The aim of this work builds up a fully automated microfluidic flow injection system from available components which can be afforded from local markets. The proposed system though to save time, regent, Improve quality, decrease cost-reducing the consumption of sample and waste, and inherited the reproducibility and accurate obtained results from flow injection analysis.

2. Experimental

2.1. Chemicals

All solutions were prepared with distilled water, where (0.01 M) was prepared by dissolving (3.199 g) of methylene blue in (1000 ml) of distilled water, then a series of concentrations were prepared through serial dilution[6].A solution of sulfuric acid at a concentration of (0.05M) was prepared by diluting the concentrated acid by adding (1.366 ml) to (500 ml) of distilled water for use as a carrier stream by drawing it with a peristaltic pump[12].

2.2. TheMicrofluidic device design for automatic injection.

2.3. Figuer1. and Table 1. show the components were used in designing the mini peristaltic pump .

Table 1. Components of the micro unit controlling the mini peristaltic pump.

Component	work	type	Origin
ARDUINO	Micro controller	UNO	Italy
Driver Motor	Micro controller	L298N	China
Buzzer	Alarm	HYDZ	China

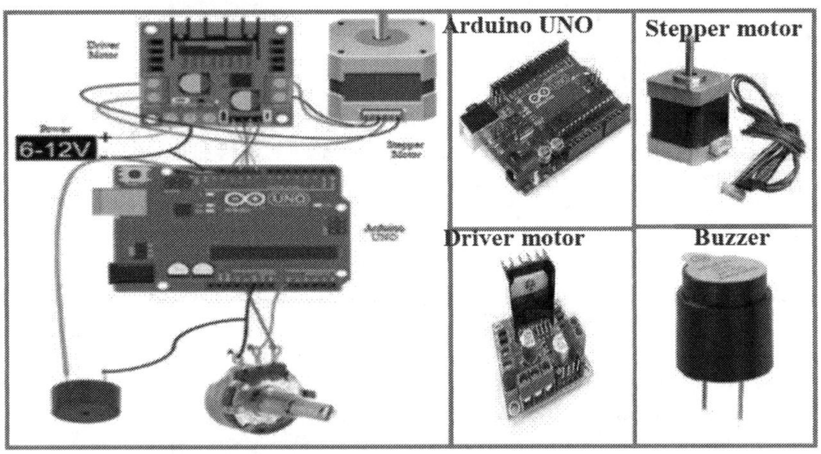

Figure 1. Components of lab – build mini peristaltic pump

Two microcontrollers types Arduino were used, one UNO to manipulating the two lab-made mini peristaltic pumps to drive the carrier reagent and to suck the right volume of sample by selects the desired speed[7,8]. The second one single–to–peak Mega was equipped with a homemade software, single–to–peak, which was used as a data logger (Fig.2) to record the signal as smooth peak height to concentration with aid of Microsoft Excel 2016[13,16].

Figure 2. Ardouino type work Mega as data logger.

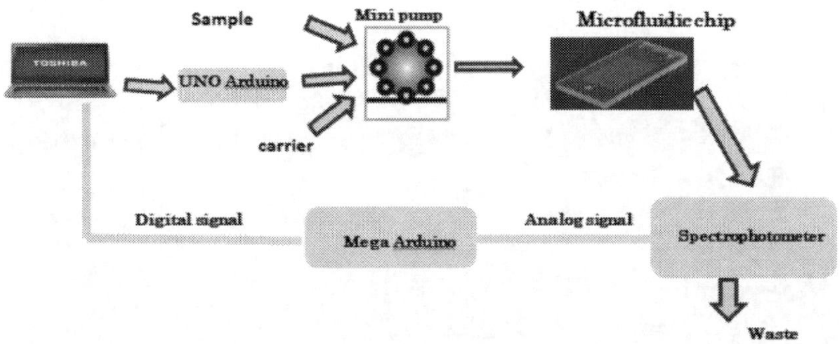

Figure 3. Shows the fully automated microfluidic Flow Injection spectrophotometric system

3. Result and discussions

4.1. Spectral characteristics

Fig 4 shows 0.075 µg/ml methylene blue spectrum which show that maximum wavelength at 665 nm which was used in all measurement of methylene blue during this work.

4.2. Optimum conditions

In order to get high performance the physical and chemical parameters affecting the results of the lab-made full automated microfluidic flow injection system were established[8].One of the important physical parameters is flow rate, especially in the microfluidic system. Figure 5 shows the effect of flow rate on the peak height of (0.075 µg/ml) methylene blue. It is indicated that peak height decrease with increasing flow rate in the range which can be attributed to the increasing the dispersion in the system [14]. So ,(1.5 mL/min) was chosen for further work.

The fully automated system was rely on sucking the right volume by the mini peristaltic pump which controlled by the UNO microcontroller. Table 2 lists the period of time and the volume of methylene blue introduce in the 0.05 M sulphuric acid relied on.

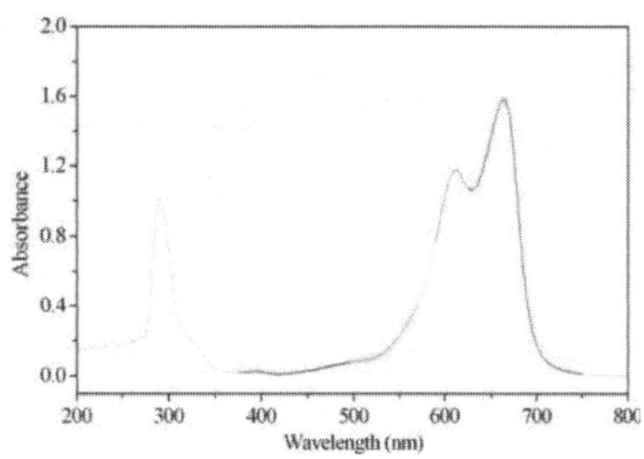

Figure 4.Shows 0.075 µg/ml methylene blue spectrum

Journal of Physics: Conference Series **2063** (2021) 012018 doi:10.1088/1742-6596/2063/1/012018

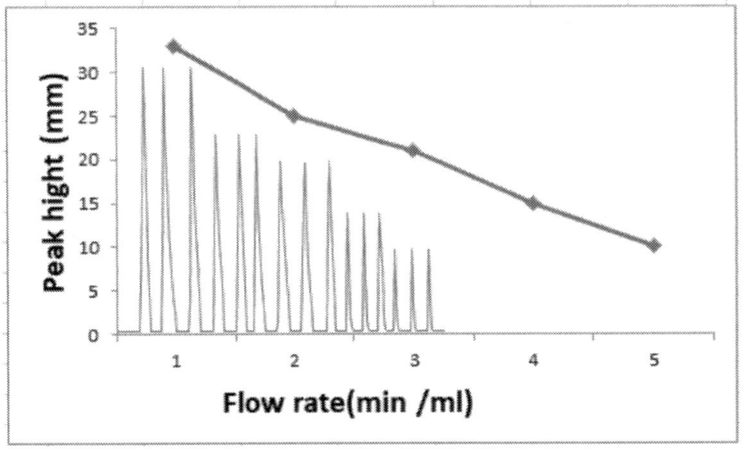

Figure 5. Effect of flow rate on the peak height of (0.075 µg/ml) methylene blue.

Table 2. List the period of time and the volume of methylene blue

No.	period of time(Sec)	volume of methylene blue (µL)
1	10	20
2	20	40
3	30	60
4	40	80
5	50	100
6	60	120

Fig 6 shows that peak height increase with increasing the injected volume which can be due to decreasing of dispersion in the flow system at the flow rate obtained previously[15]. So, (20 µL) was used in the next experiments to obtained good sensitivity and high sample throughput.

The effect of tube length on peak height was studied in tubes' in the range (3-15cm). Figure 7 shows the results obtained which inducted that peak height decrease with increasing the tube length. This tubes' length was inserted between the mixing point, of injected methylene blue with carrier stream, and the flow cell in the detector. (9Cm) was chosen for further work..

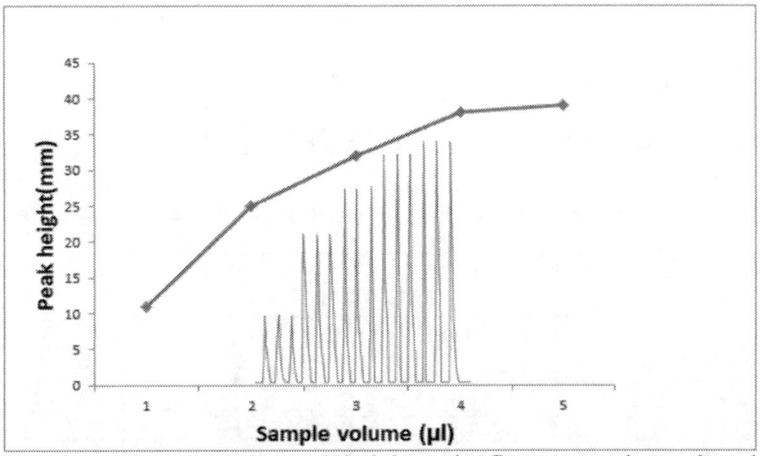

Figure 6 . Effect of sample volume on peak height at the flow rate and sample volume found previously.

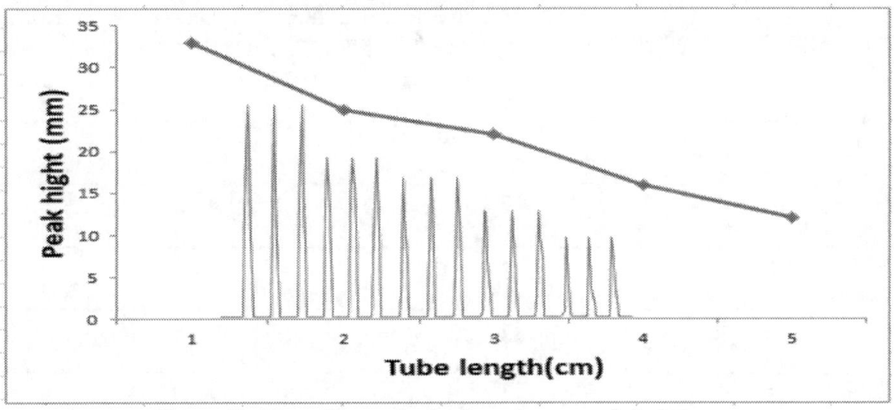

Figure 7. The effect of tube length on peak height.

4.3. Standard calibration curve

Under the established optimum condition listed in table 3 .

Table 3.The established optimum conditions for methylene blue determination

Flow rate(min\ml)	Sample volume(µl)	Tube length(mm)
15	20	12
30	40	14
45	60	16

The calibration curve for determination of methylene blue by lab builds a full automated microfluidic flow injection system. The linearity was found in the range (2.5-12.5 µg/ml) of methylene blue (Table 4 and Fig. 8) The regression coefficient (r 2) of five points was 0.9981 and the least square equation was(y = 3.461 x) and the detection limit was (0.001 µg/ml) . The sample throughput was (300sample/1hour) and each sample required (0.025 mL) of 0.05 M H2S04 in this case.

Table 4. optimum conditions for methylene blue determination.

Parameters	Values
Chip volume	20µL
Total flow rate	0.5 ml/min
Sample volume	25µL
Sulphanilamid Conc.	0.05M
Tube diameter	0.2mm

Table 5. standard calibration curve of methylene blue

methylene blue concentration (µg/ml)	Peak height(mm)	R.S.D %
2.5	9	0.0250
5	18	0.0150
7.5	25	0.0400
10	34	0.0450
12.5	44	0.0700

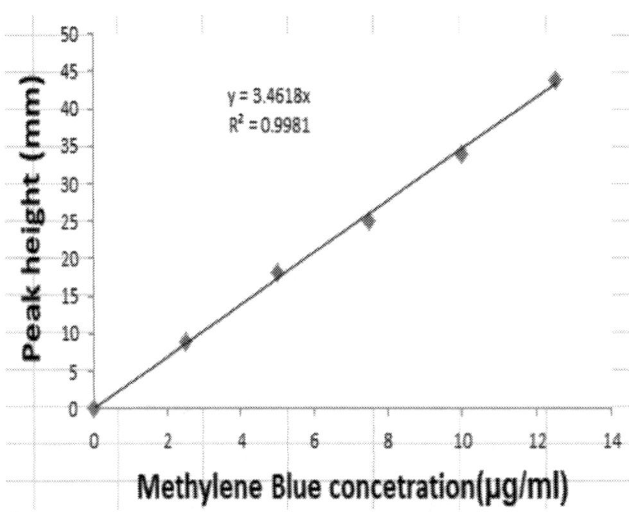

Figure 8.Calibration curve for Methylene blue

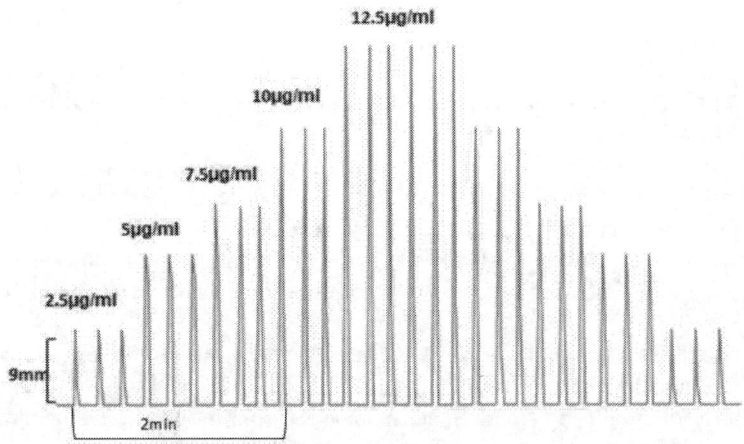

Figure 9.The The peak height of the corresponding calibration curve.

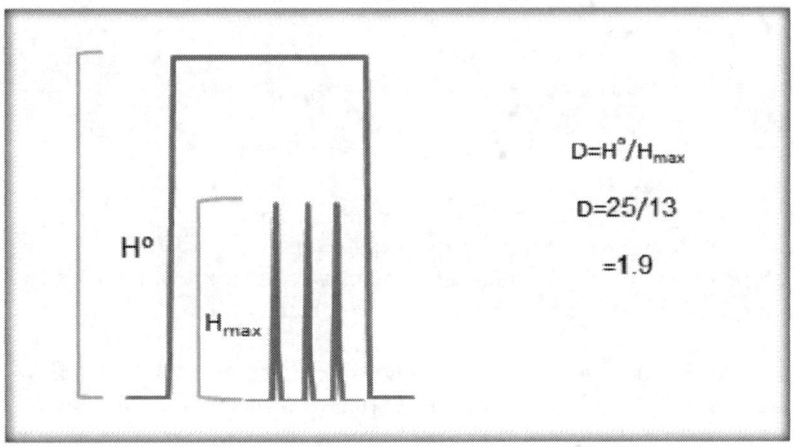

Figure 10. shows the dispersion coefficient and reproducibility which are (1.9)and (25) respectively.

4. Conclusions

The Lab made a fully automated microfluidic flow injection system that was for the first time designed in our laboratories of the Department of Chemistry at the College of Education for Pure Sciences at Basra University. It was a simple, fast, inexpensive, accurate, and reproducible system. It was successfully applied for the determination of methylene blue (as an example).In addition, the proposed system has minimized the consumption of reagents and waste, so the system is environmentally friendly and it is less hazardous due to fully automated with no contact with chemicals.

References

[1] Al-Sowdani K. H, Abdulkadhim M..H ,Lamia.A.Rusinb . 2016 Quenching-Chemiluminescence New Method for Determination of Ethylmorphine by New Design Chemiluminescence Flow Injection System,American Research Journal of Chemistry.1.1.

[2] Skeggs,L.T,.Jr.1957An automatic method for colorimetric analysis.Am.J.clin. Park,28,311-322.

[3] Trtjanowicz, M,2020. Flow chemistry in contemporary chemical sciences .A real variety of its application, Molecules.25:1434.2

Journal of Physics: Conference Series **2063** (2021) 012018 doi:10.1088/1742-6596/2063/1/012018

[4] Jean Berthier and Pascal Silberzan,2010. Microfluidics for Biotechnology 2nd.ed, Artech House, Boston, London. 503.

[5] Al-Sowdani. K. H, Mohammad .Th. K. Al-Balaawi.2019 Lab – Built semi- automated Stop – flow System for Spectrophotometric Determination of Nitrite in Different water Samples. Journal of Physics: Conf. Series 11.1294.

[6] Al-Omran, Safaa Sabri Najim K H Al-Sowdani,2009. Design and constraction A home-made semi-automated flow injection system and a baseline study for determination of human serum albumin.journalBasrahresarchers science,61.

[7] Al-Jorani .Y.Shafi , K Hussien Al-Sowdani. 2020 Designing and constructing a micro flow injection system (μFIA) for phosphate determination in soft drinks, IOP Conf. Ser.: Mater. Sci. Eng.871:012032.

[8] Al-Sowdani .K. H Mohammad Th. K. Al-Balaawi.2019 Lab -Built semi-automated Stop -flow System for Spectrophotometric Determination of Nitrite in Different water Samples Lab - Built semi-automated Stop -flow System for Spectrophotometric Determination of Nitrite in Different water Samples .Journal of Physics: Conf. Series1294.

[9] Al-Sowdani, K. H. and Noor.A.H. Al-Bahay.2016 Determination of some Nutrients By Home-made Ion Chromatography. International Journal of Science and Research,11 pp189-193.

[10] Al-Joran.Yousif Sh. J. , K. H. Al-Sowdani.2020 Designing and constructing a micro flow injection system (μFIA) for phosphate determination in soft drinks, Material science and engineering,871, 012032.

[11] Bnzi, BenziM.,2008.Getting started with Arduino, O.Reilly , Sebastopol:21.

[12] Oxer,J,H.B.2009.Mor practicalArduino Cool projects for open source Hardware,Technology in Action,43.

[13] Yebra-Biurrun, M.C, 2009. Flow injection analysis of marine samples. Nova Science Publishers, Inc. New York,21.

[14] Guo-QiZhangaYong-ShengLia,YangZhaoaYa-XiongLuobXiu-FengGaob.2020 A rapid and automated flow injection spectrophotometric determination method for pioglitazone/metformin hydrochloride in pharmaceutical preparation and to confirmation of its reaction principle using PCX column, Microchemical Journal, Volume 155, June 104698.

[15] Namea. A.S., Al-Sowdani. K.H..2020 Lab-build very simple fluorimeter flow injection system with assistance of 3D printer ,European Journal of Molecular and Clinical Medicine, 7 3818–3827.

Nano Silicate Ceramic Coating Material Applied to Increase the Emissivity and Efficiency of Primary Reformer Radiant

Basim A. Abdul Hayi [1] and Abbas. hyal [2]
Research and Development Department, The State Company for Iron and Steel Company Basra, Iraq, Email: basim.11@yahoo.com

Abstract

Nano silicate ceramic material prepared in this research for coating the walls of the primary reformer unit in the state company of fertilizers. It is applied to increase the emissivity and thermal efficiency to improve the properties of insulation materials. Moreover, it is increasing the working operation time for catalyst tubes and reducing the carbon dioxide emissions. Nano coating materials were prepared by using different refractory materials that include silica, filed spar, alumina, manganese dioxide and magnesium oxide. The preparation method was done by, calcination and firing these materials then crashing to get extremely fine powder. The results show that coating materials consist mainly from quartz crystalline and glassy phase structure with a grain size of 158 nm and penetration between 10 to 20 mm when using different concentration 0.5 to 1.0 % for total mix. The emissivity of primary reformer insulation materials was increased with 30 % comparing with insulation materials without coating.

Keywords: High emissivity coated, glassy phase, insulation, reformer, refractory.

1. Introduction

The main problems that facing reformers are include energy liberated transferred by radiation and increase in the fuel consumption. which cost the company more money and time to replace and repair the damage area of reformers. Therefore, many scientist and researchers have been focusing on this problem by using different types of refractory materials and make it as a particles to get the best specification and insulation surface[1]. Emissivity of any surface is a function of wavelength, direction, temperature and surface conditions which is defined as the radiation of materials. The effectiveness of the energy as thermal radiation includes visible radiation and infrared radiation. Also, it is defined as a measuring of the efficiency of surface radiation in which the surface emits thermal energy. Recently, researchers have been focused on the development of high emissivity coated by using ceramic fibers and refractory bricks for furnaces and reformer sections in the refining, petrochemical and iron steel plants [2] . I. Benko studied a new method which increased the radiation heat transfer of furnace refractory lining through applied a high emissivity coating material to emissivity of ceramic fiber insulations by 45% and Shamotte with 20 % [3]. Baskara Aji Nugraha was increased the production of Ammonia plant by using ceramic coating in the primary and secondary reformers. The ceramic coating was decreased the fuel consumption and the fouling effect that causes scales on catalyst tubes red-colored material [4]. P. B. Jensen et. al applied ceramic coating in furnace walls which lead to increase the emissivity in the furnace box and improve the efficiency with the mechanical properties of bricks [5]. In this research, new types of Nano refractory materials were prepared to applied in the primary reformer for the state company for fertilizing company in the south region of Iraq. The ceramic coating material that prepared was examined and testing by using different device such as XRD, EDX, SEM, XRF and FTIR with Master Sizer 2000 Laser to find the grain size.

2. Experimental Work

2.1 Materials

The main materials that used in this research are include silica oxide SiO_2, Alumina oxide Al_2O_3, Magnesium Oxide MgO, Calcium oxide CaO, Manganese oxide MnO and Calcium Florid CaF_2 with sodium silicate and potassium silicate as binding materials as shown in Table 1.

Table 1. Specification of liquid binder

Binder	% Na_2O : K_2O	SiO_2:Na_2O	SiO_2:K_2O
Sodium Silicate	12	3.5:1	-------
Potassium Silicate	15	-------	4:1

2.2 Preparation Procedure

Different percentages of fine materials that include SiO_2, MgO, MnO_2, CaO, CaF_2 and Al_2O_3 were mixed by using potassium silicate and sodium silicate with percentage of 85 % w/v solution. They were mixed for 10 minutes and calcinated in the furnace with temperatures between 700 -800 °C for 10-15 minutes as shown in Fig.1and Fig. 2.

Intensive experimental work has been made to change the oxides percentage of the composition of the ceramic coating by selecting the class of basic or acidic oxides with different temperatures at 800, 1000, 1100 and 1200 °C.

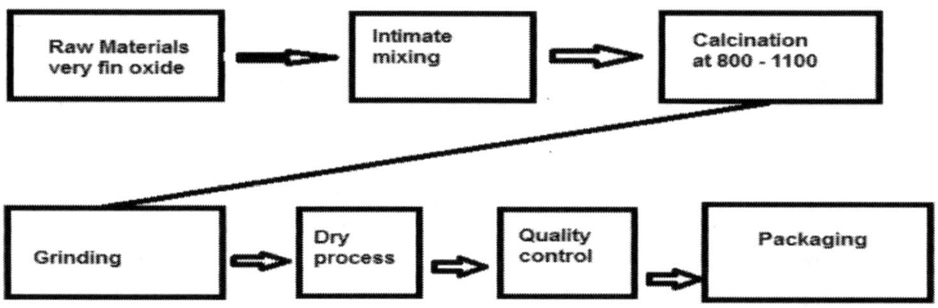

Figure 1. Schematic prepares ceramic coating.

Figure 2. Glassy phase of ceramic coating in 1100°C.

3. Results and discussion

3.1 Chemical Composition

The chemical composition of ceramic coating materials that prepared in this research shown in Table 2.

Table 2. Chemical composition of silicate ceramic

$SiO_2 - Al_2O_3$ %	MgO %	CaO %	MnO %	CaF_2 %
65	15	20	20	10

3.2 Morphology

The samples that were gold-coated were examined in a INSPECT S 50 that is supplied by FEI scanning electron micro-scope coupled to an energy dispersive spectrometer EDS of sintering sample at1100 ^{0}C has been taken in different magnification at 5000 X to 250 000 X. The scanning electron microscopic picture shows that there is a multi-lamellar quartz grains as present in figures 3 and 4. These resulted enhanced the capacity of coating by increasing the emissivity and prevented permeability acid or base gases to walls. Therefor increasing the durability of the bricks and decreased the corrosion rate. Moreover, thermal efficiency of the furnaces box, uniform temperature distribution and life time of the catalyst tubes were promoted.

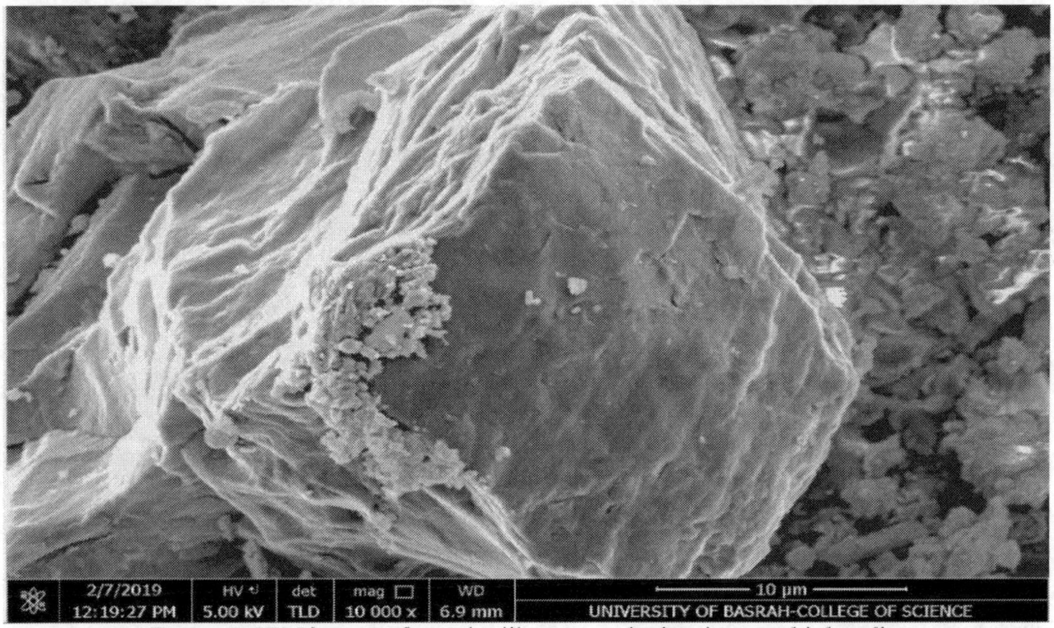

Figure 3. SEM images of sample silicate ceramic showing a multi- lamellar

Figure 4. SEM images of sample silicate ceramic showing crystalline structure

Journal of Physics: Conference Series **2063** (2021) 012019 doi:10.1088/1742-6596/2063/1/012019

The atomic percentage of the elements was found by using EDXS analysis device and the results show in Table 3 and Figure 5.

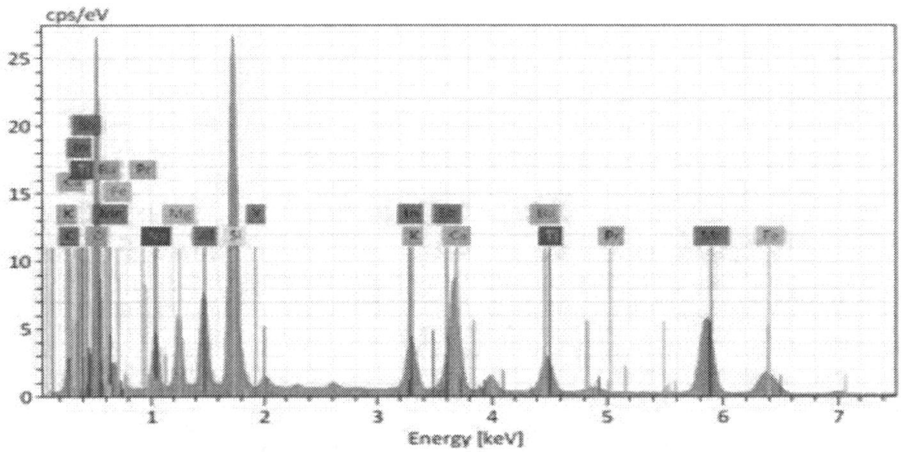

Figure 5. EDXS analysis of sintering phases showing high amount of silicon and other oxides.

Table 3. Atomic percentage of the samples found from EDXS analysis

Element	O	Si	Mn	Al	Na	Ca	Mg
Percentage %	65	12	4	2	4	4	3

Table 3 indicates that the elements found in the highest quantities were oxygen and silicon in addition to manganese, calcium, aluminum and sodium. This was expected, since silica SiO_2 alumina and Al_2O_3 are common constituents in most ceramic materials.

3.3 Effect Nano Silicate Penetration on Coating Behavior.

In this study the effect of Nano particles size by measuring penetration inside the refractory bricks was carried out by addition 0.5 to 1 % at 1100°C of Nano particles in total mix after inspection the grain size in Master Sizer AD 2000 from MALVERN INSTRUMENTS LIMITED made in U.K. The results shown in Fig. 6 the particle size distribution between 158 nm to 549 nm and the cross section between 10 to 20 mm show in Fig 7.

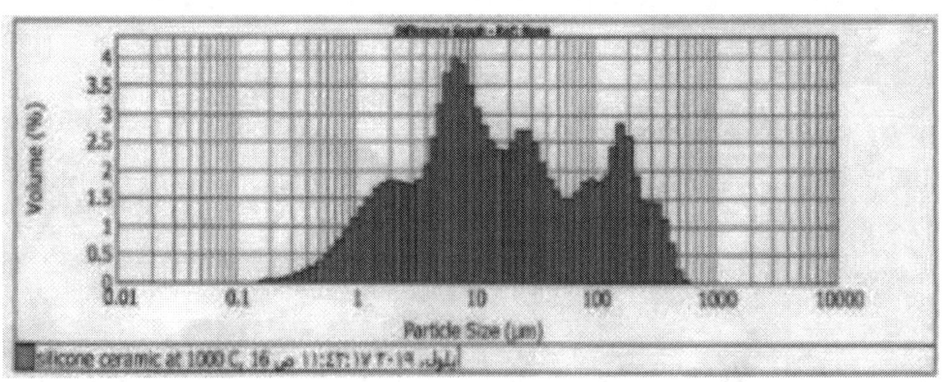

Figure 6. Distribution of grain size of glassy sample.

Journal of Physics: Conference Series **2063** (2021) 012019 doi:10.1088/1742-6596/2063/1/012019

Figure 7. Measuring nano Silicate ceramic coated penetration.

3.4 XRD and XRF analysis

The measurements were taken at room temperature in a Panalytical (Xpert-pro) pw, 3050/ 60 diffract meter equipped with a copper tube, with a 40 kV x 20 mA power supply, which is available at university of Basra. The results are shown in Table 4 and Figure 8 shown high intensity with a very high quality of quartz crystallinity in (2Ø) 26.50 of sintering degree of sintering and glassy phase. Moreover, the results in XRF were taken at room temperature in a Panalytical (Epsilon 3-XL) Dy 1540 are shown in Table 5 and in Fig. 9 show the 70 % silica as in glassy phase.

Table 4. XRD analysis for preparing ceramic coating materials

Pos. 2Ø	Height	d-spacing °A
26.641	170.46	3.34600
42.835	0.5904	2.11122

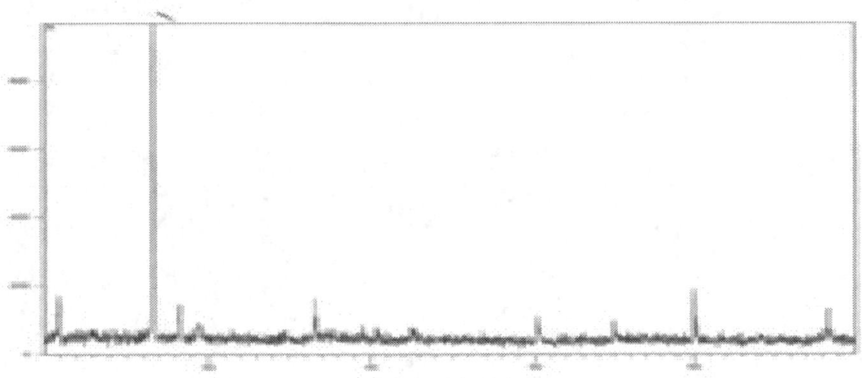

Figure 8. XRD Pattern of glassy phase sample showing high intensities of quartz.

Table 5. XRF results of glassy phase.

Element	Si	Mn	Al	Na	Ca	Mg
Percentage %	55	4.2	7	3	5	11

Journal of Physics: Conference Series **2063** (2021) 012019 doi:10.1088/1742-6596/2063/1/012019

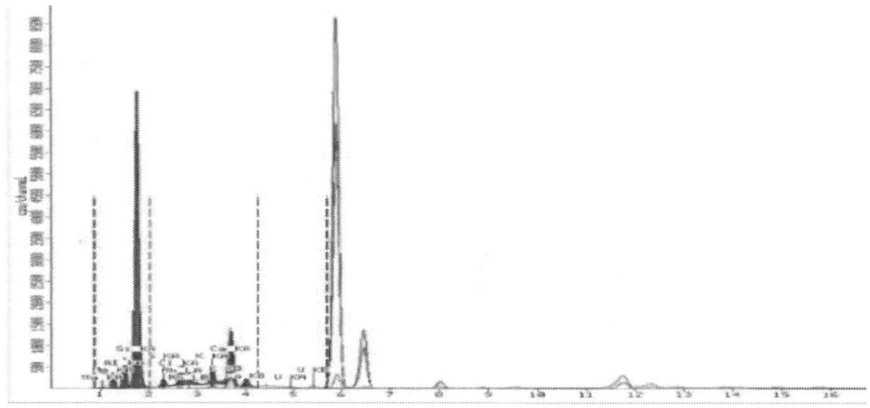

Figure 9. XRF analysis of sintering and glassy phase.

3.5 FTIR spectroscopy analyzed.

The measuring was identified from 400 to 4000 cm^{-1} using a FTIR model 8400 S Fourier transform infrared spectrophotometer of transmittance spectra of samples sintering in 1000 °C and glassy phase in 1100 °C as shown in Fig. 10 and the assignment of the bands are shown in Table 6. The characteristic bands almost at 1043, 877 and 504 cm^{-1} are corresponding to the stretching bending and out of plane of Si–O bonds, respectively. The position and the shape of the main Si–O vibrational band at 1043 cm^{-1} show a stoichiometry silicon dioxide structure. Moreover, some impurity vibrational bands are seen in the FTIR spectra which were shown in Table 6.

Table 6. FTIR spectroscopic analysis for preparing ceramic coating materials

Functional groups	Sintering sample cm^{-1}	Glassy sample cm^{-1}
Si- O deformation	458	473
Si- O bending	877	870
Si- OH stretching	940	935
Si- O- Si stretching	1042	1050
C- O bending	1622	1623

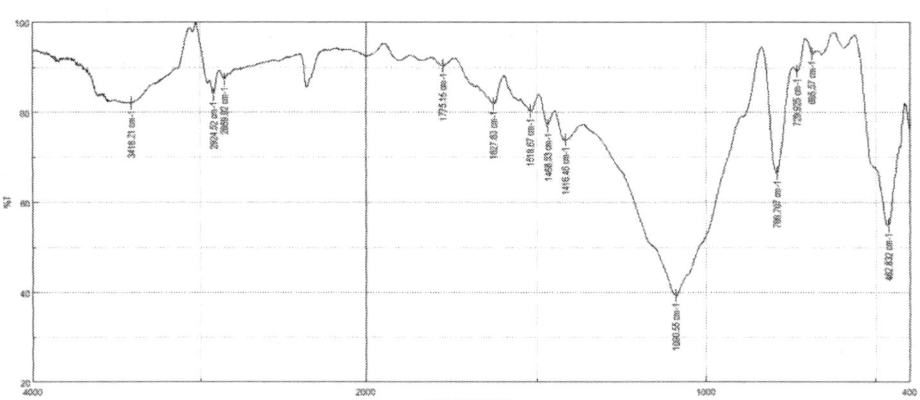

Figure 10. FTIR spectra of both sintering and glassy phase.

4. Application:

The preparation ceramic coating materials that prepared in this research was applied in the primary Reformer of the State company of Fertilizer in the south region of Iraq. It is used to coat Shamotte bricks that lined in the laboratory furnace with temperature of 1100 °C and the results shown in Fig. 11 and 12 as the glassy phase. Therefore, the ceramic coating material was applied to coat the gates bricks of reformer. Heating of walls and coated position was measured by using Infrared IR device and the result shows in Table 7. The results show that temperature in the walls and gates was decreased with 27°C.

Figure 11. Shamotte brick, coated and glassy ceramic coated

(B) (A)

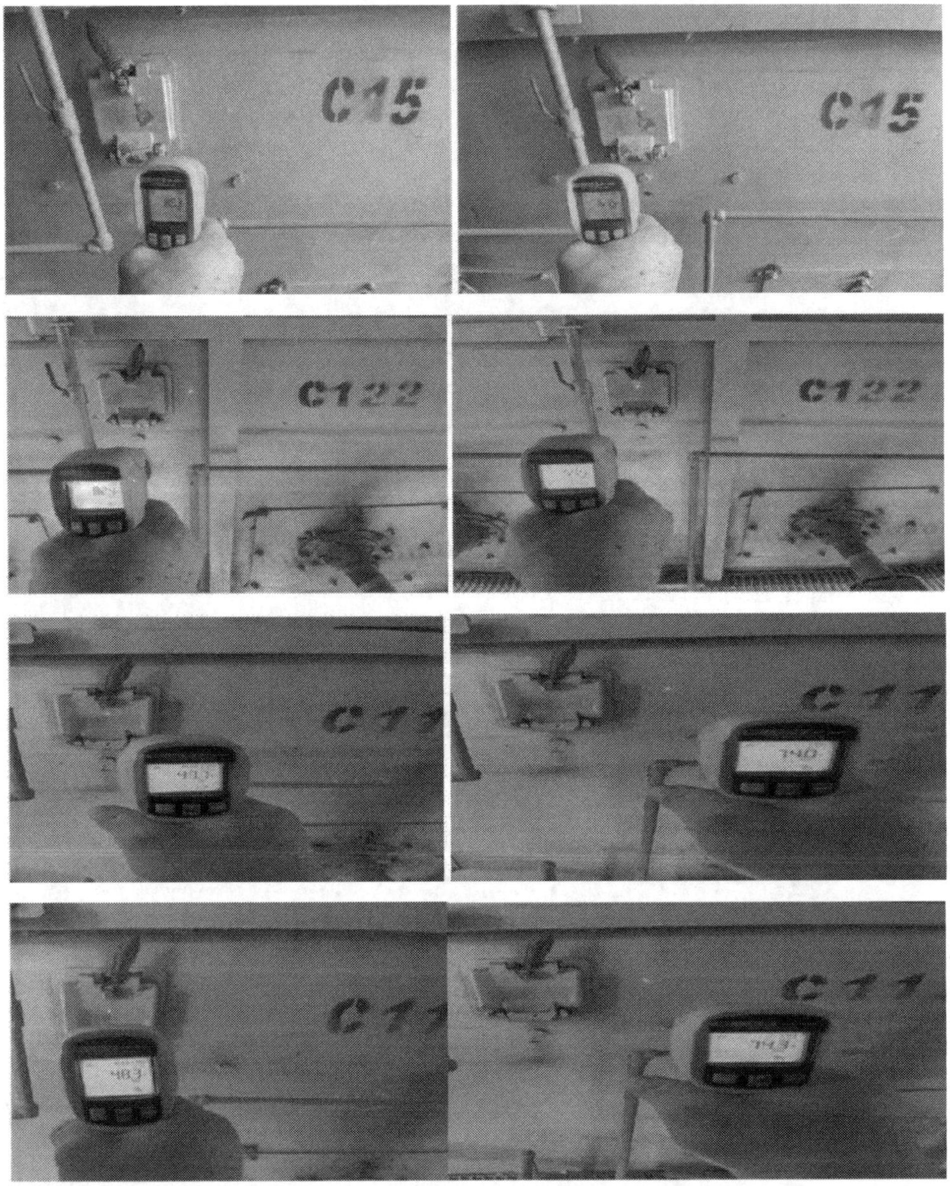

Figure 12. Heating measured by IR for walls and gates (A) without coated and (B) with coated.

Table 7. Results analysis by infrared IR device for gates reformer positions and walls

Number Position	Temperature Walls °C	Temperature without coated °C	Temperature coated °C
C 15	89.3	80	61.5
C 122	82.4	85	54.4
C 111	74.0	78	47.1
C 113	74.3	82	48.3

Journal of Physics: Conference Series **2063** (2021) 012019 doi:10.1088/1742-6596/2063/1/012019

5. Conclusion:

1. The emissivity of primary reformer insulation materials was increased by 30 % compared with insulation without coating.
2. Ceramic coating material that applied lead to produce quartz phase.
3. High amount of silicate and multi-lamellar were obtained after sintering with glassy phase.
4. Ceramic coating material was penetrated with thickness of 10–20 mm in the refractory bricks at sintering temperature of 1100°C.

References

[1] Peter Bruun Jensen, Venkat Pattabthula , John Bacon and Darius Pasaribu" High Emissivity Ceramic Coating of Furnce Walls in Tubular Reformers" Ammonia Technical Manual, January, 2012

[2] C. Ohlhorst, W.L. Vaughn and J.R.Koenig" Emissivity Results on High Temperature Coatings for Refractory Composite Materials" Materials Science. 2007.

[3] I. BENKO, " ENERGY CONSERVATION THROUGH INCREASED EMISSIVITY IN FURNACES", PERIODICA POLYTECHNICA SER. MECH. ENG, VOL.35, No. 4,PP. 235-245 , 1991.

[4] Baskara Aji Nugraha ," Coating Primary Reformers Radiant Section", Journal Rekayasa Proses, Vol. 7, No. 1, 2013.

[5] P.B. Jensen e P.B.Jensen, V.Pattabathula, S.Maule and J.Bacon " High emissivity ceramic coating of furnace walls in tubular reformers theory and practice" American Institute of Chemical Engineers in Safety in ammonia plants and related facilities symposium, pp 53-64, 2012.

[6] Navee Angsuputiphant, " Innovative high emissivity coating for energy saving in industrial furnaces" , , June 8 2016, Member of technology Business in SCG Chmicals., Co,Ltd. APMP-Midyear Symposium, DusitThaui Hall.

[7] Craig.W. Ohlhorst. Wallace. L. Vaughn. Kamran Daryabeigi Ronaid. K. Lewis. Alvaro C. Rodriguez James. D. milhoan John R. Koening" Emissivity Results on High Coating for Refractory Composite Materials" NASA Langley Research Center, MS 188A, Hampton, VA 23681, USA.

[8] J Beacon, "Technical Introduction to Cetek Ceramic Coating Technology", 2009, PETROTHCH, New Delhi, India.

Estimation and evaluation of medicinal efficacy of maximum inhibitory concentration of alkaloids extracted from Iraqi *Cordia myxa* leaves against some pathogenic bacteria.

Abbas D M Al-Maliki [1] , Maitham N Aboud [1] , Mustafa F Hameed [2]
[1] Dep. of Chemistry, College of Education for Pure Sciences,
University of Basrah, Basrah, Iraq .
[2] Basra Education Directorate, Basrah , Iraq .
corresponding author's e-mail address : mustafafadil837@gmail.com

Abstract. *Cordia myxa* is considered as one of traditional medicinal plants abundant in many countries including Iraq . Because of the healthy importance of this plant therefore the current research was established to extract the alkaloids from the its leaves with extraction percentage equal to 5.8% Preliminary qualitative detections were achieved for the alkaloids by using various chemical reagents. The medicinal efficacy of these alkaloids was determined depending on maximum inhibitory concentration (Max IC) which has value equal to 150 mg/ml and this concentration recorded inhibition zone diameters values equal to 30,22,26,25 and 30 mm against *Staphylococcus aureus*, *Escherichia coli*, *Pseudomonas aeroginosa*, *Proteus Sp.* and *klebsiella Sp.* bacteria. So *Cordia myxa* leaves alkaloids can be used as natural chemical drug to treat the most infections and inflammatory caused by these pathogens .

1. Introduction

The healthy value and medical importance of various traditional plants belong to existence of many active natural metabolites in the different parts of medicinal plants spreading in the world-wide. The chemical production of these active biochemical compounds takes place through secondary metabolism pathways [1] [2]. Various current studies proved the advantages of the environmental natural plants because of their important sources to produce drugs belonging to nature origin. Excellent therapies were isolated, purified and characterized from variety of medicinal plants depending on their big biochemical efficacy and chemical ability to treat various infections and inflammatory caused by different microorganisms. The necessary benefit of the active metabolites comes from they have no side effects opposite synthetic drugs[3][4]. The healthy and medical importance of various traditional medicinal plants results from existence of many potent chemical families as secondary metabolites are anabolized biochemically in different parts of these plants according to metabolic pathways catalyzed by required enzymes. The active metabolic compounds are represented by phenolic compounds, glycosides, terpenes, alkaloids, essential oils, flavonoids, coumarins, xanthense, tannins, steroids and Saponin[5][6][7] .

Alkaloids are considered as one of active secondary chemical metabolites abundant in various medicinal plants and they are chemically heterocyclic basic nitrogenous compounds have one or more nitrogen ring. Also the molecule of alkaloid has nitrogen atom is attached to at least two carbon atoms. The alkaloidic compounds are derived basically from primary metabolism especially from amino acids. Alkaloids have medicinal and dramatic properties and they were used successfully as antibacterial, antifungal, analgesics, miotics, antileukemic and antifungal agents [8][9] .

Cordia myxa is considered as one of medicinal traditional plants spreading in different countries in the world. It is a flowing species belongs to Boraginaceae family and the size of this plant is medium has wide leaves and the tree is deciduous. There are common names for *Cordia myxa* such as pidar, lasura, spistan, naruvilli and wanan. Also this plant is used as popular food in various countries and a good substance to produce materials such as potassium, calcium, zinc and iron [10][11] . Various studies were achieved to examine the chemical content of *Cordia myxa* parts especially fruits where glycosides, gyms ,tannins, alkaloids, coumarins, flavonoids and phenolics were tested qualitatively. Also the therapeutic importance of the fruits as anti-inflammatory and antiparasitic were evaluated [12][13] . The chemical constituents in fractions of *Cordia* oblique in methanolic extract of leaves were estimated and determined by gas chromatography- mass spectroscopy technique. A research indicated presence of sesquiterpene, amino acids, azafrin, oleic acid, phthalic acid, steroids, morphinan and galactopyranoside [14] . Also alkaloids of *Cordia myxa* leaves were extracted and they were examined on survival of mustard aphid [15] .

The present research was done for estimation, investigation and evaluation the biochemical and medicinal efficacy of alkaloids extracted from *Cardio myxa* leaves at maximum inhibitory concentration against growth of five pathogenic bacteria.

2. Materials and Methods

2.1 Preparation of plant

The Leaves belonging to *Cordia myxa* were collected from Al-Qurna distric at Basrah Governorate in Iraq. The plant was taxonomied, cleaned with distilled water, dried, powered and kept in a dark glass container for the sake of application of various analysis.

2.2 Extraction of alkaloids from Cordia myxa

Journal of Physics: Conference Series **2063** (2021) 012020 doi:10.1088/1742-6596/2063/1/012020

Twenty five grams of *Cordia myxa* leaves powder were added into 250 ml of ethanolic-acetic acid (10% v/v) and the mixture was stirred by using magnetic stirrer for eight hours. Then the contents were filtered and the precipitate was removed. The filtrate was concentrated to quarter of its volume then 5 ml of concentrated sulphuric acid was added gently and the mixture was shaken for 3 minutes. After that the acidic mixture was treated with a volume of ammonia for adjusting pH to 9 then the contents were put in separation funnel by adding 20 ml of chloroform for three times. The mixture was shaken well then two layers were formed and the aqueous layer was removed whereas organic layer hasing alkaloids, was gotten then it was dried and the weight of alkaloidic Extract was 1.45 grams[16]

2.3 Qualitative Detections Analysis of Alkaloids

The extracted alkaloidic compounds belonging to leaves of *Cordia myxa* were qualitatively tested by using various chemical reagents such as Dragendroft reagent for alkaloids, Molisch for carbohydrates, ferric chloride(1% w/v) for phenolic compounds, Benedict for glycosides ,ethanolic potassium hydroxide(5N) for flavonoids, ninhydrin(1% w/v) for amino acids, mercuric chloride(5% w/v) for saponin and lead acetate (1% w/v) for tanninic compounds [17][18] .

2.4 Culture Medium for pathogens

The pathogenic bacteria were treated with Muller-Hinton Agar (MHA) as a culture medium which was gotten from laboratories of marine science Centre in Basrah university-Iraq. It was prepared depending on all information's determined by company of manufacturing

2.4.1 Pathogenic microorganisms in the study

The pathogenic microorganisms are bacteria strains are characterized to know the identity of all bacteria represented by *Staphylococcus aureus, Proteus sp., Klebsiella sp., pseudomonas aeroginosa* and *Escherichia coli*.

2.4.2 Determination of Medicinal Efficacy of Alkaloidic Extract

The extracted alkaloids belonging to leaves of *Cordia myxa* were tested against pathogenic bacteria by using various concentrations then the maximum inhibitory concentration (150mg/ml) was carried out against these microorganisms by Agar-diffusion method by using many wells. Twenty mL of culture medium was put in petri dish then the medium was treated with 0.1 mL of bacterial suspension by using sterilized glass spreader . That 0.2 ml of Max IC belonging to alkaloids was added in each well. Finally the dishes were put in the incubator at 37 °C for one day then diameters of inhibition zones were measured [19] .

3. Results and discussion

The increasing medicinal necessity for use of various potent chemical compounds extracted and identified from variety of herbs and plants led to increase the demands to get natural therapeutic drugs from plants origin .So from these plants / *Cordia myxa* which is considered as one of feeding sources for human and animal together . Also abundarce of many effective metabolitic compounds ensured the importance of these secondary chemicals as natural therapies for different diseases caused by pathogenic microorganisms .Alkaloids were extracted successfully and purified with yield equal to 1.45 g from 25 g of *Cordia myxa* leaves therefore the extraction percentage was calculated as in table (1) .

Table 1. Extraction percentage of alkaloids extracted from leaves of *Cordia myxa*

Plants leaves weight (g)	Alkaloidic Extract weight (g)	Extraction percentage (%)
25	1.45	5.8

From above table , the extraction percentage was measured from the following law :

$$\text{Extraction percentage} = \frac{\text{Extract weight}}{\text{leaves weight}} \times 100$$

The percentage of extraction of alkaloids is somewhat very good because of existence of other secondary active compounds besides alkaloids specially phenolics, saponin, essential oils and flavonoids [20] . Therefore presence of alkaloidic compounds in *Cordia myxa* insures the bioactivity and medical efficacy of this plant . The variety of chemical reagents which were used to detect qualitatively on different secondary active metabolites was for insurance of existence of only alkaloidic compounds are present in the alkaloidic extract . Therefore orange precipitate was formed in good quantity by using Dragendroff regent and appearance of this precipitate proves abundance of alkaloids in *Cordia myxa* leaves while the other active compounds showed a negative test as in table (2):

Table 2. Qualitative detections analysis of alkaloids extracted from *Cordia myxa* leaves

Reagent type	Result of detection	Chemical result	Conclusion
Dragendroff	++	Orange precipitate	Alkaloids are present
$FeCl_3$ (1%)	-	No green- blusih colour	Phenols are absent
Molisch	-	No violet ring	Carbohydrates are absent
Benedict	-	No red p.p.t	Glycosides are absent
Lead acetate (1%)	-	No light brown p.p.t	Tannins are absent
$HgCl_2$ (5%)	-	No white p.p.t	Saponins are absent
Ethanolic KOH (5N)	-	No yellow p.p.t	Flavonoids are absent
Ninhydrin(1%)	-	No violet colour	Amino acids are absent

Journal of Physics: Conference Series **2063** (2021) 012020 doi:10.1088/1742-6596/2063/1/012020

The existence of alkaloidic compounds only in alkaloids fraction extract proves the accuracy of procedure which was carried out for extracted of alkaloids. The use of sulphuric acid in this method is for reaction with alkaloids because they have basic chemical features [18] . Because the *Cordia myxa* plant has a wide spectrum of potent chemical natural compounds therefore the medicinal efficacy of leaves belonging to this medical plant was investigated, determined and evaluated for Max IC with value equal to 150 mg/ml. So the all pathogenic bacteria used in the current research were underwent the Max IC of *Cordia myxa* leaves as indicated in table (3).

Table 3. Medicinal efficacy of alkaloids extracted from *Cordia myxa* leaves against some pathogens.

Max. inhibitory Concentration (mg/ml)	Bacterial stains	Inhibition zone diameter (mm)
150	*Staphylococcus aureus*	30
150	*Pseudomonas aeroginosa*	26
150	*Proteus sp.*	25
150	*Escherichia coli*	22
150	*Klebsiella sp.*	30

From table (3) , it was noticed that the maximum inhibitory concentration of alkaloids, has recorded the highest inhibition zone diameter against *Staphylococcus aureus* and *klebsiella sp.* Bacteria with value equal to 30 mm but the same Max IC has showed the lowest value of inhibition zone with diameter equal to 22 mm towards *Escherichia coli* bacteria . While the Max IC has given inhibition diameters with value equal to 25 and 26 mm towards growth of *Proteus sp.* and *Pseudomonas aeroginosa* respectively.

It was fourd that *Escherichia coli* showed the greatest resistance towards chemical action of alkaloids because this bacteria has dense lipidic layers in its cell wall then the alkaloidic extract was difficult to inter into the cell of this living microorganism. So the medicinal efficacy has decreased for killing this bacteria wheras the inhibition zone diameter was the highest against *Staphylococcus aureus* because this pathogen has less lipidic layers in this cell wall then the alkaloids can be inter easier into cell of this pathogenic bacteria therefore this extract could kill great number of this pathogen [21][22] .

The biochemical and medicinal potential of alkaloids can be explained by chemical bonding of these active metabolites with nucleic acids represented by DNA and RNA after that this process leads to chemical disorder in the metabolic pathways of these acids leading to inhibit the chemical roles and biological functions belonging to both acids [22][23] . Alkaloidic compounds can also destruct the enzymes which are responsible for biochemical anabolism of nucleic acids such as DNA polymerase and RNA polymerase [24] .

Various botanic studies represented the physiological and chemical properties of alkaloids as potent chemicals in inhibition the biochemical roles responsible for multi-metabolic pathways belonging to lipidic compounds, carbohydrates and proteins. The greatest functional group in alkaloidic compounds is imine which has chemical ability to destruct the pathogenic cell belonging to bacteria [25][26] .

The medicinal role of plants alkaloids can also belongs to capability of these natural metabolic compounds to do many chemical disorders in bio chemical anabolism process of proteins existing in the living cell of pathogenic bacteria leading to big deficiency in the different proteins of microorganism [27][28] .

4. Conclusions

The extracted alkaloids belonging to Iraqi *Cordia myxa* leaves showed an excellent medicinal efficacy against growth of five pathogens represented by *Pseudomonas aeroginosa, Staphylococcus aureus, Proteus sp. ,Escherichia coli* and *Klebsiella sp.* Bacteria. The maximum inhibition concentration (Max IC) proved the highest ability to inhibit or kill most pathogenic bacteria because the diameters values of inhibition zones were very good and they were insured by alkaloidic compounds which showed the high activity to finish or inhibit the biological and chemical systems of these pathogens. So the alkaloids extracted and purified from leaves of *Cordia myxa* can be carried out as natural active compounds for treating the various inflammatory caused by these pathogens instead of some synthetic drugs containing side effects.

5. References

[1] Pramila,D.M.,Xavier, R.; Marimuthu, K.; Kathiresah, S.; Khoo.,M.L.; senthilkumar, M.;Sathya, K. and Steeramanan, S. 2012 . Phytochemical analysis and antimicrobial potential of methanolic leaf extract of peppermint (*mentha piperita*:L.), *Journal of Medicinal Plant Research*, **6**(2):331-335.

[2] Kurian, J.C. 2012. Ethno-medicinal plants of India, *Thailand and Vietnam. Journal Biodiversity*, **3**(1): 61-75.

[3] Al-Maliki, A.D.M. 2011. Isolation and identification of an alkaloidic compound from Corriandrum sativum seeds and study of its medicinal activity against pathogenic bacteria of urinary tracts, *Journal of Basrah Researches (Sciences)* ,**37**(2):119-127.

[4] Britto, J.; Steena, R.S. and Mary.S.R. .and Oksman,K.M. 2011. Phytochemical analysis of eight medicinal plants of Lamiaceae. *Journal Research Plant Sciences*,**1**(1):1-6.

[5] Rehab,A.H. and Amira A.E. 2019. *Plants secondary metabolites*, Article puplication, intechopen, Cairo, Egypt.

[6] Suvi, T.H.; Anneli, R.; Heiko,R. 2012. *Medicinal plants, Engineering of secondary metabolites in cell cultures, Encyclopedia of sustain ability science and technology*, Roberts and Meyers Edit., NewYork, USA.

[7] Jean, C.C.; Maria, E.BS de O. and Fernando, CI.C., Advances and challenges on the in vitro production of secondary metabolites from medicinal plants, Horticultura Brasileira , 2019,**37**(2):124-132.

[8] Harborne, J. and Baxter, H. 1993 . *phytochemical dictionary.* Tylor and Frances publishers,

London UK.

[9] Padmini, E., V.alarmathi, A. and Usha, R.M. 2010. Comparative analysis of chamical composition and antibacterial activities of Mentha spicata and Camillia sciences. *ASIAN JOURNAL OF EXPERIMENTAL BIOLOGICAL SCIENCES*,**1**(4): 772-781.

[10] Masoud, K.; Eskandar, M. and Sayyed, F. S. 2015. preparation and evaluation of Cordia mayxa fruit topical cream .*World Journal of Pharmaceutical Research*, **4**(7):244-253.

[11] Ranjbar, M.M.; Najafzadeh, H.V.; Sabbagh, A.; Bolooki, A. and Sazmand, A.R. 2013. Study on analgesic and anti-inflammatory properties of *Cordia myxa* fruit hydro-alcoholic extract. *Pakistan Journal of Biological Sciences* ,**16**(24):2066-2069.

[12] Al-Snafi, A.E. 2016. The pharmacological and therapeutic importance of *Cordia myxa*-A review, *Journal Pharamacy*, **6**(6):47-57.

[13] Al-Snafi, A.E. 2015. Therapeutic properties of medicinal plants: a review of plants with hypolpidemic, hemostatic, fibrinolytic and anticoagulant effects(part1). *Asian Journal of Pharmacy and Technology,* **5**(4):271-284.

[14] Gupta, R. and Gupta, G.D. 2017. Determination of plant constituent in fractions of *Cordia obliquo* willd. Leaf methanol extract using GC-MS analysis. *International Journal of Pharma and Chemical Research*,**9**(9):1274-1279.

[15] Hind, S.A. 2016. The effect of alkaloids extracted from *Cordia myxa* leaves on the survival of Mustard Aphid Lipaphis erysimi(Kalt.). *Journal of Chemical, Biological and Physical Sciences*,**6**(1):84-94.

[16] Harbome,J. 1984 . *Phytochemical methods*,2nd.ed. ,Chapman and Hall,London, United Kingdom.

[17] Cowan, M. 1999. *Plant products as antimicrobial agents. Clinical microbiology reviews* ,12:564-582.

[18] Harborne, J. 1974. *Phytochemical methods*, 1st.ed. , Chapman and Hall, New York, USA.

[19] Leen,O,Ahmad,S. and Roula, M.A. 2019. *Antimicrobal activity of polyphenols and and alkaloids in middle Eastern plants. Front Microbial.* **10**:911-940.

[20] Alaa,T. 2016. Effect of some chemical elicitors on some secondary metaboliol inductions of *Cordia myxa* L. in Vitro. *Journal of Pharmaceutical and Biological Sciences* , **6**(5):2250-3013.

[21] Collee, J.; Frasher, A., Marimion, B. and Bimon, A., Practical Medical microbiology. Makie and McCarteney. 5th. Ed., Chrchill Livingston, New York, USA.

[22] Jayasurriya, H.; Nupharan, M.K.; Geahlen, R.L., McLanghlin, J.L. and Chang, C.J. 1991. Emodine, a protein Kinase inhibitor from polygonameup. *Journal of Natural Products*,**55**(5):696-703.

[23] Swapna, G. and Estari, M. 2017. In vitro antibacterial activity of alkaloids isolated from Leaves of *Eclipta alba* against human pathogenic bacteria, *Pharmacognosy Journal*, **9**(4):573-577.

[24] Tasiu, I. 2016. Anticancer Alkaloids from trees: Development into drugs. *Pharmacognosy Reviews*,**10**(20):90-99.

[25] Thamir, M.J.; Suhad,F., H.A; Ibrahim, S.A and Qabas, N.L. 2016. Phytochemical study and antibacterial activity of crude alkaloids and Mucilage of *Cordia myxa* in Iraq. *International Journal of Pharmaceutical Sciences Review and Research*, **39**(1):332-336.

[26] Pankaj, B.N.; Nayan, R.B.; Shukla, V.J. and Acharya, R. N. 2011. Antimicrobial and antifungal activities of *Cordia dichtoma*(Forster F.) bark extracts. *Ayu*, **32**(4): 585-589.

[27] Wink, M. and Twardowski, T. 1992. Allelo chemical properties of alkaloids. Effect on plants, bacteria and protein biosynthesis, *Allelopathy.*, **2**:129-150.

[28] Al-Saadi, J.H.H.; Al Maliki, A. D. M. and Aouda, M.A. 2016. Medicinal activity of alkalodic compound 1 H-pyrido{2,3-b} indole Isolated feom *Solanum melongena* cortex against some phathogenic bacteria. *Journal of Natural Sciences Research* , **6**(20): 80-85.

Isolation and identification of some pathogenic species of Fungi from Shatt-al-Arab River

Shrouk Abdulrazak Hassan AL- Ibraheem[1] , Angham O S Al- Zeadei[2]
[1]Southern Technical University / College of health and medical technology, Iraq. [2] Southern Technical University /Basra Engineering Technical College, Iraq.
[1]*Shrouk.albraheem@stu.edu.iq*, [2]*angham.omeer@stu.edu.iq*

Abstract. This study aimed to isolation and identification of pathogenic fungi from Shatt – al-Arab River in Basra city , Fourteen water samples were collected from different area from Shatt- al-Arab River (AL Ashar, AL Tnoma, AL Makal, AL Qurna, AL Karma, AL Jabiluh, AL- Hartha), from October to December in 2017, with 250 ml volume, this samples centrifuged at 5000 rpm for 10 min at room temperature , the floating was removed and then take the precipitate and pour directly into the center of the media of SDA and PDA and then incubation in a temperature range25-27c for 4 days after that the growth on the media made pure culture and each fungi species diagnosed based on the cultural and microbiological phenotypes, smear prepared with lacto phenol cotton blue stain and the results show 57.1% of growth was *Aspergillus niger*, 85.7% *Aspergillus flavus and 42.8%* was *Aspergillus candidus and* 14.2% was *Rhizopus,* while the results show 42.8% of growth was *Penicillium..*

1. Introduction :

The waters of Shatt al-Arab River suffers from serious pollution, which is evident in many types of fish that settled in its waters, and some of them have disappeared permanently, The causes of pollution in Shatt al-Arab River water vary, starting with wastewater, household waste, industrial and oil waste in addition to the salty tide flowing from the Arabian Gulf. Pathogenic and non-pathogenic microorganisms are commonly found in rivers. [1]. Infections acquired by recreational water interaction are normally mild, rendering routine surveillance systems difficult to detect. And if the disease is more serious, it can be difficult to connect it to the water. Targeted epidemiological studies, on the other hand, have linked faecally contaminated recreational water to a variety of negative health outcomes (including gastrointestinal and respiratory infections).Targeted epidemiological studies, on the other hand, have linked faecally contaminated recreational water to a variety of negative health outcomes (including gastrointestinal and respiratory infections). This can lead to a serious illness. burden and financial loss [2]. The pathogen, its type, and the number of microorganisms that can cause infection or disease determine the number of microorganisms that can cause infection or disease, the conditions of exposure, and the host's susceptibility and immune status. In the case of viral and parasitic protozoan infections, this dosage may be as low as a few viable infectious units [3]. Kelley ***et.al.,*** 2003 provided a comprehensive investigation of filamentous fungi in US distribution systems. *A.alternate, A. niger,Cladosporium sp, Epiccocum nigrum*, *Penicillium sp*. In Norway, the *Aspergillus, Penicillium and Trichoderma* are detected in drinking water, surface and ground water [4]. While in Pakistan the *Aspergillus niger* and *A. clavatus* are isolated from water and fruit juice [5].In addition the *Cladosporium, Penicillium, Aspergillus* and *Fusarium* are isolated from municipal water of Australia [6]. While in Brazil the *Penicillium, Aspergillus* and *Phoma* are detected in water treatment plant and tap water [7]. As well as the Aspergillus, Cladosporium and Penicillium are isolated from surface, spring and ground water of Portugal [8,9].

When the world's population rises, so does the demand for water, placing greater pressure on these resources [10]. Previous contributions on fungi found in the surface sediments of the Shatt Al-Arab River in Iraq, on the other hand, were based on taxonomic studies that described several new and interesting species. [11, 12, 13, 14 and 15]. Studies on the occurrence of thermophilic and thermo resistant bacteria in Shatt Al-Arab River sediments, as well as the incidence of dermatophytes and other keratinophilic fungi in Shatt Al-Arab surface sediments, were also carried out [16].
One of the most significant prerequisites for a healthy life is an ample supply of clean drinking water. When a substance is introduced to our natural world, it has a detrimental impact on nature's ability to dispose of it. A pollutant is something that has a negative impact on people's health, comfort, land, or climate. The majority of contaminants are released into the atmosphere by sewage. [17].
Filamentous fungus in drinking water has become a serious issue. One of the most important issues was the identification of isolated strains to species rather than genus level, which was a direct result of the efforts of key researchers in the field. More information about changes in biodiversity within locations and over time can be gleaned when species identification is completed. Characteristics of the isolates. [18].
According to the importance of Shatt-al-Arab River for Basra city and there is no study about biological pollutant from different source, therefore the study aimed to isolation and identification of some pathogenic fungi from different parts of it.

2. MATERIALS AND METHOD:

Fourteen samples are taken from various Shatt al-Arab River locations (AL Ashar,AL Tnoma, AL Makal,AL Qurna,AL Karma, AL Jabiluh, AL-Hartha),from October to December in 2017, with 250 ml volume under deep 15 cm of water then taken to the lab to investigation for the micro pollutant in water [19]. This samples were centrifuged at 5000 rpm for 10 min. at room temperature, the floating was removed and then take the precipitate and pour directly into the center of the mediaof SDA and PDA and then incubation in a temperature range25-27c for 4 days after that the growth on the media made,pure culture and each fungi species diagnosed based on the cultural and microbiological phenotypes, smear prepared made with lacto phenolcotton blue stain [20].

3. Results and Discussion:

The results show 57.1% of growth was *Aspergillusniger*, 85.7% *Aspergillus flavus and 42.8%* was *Aspergillus candidus and* 14.2% was *Rhizopus*, while the results show 42.8% of growth was *Penicillium*.. Table (1) show the different species of fungi isolating from different site of Shat Al-Arab, and the table (2) show the percentages occurrence of fungi species in Shatt Al- Arab River.

Table (1): fungi detected from different sites from Shat al-Arab River

Site of collection	Type of fungi
Al-Ashar	*Rhizopus , A.Flavus,A. candidus*
Al-Garma	*A. niger*
Al-Jobiluh	*A. flavus*
Al-Hartha	*A. flavus*
Al-Qurna	*A. niger, A. candidus, A. flavus*

Al-Tanoma	*A. niger, A. flavus*
Al-Makael	*A.niger, A.candidus , A.Flavus*

Table (2) show the percentages occurrence of fungi species in Shatt al- Arab River

No	Fungi	Number of appearance	Percentage of occurrence
1	*A. niger*	8	57.1%
2	*A. flavus*	12	85.7%
3	*A. candidus*	6	42.8%
4	*Rhizopus* sp.	2	14.2%
5	*Penicillium sp.*	6	42.8%

3.1 *Aspergillus niger:*

3.1.1 Colony morphology:
when culture on SDA the Conida produce the pigment aspergillin during maturation, Starting with a white to yellow felt-like layer of mycelia, paint it black, The Reverse is white or pale in color . [21] as appear in figure 1-A.

3.1.2 Microscopic Morphology:
The hyphae are Septet and hyaline , Conidiophores are long (450-3000 M) with spherical vesicles measuring 35-75 m at the apex. Metulae almost completely cover the entire surface from which the phialides extend in ***A.niger***. Conidia have a globose shape, are brown to black in colour, have a rugged base, and have a diameter of 3.5-4.5 M. hyaline septet, [21]. As shown in Figure1 – B.

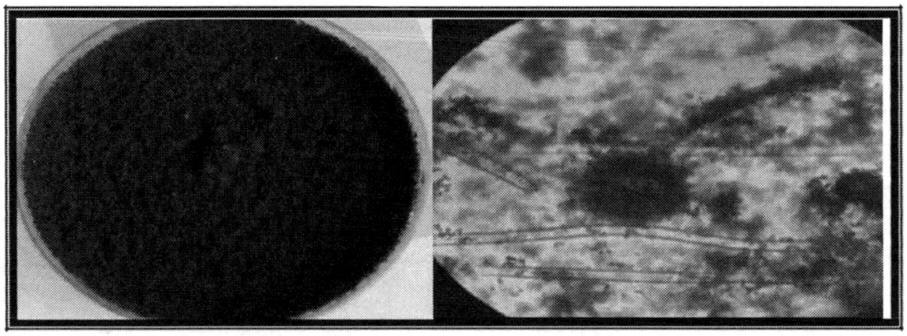

A B
Figure (1):*Aspergillus niger* on SDA(400 x)

3.2 Aspergillus flavus:

3.2.1 Colony Morphology:

Growing at a breakneck pace, In just three days, it will be fully mature. The color ranges from greenish-yellow to olive, with a white line running along it. The texture is often floccose, particularly near the middle, and can range from velvety to woolly in appearance. On Sda newspapers, reverses from cream to tan to yellowish is unremarkable. [22]. As appear in Figure 2-A.

3.2.2 Morphology at the microscopic level:

The hyphae are septate have long conidiophores (420-800 X 8.5-17 M) particularly just below the vesicle, with a rough or even spiny texture. Vesicles range in size from spherical to elongate, with a width of 20 to 45 meters. Seriation in A. flavus is variable, with most strains being around 20% biseriate, but some strains being almost entirely uniseriate. Metulae (8-10 X 5-7 m) cover three-quarters of the vesicle's surface, from which the phialides (7-12 X 3-4 m) emerge. The phialides are supported by the metulae, which together form the biseriate structure. Conidia have smooth to finely roughened walls and vary in size from globose to ellipsoidal (3-6 m). [22]. Figure 2-B.

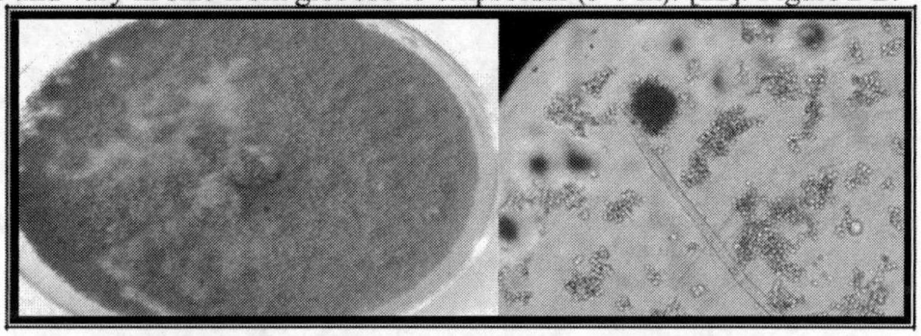

A B

Figure(2): *Aspergillus flavus* on SDA(400x)

3.3 Aspergillus candidus:

3.3.1 Colony Morphology:

A. candidus is a fast-growing fungus that matures in about three days. The surface color of colonies can range from a yellowish brown to a drab olive with grays. They may have a lighter outer lip, and droplets of purple exudates may appear on the maturing colony's surface. The back of the card is a yellowish brown color. The texture was even and velvety. [23].Figure3-A.

3.3.2 Morphology at the microscopic level:

Septate hyphae are produced by ***A. candidus***, from which smooth-walled conidiophore stipes extend. As conidiophores mature, they can turn a brownish color. Stipes are usually small (130 m to 300 m), but the ones measured on the isolate in this study were only around 150 m long. Vesicles were subspherical in form and tiny (7 to 15 m in diameter). bisect the vesicles from the vesicles. [23]. As in Figure 3-B.

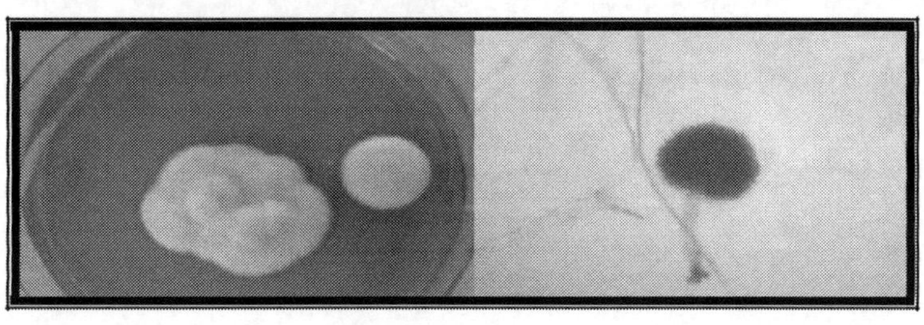

A B

Figure(3):*Aspergillus candidus* on SDA(*400x)*

3.4 Rizopus:

3.4.1Colony Morphology:

Rhizopus is a fast-growing fungus that can easily fill a petrie dish with soft, cotton-candy-like growth in as little as 5 days. The maturation of the sporangiospores within the sporangium causes growth to be whitish in color, which can turn brown with age [24].

3.4.2 Microscopic Morphology:

sporangiophores on the stolons, solitary or in tufts, diverging from the point where the rhizoids form; sporangia round; The apophysis is missing or partially visible, and the sporangiophores are ovoid.oid [24]. As in Figure 4

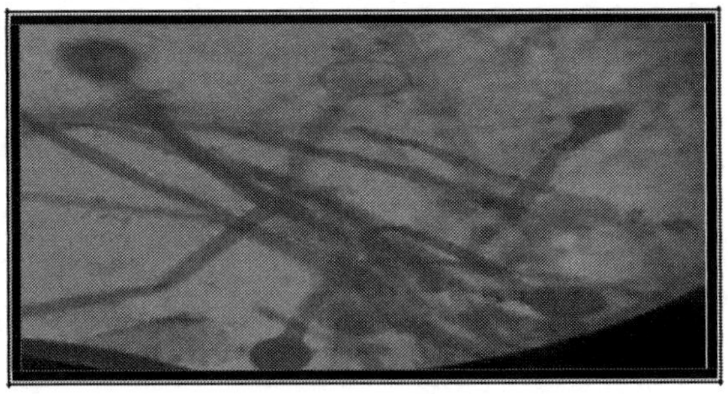

Figure (4) *Rhizopus* on SDA (400x)

From results above, the most fungi isolates belong to tow genera:*Aspergillus* and*Rhizopus*. During the analysis, 14 samples were collected from seven different locations in Shatt AL-Arab, and all of the
samples were positive. One of the most frequently isolated genera in water is Aspergillus as found in
study of [25 and 26]. Fungi, as we know, are one of the most diverse classes of living organisms on the earth , despite being understudied globally [27]. UV radiation is a common form of water treatment. In
terms of fungi,[28]. Since UV exposure is related to pigmentation in fungi, fungi with pigmented
spores, such as Aspergillus and Penicillium, have higher radiation tolerance and are less susceptible to UV. [29]. This was also found in a study conducted by [30], in which melanized, thick-walled organisms predominated following water treatment. The effect of UV dose on fungi in drinking water should be investigated. This treatment method could be used as an in-house treatment in hospitals, according to a new study on the impact of copper and silver ionization on fungal colonization in hospital water systems [31]. Internal water treatment in hospitals may be a useful tool in the fight against the fungus outbreak. Furthermore, since many of the fungal species found in drinking water are common contaminants in food and beverages, additional water treatment could help food processing plants and other businesses where water quality is critical. Fungi, on the other hand, will grow on their own. [32].

3.5Statistical analyses:

The results show, Al- Ashar, Al- Qurna and Al Makael has significant differences $p > 0.05$ compare with other sites of sample collection, and Al- Qurna and Al Makael has significant differences $p > 0.05$ compare with Al- Ashar site [33].

4. Conclusions and Recommendations:

- There are many source of pollution in Shatt al-Arab river Al-Qurna and AL-Ashar was the most pollutant which isolated many types of pathogenic fungi which is: *Aspergillus niger, Aspergillus candidus, Aspergillus flavus, Penicillium chrysogenum*
- The pathogenic fungi *A. flavus* found in highest ratio in Shatt al-Arab river reached to 85.7 %Which is isolated from six sites we collect of the samples excepting one site not isolated from .
- Conducting more studies in this field to find out the types of chemical and biological pollutants that cause pollution in the waters of the Shatt al-Arab river being a vital source for Basra residents in particular.
- Control of sewage water that extends into the Shatt al-Arab river .
- Applying environmentally-friendly biological treatments to treat pollution in Shatt al-Arab River.

5. References:

[1]US Environmental Protection Agency (USEPA)(2014). "Impaired waters and total maximum daily loads

[2]Craun, Gunther F., and Rebecca L. Calderon.(2016). "Workshop summary: estimating waterborne disea United States." *Journal of water and health* 4.S2: 241-253.

[3] Ibekwe, A. Mark, Menu Leddy, and Shelton E. Murinda.(2013). "Potential human pathogenic bacter urban watershed as revealed by pyrosequencing." *PLoS One* 8.11: e79490.

[4] Hageskal, G.; Gaustad, P.; Heier, B.T.; Skaar, I.(2007). Occurrence of moulds in drinking water. *J. Appl.Microbiol. 102*, 774-780.

[5] Nazim, S.; Dawar, S.; Tariq, M.; Zaki, M.J.(2008). Quantitative estimation of mycoflora in drinkingwater and fruit juices of Karachi. *Pakistan J. Bot. 40*, 1263-1268.

[6] Sammon, N.B.; Harrower, K.M.; Fabbro, L.D.; Reed, R.H.(2010). Incidence and distribution ofmicrofungi in a treated 9 municipal water supply system in sub-tropical Australia. *Int. J. Environ.Res. Public Health 7*, 1597-1611.

[7] Oliveira, H.M.B. (2010). Fungos filamentosos na água e em biofilmes na rede de distribuição de águapotável do sistema Alto do Céu, Recife-PE; Master's Thesis, Biologia de Fungos do Departamento de Micologia do Centro de Ciências Biológicas da Universidade Federal de Pernambuco (UFPE), Recife, Brazil.

[8] Pereira, V.J.; Basílio, M.C.; Fernandes, D.; Domingues, M.; Paiva, J.M.; Benoliel, M.J.;Crespo, M.T.; San Romão, M.V. (2009). Occurrence of filamentous fungi and yeasts in three differentdrinking water sources. *Water Res. 43*, 3813-3819.

[9] Pereira, V.J.; Fernandes, D.; Carvalho, G.; Benoliel, M.J.; San Romão, M.V.;Barreto Crespo, M.T. (2010). Assessment of the presence and dynamics of fungi in drinking water sourcesusing cultural and molecular methods. *Water Res. 44*, 4850-4859.

[10] Abdullah, S.k. and Abbas, B.A. (1994a). Occurrence of thermophilic and thermotolerant fungi in aqua of Shatt Al-Arab River and its creeks at Basrah, Iraq. Marina Mesopotamica 9:39-47.

[11] Abdullah, S.k.and Abbas, B.A. (1994b). Taxonomic study on fungi from water and surface sediments Arab River and its creeks !: Ascomycetes. Marina Mesopotamica 9:91-100.

[12] Abdullah, S.K., Al-Dossary, M.A. and Al-saad, H.T.(2000). A mycofloral study on aquatic

sediments Arab estuary and North-West Arabian Gulf. Basrah J. Science 18:1-14.
[13] Cannon, P.F., Abdullah, S.K. and Abbas, B.A. (1995). Two new species of *Monoascus* from Iraq, with known species of the genus. Mycol. Res. 99:659-662.
[14] Sivanesan, A., Abdullah, S.K. and Abbas, B.A.(1993). *Exserohilum curvisporum* sp.nov., a new hypho Iraq. Mycol. Res. 97:1486-1488.
[15] Abdullah, S.K. and Abbas, B.A. (1997). Taxonomic study on fungi from water and surface sediment o Arab River and its creeks 11:Hyphomycetes. J. Basrah Researches 13:46-55.
[16] Abdullah, S.K. and Hassan, D.A.(1995). Isolation of dermatophytes and other keratinophilic fungi from surface sediment of the Shatt Al-Arab River and its creeks at Basrah, Iraq. Myco 166.
[17] Grabińska-Łoniew, Anna, et al. (2007). "Occurrence of Fungi in Water ska Distribution System." *Poli Environmental Studies* 16.4.
[18] Gillespie, Stephen, and Peter M. Hawkey, eds.(2006). *Principles and practice of clinical bacteriology* & Sons.
[19] Clyde, Calvin G., et al. (2011). "Water Quality in Pleasant Valley, tah.".
[20] Bauer, Heidi, et al. (2002). "The contribution of bacteria and fungal spores to the organic carbon con water, precipitation and aerosols." *Atmospheric Research* 64.1-4: 109-119.
[21] Wucherpfennig, T., Hestler, T., & Krull, R. (2011). Morphology engineering-osmolality and Aspergillus niger morphology and productivity. *Microbial cell factories*, *10*(1), 58.
[22] Brown, S. Horowitz, et al. (2008). "Morphological transitions governed by density dependence and activity in Aspergillus flavus." *Applied and environmental microbiology* 74.18: 5674-5685.
[23] Balajee, S. Arunmozhi, et al. (2005). "Aspergillus lentulus sp. nov., a new sibling sp fumigatus." *Eukaryotic Cell* 4.3: 625-632.
[24] Gustafsson, Hanna, et al.(2012). "Immobilization of lipase from Mucor miehei and Rhizopus mesoporous silica—The effect of varied particle size and morphology." *Colloids and Surfaces B: Bioin* 22-30.
[25] Arvanitidou, M., et al. (2000). "High level of recovery of fungi from water and dialysate in h units." *Journal of Hospital Infection* 45.3: 225-230.
[26] Sharma, K., and Shaista Parveen. (2011). "Ecological study of fungi isolated from the surface water Dam Dhamtari, Chhattisgarh, India." *Journal of Phytology* 3.4.
[27] Grover, Ruby, et al. (2007). "Response of fungal community in the unpolluted and polluted (textile wastes) habitats." *Journal of environmental science & engineering* 49.2: 93-98.
[28] Kanzler D, Buzina W, Paulitsch A, Haas D, Platzer S, Marth E, Mascher F, (2007). Occurrence and hy relevance of fungi in drinking water. Mycoses 51: 165–169.
[29] Waipara NW, (1998). A Method to Asses the Lethal Effect of UV-C Irradiation of Fungal Spores on a Dry Surface. National Veterinary Institute, Oslo.
[30] Kelley J, Kinsey G, Paterson R, Brayford D, (2003). Identification and Control of Fungi in Distribu Awwa Research Foundation and American Water Works Association, Denver, CO.
[31] Pedro-Botet ML, Sanchez I, Sabria M, Sopena N, Mateu L, GarcI´aNu´ n˜ ez M, Rey-Joly C.(2007). Impact of copper and silver ionization of fungal colonization of the water supply in health implications for immune compromised patients. Clinical Infectious Diseases 45: 84–86.
[32] Huq, A.; Whitehouse, C.A.; Grim, C.J.; Alam, M.; Colwell, R.R. (2008). Biofilms in water, its role and impact in human disease transmission. *Curr. Opin. Biotechnol. 19*, 244-247.
[33] Al- Rawei, kh. M. and Khalafeallah, A. m.(2000). *Design and analysis of agriculture experiments*. Al-Mosul university. Book home.

Journal of Physics: Conference Series **2063** (2021) 012022 doi:10.1088/1742-6596/2063/1/012022

Synthesis of some new Rhodium complexes containing diaryl chalcogenide ligands and their uses as catalysts in hydrosilylation of alkenes

Ali Z Al-Rubaie*, Zaki N Kadhim, Majeed Y Al-Luaibi and Luma Sabri
Department of Chemistry, College of Science, University of Basrah, Basrah, Iraq

*Corresponding author: alrubaie49@yahoo.com

Abstract. Diaryl Chalcogenides (*i.e.* Ar_2E where Ar = 4-$CH_3C_6H_4$, 4-BrC_6H_4, E= S, Se and Te) were reacted with $[RhCl(CO)_2]_2$ and rhodium(III) chloride trihydrate to give compounds of type $[RhCl(CO)_2(Ar_2E)]$ and $[RhCl_3(Ar_2E)_3]$, respectively. All compounds were characterized by IR, NMR, and mass spectroscopic data. Attempts to prepare hydroxyapatite (HAp) supported rhodium catalyst by using different methods were unsuccessful. Complexes $[RhCl(CO)_2((4\text{-}R\text{-}C_6H_4)_2S)]$, $[RhCl(CO)_2((4\text{-}R\text{-}C_6H_4)_2Se)]$ and $[RhCl(CO)_2((4\text{-}R\text{-}C_6H_4)_2Te)$, where R= CH_3 or Br], were evaluated as catalysts for hydrosilylation of allyl phenyl ether and 1-decene. They showed good catalytic activities for hydrosilylation of alkenes with triethoxysilane.

1. Introduction

In recent years, due to their exceptional properties calcium apatites, $Ca_{10}(PO_4)_6(OH)_2$ (HAp), has wide importance in the fields of biomedicine, dental implants, catalysis, and drug delivery [1-8]. Several papers described the preparation of heterogeneous catalysts from transition metal ions and HAp. It was found that Pd and Ru supported HAp are the suitable heterogeneous catalyst for oxidation, aldol reaction, and Diels-Alder reactions [9-11]. On the other hand, metal-organic coordination compounds have drawn great research interest in recent years because of their potential applications in catalysis [12]. It is worth noting that our previous work described the synthesis of some rhodium-tellurium heterocycle complexes [13]. Rhodium complexes containing organotellurium compounds, namely diphenyl telluride, and dimethyl telluride, as donors have also been previously reported [14-16]. Thus, it is therefore of interest to study rhodium(I) carbonyl and rhodium (III) complexes containing monodentate chalcogenide ligands since such complexes are very rare in literature and not used as catalysts, as far as we are aware. Rhodium(I) carbonyl complexes containing monodentate phosphine chalcogenides were reported and their catalytic activity for carbonylation of methanol was studied [17,18]. The present work describes the preparation of some new rhodium-chalcogenide complexes. Our recent work [19] described the catalytic activity of platinum(II) complexes of hexenyl aryl tellurides in the hydrosilylation of olefins promoted us to support these new complexes on hydroxyapatite to investigate their catalytic activity. Furthermore, the catalytic activity of these new complexes in the hydrosilylation of allyl phenyl ether and 1-decene will be examined at different temperatures.

2. Experimental

All the preparations were performed in a dry nitrogen atmosphere. All chemicals were reagent grade and used without further purification. Bis(4-methylphenyl) sulfide, bis(4-methylphenyl) selenide, bis(4-bromophenyl) sulfide and bis(4-bromophenyl) selenide were prepared by reported methods [19-21]. Bis(4-methylphenyl) telluride and bis(4-bromophenyl) telluride were prepared according to a literature

Journal of Physics: Conference Series **2063** (2021) 012022 doi:10.1088/1742-6596/2063/1/012022

method [20-21]. $[Rh_2(\mu\text{-}Cl)_2(CO)_4]$ was prepared according to a literature method [22] by bubbling carbon monoxide through a refluxing ethanolic solution of rhodium trichloride trihydrate. All diaryl Chalcogenides (*i.e.,* Ar_2E where Ar = 4-$CH_3C_6H_4$, 4-BrC_6H_4, E= S, Se, and Te) were characterized by spectroscopic methods and CHN analysis and are not reported.

2.1. Synthesis of [Rh(CO)₂ClL] complexes

Complexes **1-6** (*i.e.* $[Rh(CO)_2ClL]$, L = $(4\text{-}CH_3C_6H_4)_2E$ and $(4\text{-}BrC_6H_4)_2E$, where E= S, Se and Te) were prepared by the following general method:

To a solution of $[Rh(CO)_2Cl]_2$ (150 mg, 0.65 mmol) in 30 mL of THF was added a solution of the respective ligand (1.35 mmol) in 20 mL of THF. The reaction mixture was stirred for 6 h at room temperature and the solvent was removed under vacuum by a rotatory evaporator. The remaining solid was washed with pentane (3 x 10 mL) and dried under vacuum over silica gel in a desiccator.

All these new complexes along with their melting points and elemental analyses, which were prepared by the above general route, are listed in Table 1.

2.1.1. Synthesis of [RhCl₃(Ar₂E)₃] complexes

Complexes **7-12** (i.e. $[RhCl_3(Ar_2E)_3]$, where Ar = $(4\text{-}CH_3C_6H_4)_2E$ and $(4\text{-}BrC_6H_4)_2E$, E= S, Se and Te) were prepared by the following method: Rhodium trichloride trihydrate (0.79 g, 3.0 mmol) in dry ethanol (50 mL) was added to an ethanolic solution of diaryl chalcogenide (9.2 mmol). The mixture was stirred at room temperature for 6 h. A brownish-red solid was obtained, collected by filtration, washed with ethanol, and dried in *vacua*. The rhodium(III)-diaryl chalcogenide complexes which were prepared by the above general route (along with their melting points and elemental analyses) are listed in Table 2.

2.1.2. Attempts to prepare the Rh-complex/HAp catalysts.

The following methods were used to prepare Rh-complex/HAp catalysts:

a) To 0.20 mmol of the corresponding rhodium complex (i.e. $[Rh(CO)_2ClL]$) in 100 mL of acetone was added to 2.5 g of HAp. The light-yellow mixture was stirred at room temperature for 12 h and the liquid became colorless. The mixture was filtered. The solid product was collected by filtration and washed with several amounts of acetone and water, respectively. The product was dried under a vacuum. The obtained solid was grid into powders and divided into two parts which were then calcined in a muffle oven under static air at 300 and 500°C respectively, for 12 h.

b) Three rhodium concentration (10 mM, 20 mM, and 50 mM) solutions were prepared by dissolving $[Rh(CO)_2ClL]$ in ethanol. To each of the above solutions (40 mL), and 2.0 g of HAp was added respectively. The solutions were stirred at 70°C. The solid was collected by filtration. The product was washed several times with a mixture of ethanol and water. The dried samples were kept in an oven maintained at100°C for 12 h. Finally, the dried product was divided into two parts and calcined in the air at 300°C and 500°C for 12 h. The above methods were also unsuccessful to support $[RhCl_3(Ar_2E)_3]$ complexes on hydroxyapatite (HAp).

All samples prepared by the above methods showed no presence of rhodium, sulfur, selenium, or tellurium by the powder X-ray diffraction analysis. In general, no significant variation is observed in the diffraction pattern of samples prepared with different rhodium ligands and no characteristic peaks of sulfur, selenium, tellurium, or rhodium were observed. The peaks obtained were compared with standard references [23].

2.2. Catalytic activity

2.2.1 Hydrosilylation of olefins

Hydrosilylation was carried out in a round-bottomed flask equipped with a magnetic stirrer and reflux condenser. A drying system was attached to the upper of the condenser. Olefin and platinum complex were stirred at the reaction temperature (50, 70, and 90°C) for 60 min before triethoxysilane was added. The structure and the yield of hydrosilylation products were determined by GLC using a standard sample and a standard curve at regular intervals.

Typical conditions were by using 10 mmol of $HSi(OEt)_3$, 10 mmol of olefin, and 10^{-3} – 10^{-4} mmol of rhodium complex.

2.3. Physical measurements

IR spectra were recorded as CsI discs in the range 4000-200cm^{-1} using a Pye-Unicam SP3-300s spectrophotometer. ^1H NMR spectra were recorded on a Bruker AVANCE 300 (300.13 MHz) or a Bruker 400(400.13 MHz) spectrophotometer. ^{13}C NMR (^1H decoupled) spectra were recorded on a Bruker AVANCE 300 (75.47 MHz) or Bruker DRX 400 (100.63 MHz) spectrometer. Spectra were recorded in deuterated chloroform ($CDCl_3$) or CD_2Cl_2 unless otherwise stated and chemical shifts are reported in parts per million downfield from tetramethylsilane (TMS) for ^1H NMR spectra and relative to the central signal of deuterated solvent for ^{13}C NMR spectra. Coupling constants are given in Hertz. Microanalysis of carbon, hydrogen, and nitrogen was carried out by a Carlo Erba EA 1108-Elemental Analyzer instrument. Melting points were determined on a Gallenkamp melting point apparatus and are uncorrected. GLC determinations were carried out on a Hewlett Packard-5840A chromatograph equipped with a 6mX1/8 column of OV-17 on chromosorb w. Mass spectra were measured with an MAT 1125 Finnigan mass spectrometer.

3. Results and Discussion

The dimeric precursor $[Rh(CO)_2Cl]_2$ reacts with various diaryl chalcogenide (Ar_2E) ligands in 1 : 1 mole ratio afforded the new *cis*-dicarbonyl complexes of the type $[Rh(CO)_2ClL]$ (**1-6**) (where L= Ar_2E; Ar = 4-$CH_3C_6H_4$, 4-BrC_6H_4, E= S, Se and Te), Scheme 1. Reaction of rhodium (III) chloride trihydrate with an excess amount of Ar_2E gave new complexes of the type $[RhCl_3(Ar_2E)_3]$ (*i.e.*, compounds **7–12**), Scheme 1. All these complexes (**1** – **12**) gave a satisfactory CHN analysis, Tables 1 and 2.

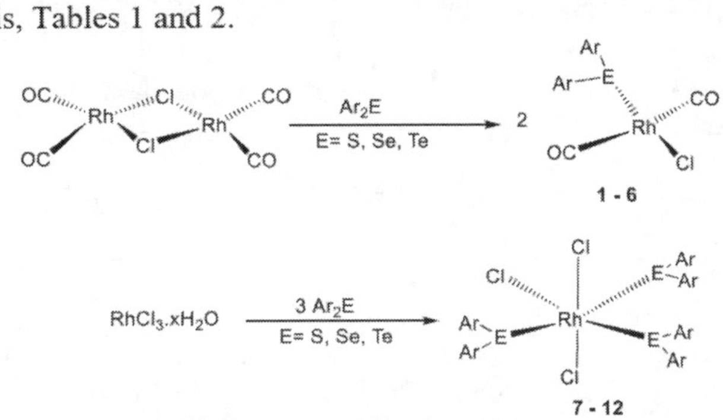

Scheme 1. Preparative methods for complexes **1** – **12**.

Table 1. Physical properties of $[RhCl(CO)_2(Ar_2E)]$ complexes

Ar_2E	M.p.°C	Colour	Elemental analyses Found(calc.)	
			C	H
$(4\text{-}CH_3C_6H_4)_2S$ (**1**)	172	deep brown	46.86(47.02)	3.52(3.45)
$(4\text{-}CH_3C_6H_4)_2Se$ (**2**)	156	red–brown	42.03(42.18)	2.95(3.10)
$(4\text{-}CH_3C_6H_4)_2Te$ (**3**)	145	dark red	37.86(38.11)	2.76(2.80)
$(4\text{-}BrC_6H_5)_2S$ (**4**)	180	red–brown	30.85(31.23)	1.24(1.50)
$(4\text{-}BrC_6H_5)_2Se$ (**5**)	160	dark red	28.45(28.73)	1.23(1.38)
$(4\text{-}BrC_6H_5)_2Te$ (**6**)	149	deep red	26.48(26.52)	1.21(1.27)

Journal of Physics: Conference Series **2063** (2021) 012022 doi:10.1088/1742-6596/2063/1/012022

Table 2. Physical properties of $[RhCl_3(Ar_2E)_3]$ complexes

Ar_2E	M.p.°C	Colour	Elemental analyses Found(calc)	
			C	H
$(4\text{-}CH_3C_6H_4)_2S$ (7)	150(d)	red–brown	58.76 (59.19)	5.05 (4.97)
$(4\text{-}CH_3C_6H_4)_2Se$ (**8**)	120(d)	red–brown	50.91 (50.81)	4.02 (4.26)
$(4\text{-}CH_3C_6H_4)_2Te$ (**9**)	68	red–brown	43.86 (44.30)	3.58 (3.72)
$(4\text{-}BrC_6H_5)_2S$ (**10**)	76	dark brown	35.04 (34.83)	2.03 (1.95)
$(4\text{-}BrC_6H_5)_2Se$ (**11**)	110	dark brown	31.07 (31.28)	1.66 (1.75)
$(4\text{-}BrC_6H_5)_2Te$ (**12**)	121	dark brown	27.86 (28.30)	1.52 (1.58)

The IR spectra of the complexes **1-6** showed two almost equally intense terminal ν(CO) bands in the range 2003-2093 cm^{-1} indicating two carbonyl groups are mutually cis to one another [24,25], Table 3. The positions of the ν(CO) bands follow the following order:

$$Ar_2S > Ar_2Se > Ar_2Te$$

This may be explained in terms of 'hard-soft' interactions between the metal and the chalcogenide donor [26]. The IR spectra of complexes **7-12** show some characteristic bands between 320 and 385 cm^{-1} due to the terminal ν(Rh-C1)[13,27], Table 3. The appearance of a medium band near 345 cm^{-1} typical of asymmetric stretching vibration of a linear Cl-Rh-Cl group, indicates a *trans* arrangement of chlorine atoms[28-30]. Thus, based on the above evidence, square planar and octahedral structures can be assigned for complexes **1** – **12**, Scheme 1.

Table 3. Characteristic IR frequencies (cm^{-1}) of complexes **1** – **12.**

Complex	υ(C–H) Phenyl ring	ν(CO)	ν(Rh-Cl)	Asy. ν(Cl-Rh-Cl)
1	3056	2062, 1995	295, 285	
2	3060	2055, 1945	290, 282	
3	3057	2058, 1938	302, 275	
4	3053	2068, 2001	284, 270	
5	3056	2058, 1985	296, 280	
6	3056	2047, 1956	300, 276	
7	3063		380, 322	345
8	3065		385, 322	337
9	3058		380, 322	345
10	3061		385, 322	342
11	3060		370, 318	345
12	3058		385, 320	343

^1H and ^{13}C spectra for all complexes (**1–12**) agree well with their structure (Table 4). The ^1H NMR spectra of **1-6** exhibit two doublet resonances in the range δ 7.32-8.83 ppm for the disubstituted benzene rings. The ^{13}C NMR spectra of 1-**6** show characteristic resonances of terminal carbonyl groups in the range δ 182-188 ppm and four signals in the region δ 112-140 ppm for carbon atoms of a benzene ring, Table 4.

Journal of Physics: Conference Series **2063** (2021) 012022 doi:10.1088/1742-6596/2063/1/012022

Table 4. ^{1}H and ^{13}C NMR spectroscopic data

Complex	^{1}H NMR (δ, 300 MHz, $CDCl_3$)	^{13}C (δ, 75 MHz, $CDCl_3$)
1	7.35 (d, J = 7.8 Hz, 4H, H-Ar), 4H), 7.06 (d, J = 8.1 Hz, 4H, H-Ar),), 2.14 (s, 6H, CH_3).	186.7, 182, 139.00, 132.50, 131.66, 130.29, 21.19
2	7.47–7.44 (m, 4H, H-Ar), 7.35 (d, J = 8.3 Hz, 4H, H-Ar), 2.38 (s, 6H, CH_3).	188.1, 185, 137.4, 137.1, 129.4, 128.7, 21.6.
3	7.48 – 7.36 (m, 4H, H-Ar), 7.15 (d, J = 8.1 Hz, 4H), 2.38 (s, 6H, CH_3).	186.2, 182.6, 137.4, 137.1, 129.4, 109.3, 21.5
4	7.77 – 7.67 (m, 4H, H-Ar), 7.54 – 7.43 (m, 4H, H-Ar).	187.1, 184.3, 134.5, 132.6, 130.7, 124.6.
5	7.55 (d, J = 7.8 Hz, 4H, H-Ar), 7.25 – 7.15 (m, 4H).	187.6, 183.8, 139.4, 132.60, 128.8, 122.5.
6	7.44 (d, J = 8.3 Hz, 4H, H-Ar), 7.34 (d, J = 7.9 Hz, 4H, H-Ar).	186.5, 182.5, 139.4, 132.6, 122.5, 112.4.
7	7.41 (d, J = 7.8 Hz, 4H, H-Ar), 4H), 7.22 (d, J = 8.1 Hz, 4H, H-Ar),), 2.34 (s, 6H, CH_3).	138.3, 132.7, 131.5, 130.3, 22.4
12	7.35 (d, J = 8.1 Hz, 4H, H-Ar), 7.27.15 (d, J = 8.3 Hz, 4H, H-Ar).	138.5, 133.1, 131.2, 119.3

Attempts to support these new complexes (**1–12**) on hydroxyapatite were unsuccessful. The IR spectra of all products exhibited only the characteristic absorption peaks of hydroxyapatite [28,29], Fig. 1.

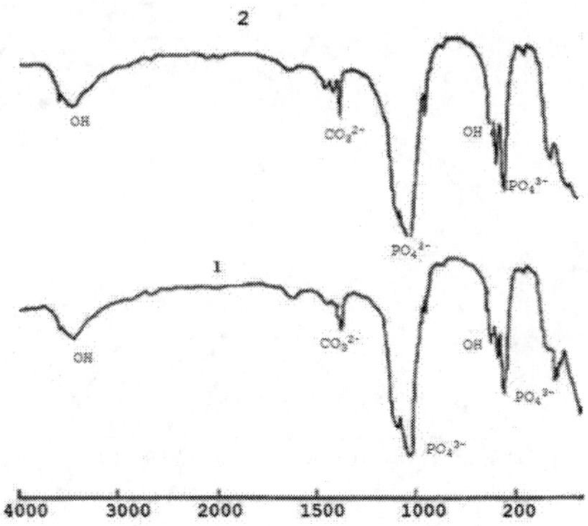

Figure 1. IR spectra of Hap and Rh complexes at **1**) 300°C and **2**) 500°C.

All samples obtained for hydroxyapatite-rhodium complexes by different methods showed no presence of rhodium, sulphur, selenium, or tellurium by the powder X-ray diffraction analysis, Fig. 2 and no significant variation was observed in the diffraction pattern of samples prepared with different rhodium ligands. Furthermore, no characteristic peaks of sulphur, selenium, tellurium or rhodium were observed. The peaks obtained were compared with a standard reference [23].

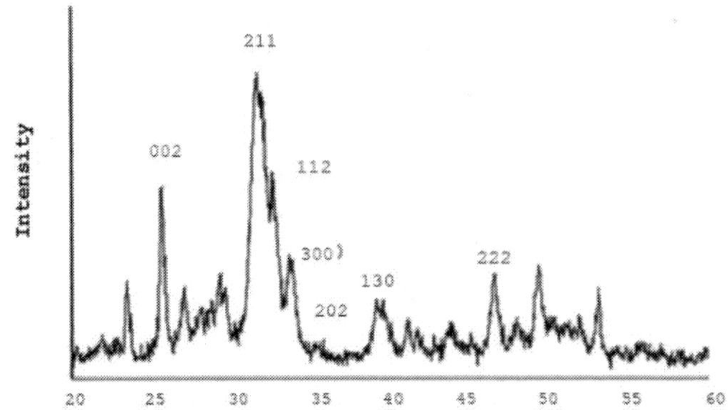

Figure 2. XRD of HAp and Rh complexes

The unsuccessful attempts to support the rhodium (I) or rhodium (III) complexes on hydroxyapatite, prompted us to investigate the catalytic activities of complexes **1-6** on hydrosilylation of allyl phenyl ether and 1-decene. It is worth noting that our previous work[31] described the hydrosilylation of olefins using platinium (II) complexes containing hexenyl aryl telluride ligands.

The catalytic activity of complexes **1**, **2**, and **3** were examined between 50 and 90°C using 1-decene and triethoxysilane as a model reaction, Fig. 3. The experimental results indicated that no remarkable induction period was observed and the reaction rate became faster. In the range 50 to 90°C, the catalytic reaction finished within 50 min, see Fig. 3.

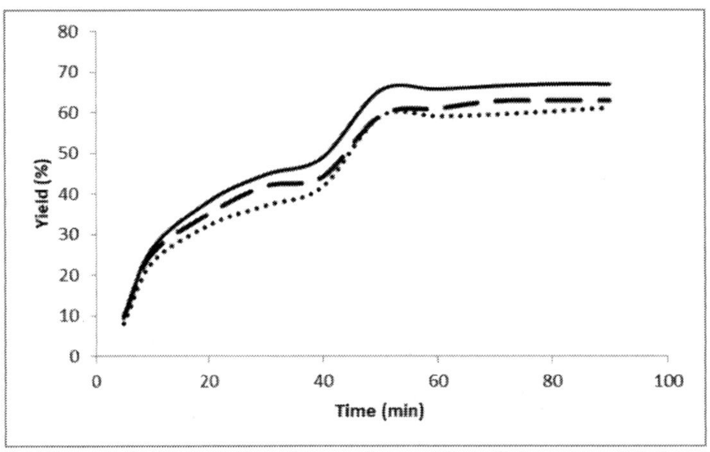

Figure 3. The catalytic activity of complexes 1(—), **2** (— —) and **3** (.....) between 50 and 90°C using 1-decene and triethoxysilane as a model reaction

Thus, the catalytic behavior of complexes 1, **2**, and **3** were examined by using allyl phenyl ether and 1-decene as substrates. The substrates were hydrosilylated with $HSi(OEt)_3$ by stirring the olefin and rhodium(I) complexes for 60 min followed by the addition of triethoxysilane to give the saturated products between 59- 65% yields, Table 5. The catalytic activity of complexes 4, **5**, and **6** were also examined with allyl phenyl ether and 1-decene by the same procedure and gave the saturated products in the range 48 - 60%, Table 5.

The efficacy of complexes **1**, **2**, and **3** towards the hydrosilylation was found to follow the order **1** > **2** > **3**. This may be due to the difference in bond lengths of Rh-S, Rh-Se, and Rh-Te. In general, complexes **1**, **2**, and **3** showed moderate catalytic activity. On other hand, Complexes **7**, **8**, and **9** showed no catalytic activity in comparison with Rh(I) complexes.

4. Conclusion

In conclusion, some new Rh(I) and Rh(III) complexes were prepared and characterized by various spectroscopic techniques. Attempts to support Rh(I) and Rh(III) complexes on hydroxyapatite by different methods were unsuccessful. Complexes $[RhCl(CO)_2((4\text{-}CH_3C_6H_4)_2S)]$, $[RhCl(CO)_2((4\text{-}CH_3C_6H_4)_2Se)]$ and $[RhCl(CO)_2((4\text{-}CH_3C_6H_4)_2Te)]$ showed moderate to good catalytic activity for the hydrosilylation. Complexes of type $[RhCl_3(Ar_2E)_3]$ *(i.e.* compounds 7, 8, and 9) showed no catalytic activity in comparison with Rh(I) complexes.

Acknowledgements

The financial support of this work by the Ministry of Higher Education and Scientific Research/ Research and Development is hereby gratefully acknowledged.

References

[1] Moteki T, Rowley AT, Bregante DT, and Flaherty DW. 2017 *ChemCatChem*. **9**(11) 1921.
[2] Guo J, Yu H, Dong F, Zhu B, Huang W, and Zhang S. 2017 *RSC Advances* **7**(72) 45420.

[3] Epple M. 2018 *Acta Biomaterialia* **77** 1.
[4] D. Choi, K. Marra, P.N. Kumta, 2004 *Materials Research Bulletin* **39** 417.
[5] Jarcho M, Bolen CH, Thomas MB, Bobick J, Kay JF, Doremus RH. 1976 *J. Mater. Sci*. **11** 2027.
[6] Gomes DS, Santos AMC, Neves GA, and Menezes RR. 2019 *Cerâmica* **65** 282.
[7] Kołodziejska B, Kaflak A, Kolmas J. 2020 *Materials (Basel)* **13**(7) 1748.
[8] Nayak AK. 2010 *Int. J. Chem. Tech. Res*. **2** 903.
[9] Mori K, Hara K, Mizugaki T, Ebitani T, Kaneda K. 2004 *J. Am. Chem. Soc*. **126** 10657.
[10] Yamaguchi K, Mori K, Mizugaki T, Ebitani K, Kaneda K. 2000 *J. Am. Chem. Soc*. **122** 7144.
[11] Mori K, Hara T, Mizugaki T, Ebitani K, Kaneda K. 2003 J. *Am. Chem. Soc*. **125** 1146.
[12] Ghasemzadeh MA, Mirhosseini-Eshkevari B, Tavakoli M, Zamani F. 2020 *Green Chem*. **22** 7265.
[13] Al-Rubaie AZ, Al-Obaidi YN, and Yousif LZ. 1990 *Polyhedron* **9** 1141.
[14] Sato M, and Yoshida T. 1973 *J. Organomet. Chem*. **51** 321.
[15]Anderson SJ, Barnes JR, Goggin PL, and Goodfellow RJ. 1979 *J. Chem. Res*. (**S**), 236; 1978 (**M**) 3601.
[16] Gardner SA, 1980 *J. Organomet. Chem*. **190** 289.
[17] Das P, Konwar D, Sengupta P, and Dutta DK. 2000 *Trans. Met. Chem*. **25** 426.
[18] Moosun S, Laulloo SJ, Bhowon MG. 2013 *J. Mol. Struct*. **1053** 106.
[19] Jung HS, Suh N, and Jeon HB, 2016 *Bull. Korean Chem. Soc*. **37** 779.
[20] Perin G, Alves D, Jacob RG, Barcellos AM, Soares LK, and Lenardão EJ. 2016 *ChemistrySelect*. **1** 205, and references therein.
[21] Lin HM, Tang Y, Li ZH, Liu KD, Yang J, and Zhang YM. 2012 *ARKIVOC* **viii** 146.
[22] McCleverty JA, and Wilkinson G. 1966 *Inorg. Synth*. **8** 221.
[23] Izumi F and Ikeda T. 2000 *Mater. Sci. Forum* **321** 198.
[24] Vallarino LM and Sheargold SW. 1979 *Inorg. Chim. Acta* **36** 247.
[25] Dutta DK and Singh MM. 1994 *Trans. Met. Chem*. **19** 290.
[26] Pearson RG, 1968 *J. Chem. Educ*. **45** 581.
[27] Khan MMT and Rao ER. 1988 *Polyhedron* **7** 29.
[28] Bennett MA, Clark RJH and Milner DL, 1967 *Znorg. Chem*. **6** 1647.
[29] Browing J, Goggin PL, Goodfellow RJ, Norton MG, Rattary AJM, and Taylor BF. 1977 *J. Chem. Soc., Dalton Trans*. 2061.
[30] Kohls E, and Stein M. 2017 *Sec. Nat. Math. Biotech. Sci., MASA* **38** 43.
[31] Al-Rubaie AZ, and Al-Khazragi ZK. 2020 *IOP Conf. Ser.: Mater. Sci. Eng*. **928** 052031

Journal of Physics: Conference Series **2063** (2021) 012023 doi:10.1088/1742-6596/2063/1/012023

Preparation, Characterization and Analytical Studies of novel Azo dyes and Diazo dyes

Muthanna M Mutar, Hanan M Ali

Depart. of Chemistry, College of Education for pure Sciences, University of Basrah

taledmmm11@gmail.com

Abstract. The amino {4-hydroxy-3-[(*E*)-{4-[(5-methylisoxazol-3-yl) sulfamoyl] phenyl} diazenyl] phenyl} acetic acid (1) and the amino {4-hydroxy-3,5-*bis*[(*E*)-{4-[(5-methylisoxazol-3-yl) sulfamoyl] phenyl} diazenyl] phenyl} acetic acid (2) were Prepared. The resulting azo and diazo dyes were then characterized using m.p., IR, UV-visible, mass spectrum and ^{1}H-NMR spectrum. Analytical studies were carried on the azo dye (1) and the diazo dye (2). The best solubility of (1) and (2) was in methanol and ethyl acetate respectively, with no deviation from the linear relationship in each dye, which is due to the fact that the effect of the dielectric constant is the main factor that can control the shift of the absorption beaks. Though, the pH effects of (1) and (2) in a range of buffer solution were showed one and two isopiestic points respectively. Calculation of the ionization (pK_a) and the protonation (pK_p) constants by using the half height method was associated to the nitrogen atom and the OH-groups respectively. The pK_{p1}, pK_{a1} and pK_{a2} were also intended. But, the value of the pK_{p2} was absent in the azo dye (1) and seems to be equal to 2.5 in the diazo dye (2). These results were indicated the suggested ionization scheme in acidic and basic medium of each dye. Furthermore, the dyes with different concentrations have harmless, nontoxic and no haemolysis effects.

1. Introduction

Scientific research are regarding the azo dyes [1] [2] [3] [4], besides expended a great rank in chemical analysis. the powerfully coloured dyes, depending on their exact molecule structure were created as very significant compounds as pigments for an extensive period[5]. The structural features in the organic compounds, that typically display colour are C=C, N=O, N=N, aromatic rings, C=O and NO_2. However, the azo (–N=N–) and nitroso (–N=O) groups can confer colour, while the other groups actually dose in convinced conditions[6]. one or more azo groups, that are linked to SP_2 hybridized carbon atoms also founded on same dyes [7]. dyes can hold an extra active group able to express chelatic coordinational complexes with metal ions. Essentially reactive dyes, that reported for their pharmaceutical significance as antidiabetic[8], antineoplastic[9], antibacterial [10], and anticancer agent[11]. Analytical studies were carried on some diazo dyes [12]. The solvents and the pH effects of these dyes displayed high solubility in ethanol and distil water, and appropriate values, (pH= 12) in each dye were established. Three isopiestic points were gained in these dyes owing to envision their pK_a and pK_b constants. But, the pK_{a1} was disappeared. Theoretical studies were also carried on the structure of each synthetic dye in order to explain the relation between the structure and their properties. The structure of each synthetic diazo dye was accepted by means of the internal coordinate mechanics (ICM), the conformational analysis and the molecular mechanics (MM2) studies [12]. Furthermore, the best minimization of these dyes was attended successfully.

Journal of Physics: Conference Series **2063** (2021) 012023 doi:10.1088/1742-6596/2063/1/012023

2. Experimental section

The melting points of azo and diazo dyes were attended using Buchi B190K. The IR spectrum was carried out on aFT-IR-8400S.Fourier Transform Infrared Spectrophotometer Shimadzu (Japan) by using a KBr disc in the range (600 – 4000) cm^{-1}. The UV-Visible spectrum was done using ethanol (1 x 10^{-4} M). The IR, UV-Visible spectrophotometer and melting point completed by Chemistry Department–Education College of pure science– Basrah University, Iraq. Correct mass spectrum and the ^{1}H NMR were measured in Tehran University.

2.1 Preparation of amino {4-hydroxy-3-[(E)-{4-[(5-methylisoxazol-3-yl) sulfamoyl] phenyl} diazenyl] phenyl} acetic acid (1=L_3 in spectrum figures)

The azo dye was prepared using each of sulfamethoxazole (0.005 mol., 1.266 g), with (1.75 mL) conc. HCl in separate beaker followed by add 10 mL of distilled water. Then, the solution of $NaNO_2$ was prepared by dissolving (0.468 g) in (5 mL) of distilled water, the $NaNO_2$ solution was then added to the 1st beaker. The resulting diazonium salt was then added to 4-hydroxyphenylglycine (HPG) (0.005 mol., 0.836 g) in 25% sodium hydroxide solution. The resulting crude was recrystallized in ethanol and hexane to yield (63%) from the titled dye; m.p.: (177-179) °C This showed δ_H (500 MHz, DMSO): 2.32 (3H, CH_3);4.66 (H, CH); 6.20 (1H, CH, isoxazole); 7.01-8.07(7H, Ar-H); 9.41(1H, OH); 10.56 (1H, NH). υ :(3223.05, 3157.05, 1433.11, 1506.41, 1465.90, 1168.86, 1334.74, 1168.86, 3091.89, 2993.52, 2889.37, 1614.42) cm^{-1}; λ_{max}: 360 nm and 390 nm.

2.2. Preparation of amino {4-hydroxy-3,5-bis[(E)-{4-[(5-methylisoxazol-3-yl) sulfamoyl] phenyl} diazenyl] phenyl} acetic acid (2=L_4 in spectrum figures)

The diazo dye was prepared using sulfamethoxazole (0.005 mol., 1.266 g), with (1.75 mL) conc. HCl in separate beaker followed by add 10 mL of distilled water. Then, the $NaNO_2$ were prepared by dissolving 0.468 g in 5 mL of distilled water, the $NaNO_2$ solution was then added and the resulting diazonium salt then was added to azo dye (1) (0.005 mol., 3.478 g) in (25%) sodium hydroxide solution. The resulting crude was recrystallized in ethanol and hexane to yield (62%) from diazo dye; m.p: (232-233) °C; This showed δ_H (500 MHz, DMSO): 2.33 (6H, $2CH_3$);4.62 (1H, CH); 6.21 (2H, 2CH, isoxazole); 7.01-8.23(10H, Ar-H); 9.42(1H, OH); 10.58 (2H, 2NH);11.63(1H, OH). υ :(3442.94, 3161.33, 1494.83, 1494.83, 1465.90, 1168.86, 1334.74, 1168.86, 3091.89, 2985.81, 2893.22, 1616.35) cm^{-1}; λ_{max}: 340 nm and 380 nm.

2.3 The pH effect

The solutions of each azo and diazo dye was prepared by dissolving weights (0.0215 g) and (0.0347 g) respectively in (50 mL) of ethanol to give the stock solution (1 x 10^{-3} M) from each. Then, the (0.5 mL) of each dye was took from their stock solution, (1 x 10^{-3} M) and diluted with (5 mL) using range of buffer solutions, (2-12), to give (1x10^{-4} M) concentration from each in each buffer solution.

2.4 The Solution of the azo and diazo dye in ethanol

The stock solution of each azo dye and diazo dye was prepared by dissolving weights (0.0215 g) and (0.0347 g) respectively in (50 mL) of ethanol to give the resulting concentration (1 x 10^{-3} M) from each dye. Then, the (0.5 mL) of each dye was took from their stock solution, (1 x 10^{-3} M) and diluted with (5 mL) of ethanol, to give (1x10^{-4} M) concentration.

2.5 The solvent effect

The stock solution of each azo dye and diazo dye was prepared by dissolving weights (0.0215 g) and (0.0347 g) respectively in (50 mL) of solvents, (ethanol, methanol, water, DMSO, DMF, chloroform and ethyl acetate) to give the resulting concentration (1 x 10^{-3} M) from each dye in each solvent. Then, the (0.5 mL) of each dye was took from their stock solution, (1 x 10^{-3} M) and diluted with (5 mL) of each solvent, to give (1x10^{-4} M) concentration.

2.6 The toxic effect of azo and diazo dyes in the haemolytic red blood cells

A stock solution (200 mg / mL) was prepared, then a series of diluted (1000, 500, 250, 125,62.25) $\mu g/mL$ solutions were setup individually (0.8 mL) with red blood cells (0.2 mL) in Eppendorf tubes. Two tubes then equipped. At that point, Ringer solution (0.8 mL) was added to the first tube as a negative control, then the tap water used as a positive control in the second tube. Followed by adding 0.2 mL of red blood cells to each tube. These tubes kept for 3 hours at 37ºC in a special incubator, and the results recorded.

3. Result and Discussion

The azo dye (1) and the diazo dye, (2) were prepared as seen in Scheme (1) below.

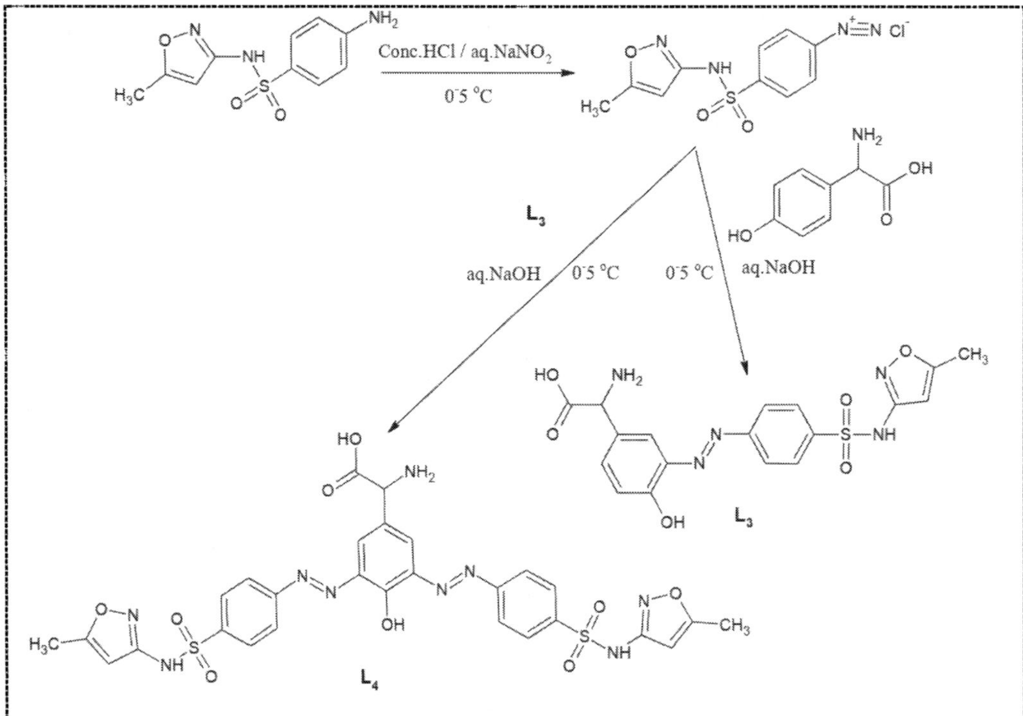

Scheme (1): The synthesis of the new azo dye (1=L_3) and diazo dye (2=L_4).

The synthetic dyes, (1) and (2) were derived from the amoxicillin drug using with optimize stoichiometry and conditions of the reactions. Then, the azo and diazo dyes were characterized by IR spectrum, mass spectrum, ^{1}H NMR spectrum and the UV-visible spectrum. The IR spectrum of the (1) and (2), (Figures 1 and 2) were showed the stretching vibration of the ν (O-H) and (N-H) groups in the regions 3223.05 cm^{-1} and 3442.94 cm^{-1} respectively. But, the (N-H) groups of NH_2 were in the regions 3157.05 cm^{-1} and 3161.33 cm^{-1} respectively. the stretching vibration band of ν (N=N) was looked at 1465.90 cm^{-1} and 1465.90 cm^{-1} respectively [2][3]. Additional bands considered as skeletal vibrations, the (C=C) seems at 1506.41 cm^{-1} and 1494.83 cm^{-1} respectively[2][3]. The (C=O) and the aromatic CH bands were appeared in the regions (1614.42 and 3092.83) cm^{-1} and (1616.35 and 3091.89) cm^{-1} respectively [3]. Though, the C-N bands were appeared in the regions1168.86 cm^{-1} and 1168.86 cm^{-1} respectively. The (O=S=O) asymmetric w. and symmetric s. bands were appeared in the regions (1334.74 and 1168.86) cm^{-1} and (1334.74 and 1168.86) cm^{-1} respectively.

Journal of Physics: Conference Series **2063** (2021) 012023 doi:10.1088/1742-6596/2063/1/012023

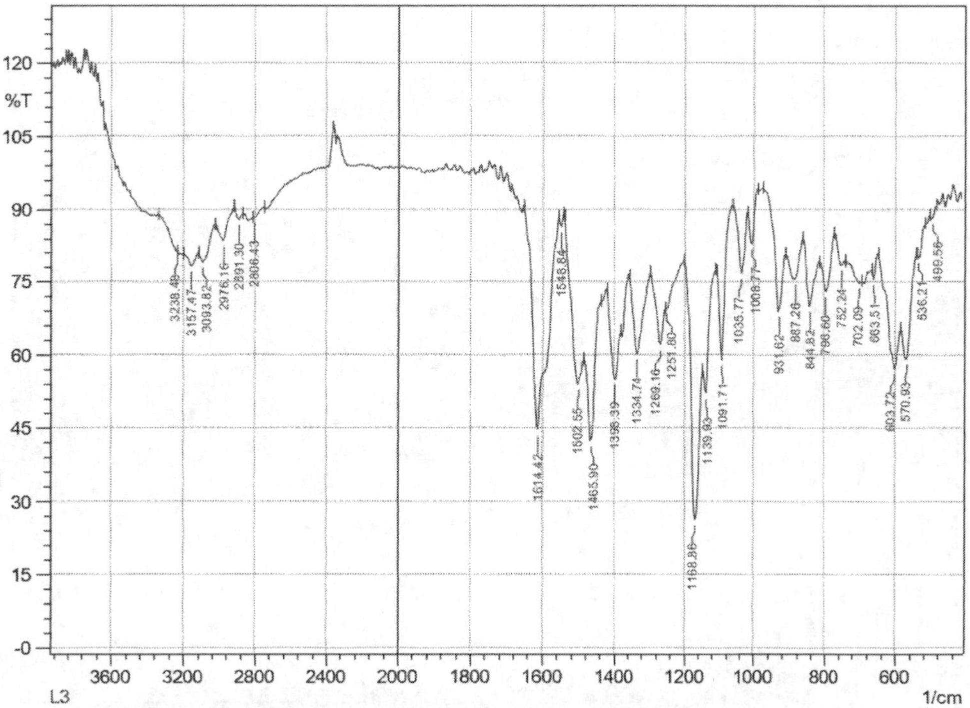

Figure 1. FT-IR spectrum (1=L_3).

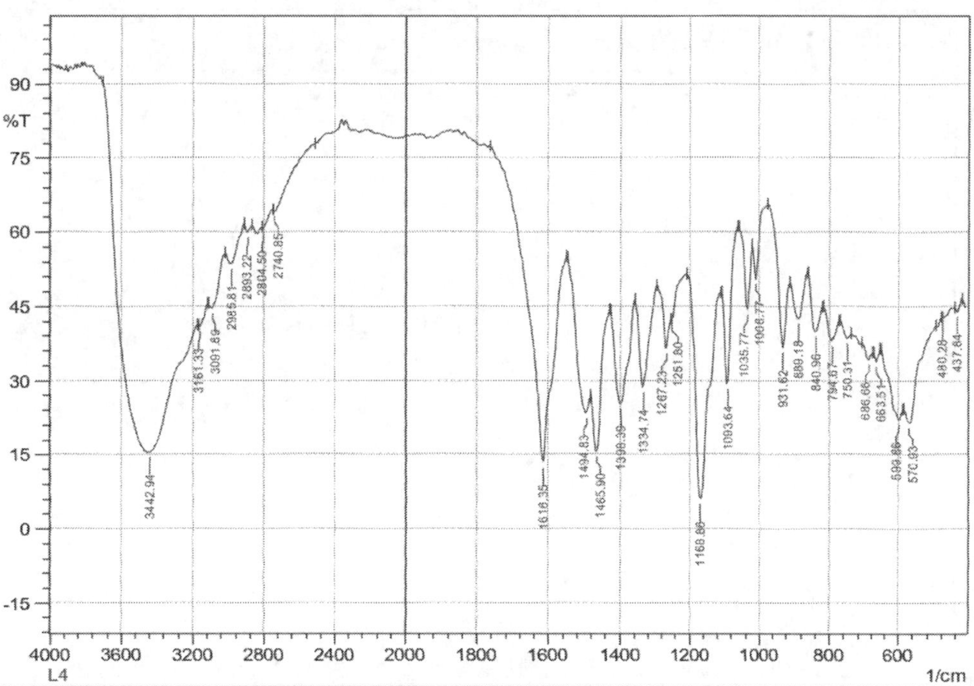

Figure 2. FT-IR spectrum (2=L_4).

Therefore, the mass spectrum was showed that the peak of (1) and (2) at *m/z* were equal to 431 and 695 respectively as seen in figures (3) and (4).

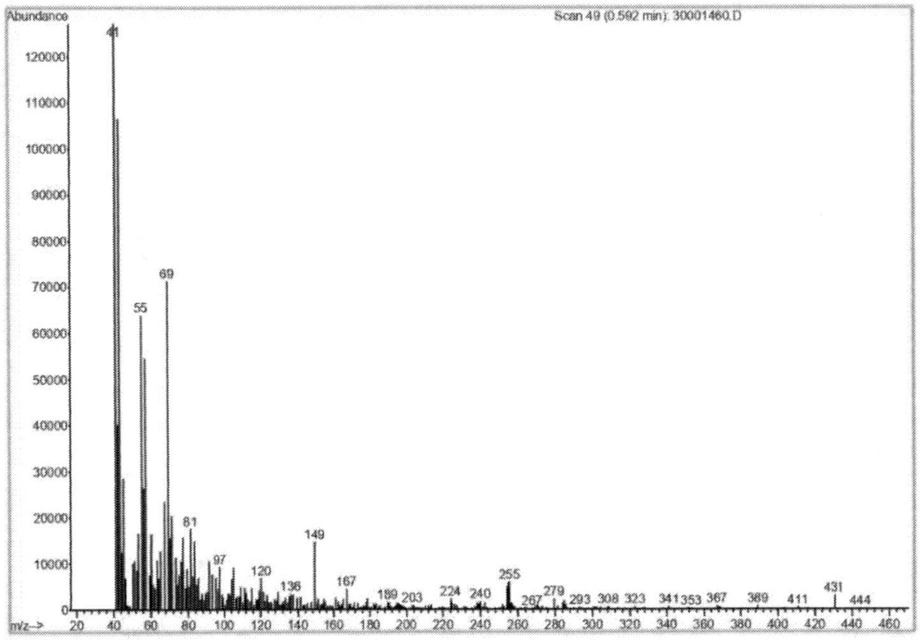

Figure 3. The mass spectrum of the azo dyes (1=L$_3$).

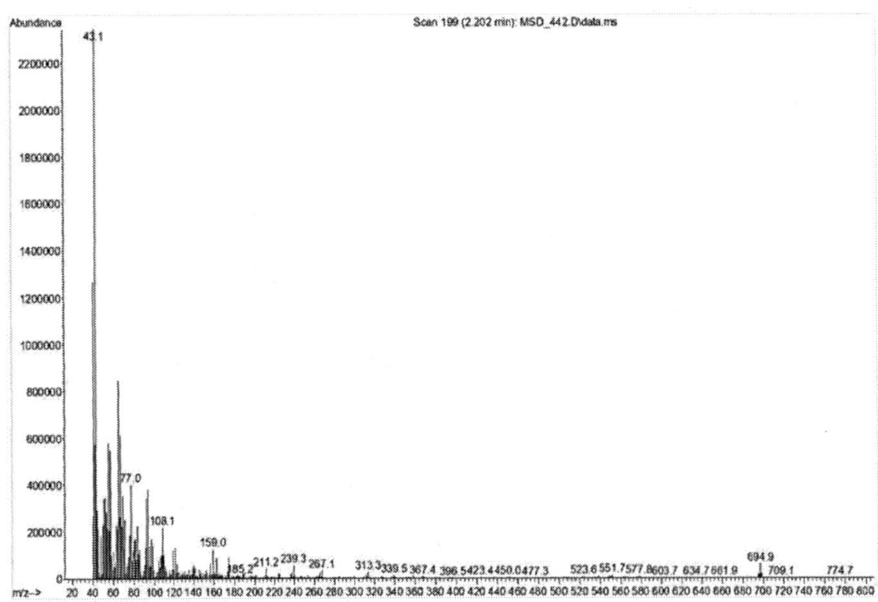

Figure 4. The mass spectrum of the diazo dyes (2=L$_4$).

Add to which, the ^{1}H NMR spectrum was confirmed the formation of the synthetic azo dyes (1) and diazo dye (2), (figures 5 and 6).

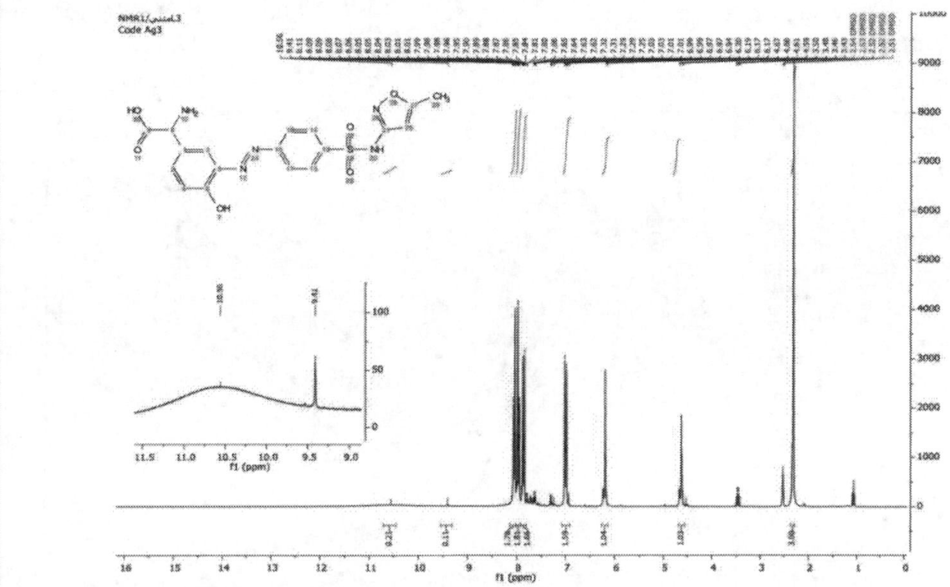

Figure 5. The ^{1}H NMR spectrum of the azo dyes ($1=L_3$).

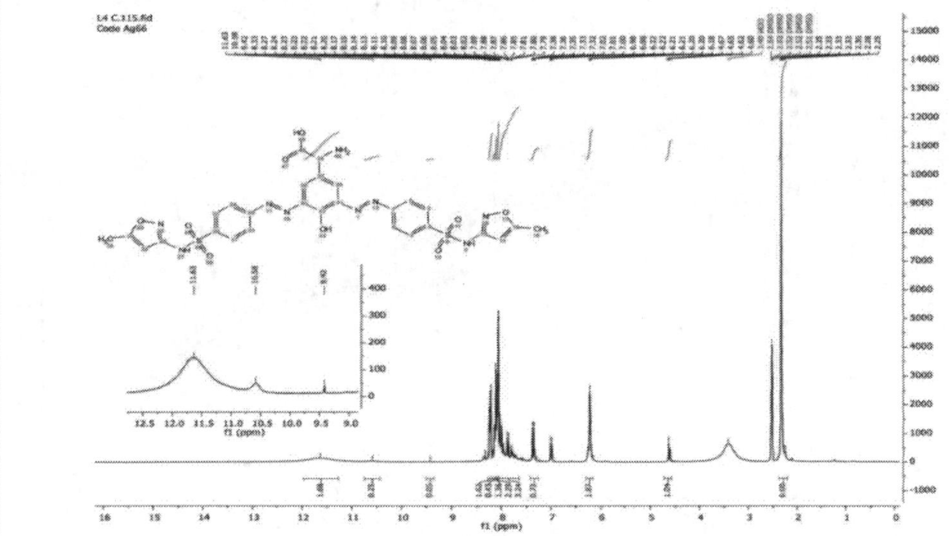

Figure 6. The ^{1}H NMR spectrum of the azo dyes ($2=L_4$).

The UV-visible spectrum was documented at the range (320-520) nm in ethanol for each synthetic dye. The absorption spectrum of synthetic azo dye (1) and diazo dye (2) were showed bands at (360 nm and 410 nm) and (330 nm and 380 nm) related to (π -π*) and (n-π*) respectively. Analytical studies were also curried on (1) and (2), first of all, the solvent effect of (1) and (2) were intended, (Figures 1 and 2) using set of different solvents. The results displayed that the best solubility of (1) in methanol and (2) in ethyl acetate.

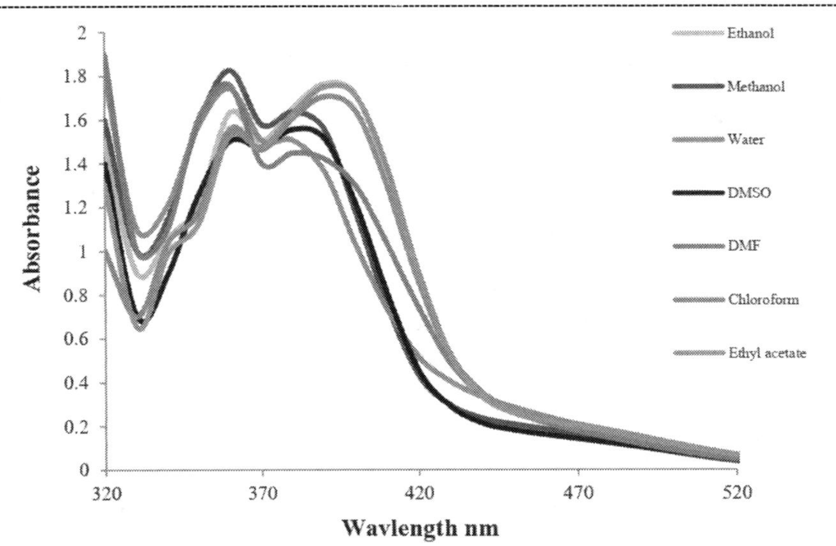

Figure 7. The UV-vis spectrum of the novel azo dyes (1=L_3) in a range of solvents.

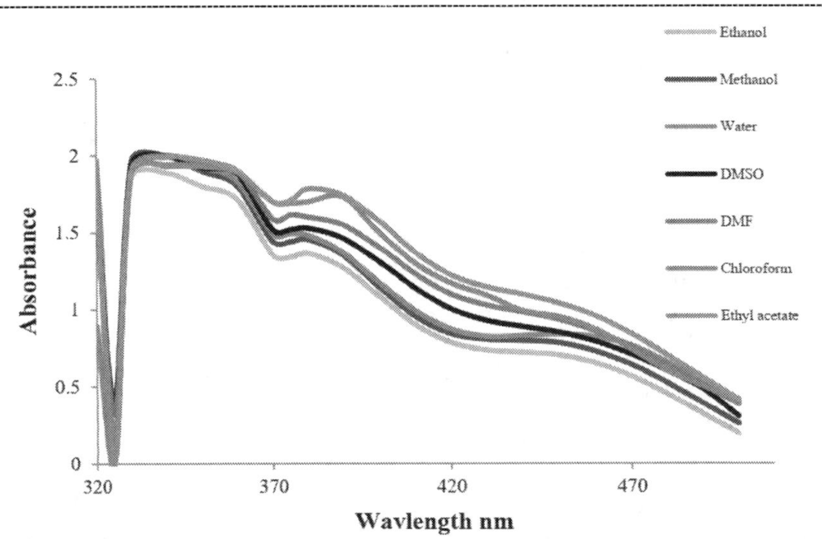

Figure 8. The UV-vis spectrum of the novel azo dyes (2=L_4) in a range of solvents.

Each of azo dye (1) and diazo dye (2) was gave different values of λ_{max}, (Table 1) attributed to π -π* transition of the azo group and indicates the absence of the hydrazone formula within the studied region.

Journal of Physics: Conference Series **2063** (2021) 012023 doi:10.1088/1742-6596/2063/1/012023

Table 1. The UV-visible spectrum of the azo dye (1) and the diazo dye (2).

Solvent	(1)		(2)	
	λ_{max}(nm)	$\varepsilon_{max}(\times 10^{-4})$	λ_{max}(nm)	$\varepsilon_{max}(\times 10^{-4})$
Methanol	390	1.76	380	1.36
Ethanol	380	1.64	380	1.45
Water	380	1.51	375	1.49
DMSO	380	1.56	380	1.53
DMF	380	1.45	375	1.62
Chloroform	390	1.71	390	1.73
Ethyl acetate	390	1.75	380	1.78

The table shows the λ_{max}(nm) in the absorption spectrum for each dye. The azo dye (1) was displayed a slight red shift towards a higher wavelength using methanol and ethyl acetate in contrast with the diazo dye (2). These results indicated that the synthetic dyes were affected by the solubility and dielectric constant (D), which can be expressed in relation to Gati and Szalay [15] as follows:

$$\Delta\tilde{V}= [(a\text{-}b)(n^2\text{-}1\ /\ 2n^2+1)]+b(D\text{-}1\ /\ D+1)$$

The F(D) and Φ(D) were also calculated, (Table 2) which gave a linear relationship when the dielectric constant is the only effect controlling the beak shift.

$$F(D)= 2(D\text{-}1)/(2D+1)$$

$$\Phi(D) = (D - 1)/\ (D + 2)$$

Table 2. The values of the solvents of the dipole moment constants with the maximum wavelengths of the synthetic dyes.

The Solvent	D	(D-1)/(D+1)	D(Θ)	F(D)
Ethanol	24.55	0.92	0.887	0.94
Methanol	32.6	0.94	0.914	0.955
Water	78.3	0.975	0.963	0.981
DMSO	46.7	0.958	0.938	0.968
DMF	36.71	0.947	0.923	0.96
Chloroform	4.7	0.649	0.552	0.711
Ethyl acetate	6.02	0.715	0.612	0.776

The results, (figures 9 and 10) also indicated, that there is no deviation from the linear relationship, which is due to the fact that the effect of the dielectric constant can control the shift of the absorption beaks.

Journal of Physics: Conference Series **2063** (2021) 012023 doi:10.1088/1742-6596/2063/1/012023

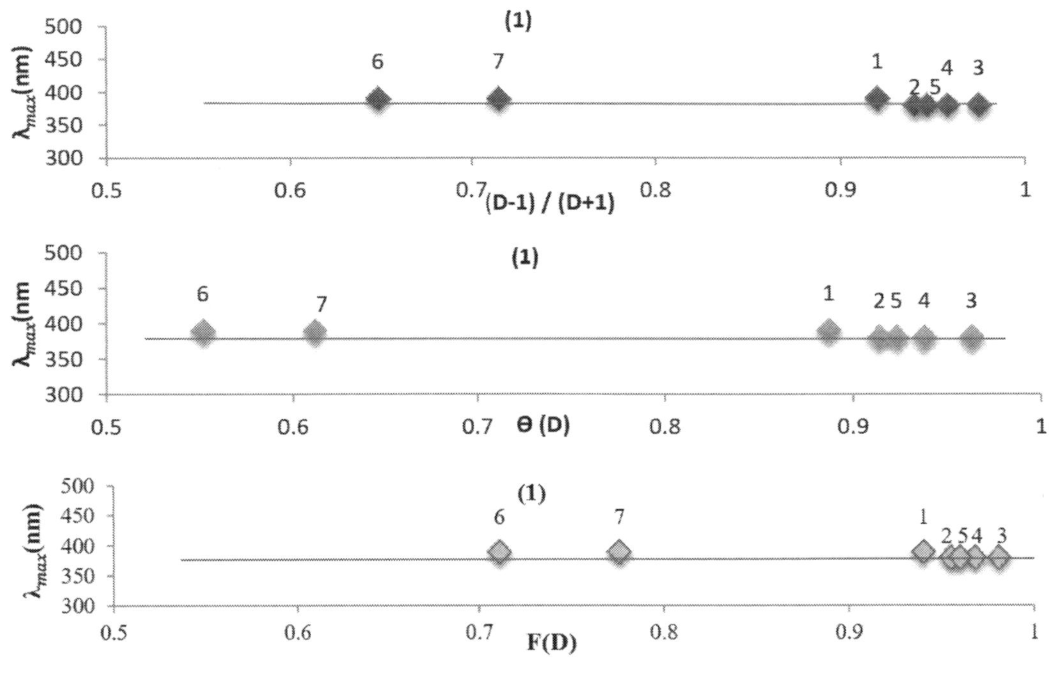

1- Ethanol , 2- Methanol , 3- Water , 4- DMSO , 5- DMF , 6- Chloroform , 7-Ethyl acetate .

Figure 9. The relation between the dipole moment constants and the maximum wavelengths using different solvent.

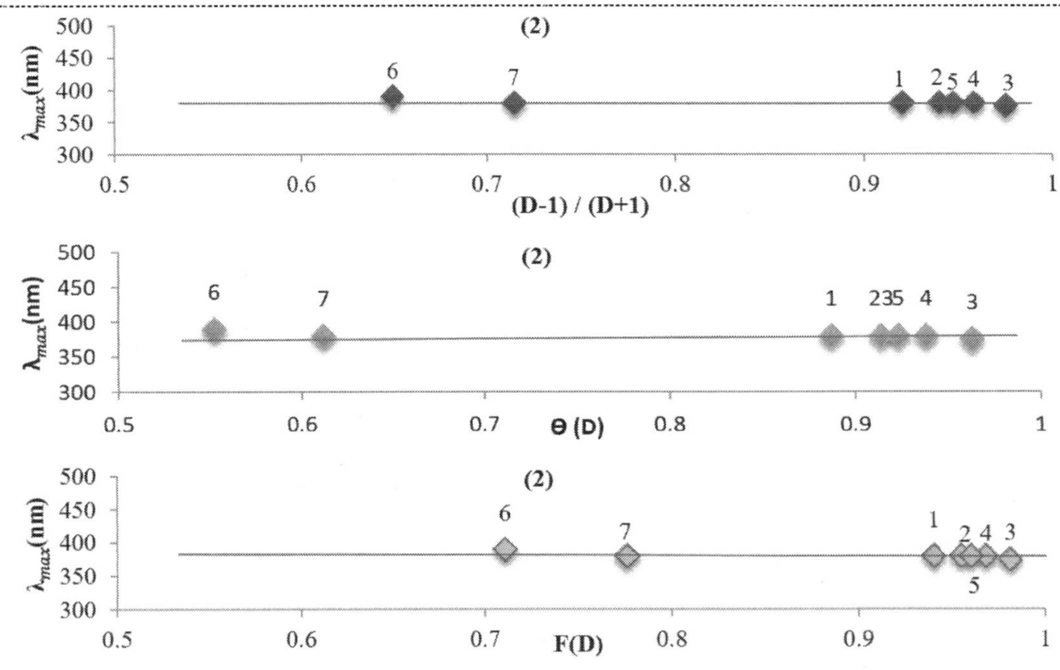

1- Ethanol , 2- Methanol , 3- Water , 4- DMSO , 5- DMF , 6- Chloroform , 7-Ethyl acetate .

Figure 10. The relation between the dipole moment constants and the maximum wavelengths using different solvent.

Further, the pH effect in the range of λ (320-620) nm was also studied for each dye (1) and (2) in pH (1-12) using 1×10^{-4} M concentration as seen in figures (11) and (12) below.

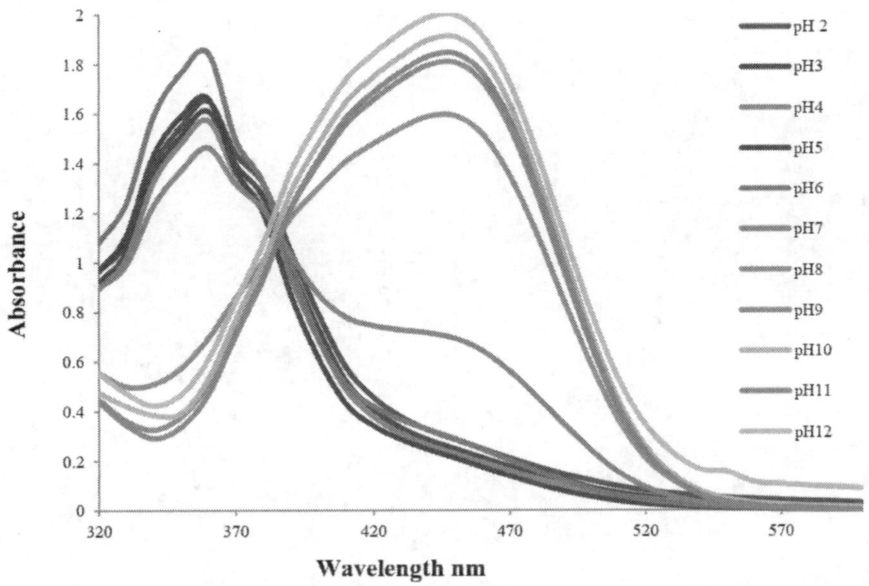

Figure 11. The pH effect in the novel azo dyes (1=L_3).

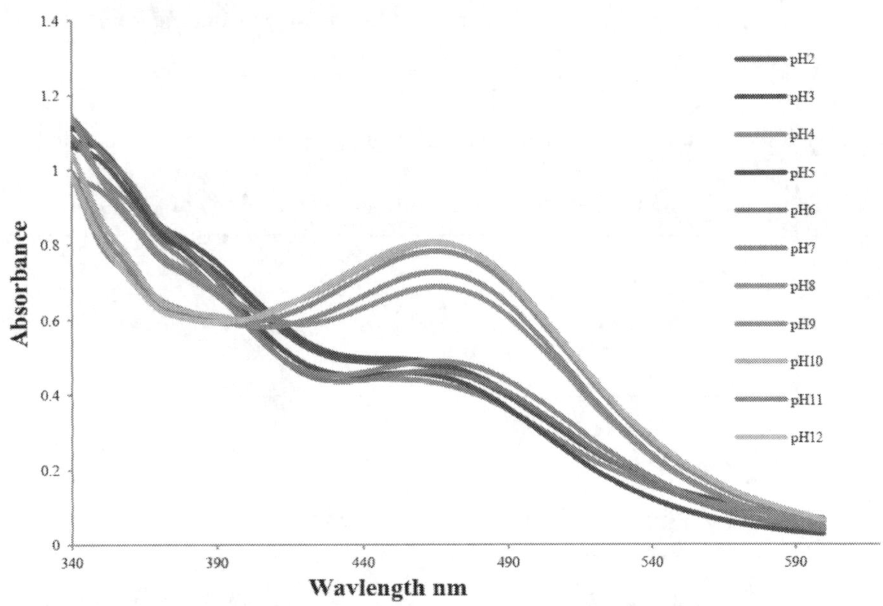

Figure 12. The pH effect in the novel azo dyes (2=L_4).

Journal of Physics: Conference Series 2063 (2021) 012023 doi:10.1088/1742-6596/2063/1/012023

The results revealed suitable pH values at pH12 in each dye with one and two isopiestic points as seen in Figures (11 and 12) above respectively. Due to calculate the pK_a and pK_b in each of (1) and (2) applying half height method [16], (Figures13 and 14).

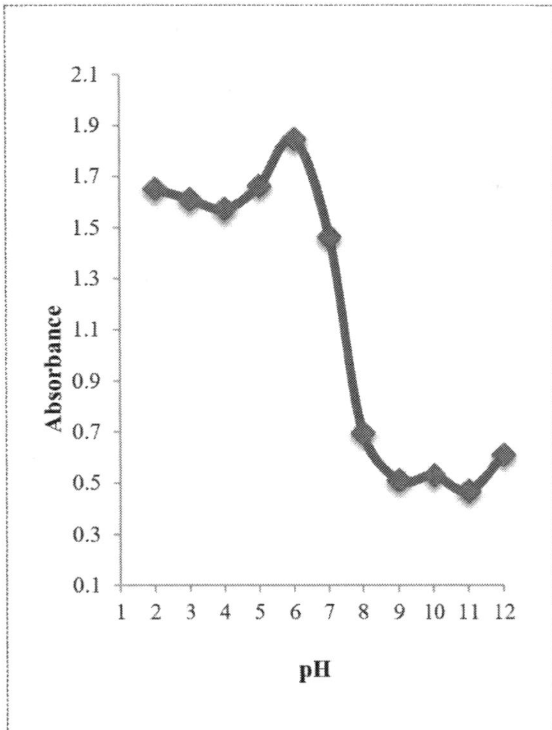

Figure 13. Absorbance-pH effect of ($1=L_3$) at λ_{max} 360 nm.

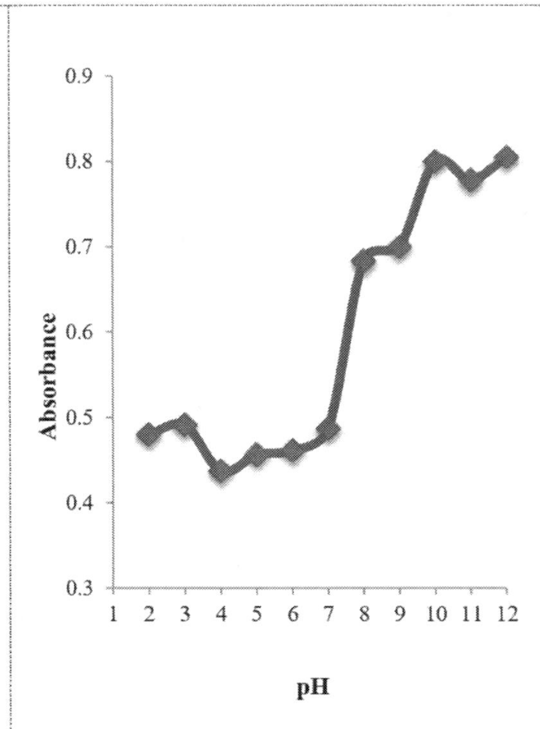

Figure 14. Absorbance-pH effect of ($2=L_4$) at λ_{max} 460 nm.

The results obtained from the absorbance-pH curve in figures above presented in table (3) below.

Table 3. define the ionization and protonation constants of (1) and (2).

The synthetic dye	$\lambda_{max\ (nm)}$	$A_{1/2}$	pK_{p1}	$A_{1/2}$	pK_{p2}	$A_{1/2}$	pK_{a1}	$A_{1/2}$	pK_{a2}
Azo dye (1)	360	1.709	5	....	....	0.521	9.5	0.539	11.5
Diazo dye (2)	460	0.449	5	0.485	2.5	0.75	9.5	0.791	11.5

The following equilibrium schemes, (2 and 3) displays the suggested ionization of azo and diazo dye respectively in acidic and basic medium.

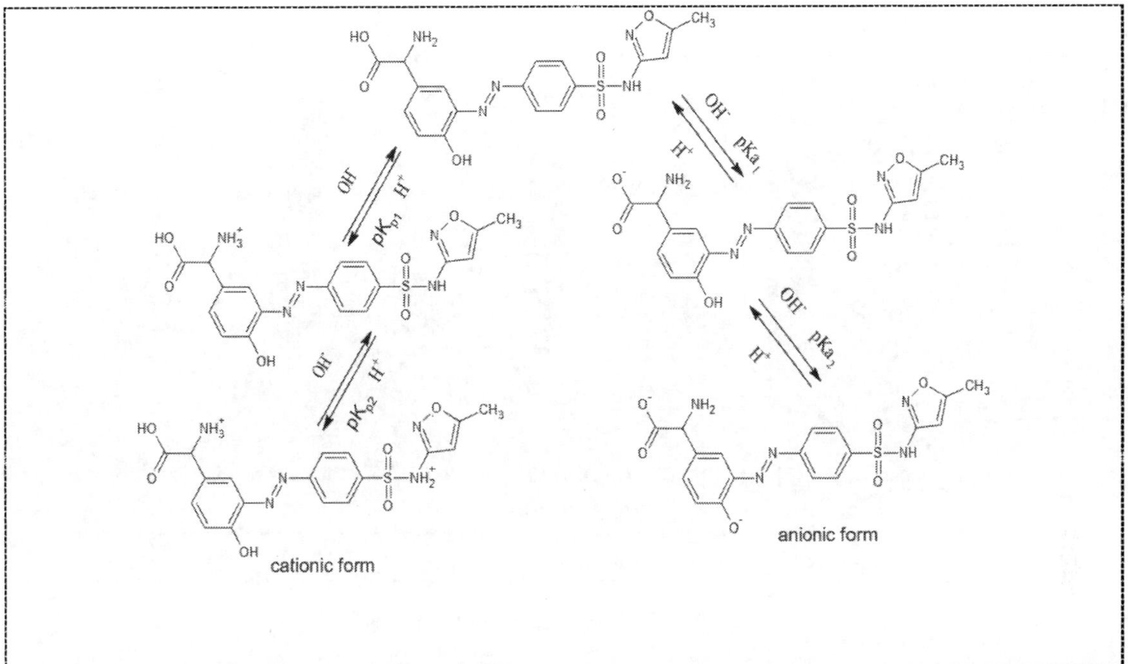

Scheme 2. Optional ionization of (1) in acidic and basic medium.

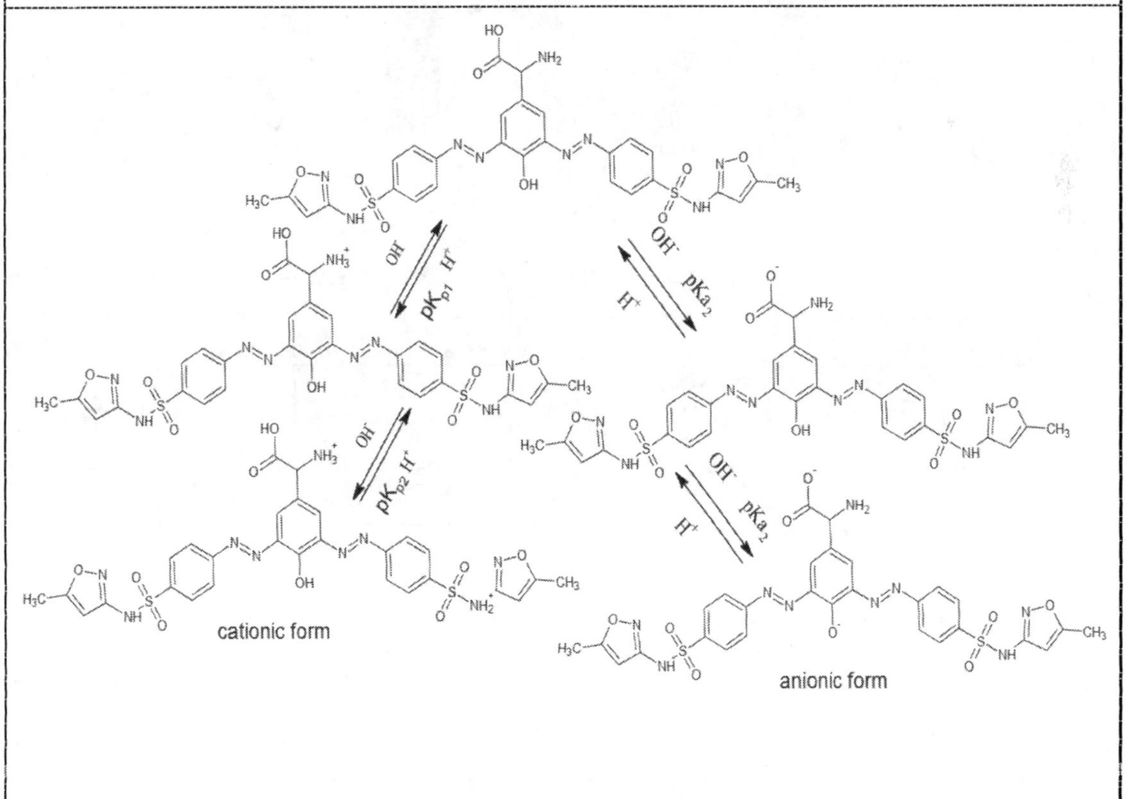

Scheme 3. Recommended ionization of (2) in different media.

Journal of Physics: Conference Series **2063** (2021) 012023 doi:10.1088/1742-6596/2063/1/012023

Different buffer solutions effect each dye, (Figures 15 and 16).

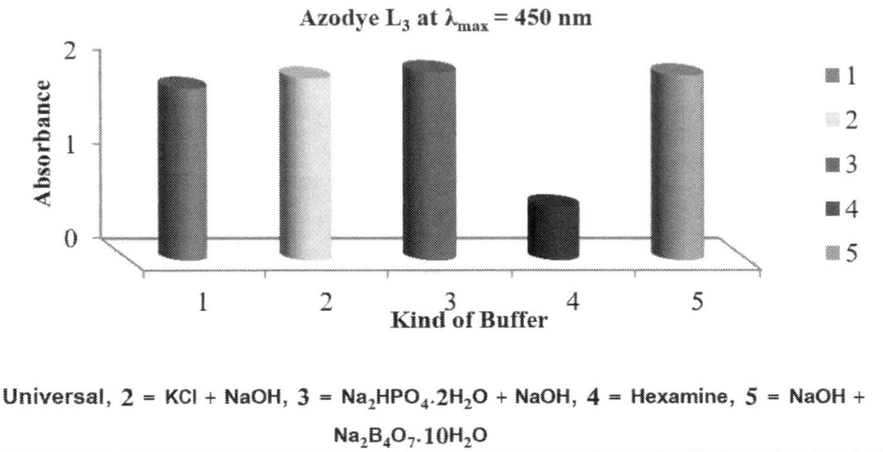

Figure 15. The effect of different buffer solutions in azo dye (1).

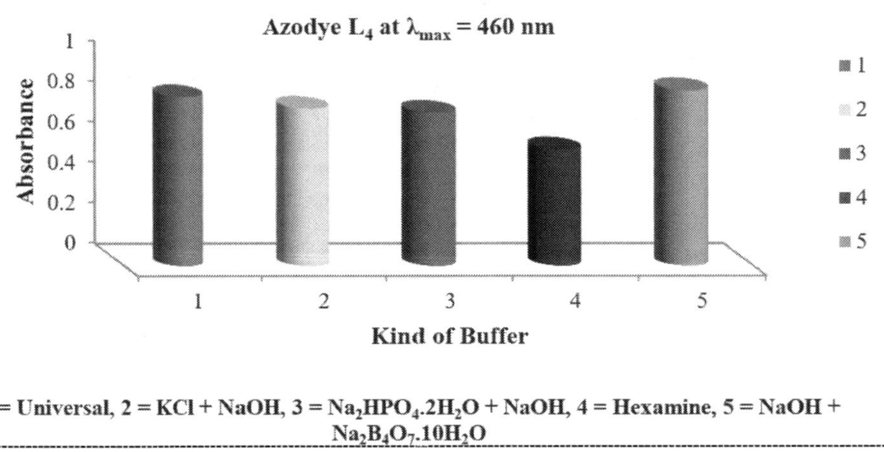

Figure 16. The effect of different buffer solutions in diazo dye (2).

Furthermore, the results were displayed effect of different buffer solutions on (1) and (2), (Tables 4).

Table 4. The effect of different buffer solutions on azo dye (1) and diazo dye (2).

No.	Buffer Solutions	The absorption at λ_{max}	
		L_3 (450)nm	L_4 (460)nm
1	Universal	1.82	0.834
2	KCl + NaOH	1.926	0.775
3	$Na_2HPO_4.2H_2O$ + NaOH	1.99	0.754
4	Hexamine	0.567	0.574
5	NaOH + $Na_2B_4O_7.10H_2O$	1.965	0.867

The method [14] applied for toxity measurement using haemolytic red blood cells in vitro. The results were showed that all prepared compounds harmless and didn't display haemolysis effect in the cells, using different concentrations.

4. Conclusion

This study focused on the synthesis, characterization and analytical studies of novel synthetic azo and diazo dyes. These dyes were prepared cheaply because the starting materials are obtainable and most of the chemistry is prepared at or below room temperature. Moreover, the synthetic dyes gained good colour and revealed non-toxic properties.

5. References

[1] Kirkan, B. and Gup, R. "Synthesis of New Azo Dyes and Copper (II) Complexes Derived from Barbituric Acid and 4-Aminobenzoylhydrazone", *Turk. J. Chem.*, 2008, 32, 9-17.

[2] H. Majeed, Synthesis, Characterization and study of the spectral and electronic properties of a New Azo Dyes Compounds, *J. Thi-Qar Sci*, 2013, 4, 91 – 101.

[3] Ali H., H. Majeed, A. Hussain, Synthesis, Analytical and Theoretical studies of (Z)-4amino-3-hydroxy-2-((4-(N-(5-methyl isoxazol-3yl) sulfamoyl) phenyl) diazenyl) naphthalene-1-Sulfonic Acid, *Journal of Natural Sciences Research*, 2017, 7, 81 – 88. 16.

[4] Ali H., H. Majeed, I. Al-Asadi, A. Abdulredha, A. Hussain, Structures effect of two azo dyes associated with their antimicrobial activity, *Journal of Chemical, Biological and Physical Sciences*,2018, 8, 171-185.

[5] Otutu, J. O., " Synthesis and application of azo dyes derived from 2-amino-1, 3,4-thiadiazole-2-thiol on polyester fibre", *J. IJRRAS*, 2013, 15, 292 – 296.

[6] Fayadh, R. H. F., Ali, A. A. and Al –Jabri, F. M., Synthesis and Identification Symmetrically Azo Dyes Derived from Sulfa Compounds and Spectrophotometric study of Nickel (II) Complexes with Prepared Dyes, *International Journal of Engineering and Technical Research* (IJETR), 2015, 3, 25-28.

[7] Zollinger, H., "Color chemistry; synthesis, properties and Application of organic Dyes and Pigments" ,*VCH*,1991.

[8] Garg H. G. and Praksh C., "Preparation of 4-arylazo-3,5-disubstituted-(2H)-1,2,6-thiadiazine1, 1-dioxides," *Journal Medicinal Chemistry*,1972, 15,.435–439.

[9] Child R. G., Wilkinson R. G., and Tomcu-Fucik A., "Effect of substrate orientation of the adhesion of polymer joints," *Journal of Chemical Society*, 1977, 87, 6031–6038.

[10] Ali H., Majeed H., Al-Asadi I., Abdulredha A. and Hussain A., Structures effect of two azo dyes associated with their antimicrobial activity, *Journal of Chemical, Biological and Physical Sciences* 2018, 8, 171-185.

[11] Farghaly T. A. and Abdallah Z. A., "Synthesis, azo-hydrazone tautomerism and antitumor screening of *N*-(3-ethoxycarbonyl4,5,6,7-tetrahydro-benzo[b]thien-2-yl)-2-arylhydrazono-3-oxobutanamide derivatives," *Archieve for Organic Chemistry*, 2008, 17, 295–305.

[12] Ali H., Majeed H., Hussain A. and Mohsin R., Journal of Global Pharma Technology Analytical and Theoretical Studies of Some Diazo Dyes, *Journal of Global Pharma Technology*, 2019 11, 353-362.

[13] Xian-guo H. and Ursula M., Determination of hemolysis against fresh sheep RBCs, *J. Ethnopharmacol*, 1994, 43, 173-177.

[14] Ali, A. A., Fahad, T. A., & Al-muhsin, A. A. (2020, November). Preparation and Spectroanalytical Studies of Two New Azo Dyes Based on Luminol. In *IOP Conference Series: Materials Science and Engineering* (Vol. 928, No. 5, p. 052007). IOP Publishing.

[15] Issa RM, Zewail AH (1971) J. Chem. U.A.R., (Egypt J. Chem.), 14, 461

Preparation, analytical Studies and application of a New Azodye Derived from Pharmaceutical Materials (Procaine Hydrochloride & Salicylic acid)

Huda T Thuiny, Tarek A Fahad and Asaad A Ali

Department of Chemistry, College of Education for Pure Sciences, University of Basrah, Basrah, Iraq
Corresponding author's e-mail address: huda.ta78@gmail.com

Abstract. Procaine hydrochloride was used as an aromatic amine, and salicylic acid (5-[2-(diethylamino)-4-benzoateazo]-2-hydroxybenzoic acid) was used to make a new azodye derived from pharmaceutical materials (PS).C.H.N., H1-NMR, I.R., and visible spectroscopic techniques have also been used to characterize dye.At different pH values (2 - 12), the electronic spectra of this azo dye was investigated in terms of acid-base properties, which included defining isobestic points and determining protonation and ionization constants.The effect of different polarities of solvents on the electronic spectra was the subject of another study.The dye has been used in a variety of applications, including as an indicator for strong acid with strong base and for nitrite determination.

1. Introduction

Azo compounds are commonly used as dyes because of their numerous uses in fields such as synthetic fabrics, material coloring, biomedical research, and organic synthesis [1-2].Azo compounds are often used as markers of biomedical research.Furthermore, azo dyes have been documented to have a wide range of biological applications, including antineoplastics, antidiabetics, antiseptics, anti-inflammatory, and other chemotherapeutic agents [3-6].Since most azo dyes have acid-base properties and have fixed isobestic points (which reflect the number of equilibriums in such an azo dye), they are used as acid-base indicators [7] and some can also be used as metallochromicindicators.Azo compounds are commonly used as dyes because of their numerous uses in fields such as synthetic fabrics, material coloring, biomedical research, and organic synthesis [8-12].Azo compounds are often used as markers of biomedical research.Furthermore, azo dyes have been documented to have a wide range of biological applications, including antineoplastics, antidiabetics, antiseptics, anti-inflammatory, and other chemotherapeutic agents [13-17].Since most azo dyes have acid-base properties and have fixed isobestic points (which reflect the number of equilibriums in such an azo dye), they are used as acid-base indicators [18] and some can also be used as metallochromic indicators. To the spectrophotometric determination of nitrite, the azo dye was used by forming daizonium salt. Such as p-nitroaniline in acidic medium with frusemide at λmax.680nm[19], and also the azo dye formed form coupling of 2- thioparbutric acid with diazonium salt of p-aminobenzoic acid at λ_{max}.420nm [20] and the diazonium salt of luminal with 4-hydroxycumarine [21].

2. Experimental

The reagents and solvents were all of the reagent-grade variety. TLC was used to monitor the reaction's progress on a silica gel-coated plate, and spots were visible under UV light. The FT.IR-

Journal of Physics: Conference Series **2063** (2021) 012024 doi:10.1088/1742-6596/2063/1/012024

8400S shimadzu was used to record infrared spectra (in KBr pellets). Melting points were measured using a melting point apparatus, element analysis (C.H.N.) was performed using a Perkin element 2400 element analyzer, and UV/V absorption spectra were reported using a Perkin Elmer Lampda EZ 210 UV/V Spectrophotometer. PH-Meter was used to take the readings (H. Jurgons Co. Bremen,L.Puls Munchen15).

A sequence of acetate and universal buffer solutions with different pH values (2-12) were prepared to measure the ionization and protonation constants for hydroxyl and nitrogen groups [22].

2.1. Preparation of azodye

The azo dyes reagentsPS were made by dissolving 0.004 mole of procaine hydrochloride in 1.4 ml concentrated HCl, 20 ml distilled water, and then adding a drop-by-drop solution of (0.304 g sodium nitrite in 20 ml distilled water) to form diazonium salt at a temperature of less than 5 o C.

To the above diazonium salt, add an alkaline solution containing 0.004 mole of salicylic acid in 1.2 percent v/w NaOH.Adding dilute HCl to the prepared dye to convert it from sodium salt to hydrogen type.Recrystallization from metanol was used to purify the dye [23]. Figure (1).

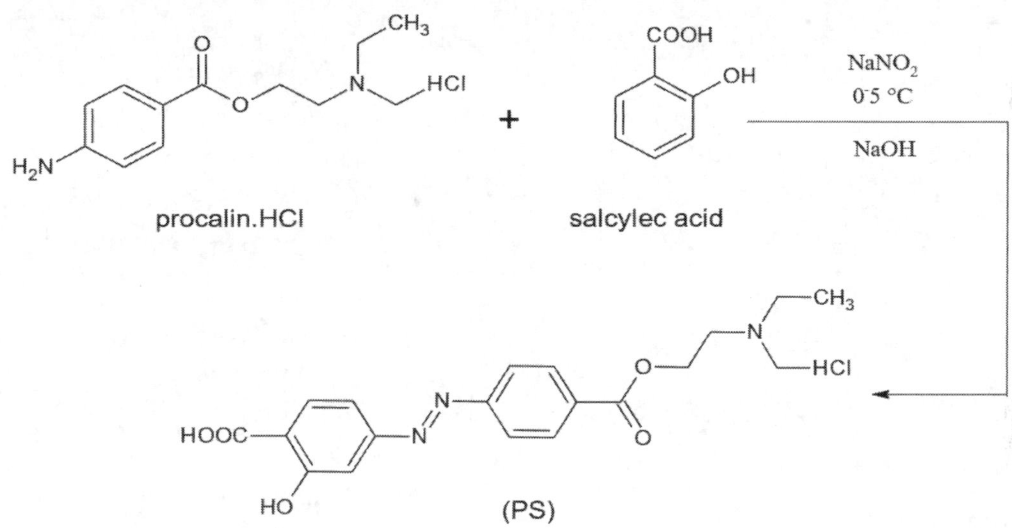

Figure 1. PS 5-(Procaine Azo) - Salicylic acid =5-[2-(diethylamino)-4-benzoateazo]-2-hydroxybenzoic acid

2.2. Solutions

* 1 x 10^{-3}M stuck solution of azo dye PS
* Universal buffer solutions (pH 2-13) [24]
* 0.1 M solution of each sodium carbonate, sodium hydroxide and hydrochloric acid. Thesolutions were prepared and standardized by recommended methods [24]
* For nitrite solution
  1-100 ml Acidic solution of 8 x 10^{-3} M Procaine hydrochloride (P) (0.2183 g of P + 0.28 ml conc.HCl).
  2- 100 ml Basic solution of 8 x 10^{-3} M Salicylic acid (S) (0.1105 g of S + 0.25 g NaOH)
  3- 100 ml solution of 8.8 x 10^{-3} M NaNO2 (N)

2.3 Procedures

- For acid-base, the absorbance of PS dye solutions at concentrations of 6x10-5 M were calculated in the range of 350- 560 nm at different pH buffer solution values of (2-12).
- The absorbance of PS dye solutions with a concentration of 4x10-5 M in different solvents with different polarities was calculated in the range of 320-600nm for the solvent effect.
- To investigate the best conditions for determining nitrate.

Journal of Physics: Conference Series **2063** (2021) 012024 doi:10.1088/1742-6596/2063/1/012024

Mix equal volumes of cold P and N solutions at 5° C, then add an equal volume of S solution and calculate the absorbance in the 320–600 nm range.

3. Results and discussion

The compounds were prepared in good yield. The compounds are stable in air and light. Elemental Analysis data and physical properties of compounds are summarized in table (1)

3.1. Infrared radiation

Table 1 shows the spectral data of azo dye PS .The (N-H) and (O-H) bands in the IR spectrum of free ligand are responsible for the medium band about 3234 cm^{-1} . There is a high intensity band in the range of 1300-1270cm-1 due to (C=N) in the novel azo dye. In the azo spectrum, the azo (-N=N-) group can be seen at 1427.32s cm-1, the bands at 1664 cm-1 for (C=O), the bands around 1309 cm-1 observed due to C-N stretching . Figure. (2) [22]

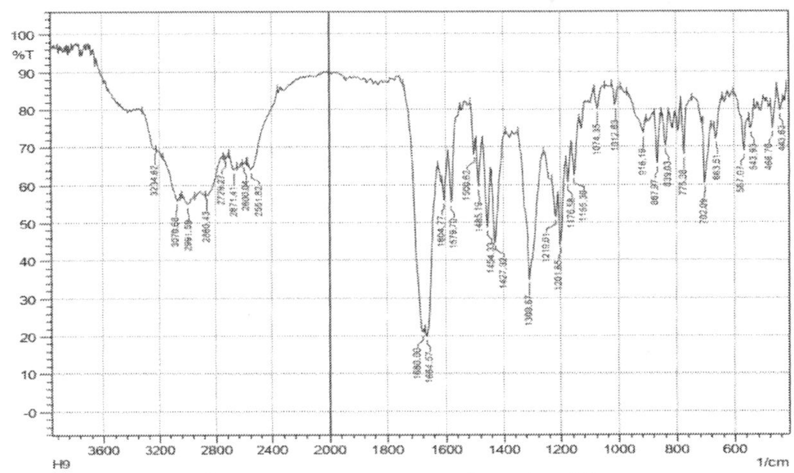

Figure 2. IR-Spectrum of azo dye PS

3.2. Spectra of ^{1}H-NMR

In deuterated dimethyl sulfoxide solutions with tetramethylsaline as an internal norm, 1H-NMR spectra of the prepared azo dye PS were obtained. The DMSO solvent caused a peak at 2.5 ppm in the spectra. Figures (3) show the 1H NMR data.

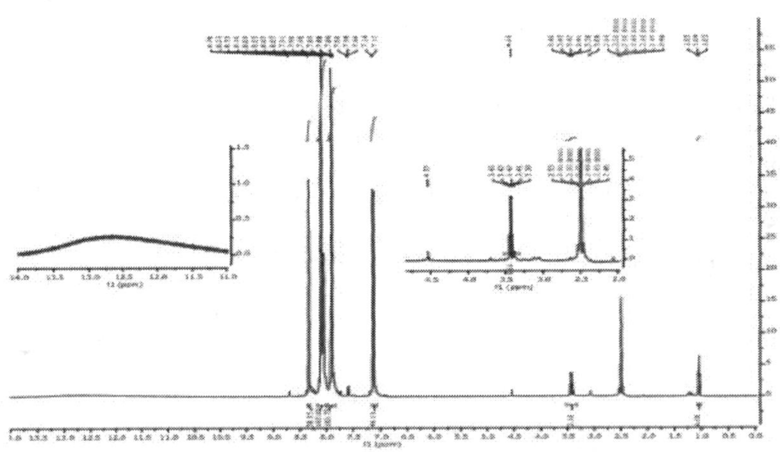

Figure 3. NMR-Spectrum of azo dye PS

Journal of Physics: Conference Series **2063** (2021) 012024 doi:10.1088/1742-6596/2063/1/012024

Table 1. Chemical analysis (CHN, IR and NMR) of azo dye PS

CHN	C%		H%		N%	
	Calculated	Found	Calculated	Found	Calculated	Found
	62.34	62.66	5.97	6.01	10.91	11.01
IR	Wavenumber (cm^{-1})					
	OH&NH	C=O	N=N	C-N	C-O	C=C
	3234.62w	1664.57s	1427.32s	1309.67	1219.01s	1579.70s
	Chemical shift (ppm)					
H^1NMR	S,3H,CH_3	S,CH2,	S,6H,DMSO	m,7H,Ar-H	S.H.OH	S NH
	1.2	3.5	2.5	7.3-8.2	11-12	7.2

3.1. Some of spectrophotometric studies of azo dye PS

3.1.1. Ionizationscostants

The absorption spectra of 6 x10^{-5} M solution of dye in the range of wavelength (350- 560 nm) of varying PH values (2 – 12) were represented graphically (Figure. 4).

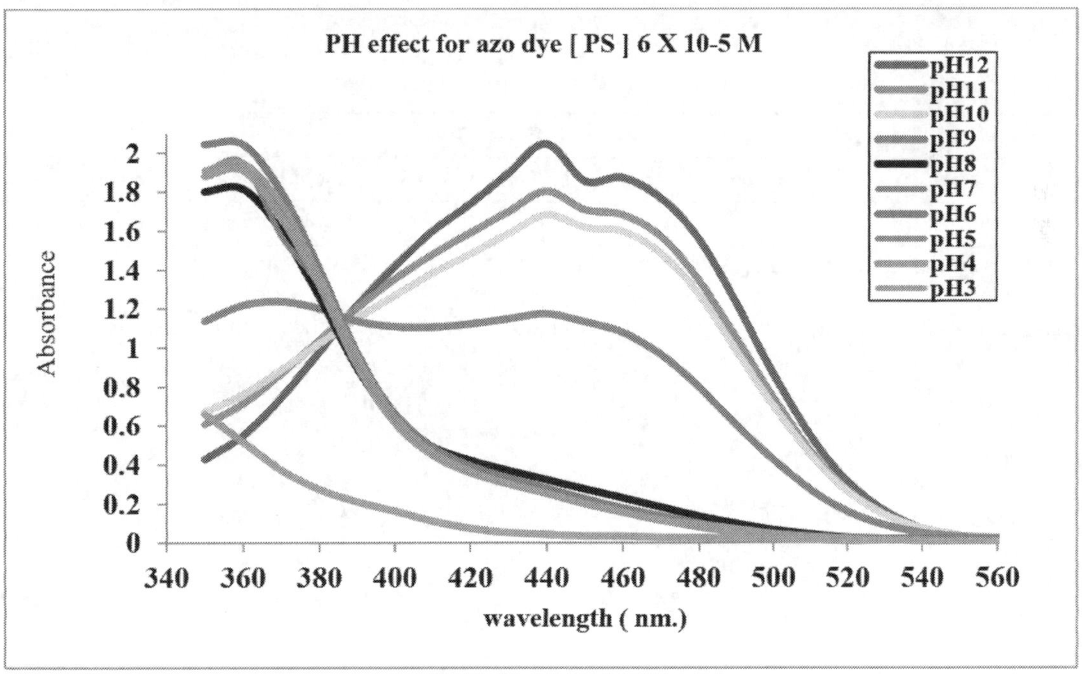

Figure 4: Electronic spectrum of dye PS at different value of PH.

The spectrum had three bands, the first of which was at 360 nm and was in the pH range (3-8) due to dye protonation.In the pH range, the second and third bands (440 and 460 nm,) of high intensity (8-12).At 390 nm, the spectrum also revealed an isobestic point (a wavelength with constant absorbance despite changes in pH value medium).

Journal of Physics: Conference Series **2063** (2021) 012024 doi:10.1088/1742-6596/2063/1/012024

At pH 2, the spectrum revealed no and band. To determine the protonation and ionization constants of azo dye PS, the wavelength 440nm. Is chosen as best wavelength (from Figure. 4) to draw a curve between pH values as x-axis with corresponding their absorbance (Figure. 5).

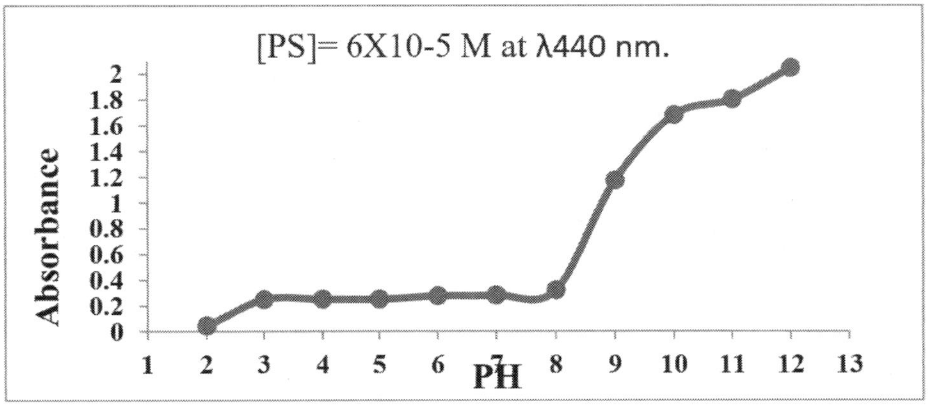

Figure5. Absorbance – pH curve of the azo dye PS

From (Fig.5) the ionization and protonation constants were calculated (Table 2). From Absorbance–PH curve and by the aid of half-height method [25], the PK values were obtained by this relation : PK = PH (at $A_{1/2}$) where ; $A_{1/2} = (A_L + A_{min.}) / 2$
And A_L and A_{min} are limiting and minimum absorbances respectively.

Table (2): Protonation and ionization constants of azo dye PS

PS Azo Dye at λ_{440} nm.											
A_m	A_l	$A_{1/2}$	PKp	A_m	A_l	$A_{1/2}$	PKa_1	A_m	A_l	$A_{1/2}$	PKa_2
----	0.25	----	< 3	0.3	1.7	1	8.8	1.8	----	----	> 11

P = protonation of H of –COOH group, a_1= ionization of H of –COOH group,
a_2 = ionization of H of –OH group
The suggested chemical mechanism of protonation and ionization was shown in Figure. (6)

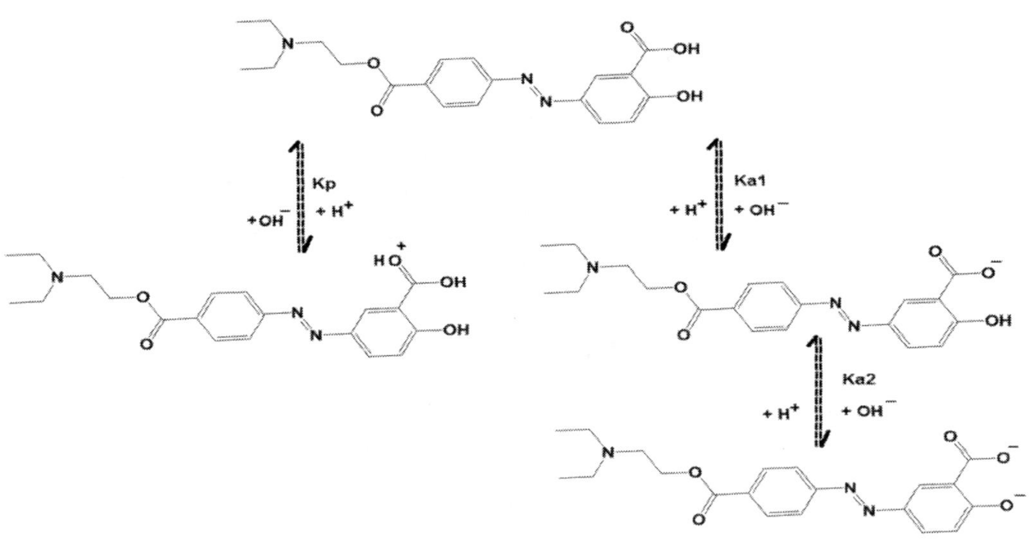

Figure 6. Suggested mechanism of protonation and ionization of azo dye PS

Journal of Physics: Conference Series **2063** (2021) 012024 doi:10.1088/1742-6596/2063/1/012024

3.2. Effect *of solvents of different polarities*

Figure 7 depicts the spectrum of the azo dye PS (concentration 4x10-5 M) in various polarity solvents (Acetone , chloroform , ether , DMF , DMSO, ethyl acetate , water , ethanol , methanol and 1,4-dioxane).In all solvents, the dye has only one band, as shown in Fig.7. When polar solvents (ether, methanol, chloroform, and ethanol) are compared to non-polar solvents (benzene) at 350 nm, there is a small blue change at 340 nm. The broad red shift at 380 nm in comparison to solvents (DMF and DMSO) was due to hydrogen bonding between dye molecules (solute) and solvent molecules, as shown in Fig.8. The dielectric function of solvents and the maximum wavelength of dye PS in various solvents are shown in Table 3.

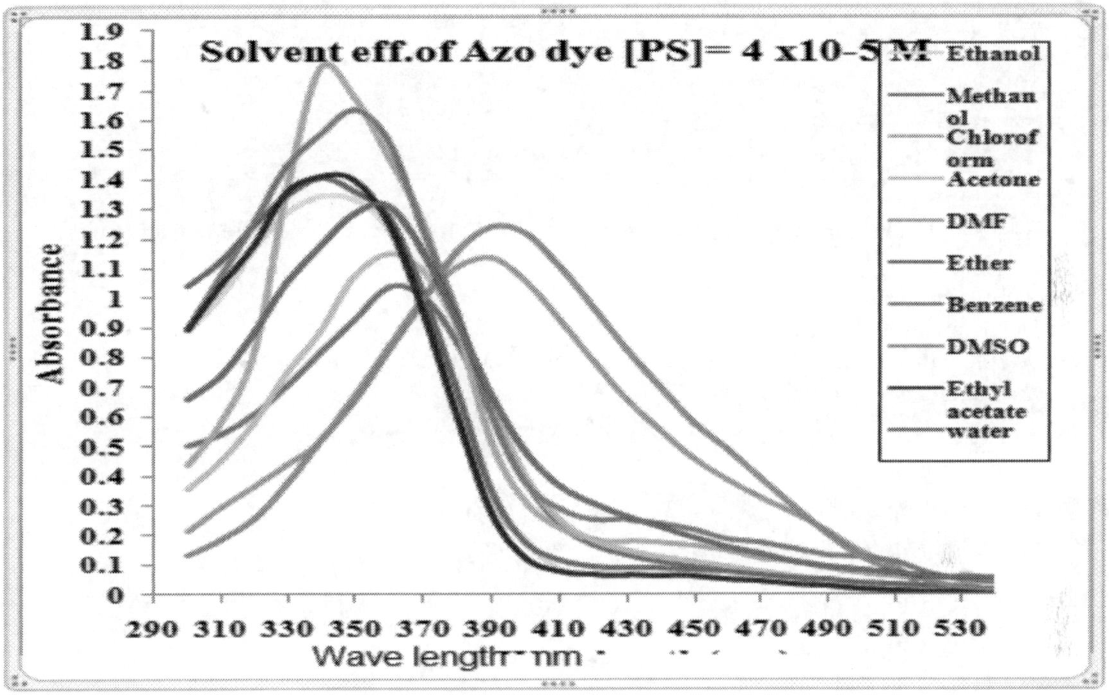

Figure7. Absorbance spectra of dye PS at different solvents with different polarities

Table (3): values of dielectric function of solvents and maximum wavelength of dye PS at different solvents

Symbol	Solvent	Dielectricco (nst.D)	(D – 1) / (D + 1)	PS	
				λ max nm	ε_{max} x10^4 l.mol^{-1} .cm^{-1}
A	Acetone	20.60	0.907	360	2.875
B	Benzen	2.30	0.394	350	4.083
C	Chloroform	4.80	0.655	340	4.433
DF	DMF	36.71	0.923	390	2.833
DO	DMSO	47.00	0.958	390	3.100
Et	Ethanol	24.30	0.921	345	3.350
E	Ether	4.33	0.625	340	3.503
Ea	Ethyl acetate	6.02	0.715	340	3.520
M	Methanol	32.70	0.941	360	2.588
W	Water	78.40	0.975	360	3.270

By plotting a curve between dielectric function (D-1)/(D+1) and λ_{max} of dye at different solvents (Figure. 8) , the linearity of all solvents which means the dielectric constant that affected except for the DMF and DMSO gave deviation from the linearity because of hydrogen bonding between dye molecules (solute) and solvent molecules.

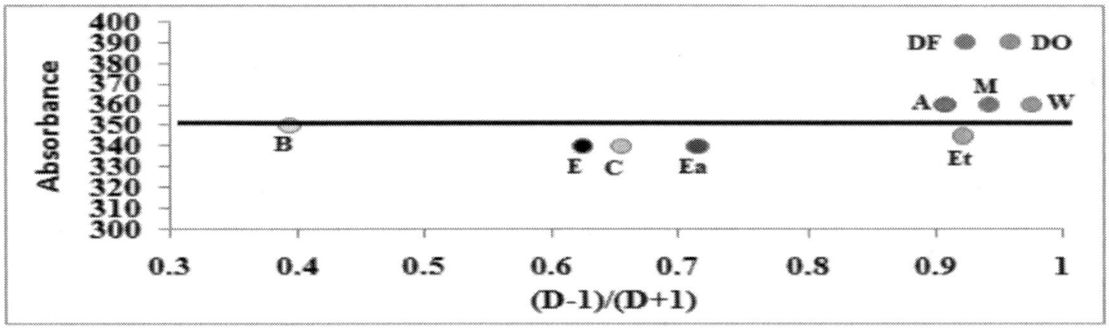

Figure 8: The relation between λ_{max} and (D-1)/ (D+1) for of dye PS at different solvents

3.2. *Some applications of the azo dye PS*

Determination of nitrite

To select the maximum wavelength of azo dye that formed, a curve of absorbance and wavelength was plotted (Figure. 9).

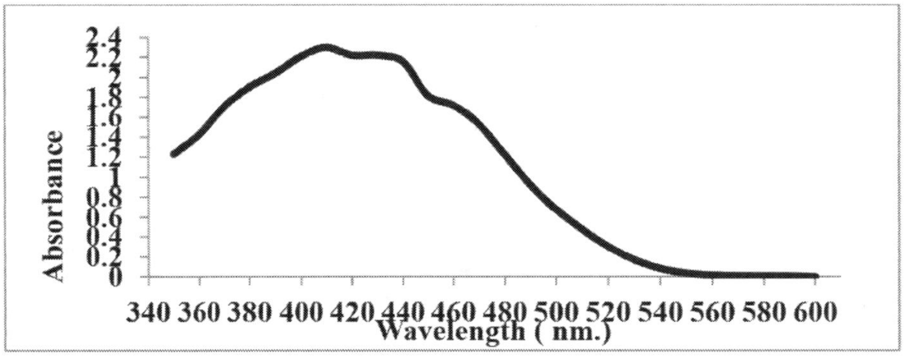

Figure 9 .Absorption curve of dye PS ([P] =[S] = $1.6x10^{-3}$ M and [N] = 1.76×10^{-3} M)

The optimum time for forming the dye PS was studied at λ_{max} = 420 nm, which found to be in the range of (20 – 25 sec) (Table 4 and Figure. 10).

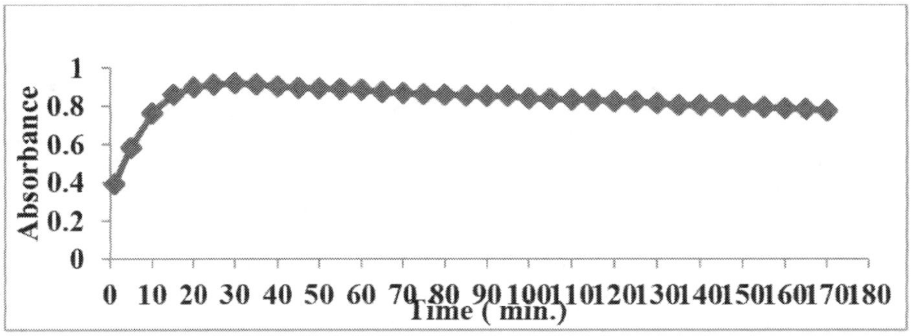

Figure10.Time effect for forming stable azo dye PS

By using of calibration curve (Beer's law) for calculating precession, sensitivity of the method and limits of Beer's law were calculated (Figure.11).

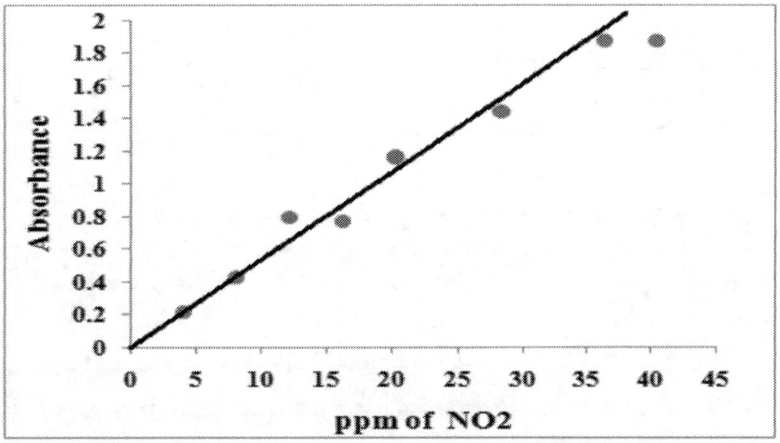

Figure11.Calibration curve for determination nitrite ([PS] = 4.8x10^{-4}M)

Table (4): beer's law of the dye PS.

ε x10^4 l.mol^{-1}.cm^{-1}	a	S μg.cm^{-2}	R^2	S.D.	D.L x10^{-5} μg.cm^{-1}	Beer's law up to ppm
0.053	0.012	0.083	0.9800	0.0085	0.048	40

Where ε, a, S, R^2, S.D. and DL are molar absorptivity coefficient, specific absorptive, Sandel sensitivity, standard deviation and detected limit respectively.
Titration of strong acid (HCl) with strong base (NaOH) uses the azo dye as an acid–base indicator. The base was titrated spectrophotometrically with standard HCl using the new azo dye PS as an indicator (Fig. 12) after HCl and NaOH were standardized to be 0.0917 N and 0.0961 N, respectively. The relative error for obtaining the best PS concentration was discovered (Table 5).

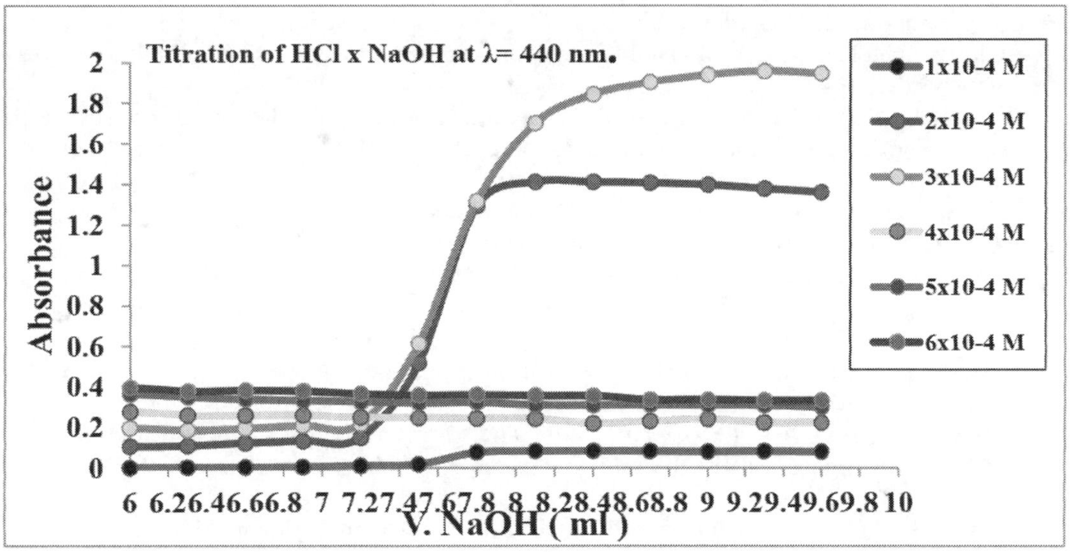

Figure12: The titration curves of (HCl and NaOH) at different concn. Of PS

Table (5): Relative error for different concn. Of dye PS

[PS] M	(e.p.) ml NaOH	[NaOH] M Found	[NaOH] M True	Jump / 0.9 ml	±R.E%
1×10^{-4}	7.65	0.0959	0.0961	0.072	0.21
2×10^{-4}	7.60	0.0965		1.262	0.42
3×10^{-4}	7.65	0.0959		1.483	0.21

Using 8 mL of 0.0917 M HCl
Table (5) and Figure. (12) Revealed that the concen.of PS in the range of (0.00001- 0.00003 M)
The best concn. Was 0.00003M of PS, which resulted in more jumbling (that means more sensitivity)
When the concn. Is greater than 0.00003, the titration fails due to minor dissuasions of dye PS at high concentrations.

References

[1] Towns, A. D., (1999) Developments in azo disperse dyes derived from heterocyclic diazo components. *Dyes and Pigments*, 42:3-28
[2] Zollinger,H., (2003). "Synthesis, properties and application s of organic dyes and pigments, Color Chemistry", third revised edition, Wiley-VCH., Weinheim
[3] Bae,J.S., Freemanand,H.S., El-Shafei,A. (2003). Metallization of non-genotoxic direct dyes, *Dyes and Pigments*, 57(2):121-129
[4] Sanjay,F.T., Dinesh,M.P., Manish,P.P., and Ranjan,G.P. (2007) Synthesis and antibacterial activity of new pyrazolo [3,4-b] quinoline based heterocyclic azo compounds and their dyeing performance),*Saudi Pharm. Journal*, 15(1):48-54).
[5] Child, R.G., Wilkinson,R.G., and Fucik, A. (1977),*Chem. Abstr,*87: 6031.
[6] Ggarg, H., and Prakash,C. (1972) Imidazo[1,2-a]pyridine anthelmintic and antifungal agents.*J Med. Chem*., 15(4): 435-436
[7] Ali ,A. A., (2014) Synthesis and Spectroanalytical Studies of a New Azodye Derived From 2-Amino-6 -ethoxybenzothiazole and 4-Chloro-3,5-dimethylphenol and its Complexes With Fe (III) Ion,*Ibn Al-Haitham Jour. for Pure & Appl. Sci*., 27 : 196
[8] VIDYA, V.G., and SADASIVAN, V., (2018) Synthesis, Characterization and Applications of SomeAzo Dyes Derived from Various Aromatic Amines, *Asian Journal of Chemistry*, 30,: 2049-2053
[9] Gupa,R., Giziroglub,E., Kırkan,B.,(2007) Synthesis and spectroscopic properties of new azo-dyes and azo-metalcomplexes derived from barbituric acid and aminoquinoline, Dyes and Pigments,73: 40-46
[10] Kadim,K.H., Mohammed,M.S., (2014) Spectrophotometric determination of Procaine Hydrochloride in Bulk and Pharmaceutical by Diazotization-Coupling Reaction, *Iraqi National Journal Of Chemistry* ,56:368-380
[11] Afroze S, Sen TK (2018) A review on heavy metal ions and dye adsorption from water by agricultural solid waste adsorbents. *Water Air Soil Pollut* 229:225.
[12] Badrey, M.G., Gomha, S.M.andMashaly,H.M.,(2016)Synthesis and studies on Chromene base dreactive azodyes and their applications on different fabrics. *European Journal of Chemistry*, 7(2),:146-151.
[13] El-Harfi,S.and El-Harfi,A.(2017) Classifications,properties and applications of textiledyes :Areview.Applied Journal of EnvironmentalEngineeringScience,3(3):00000-3
[14] Nuha, W.A and AsaadA.A, (2016), preparation and spectroanalytical studies of two new azodyes derived from procaine hydrochloride and metacloprimide, *World Journal of Pharmaceutical Research*, 5, 127-140
[15] Ali A.A., Fahad,T.A and Al-muhsin,A.A.(2020) Preparation and pectroanalytical Studies of Two New Azo Dyes Based on Luminol IOP Conf. Series: Materials Science and Engineering 928 (2020) 052007,IOP Publishing,doi:10.1088/1757-899X/928/5/052007
[16] Clissold, S.P, (1986) Aspirin and related derivatives of salicylic acid, Drugs, 32: 8–26

[17] Dhungana,B., Peng,H., Kutarna,S., Umbuzeiro, G.,Shrestha,S., Liu,J.,Jones, P.D.,Subedi,B., Giesy,J.P. and Cobb,G.P., (2019) Abundances and concentration soft brominated azo dyes detected in indoor dust. Environmental pollution,252:784-793.

[18] Snigur, D.V., Chebotarev, A.N. and Bevziuk, K.V. (2018) Acid–Base Properties of Azo Dyes in Solution Studied Using Spectrophotometry and Colorimetry. *J Appl Spectrosc* **85,** 21–26 https://doi.org/10.1007/s10812-018-0605-9

[19]Veena,K andNarayana,B.,(2009) Spectrophotometric determination of nitrite using new coupling agents.,*Indian Journal of chemical Technology,16:*89-92

[20] Otutu,J.O., Okoro,D., and Ossai,E.K.,(2008) Preparation of Dis-Azo Dyes Derived from p-Aminophenol and Their Fastness Properties for Synthetic Polymer-Fibres. *Journal of Applied Sciences,* 8: 334-339.

[21] Yazdanbakhsh ,M.R., Ghanadzadeh, A. Moradi,E., (2007)Synthesis of some new azo dyes derived from 4-hydroxy coumarin and spectrometric determination of their acidic dissociation constants,*Journal of Molecular Liquids,*136,:165-168

[22] Abdallah,S.M., (2012 Metal complexes of azo compounds derived from 4-acetamidophenol and substituted aniline) *Arabian Journal of Chemistry*, 5: 251–256

[23] Adil Muala Dhumad, Qusay MA Hassan, Tarek Fahad, CA Emshary, Nabeel A Raheem, HA Sultan (2021) Synthesis, structural characterization and optical nonlinear properties of two azo-β-diketones, Journal of Molecular Structure 1235:130196

[24] J.A.Dean,J.A., (2012)."*Lange's Hand book of Chemistry*", Mcgrawhill, ING.New York, 15th ed.

[25] Fahad, T.A.,Ali,A.A., and Baty,A.H., (2019) SYNTHESIS, CHARACTERIZATION AND ANALYTICAL STUDIES OF SOME NEW AZODYES DRIVEN FROM O-VANILLIN,*World Journal of Pharmaceutical Research,* 8: 707

Study the fluorescent and conductivity properties of a 4-hydroxy-3-nitro-1,8-naphthalic anhydride

Roza Al-Aqar

Chemistry Department – College of Science – Basrah University

rozachemistry@yahoo.co.uk

Abstract. The compound of (4-hydroxy-3-nitro-1,8 -naphthalic anhydride) and the dopant material (4-hydroxy-m-benzene-disulfonic acid) were synthesized. The UV- Vis absorption and fluorescence spectra of the compound were recorded. The bathochromic shifts in absorption due to the changes in the solvent polarity was observed, in DMSO solvent, the compound shows different peaks, three bands in the UV region and one band in the visible region, one absorption peak at 402 nm, which might be attributed to the (ICT) band. The emission spectra of this compound were sensitive to the solvent polarity. They moved to lower energies with increasing the polarity of the solvent. The influence of the solvent polarity on the fluorescence maximum was more pronounced compared with the absorption maximum, showing *ca.* 3 nm red shift on moving from butanol to DMSO. The effect of the dopant material on the conductivities (ionic and specific) of the compound was studied, the ionic conductivity was increased as the weight of the dopant material increases, and the ionic conductivity (G) was measured in water as a solvent. the (G) of the compound was $2.4X10^{-6}$ Siemens at 0.01 gm of the dopant material, and was $2.62X10^{-4}$ Siemens after the dopant material was added (0.1 gm).

1. Introduction

There are many photophysical properties for the 1,8-naphthalimide compounds for example, the stability of these compounds (i.e. optical, thermal and chemical stabilities) in addition to high quantum yield and the medical applications for example, anticancer operators, these compounds are used in the fields of high-tech , such as liquid crystal displays, and transformation of solar energy [1]. The internal charge transfer (ICT) obtains when an amine or hydroxyl is attached in the 4-position of 1,8-naphthalimide, which gives the efficient fluorophore. The 4-substituted derivatives show excellent properties such as, large Stokes Shift and cell permeable. The compound of 4-hydroxy-1,8-naphthalimide fluorophore is used as a fluorescent for the cellular imaging of nitroxyl. The ratiometric probes can be designed by the esterification of the hydroxyl analogues, which give the hypsochromic shift (is a change of spectral band position in the absorption, or emission spectrum of a compound to a shorter wavelength of the fluorescence maximum [2].

The 1,8-naphthalimides compounds are used for many applications for example optoelectronic and organic electronics, therefore it is important to prepare new derivatives compounds which are essential to use in the photo-physics, pH/metal/anion sensing and chemical photostability. The other applications of naphthalimide derivatives in the organic electronic area, the photovoltaic cells which require the compounds show good charge separation capability, in addition to these derivatives can be used as energy donor or acceptor in many energy transfer systems. The extending of the π- conjugation results to the shift to longer wavelengths [3]. There are many potential applications of the derivatives of the naphthalimide which are related to organic production of electric current which is defined as a photovoltaics. The using for the example the compound 4-bromo- naphthalic anhydride which is used as a starting material, and the push-pull chromophores are obtained by the substituents. Naphthalimide-setup systems are utilized as fluorescent signs in the imaging applications and molecular biology [4].

The 1,8- naphthalimide derivatives have been used as fluorescent dyes, and metal sensors due to the good fluorescence properties, such as excellent emission and quantum yields and high absorption coefficients, large Stokes Shifts, and good stability. The naphthalimide derivatives contain an O,N,S-substituted group, or heterocycle with five-member in the C-4 position contain an electron donating group, which it is important to extend the materials, including different groups, such as (electron-donating), for example hydroxyl or ester group to explore the fluorescence property [5]. The mechanism of intramolecular charge transfer is obtained by the 1,8-naphthalimide compound which contains the groups of an electron donor, an acceptor and its derivatives for example, 4,5-diamino-1,8-naphthalimide and 3,4-diamino-1,8-naphthalimide [6]. The 1,8-naphthalimide derivatives have been used in the DNA intercalates and cellular imaging agents therefore they appeared the good antitumor activities, the 4-Bromo-1,8-naphthalimide exhibited a good inhibitor in the blood products [7].

The compound of 6-Dimethylamino-2,3-naphthalimide (6-DMN) is displayed intramolecular charge transfer fluorophore which appears positive solvatofluorochromism, the fluorescence of the compounds 5- and 6-amino-2,3-naphthalimides is affected by the solvent which the colour changed covering the visible region [8]. The compound of 1,8-Naphthalimide appears high photostability which the fluorescence wavelength in the visible region. Therefore, the naphthalimide- markers fluorescent has many excellent properties for example solubility is low in water, and low changes in the environment for example temperature and oxygen concentration [9].

The 1,8- naphthalimide derivatives show good emission, therefore they are used as colorants in the industry of polymer, emission probes for biological applications, and in the organic light-emitting diodes. The 1,8-naphthalimide possesses a good fluorescent properties, depend on the groups which are strong electron-donating, for example alkoxy and alkyl amine at the C-4 position [10]. The naphthalimide compounds are used as colourful dyes and absorbing dyes, and in the artificial light harvesting arrays, in addition to in the fluorescent chemical probes for the sensing in the biological applications [11].

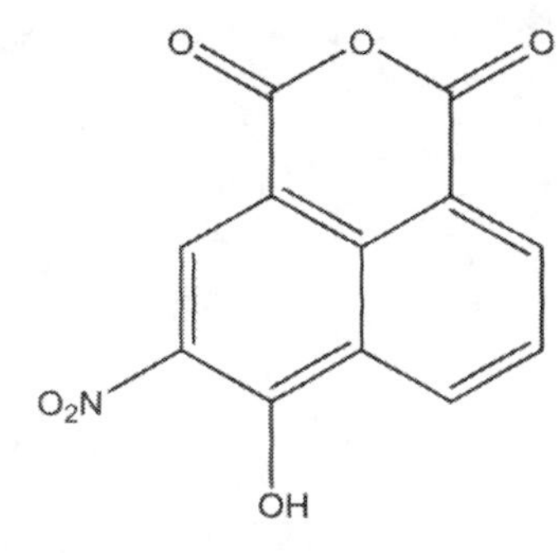

Figure 1. The structure of the compound

In the current work, we report the synthesis of 4-hydroxy-3-nitro,1,8-naphthalic anhydride, figure 1 explains the structure of the prepared compound, the dye was synthesized by a direct method. The UV- visible absorption and fluorescence properties and ionic conductance were elaborated. Application in conductance by using dopant material (4-hydroxy-m-benzene-disulfonic acid) was investigated.

Journal of Physics: Conference Series **2063** (2021) 012025 doi:10.1088/1742-6596/2063/1/012025

2. Material and methods:

2.1. Synthesis:

1. The synthesis of 4-hydroxy-3-nitro-1,8-naphthalic anhydride: 4- bromo- 3- nitro-1,8 -naphthalic anhydride (32.2 gm, 0.1 mole) was stirred in 200 ml of aqueous sodium hydroxide (20%) for eight hours at 80-85°C. The solution (dark red) was cooled , poured into ice-cold (500 ml) of 10% hydrochloric acid, stirred and filtered to give the sallow yellow deposit [12].

2. Synthesis of 4- hydroxy - m- benzene – disulfonic acid: In a dry 500 ml flask, phenol 31 gm (0.33 M) and concentrated sulphuric acid 116 gm were mixed. A boiling water bath was used to heat the mixture, in three hours by using the mechanical stirrer. The mixture was cooled to room temperature or below in ice water, and a solution of (95 gm of NaOH in 235 ml water) was added slowly, the solid product was separated [13].

3. Results and discussion:

3.1. The photophysical properties (UV- Visible absorption)

The UV- Visible absorption of the compound 4- hydroxy- 3- nitro - 1,8- naphthalic anhydride in different solvents, for example dimethyl sulfoxide DMSO, ethanol, and butanol was listed to study its spectral features in many solvents of various polarity figures 2, 3 and 4. The wavelengths identical to the spectral peaks are shown in table (1).

Table(1). Absorption and emission data for the 4-hydroxy-3-nitro-1,8-naphthalic anhydride compound in different solvents.

Solvents	Absorption maxima (nm)	Fluorescence maxima (nm)
ethanol	244	362
butanol	220	356
DMSO	284, 292, 308, 402	359, 302
DMF	-	356

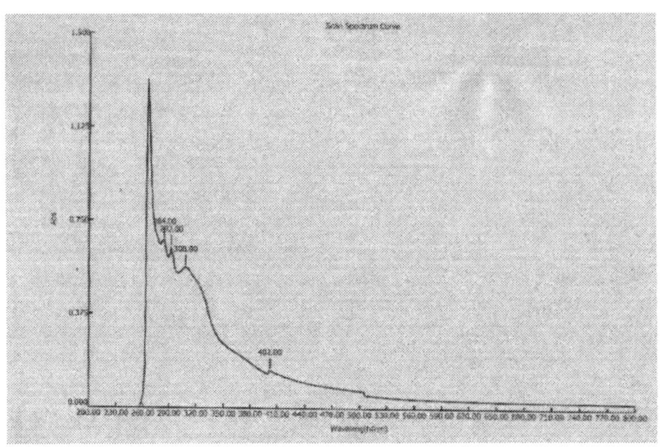

Figure 2. The UV- Visible in DMSO solvent for the 4-hydroxy-3-nitro-1,8-naphthalic anhydride compound

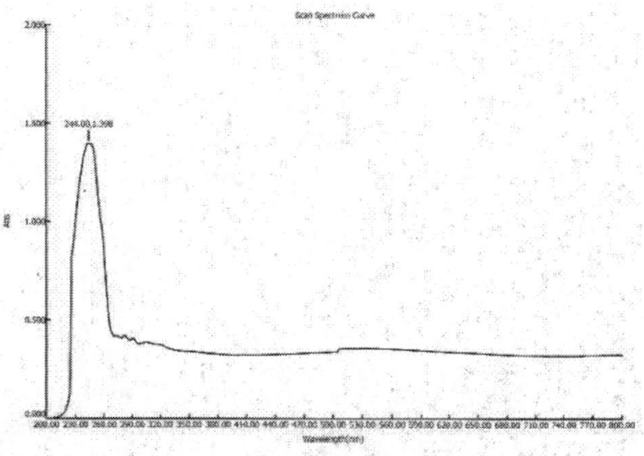

Figure 3. The UV- Visible in ethanol solvent for the 4-hydroxy-3-nitro-1,8-naphthalic anhydride compound

Figure 4. The UV- Visible in butanol solvent for the 4-hydroxy-3-nitro-1,8-naphthalic anhydride compound

As shown in the absorption spectra, the compound has a normal high-energy broad band at 244 nm in some solvents, such as ethanol. In butanol solvent the absorption maxima is 220 nm compared with the other solvents such as ethanol and DMSO (244, 284) nm respectively. Therefore, the bathochromic shifts in absorption due to the changes in the polarity of the solvent was shown, such as in ethanol the compound had a λ_{max} value at 244 nm, and a red shift in absorption for the solvent that was more polar, DMSO at 284 nm. The impact of the substituents groups at C-4 was studied, it can be seen that the ability of the methoxy ethoxy groups as a stronger electron-donating than methoxy [10]. In DMSO solvent, the compound shows different peaks, three bands in the UV region and one band in the visible region, one absorption peak at 402 nm, which might be attributed to the (ICT) band [11]. The peak of this absorption band shows at about 244 nm in ethanol solvent, which red shifted for about 24 nm in proportional to that of this compound in butanol solvent. In DMSO solvent, the compound shows the absorption peak at 402 nm, which can be attributed to the ICT, and another band in the UV region 284 nm, which might be attributed to the absorption of naphthalic ring transition. The replacement of ethanol with DMSO leads to a 40 nm red-shift on the band in the UV-region, depends on the solvent polarity, while the replacement of ethanol with butanol, leads to a 24 nm blue shift.

The red shift on the maximum in the absorption spectrum band of this compound in polar solvents compared with less polar-solvent is due to the stabilization of the polar ICT excited states which leads to reduce the energy gap between the states i.e. ground and excited, and leads to the red-shift on the maximum absorption band [11].

It could be easily followed from the data that there is a significant downward shift in the wavelength of the absorption with decrease in the polarity of the solvents, for example ($\lambda_{max, DMSO} = 284$ nm, $\lambda_{max, ethanol} = 244$ nm, $\lambda_{max, butanol} = 220$ nm). The absorption maximum of the compound in DMSO solvent is 64 nm bathochromic shifted than that of this compound in butanol. The position of the electron-donating group, such as amino group leads to increase the intramolecular hydrogen bond strength [14].

The fluorescence spectra:

The fluorescence spectra of the compound 4-hydroxy-3-nitro,1,8-naphthalic anhydride are displayed in figures 5,6,7 and 8. It can be showed from the figures that the emission spectra of this compound were sensitive to the polarity of the solvent. They moved to lower energies with increasing the solvent polarity (table 1). The influence of the solvent polarity on the fluorescence maximum was more pronounced compared with the absorption maximum, showing *ca.* 3 nm red shift on moving from butanol to DMSO, the compound of 4-amino-1,8-naphthalimides have ICT character in the excited states of these compounds [15]. The fluorescence maxima is equal to 356 nm in butanol compared to other solvents such as DMSO and DMF (359, 356 nm) respectively. The fluorescence maximum shifts to red in polar solvent, such as in DMSO ($\lambda_{max} = 359$ nm) while in less polar solvent, such as DMF ($\lambda_{max} = 356$ nm). The three peaks (301, 350, 600) nm in butanol solvent so it is expected that it might make CT complex with the electron donating in polar solvent.

This can be qualified to the polar solvents solvation to the polar ICT excited states, therefore the energy of the excited states is decreased, and then caused the significant red-shift of the fluorescence maximum peak [11]. Therefore, the polarity of the solvent is caused the red shift at the luminescence wavelength. In general, the emission intensity is shifted to a longer wavelength as the solvent polarity is increased [16]. The red-shifted might be result from the capabilities of the electron-withdrawing and electron-donating groups of the substituents of the benzene. The hydroxyl group is exerted a steric effect, and influenced in the fluorescence maximum and resulted in a bathochromic shift , such as in 1-butanol and DMF ($\lambda_{max} = 356$ nm).

The presence of electron-deficient substituent or less electron- rich is caused a ICT-induced blue shift in the fluorescence peak [17]. The solvent polarity trigged a significant bathochromic shift 3 nm of the emission wavelength of this compound with increasing solvent polarity from DMF to DMSO, and 6 nm from butanol to ethanol, in DMSO it is expected that it might make CT complex with the electron donating. The fluorescence maximum is correlated with solvent polarity, which is result in the stabilization in the less polar solvent, such as ethanol $\lambda_{max} = 362$ nm, the π- π^* is responsible for its fluorescence which is the lowest singlet excited state [18].

FT-IR spectroscopy:

The FT-IR was used for the characterization of the compound (4-hydroxy-3-nitro,1,8-naphthalic anhydride), figure 9, the FT-IR data showed stretching band at 1554 cm^{-1} for C=C, and the band at 3070 cm^{-1} for the stretching (C-H) (19), the bands at 1438, 1388 cm^{-1} are due to the NO_2 stretching (asymmetry and symmetry) (20). The bands at 1604, 1635 cm^{-1} are due to the C=O (carbonyl stretching) [19], the aromatic C-H bending at about 1033 cm^{-1} [20]. The C-N band at 1195 cm^{-1}, the OH band at 3421 cm^{-1} and the band C-OH cm^{-1} at 1033 cm^{-1} [21].

The ionic conductivity:

The ionic conductivity (G) was measured in water as a solvent, and the cell constant is equal to 9.5 cm^{-1}. The effect of the dopant material (4-hydroxy-m-benzene disulfonic acid) on the conductivities for this compound is shown in figure 10.

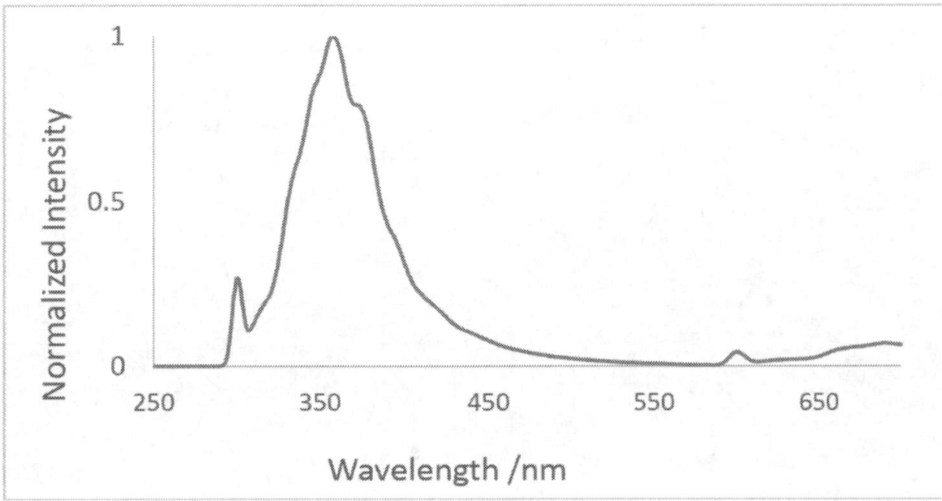

Figure 5. The fluorescence spectrum of the compound in DMSO for the 4-hydroxy-3-nitro-1,8-naphthalic anhydride compound

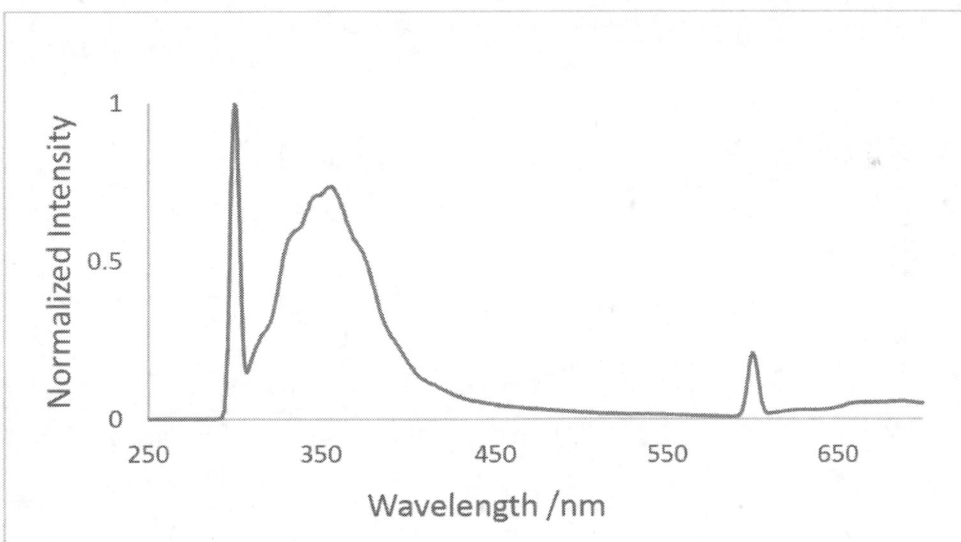

Figure 6. The fluorescence spectrum of the compound in DMF for the 4-hydroxy-3-nitro-1,8-naphthalic anhydride compound

It is apparent that the conductivity increases as the increasing of the dopant material weights, the (G) of the compound was $2.4X10^{-6}$ Siemens at 0.01 gm of the dopant material, and was $2.62X10^{-4}$ Siemens after the dopant material was added (0.1 gm), therefore the ionic conductivity depends on the concentration of this material, figure 11 shows the relationship between the specific conductivity and the weights of the dopant material, it can be showed the same behaviour which was seen in the ionic conductivity, (table 2).

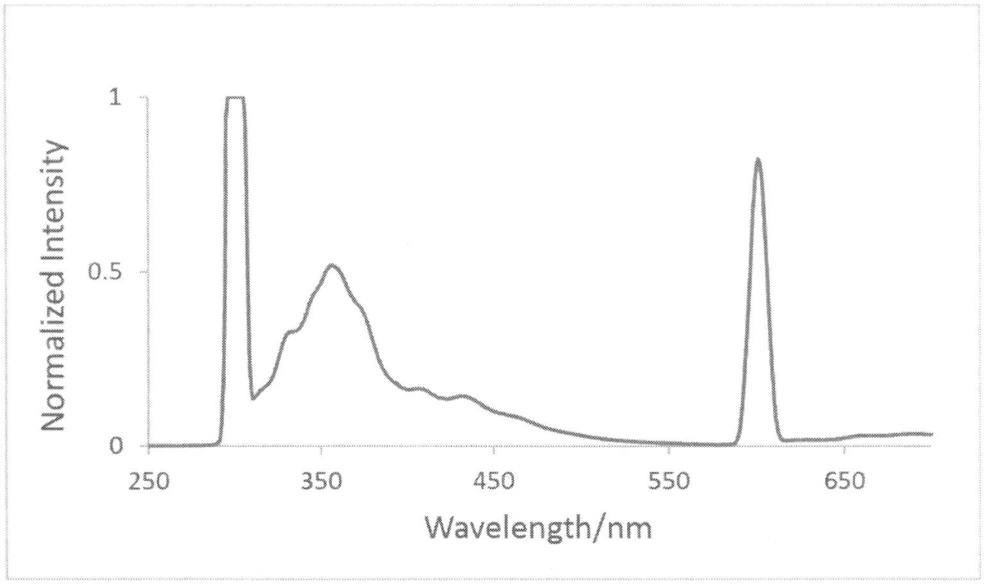

Figure 7. The fluorescence spectrum of the compound in butanol for the 4-hydroxy-3-nitro-1,8-naphthalic anhydride compound

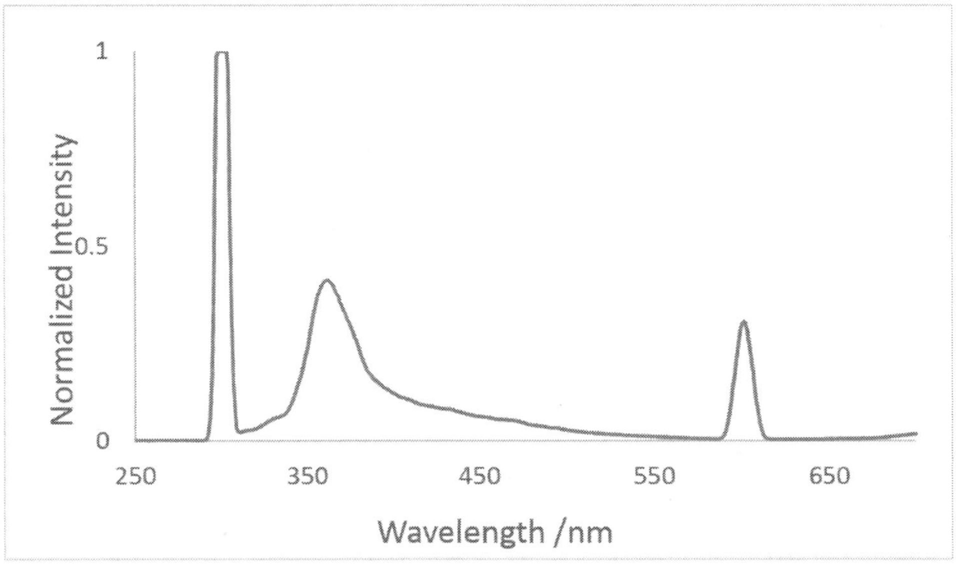

Figure 8. The fluorescence spectrum of the compound in ethanol for the 4-hydroxy-3-nitro-1,8-naphthalic anhydride compound

4. Conclusions

In this paper, the compound of (4-hydroxy-3-nitro-1,8 -naphthalic anhydride) and the dopant material (4-hydroxy-m-benzene-disulfonic acid) were prepared. The UV- Vis absorption and fluorescence spectra of the compound were recorded. The emission spectra of this compound were sensitive to the polarity of the solvent. They moved to lower energies with increasing the solvent polarity. The

Journal of Physics: Conference Series **2063** (2021) 012025 doi:10.1088/1742-6596/2063/1/012025

influence of the solvent polarity on the fluorescence maximum was more pronounced compared with the absorption maximum, showing *ca.* 3 nm red shift on moving from butanol to dimethyl sulfoxide DMSO.

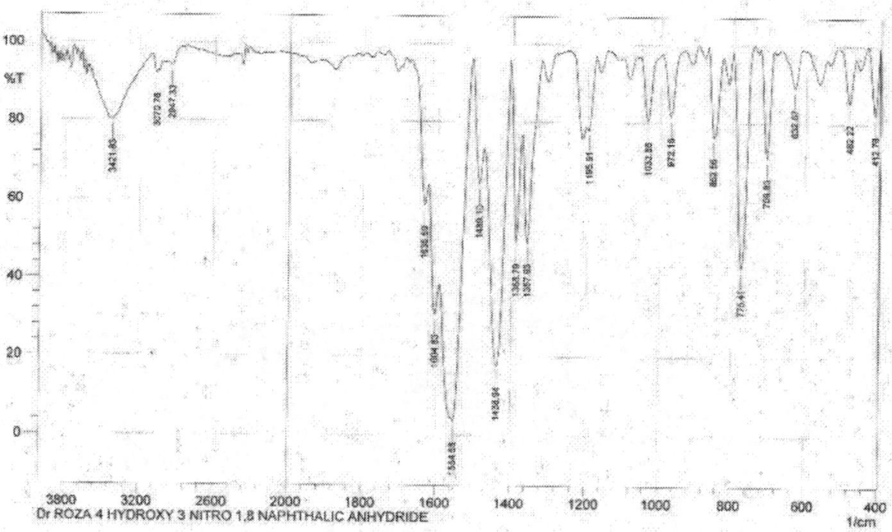

Figure 9. The FT-IR spectrum for the 4-hydroxy-3-nitro-1,8-naphthalic anhydride compound

Table (2). The values of conductivity and specific conductivity for the 4-hydroxy-3-nitro-1,8-naphthalic anhydride compound after doped with the dopant material

Weight /gm	conductivity /ohm^{-1}	Specific conductivity / $ohm^{-1}.cm^{-1}$
0.01	2.40E-05	1.39E-04
0.02	5.10E-05	2.96E-04
0.03	8.70E-05	5.05E-04
0.04	1.17E-04	6.79E-04
0.05	1.32E-04	7.66E-04
0.06	1.27E-04	7.37E-04
0.07	1.82E-04	1.06E-03
0.08	2.02E-04	1.17E-03
0.09	1.87E-04	1.08E-03

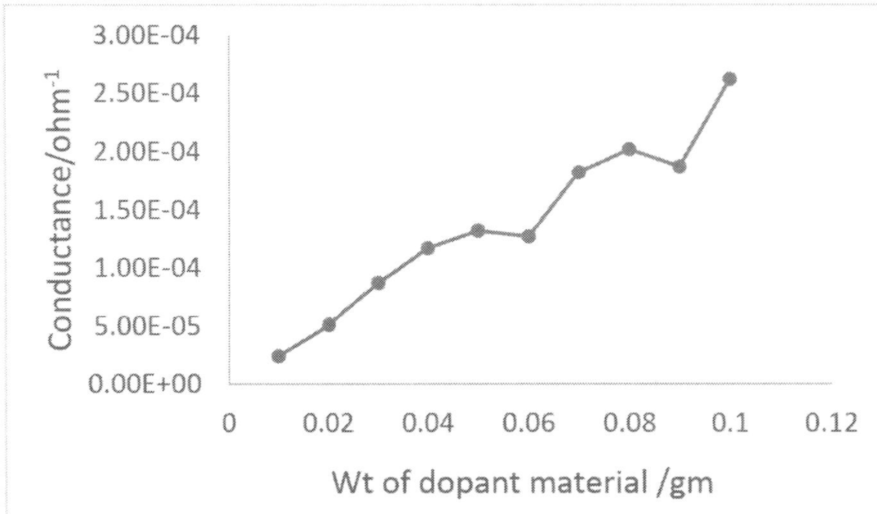

Figure 10. The ionic conductivity curve for the compound

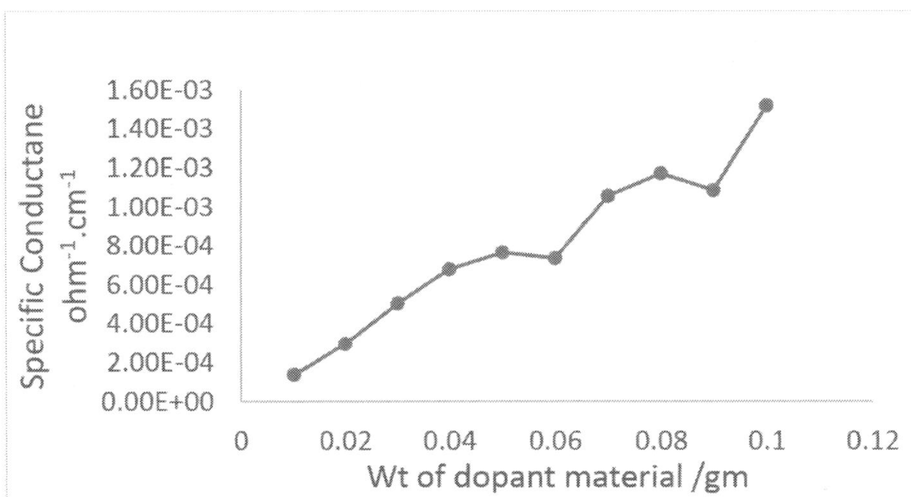

Figure 11. The specific conductivity curve for the compound

The effect of the dopant material on the conductivities (ionic and specific) of the compound was studied, the ionic conductivity was increased as the weight of the dopant material increases.

Acknowledgements:

This research was undertaken with the assistance of College of Science at the University of Basrah, Chemistry department.

References:

[1] Sánchez S, Woo AYY and Baumgartner T. Electron-accepting π-conjugated species with 1,8-naphthalic anhydride or diketophosphanyl units 2017 *Mater Chem. Front.* **1(11)** 2324.

[2] Fleming CL, Nalder TD, Doeven EH and Barrow CJ Synthesis of N-substituted 4-hydroxynaphthalimides using palladium-catalysed hydroxylation 2016 *Dyes Pigm.* **126** 118.

[3] Atahan A and Orhan E Photophysics, pH sensing, and hydrolysis study of a novel 1,8-naphthalimide derivative 2018 *JOTCSA*. **5(2)** 775.

[4] Hollauf M, Cajlakovič M, Tscherner M, Köstler S, Knall A-C and Trimmel G. Synthesis and characterization of naphthalimide-functionalized polynorbornenes 2017 *Monatsh Chem*. **148** 121.

[5] Fu Y, Pang X-X, Wang Z-Q, Qu H-T and Ye F. Synthesis and fluorescent property study of novel 1,8-naphthalimide-based chemosensors.2018 *Molecules*. **23(376)** 1.

[6] Zhang C, Ren A-M, Guo J-F, Wang D and Yu, L-Y. Theoretical design and investigation of 1,8-naphthalimide-based two-photon fluorescent probes for detecting cytochrome P450 1A with separated fluorescence signal.2018 *Phys. Chem. Chem. Phys*. **20(19)** 13290.

[7] Sharma H, Sidhu JS, Hassen WM, Singh N and Dubowski JJ. Synthesis of a 3,4-disubstituted 1,8-naphthalimide-based DNA intercalator for direct imaging of legionella pneumophila. 2019 *ACS Omega*. **4** 5829.

[8] Wang L, Fujii M and Okamoto H. Fluorescence behaviour of 2-, 3- and 4-amino-1,8-naphthalimides: effects of the substitution positions of the amino functionality on the photophysical properties.2018 *Photochem. Photobiol. Sci*. **17** 1319.

[9] Choi S-A, Park CS, Kwon OS, Giong H-K, Lee J-S and Ha TH, *et al.* Structural effects of naphthalimide-based fluorescent sensor for hydrogen sulfide and imaging in live zebrafish. 2016 *Sci. Rep*. **6(1)** 1.

[10] Xiao H, Chen M, Shi G, Wang L, Yin H and Mei C. A novel fluorescent molecule based on 1,8-naphthalimide: synthesis, spectral properties, and application in cell imaging. 2010 *Res Chem Intermed*. **36** 1021.

[11] Wang L, Shi Y, Zhao Y, Liu H, Li X and Bai M. ''Push-pull'' 1,8-naphthalic anhydride with multiple triphenylamine groups as electron donor. 2014 *J. Mole. Stru.* **1056-1057** 339.

[12] Tan S, Yin H, Chen Z, Qian X and Xu Y. Oxo-heterocyclic fused naphthalimides as antitumor agents: Synthesis and biological evaluation. 2013 *Eur. J. Med. Chem.* **62** 130.

[13] Kahol PK, Satheesh Kumar KK, Geetha S, Trivedi DC. Effect of dopants on electron localization length in polyaniline. 2003 *Synth. Met*. **139** 191.

[14] Bojinov V and Panova I. Synthesis of novel fluorophores-combination of hindered amine and UV absorber in the molecule of benzo[*de*] isoquinoline-1,3-dione. 2006 *J. Univ. Chem. Technol. Metallurgy*. **41(3)** 277.

[15] Banerjee S, Kitchen JA, Gunnlaugsson T and Kelly JM. Synthesis and photophysical evaluation of a pyridinium 4-amino-1,8-naphthalimide derivative that upon intercalation displays preference for AT-rich double-stranded DNA. 2012 *Org. Biomol. Chem.* **10**: 3033.

[16] Orhan E. Synthesis of novel diarylethenes bearing naphthalimide moiety and photochromic fluorescence behaviors. 2017 *JOTCSA*. **4(2)** 501.

[17] Li C and Liu S. Responsive nanogel-based dual fluorescent sensors for temperature and Hg^{+2} ions with enhanced detection sensitivity. 2010 *J. Mater. Chem*. **20** 10716.

[18] Ventura B, Bertocco A, Braga D, Catalano L, d'Agostino S and Grepioni F. Luminescence properties of 1,8-naphthalimide derivatives in solution, in their crystals, and in co-crystals: toward room-temperature phosphorescence from organic materials.2014 *J. Phys. Chem. C*. **118** 18646.

[19] Panah HS, Khosravi A, Gharanjig K, Khorassani M, Zadeh MK and Taromi FA. Synthesis and characterization of new fluorescent polymerizable dyes based on naphthalimide. 2010 *Iran. Polym. J.* **19(7)** 491.

[20] Shaki H, Gharanjig K, Rouhani S and Khosravi A. Synthesis and photophysical properties of some novel fluorescent dyes based on naphthalimide derivatives. 2010 *J. Photochem. Photobiol. A: Chem.* **216** 44.

[21] Saito G, Velluto D and Resmini M. Synthesis of 1,8- naphthalimide-based probes with fluorescent switch triggered by flufenamic acid. 2018 *R. Soc. Open sci.* **5(172137)** 1.

Evaluation of synergistic medicinal efficacy of four alkaloidic compounds isolated from Iraqi *Allium porrum* seeds against staphylococcus aureus bacteria

Abbas D M Al-Maliki [1] , Ala'a A H AI- Maliki[1] , Maha A A Hameed[2]
[1] Dep. of Chemistry, College of Education for Pure Sciences, University of Basrah, Basrah, Iraq
[2]Dep. of Applied Sciences -College of Marine Sciences - University of Basrah - Basrah ,Iraq.
corresponding author's e-mail address : abbasmatterchem@gmail.com

Abstract : Medicinal plants are necessary benefit natural sources for chemical production of different therapies from secondary metabolism .*Allium porrum* is considered as one of sources of foods and drugs at the same time . So the current work focused on isolation and characterization of four alkaloids from seeds of this plant . These compounds are 2-piperidinone , 2-pyrrolidinone-4-methyl , adenosine -2-deoxy–N- methyl –N- (triflouro acetyl) and maymarsine by using Gas Chromatography – Mass Spectroscopy (GC-MS) technique. The medicinal activity was achieved to investigate and evaluate the synergistic action of the mixture of the four alkaloidic compounds towards *Staphylococcus aureus* pathogen . The concentrations of alkaloidic mixture were prepared with values equal to 50,100,125,150,200,250,300,375,400 and 500 mg/ml and they revealed inhibition zones with diameters represented by 31,31,32,35,35,37,41,43,43and 45 mm . So the mixture of the four alkaloidic compounds can be used as natural therapy to treat various inflammatory caused by this pathogenic bacteria .

1. Introduction

The social and healthy need to discover natural drugs more safe than synthetic antibiotic led to development and creation in science of medicinal plants . Pre studies about the natural chemical compounds insured and proved the medical value for the chemical components abundant in the different parts of plants so the active materials were isolated , characterized and carried out successfully in treat various diseasess caused by different microorganisms [1][2][3] . The medical importance of all active metabolites is absence of the side effects causing to other infections and ,or inflammatory leading to a great disorder in human health. So the centers of

Journal of Physics: Conference Series **2063** (2021) 012026 doi:10.1088/1742-6596/2063/1/012026

scientific researches established many studies about the chemical structure of active natural metabolites existing in the various of medicinal plants in order to get new drugs have no side effects , cheap , economic and available so much in the environement [4][5][6] . Antibacterial activities were achieved for various medicinal plants containing a wide range of active chemical metabolites in the different parts of these plants . Also these studies insured the great potential of active secondary metabolites for inhibition the biological and biochemical functions of living systems belonging to the pathogenic microorganisms [7][8][9] . *Alluim porrum* (leek) is one of medical plant belongs to Brassicaceae family . This plant is traditionally used as a food for human in the world – wide because of its nutrition value for the human health .There are 500 genus are available in various countries but they differ in some features such as colour, maturing time and taste .Also these botanic genera are similar in chemical composition and neutral ceuitical components . Different phytochemical studies were achieved about various parts of *Allium porrum* and they indicated and ensured xistence of active metabolites which included phenolics , active antioxidants sulphuric phenols and flavonoids [10][11] .

Alkaloids are secondary active metabolites are found in most medicinal plants and these chemical compounds are anabolized in plants from the starting material is amino acid by different biochemical pathways by the support of required enzymes . Chemical alkalolidic metabolites are heterocyclic compounds containing nitrogen atom(s) and they have basic properties but some of them have weak acidic features whereas a few of them have neutral properties . In addition to presence of nitrogen atom(s) in the chemical structure contain sulpher , chloride , bromine and phosphorus atoms also are existing in various multi-alkaloids .Because alkaloidic metabolites are active compound so they have high medicinal efficacy to treat different diseases caused by multi- pathogens such as bacteria , virus and fungi [12][13][14] .Therefore the following research was achieved to estimate and evaluate the medicinal efficacy of synergistic interaction of mixture of four alkaloidic compounds isolated and identified from Iraqi *Allium porrum* seeds against pathogenic bacteria represented by *Staphylococcus eureus*.

2. Materials and Methods

2.1 Preparation of plant

Leek (*Allium porrum*) seeds were purchared from the local market in Al-Ashar district at Basrah Governorate in Iraq ,then the plant was taxonomied in biology department –college of education for pure sciences at University of Basrah –Iraq by specialist botanist . The seeds were washed with distilled water , dried under the shadow ground as powder with weight equal to 100 gm by electrical mill and placed in dark containers until the day of use .

2.2 Preparation of the cold aqueous extract of Allium porrum

The aqueous extract belonging to seeds of *Allium porrum* was prepared by mixing of 50 gm of seeds with 500 ml of distilled water as solvent in a conical flask . The contents were stirred on magnetic stirrer for 10 hours at room temperature , then the mixture was filtered by using Buchner funnel . The precipitate was removed and the filterate was dried carefully to get a solid crude [15] with weigh equal to 6.25 gm .

Journal of Physics: Conference Series **2063** (2021) 012026 doi:10.1088/1742-6596/2063/1/012026

2.3 Qualitative detections of alkaloidic compound

The cold aqueous extract of *Allium porrum* seeds was underwent Dragendroff reagent to detect the alkaloids existing in the plant . Fifty mg of extract was dissolved in 1 ml of distilled water in a test tube , then 2 ml of Dragendroff reagent was added into tube then orange precipitate was fromed [16] .

2.4 Isolation of alkaloids from Allium porrum

Fifty grams from seeds powder of *Allium porrum* was added into 500 ml of ethanolic acetic acid has concentration equal to (10% v/v) in conical flask and the contents were mixed and then stirred by using magnetic stirrer for 12 hours . The mixture was filtered by Buchner funnel and the precipitate was canceled then the filterate was concentrated to a quarter of its volume. After that 5 ml of concentrated sulphuric acid was added into the filterate and the contents were shaken well then the pH was adjusted to 9 by adding ammonia .The mixture was put in separation funnel and 20 ml of chloroform was added into mixture and contents were shaken well. Then extraction process was used and two layers were formed then organic layer volume was collected and aqueous layer was kept . This process was repeated by adding chloroform (20*3) to aqueous layer and organic layer was collected other time then the total volume was dried and the solid crude was gotten [15] with yield equal to 2.35 gm.

2.5 Gas chromatography –Mass Spectroscopy Technique

The alkaloides isolated from *Alllium porrum* seeds were separated then characterized by using gas chromatography – mass spectrum procedure in GC- MS laboratory in college of agriculture in Basrah university by instrument type shimadzu-GC-MS-QP- 2101 Ultra.

2.6 Stain of pathogenic bacteria

The pathogenic bacteria isolate represented by *Staphylococcus aureus* was gotten and identified by a microbiologist in microbiology laboratory –marine science center in University of Basrah Iraq.

2.7 Preparation of culture medium

The pathogenic bacteria used in this study was treated with Muller Hinton Agar (MHA) as a culture medium for growth of this pathogen.

2.8 Estimation of Medicinal efficacy of alkaloidic compounds

To evaluate the medicinal and biochemical effect of alkaloids isolated from *Allium porrum* seeds ,the method of Agar well diffusion was carried out . Two ml of Muller Hinton Agar was added into each petri dish then the medium was treated with 0.1 ml of pathogenic suspension (*Staphylococcus aureus*) has optical density equal to 0.1 with wave length is 450 nm by using spectrophotometer by using a sterilized glass spreader. Then the dishes were let for 25 minutes after that many wells were done and alkaloidic mixture was added with concentrations equal to 50,100,125,150,200,250,300,375,400 and 500 mg/ml and dishes were incubated at 37□C for 24 hours in the incubator. Finally the diameters values of inhibition zones were measured [16]

3. Results and Discussion

The health importance of *Allium porrum* plant indicates and proves the medicinal benefit resulting from existing of various active metabolites as natural drugs were used as antibacterial, antifungal, anticancer, antitumor and antiphrastic agents[17][18]. Aqueous and alkaloidic extracts were prepared with weights equal to 6.25 and 2.35

Journal of Physics: Conference Series **2063** (2021) 012026 doi:10.1088/1742-6596/2063/1/012026

gm respectively therefore the extraction percentages were 25 and 9.4% as showin in table 1. This insures that the percentages were excellent and this chemical case proves abundance of different active chemical compounds in the both extracts.

Table 1. Extraction percentages of cold aqueous and alkaloidic extracts of *Allium porrum* seeds

No.	Extract Sort	Plant powder weight (gm)	Extract weight (gm)	Extraction percentages (%)
1	Cold aqueous	25	6.25	25
2	Alkaloidic	25	2.35	9.4

It is noticed that the extraction percentage of cold aqueous extract is higher than the the alkloidic extract so this biochemical state illustrates presence of more active metabolitic compounds in the aqueous extract than alkaloids .

Table 2. shows the primary qualitative analysis of cold aqueous extract of *Allium porrum* seeds by using various chemical reagents depending on the chemical nature of secondary metabolites .

Table 2 . Primary detection on active metabolites in cold aqueous extract of *Allium porrum* seeds

Reagent sort	Test result	Test Indication	Conclusions
Molisch	+	Violet ring	Existence of carbohydrates
Benedict	+	Red precipitate	Existence of glycosides
Dragendroff	+	Orange precipitate	Existence of alkaloids
$FeCl_3$(1%)	–	No bluish-green colour	Absence of phenols
Lead acetate(1%)	–	No white- light brown precipitate	Absence of tannins
Ethanolic KOH (5N)	_	No yellow precipitate	Absence of flavonoids
Liberman-Burchard	+	Deep blue –greenish colour	Existence of steroids
Ninhydrin(1%)	_	No violet colour	Absence of amino acids

From table 1. ,it was noticed that the cold aqueous extract belonging to *Allium porrum* contains carbohydrates , alkaloids, glycosides and steroids but phenolics , tannins and amino acids are absent. Many researchers carried out different experiment and analysis about phytochemical content of various parts of *Allium porrum* including seeds and they found different chemical metabolitic classes such as phenolic compounds , alkaloids, flavonoids and glycosides [19[20] . Alkaloids as secondary active materials were isolated by using different polar solvents and these potent compounds were gotten from the organic layer and this process insures existence of non- polar alkaloidc compounds. Many researches were established for isolation ,separation and characterization of plant alkaloids in a high purity [21][22] .

Gas chromatography –mass spectroscopy technique was applied for chemical identification of alkaloidic compounds abundant in the alkaloidic extract were chemically separated by gas chromatography (GC) and four peaks were appeared in alkaloidic extract. At the beginning the active alkaloids with different retention times then these active metabolites were inserted to mass spectrophotometer instrument and each alkaloidic compound was indicated with mass spectrum has various peaks depending on structural groups and relation between mass and charge. The mass spectra insured that the active alkaloidic compounds are 2-piperidinone, 2-pyrrolidnone , 4-methyl ,2-deoxy –N-methyl (-n-(tri-fluoroacetyl) and maymyrsine as shown in figures 1 , 2 , 3 and 4 .

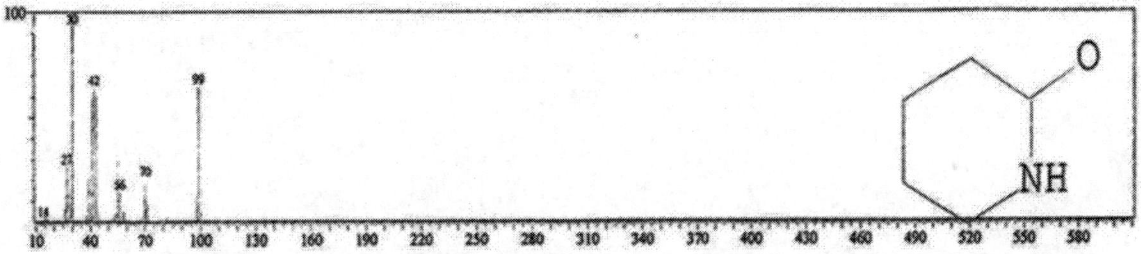

Figure 1. Mass spectrum of 2-piperidinone

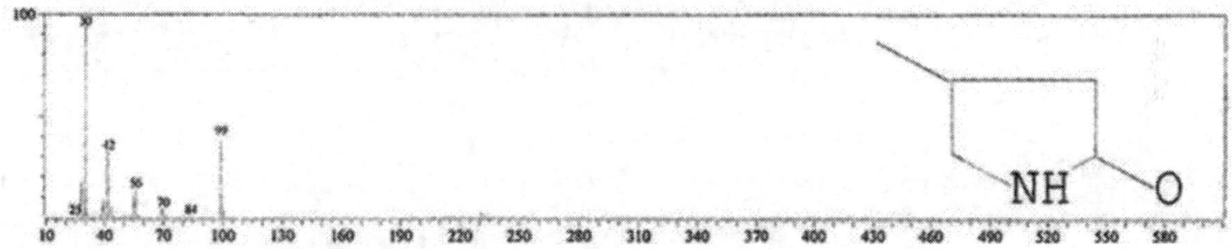

Figure 2. Mass spectrum of 2-pyrrolidinone-4-methyl

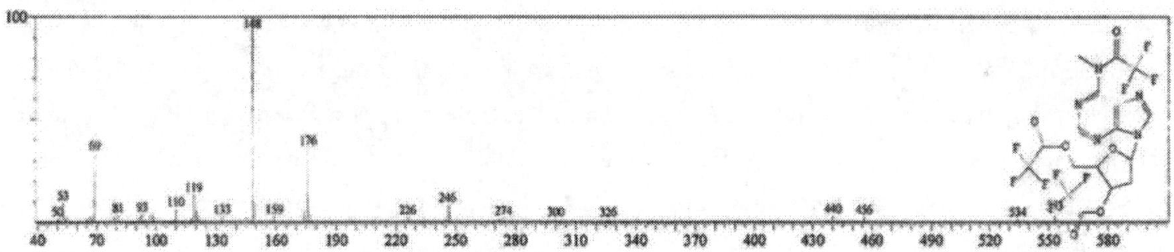

Figure 3. Mass spectrum of adenosine 2-deoxy –N- methyl –N- (tri fluoro-acetyl)

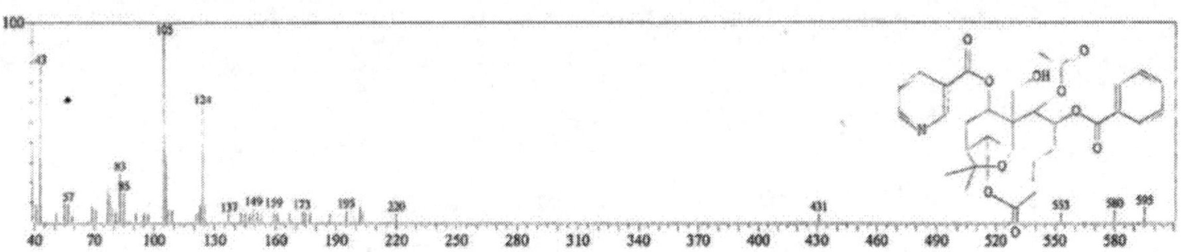

Figure 4. Mass spectrum of maymyrsine

Because the medicinal efficacy and biochemical potential of alkaloids therefore the synergistic interaction of the four alkaloidic compounds from *Alllium porrum* seeds was achieved and studied to investigate, estimate and evaluate the medical effect of the mixture of these alkaloids against growth of *Staphylococcus aureus* pathogen. Table (3) represents the biochemical role and medicinal activity belonging to multi- concentrations were prepared from mixture of four alkaloidic compounds of seeds of *Allium porrum* .

Table 3. Medicinal efficacy of synergistic interaction of alkaloidic mixture isolated from *Allium porrum* seeds.

Pathogenic bacteria Kind	Alkaloidic compounds mixture	Concentration (mg/ml)	Inhibition zone diameter(mm)
Staphylococcus Aureus	2-piperidinone + 2-pyrrolidinone-4-methyl + adenosine2- deoxy-N-methyl-N-(tri-flouroacetyl) + Maymyrsine	50	31
		100	31
		125	32
		150	35
		200	35
		250	37
		300	41
		375	43
		400	43
		500	45

It is noticed from table 3. , the increase of concentration of alkaloidic mixture showed an increasing in diameters values of inhibition zone where the concentrations (50,100,125,150,200,250,300,375,400,and 500 mg/ml) recoded inhibition diameters equal to 31,31.32,35,35,37,41,43,43,and 45 mm. The biochemical explanation of this status belongs to high antibacterial activity of alkaloids mixture with increase of concentration leading to inhibit or kill most *Staphylococcus aureus* bacteria [23] . plant alkaloidic metabolites have chemical ability for bonding with nucleic acids are DNA and RNA existing in the living cell of pathogenic bacteria leading to destruct or inhibit the chemical and biological roles of these acids. Also multi studies insured and proved chemical capability of natural alkaloids to inhibit the metabolism of amino acids , lipids, carbohydrates and proteins through the hydrogen bonding between nitrogen atom present in alkaloids with hydrogens of these biochemical compounds [24][25][26] .

Also the synergistic interaction of natural alkaloidic compound is chemically very powerful against the action of multi-enzymes abundant in the living cell of pathogens including *Staphylococcus aureus* by hydrogen bonding with thiol (-SH) group present in these enzymes leading to break the sulphuric bridges then denaturation of chemical nature belonging to these enzymes[27][28] .

4. Conclusions

The synergistic interaction of the natural alkaloidic mixture containing four active chemical compounds represented by 2-piperidinone, 2-pyrrolidinone-4-methyl, adenosine-2- deoxy-N-methyl –N-(triflouro acetyl) and maymyrsine isolated from Iraqi *Allium porrum* seeds showed the high chemical ability and medicial efficacy for inhibition of biological and physiological activities of *Staphylococcus aureus* pathogen causing multi-diseases . Also it was proved that the increase of concentration of alkaloids gave more inhibition diameters values leading to kill most of the colonies of this bacteria. So the natural mixture of alkaloidic compounds of leek can be carried out as plant therapy against this pathogenic micro-organism and comparison of the medicinal efficacy of alkaloids characterized from this plant with antibiotics which target *Staphylococcus aureus* bacteria.

5. References

[1] Ghonson,S.S. 2015. Chemical Detction of some active compounds in Egg plant (solanum melongena) callus as compared with fruit and root content *.Int.J.Curr.Microbiol. App.sci* ,**4**(5):169-165.

[2] Mohaddese,M. and Atefeh,M. 2014.Antimicrobial activity of Cappais spinosa as its usages in traditional medicine, *Herba Polonica* , **60**(1):1-9.

[3] Nazia,M.A.C. and Perween,T. 2008.Invitro antibacterial activities of kalonji, cumin and poppy seeds pak*.J.Bot* ,**40**(1):461-467.

[4] Padmini,E.,Valarmathi ,A. and Rani,M.U. 2010.Comparitive analysis of chemical composition and antibacterial activities of Mentha spicata and Camellia sinensis. *Asian J.Exp. Biol .Sci* , 4:772-781.

[5] Al-Maliki, A.D.M ; Abd,Al-Majeed, M.I;Al-Abdal, M.A.J. and Esmaeel, B.A. 2017. Characterization, estimation and evaluation of antifungal activity of lipids isolated from Iraq capparis spinosa leaves buds, *J.Med. sci. and clin.Res* ,**5**(11):29797-29808.

[6] Jeeva,S,;Johnson,M,; Aparna, J.S. and Irudayaraj,V. 2011. Preliminary phytochemical and anti-bacterial studies on flowers of selected medicinal plants , *Int.J.Med.Arom. plants* ,**1**(2):107-114.

[7] Nassem,u., Khurram, M.U; Afridi,H.H;Khan,F.A.;Khayam,M.U.;saleem,U.;Najeeb, U.; Hussain ,J. and Khan, M.A. 2011. Composition of phytochemical contituents and antimicrobial activities of Mentha spicata from four northern districts of Khyber. Pakh.tunkhwa. *J.appl. pharm. Sci* , **1**(7):72-76.

[8] Mahmoud, I.N.;El-Sayed,A.A. and Amany, A.S. 2010. Secondany metabolites and bioactivities of Myrtus communins. *Phatamaco. Res.* ,**2**(6):325-329.

[9] Al-Maliki,A.D.M. and Bader , S.Q. 2012.Isolation and identification of tannins from Cuminum cyminum L. seeds and study of their medicinal activity against selective types of pathogenic fungi *.J.Missan Res.*,**8**(16):147-161.

[10] Irkin, R.; Koru kiuoglu, M. 2007 . Control of Aspergillus niger with Garlic , Onion and Leek Extracts. *African J. Biotech.* ,**8**(4):384-387.

[11] Hero, F.S.A. and Jwan, D.T. 2012. Antibacterial activity of Lepidium sativum and Allum porrum extracts and Juices against some Gram positive and Gram negative bacteria . *Med .J. Islamic world Acad . sci.* ,**20**(1):10-16.

[12] Kabera, J.N.; Semana, E., Mussa , A.R. and He,X. 2014. Plant secondary metabolites, biosynthesis, classification , function and pharmacological properties. *J.pharm and pharmacol*.,2:377-392.

[13] Mc Brien, N.A.; Stell, W.K. ands Carr, B. 2013 . How does atropine exert its anti-myopia effects , *opthalmic and physio optics*.,**33**(3):373-378.

[14] Simera, M.; Pollacek, J. and Tak us,G. 2010 . Central antitissusive effect of codeine in the anesthentized rabbits. Europ. *J. Med. Res.* ,115:184-188.

[15] Harborne ,J.B. 1984 . *Phytochemical methods*, 2 nd.ed., Chapman and Hall London,Ulk.

[16] Cowan,M.M. 1999. Plant products as antimicrobial agents *.J. Clin. Microb. Rev.* ,124:564-582.

[17] Mehrabian,S. and Larry-Yazdy, H. 1992 . Antimicrbial activity of Allium sativum, Allium cepa and Allium porrum (Liliaceae) against enteric pathogens (Entero bacteriacea), *Int. Soc. For Horticult. Sci*.,**319**(4):177-182.

[18] Cvetnic, Z; Pepeljnjak, S.;Stjepan, V. 2003 . Knezevic of leek extract ,Allium porrum L.(Liliaceae) *Microb.soc.* ,7:4-7.

[19] Packia, L.N.C., Viveka, S.J.; Jeeva,S. and Raja,B.J. 2015 . phytochemicals in Allium species and its analytical Methods- A review. *Int. J. Inst. Pharm. And life sci.*,**5**(1):1-22.

[20] Ernesto, F.; Virginia, L.;Orazio, T.S. and Carla,C. 2001. The flavonoids of leek, Allium porrum *.J. phytochemistry*,**57**(4):565-569.

[21] Dima, M.; Anne-Sylvie, F.; Emmanuel, P.; Tayssir, H.; Naneg, N.; Christine, F.; Xavier, F. and Farid, C. 2014. Chemical composition, antibacterial and antioxidant activities of six

essentials oils from Alliaceae family. *Molecules*,19:20034-20053.

[22] Al-Saadi, J.H.; Al-Maliki,A.D.M.and Aouda, M.A. 2016 . Medicinal activity of alkoloidic compound 1-H-pyrido(2,3-b)indole isolated from Solanum melongena cortex againt some pathogenic bacteria. *J. Nat. Sci Res.* ,**6**(20)80-85.

[23] Wisseni, A.W. and Brahimm M. 2016 . Charactenzation of myrlte seeds (Myrtus communin var. bacteria) as a source of lipids,phenolic and antioxodant activities . *J. food and drug analysis*,**24**(2)316-322.

[24] Mohammad, S.R.; Mohammad, Z.R.;Md, A.; Rasheduzzaman , C. and Mohammad,A.R. 2019 . Antimicrobial activity of some indigenous plant of Bangladesh. *Res.J.Biotech* ,**14** (6):88-95

[25] Mona, Gh.;Sina,O. and Mohammad, B.O. 2016 . Review of anti-intlammatory hebal medicines. *Adv. In pharmacol. Sci* ,11:1-110.

[26] Packia, L.; Vevika,S.; Madepalli, B.V. and paja, B.J. 2015 .Phytochemical screening and in vitro antibacterial activity of Allium sativum extract agairst bacterial pathogen, *J.Sci.* ,**5**(5):281-285.

[27] Hamomoouchi, M.; Elarak, K.; Tantoui, N.E. and Agoumi,A. 2004.Eluicidation of antibacterial and antifurgal properties of essential oils of Eucalyptus. *Plants Med. Ethphyto.* ,24:278-289.

[28] Leen,O,Ahmad,S. and Roula,M.A. 2019. Antimicrobial activity of polyphenols and alkaloids in Midlle Eastern plants. *Front. Microbiol* ,**10** (911):1-28.

Journal of Physics: Conference Series **2063** (2021) 012027 doi:10.1088/1742-6596/2063/1/012027

Production of biodiesel from heat-treated edible oil

Hamza M Ahmed[1*], Siddig O Abdeldaim[1] and Ibrahim Y Erwa[1]

[1]Department of Applied and Industrial Chemistry, Faculty of Pure and Applied Sciences, International University of Africa, P.O. Box 2469, Khartoum, Sudan.

[*]Corresponding author's e-mail barbaraawi2010@gmail.com

Abstract. Among the urgent needs of developing countries are renewable sources of energy and a pollution-free climate. Acid-catalyzed transesterification for production of biodiesel from heat-treated peanut oil was investigated, and the produced biodiesel was characterized. The change in oil properties during the heating process was studied. The results showed an increase in the saponification number with heating time (from 115.47 mg KOH/g for crude oil, to 156.19, 172.7 and 184.68 mg KOH/g after 30, 60 and 120 minutes of heating respectively), a slight decrease in the acid value was observed after heating (from 5.91 to 5.87, 5.87 and 5.88 mg KOH/g after 30, 60, and 120 minutes of heating respectively) There is no obvious change in the refractive index during heating. The peroxide value increased from 2.87 to 3.95 meq O_2/kg as the heating time was increased up to 60 minutes. The characteristics of the produced biodiesel were compared to ASTM D 6751 biodiesel standards and ASTM D 975 fossil diesel standards to validate their acceptability as a fuel in diesel engines, and a good agreement was observed. In contrast, to international standards, we notice that ethanol-based fuel has properties that are similar to those of global fuels. Peanut oil can be used to make renewable, low-cost biofuels that can meet a significant portion of the world's energy needs.

1. Introduction

Diesel is one of the most important contributors to global environmental pollution; transitioning to more environmentally efficient and renewable fuels is now a must in the battle against increasing greenhouse gas (GHG) levels and climate change. Biofuels attracted a lot of attention over the last two decades because of their sustainable feedstock and short life cycle [1]. Biodiesel and its blends with diesel are being studied as a potential alternative to the issues of fossil fuel depletion and environmental destruction [2, 3]. In the face of environmental emissions, declining crude oil supplies, and accelerating price rise, biodiesel is the best choice among biofuels. Biodiesel processing is thus a green, biodegradable, non-toxic, and sustainable fuel alternative [4, 5]. Because the world's fossil oil reserves are finite, renewable energy sources like biodiesel are becoming much more important [6].

Biodiesel (a mixture of mono-alkyl esters of long chain fatty acids) is an alternative diesel-equivalent fuel which is derived from vegetable oils or animal fats, designated B100, synthesized via transesterification reaction of triglycerides with alcohol. At mild operating conditions, a catalyst is usually used in the transesterification process because it speeds up the chemical reaction by lowering the amount of energy required to complete the process [7]. Vegetable oil has become a major raw material for the production of biodiesel. However, its high density, low volatility, high viscosity and

Journal of Physics: Conference Series **2063** (2021) 012027 doi:10.1088/1742-6596/2063/1/012027

particulate accumulation in engines limit its direct use as a fuel [8]. Currently, biodiesel is mainly prepared from conventionally grown edible oils such as rapeseed, soybean, sunflower and palm, thus leading to competition with foodstuff and raising the cost of biodiesel [9]. Gebremedhin et al. [10] were analyzed Ethiopian peanut and their results revealed that the peroxide value, iodine value, saponification value, refractive index, density, dynamic viscosity and flashpoint were; 1.56 meq O_2/kg, 89.23 gI/100g, 182.37 mg KOH/g, 1.45, 0.91 g/cm^3, 55.72 mPa.s and 230°C respectively. In another study done by Angaye et al. [11] various peanut oil samples were analyzed for their relative densities, viscosities, refractive index, iodine values and peroxide values, relative density increased from 0.9115 ± 0.0005 g/cm^3 (for oil from fresh groundnuts) to 0.9241 ± 0.0002 g/cm^3 (for oil from the market roasted sample). The viscosities of the various samples also showed an increase from 33.02 m^2/s (for oil from fresh groundnuts) to 61.6 m^2/s (for oil from the market roasted seeds). The authors in [12] investigated chemical characteristics of peanut oil the results revealed that acid value and iodine value were 1.489 mg KOH/g and 0.5030 gI/100g respectively. This study aimed to examine the changes in the physical and chemical properties of crude peanut oil following heat treatments, as well as for the synthesis of biodiesel from heat-treated peanut oil at a laboratory scale using a catalytic transesterification technique. Qualitative and quantitative characterization of the resulting biodiesel was performed, and the properties of the produced biodiesel were compared with those of standard diesel and biodiesel.

2. Materials and methods

2.1. Samples collection and preparation
Peanut *(Arachis hypogea)* oil was purchased from the farm of the International University of Africa – Khartoum, Sudan. All chemicals used in this study were of analytical grade. The samples were heated in an electric hot plate at temperature of 200°C for different periods of time separately (30, 60 and 120 min), the fourth sample was kept without heating. The samples were then kept in a clean, dry dark container at room temperature.

2.2 Characterization of the oil
The density, refractive index and kinematic viscosity the samples were assessed using the methods described by (ASTM International) [13–15]. The peroxide value (PV) and saponification value of oil samples were determined by titration according to the official procedures described by American Oil Chemist's Society AOCS [16, 17]. The AOCS Official Procedure Cd 3d-63 [18] was used to evaluate the acid value of the samples.

2.3 Synthesis of biodiesel
The transesterification reaction was determined according to the method descried by Priya et al. [29] the heat-treated sample was heated to the desired temperature (45, 50, 55 and 60°C) before adding the sulfuric acid and methanol. The batch reactions were allowed to proceed at the following conditions: the amounts of catalyst added were (0.5 – 2.5% by weight), the ethanol to oil ratio (3:1 – 6:1 volume/volume) and reaction time was (30 – 120 minutes). After the specified time, the mixture was cooled to room temperature and transferred into a separating funnel. The glycerol-containing bottom layer was drained and the biodiesel rich phase separated after washing with water.

2.4 Biodiesel properties
The physicochemical fuel properties of the biodiesel were measured using appropriate ASTM standards. The following fuel properties were determined, namely density at 15°C, viscosity at 40°C, color, flash point, pour point, cloud point, water content, sulphur content, carbon residue and copper strip corrosion. The characteristics of the produced biodiesel were compared with ASTM D6751 [19] biodiesel standard and ASTM D975 [20] fossil diesel standards to confirm their acceptability as a fuel in diesel engines as shown in table 3.3.

3. Results and discussion

The peanut oil was heated at a temperature of 200 °C and with the change of heating time, the first sample was heated for 30 minutes, the second sample was heated for an hour and the third sample for two hours while the last sample was raw oil as shown in table 1. The determination of physical and chemical values is often used as a general indicator of the condition of the oil. To assess the quality of the crude and heat treated peanut oil and biodiesel produced from the heat-treated peanut oil, the content and formulations of the oil and biodiesel were subjected to a physicochemical analysis, the results obtained are shown in table 2 and table 3.

3.1 Physicochemical properties of a raw and heat-treated peanut oil

The acid value is the amount of potassium hydroxide in milligrams required to neutralize the free fatty acids in one gram of oil or fat is known as the acid value. Since free fatty acids are produced during the decomposition of oil glycerides, it is a relative predictor of rancidity. The ratio of free fatty acids measured as oleic acid is also included [22]. The acidic nature of peanut oil was decreased slightly as heating time and temperature consistency increased, peanut oil showed that the acid number of raw and heat-treated oil as shown in table 1. Mandloi et al. [12] reported that the acidic number of crude peanut oil was 1.489 mg KOH/g while our results showed a great difference from the previous study. Variation in physicochemical properties may be due to environmental and soil related factors. In another previous study conducted by Ayoola and Adeye [23] for three samples of peanut oil (raw, sun-dried and roasted) the acid value was 2.35, 1.79 and 2.52 mgKOH/g, the peroxide values were 0.740, 0.603 and 0.470 meqO_2/kg, while the refractive index was 0.247, 0.256 and 0.147 respectively. The saponification values for raw, sun-dried and roasted peanut oil were 201, 195 and 170 mg KOH/g, respectively. The results obtained in this study showed that the acid value of crude oil was 5.91 mg KOH/g, and this value decreased slightly in the case of heated oil for 30 minutes and for one hour, where the same value was recorded, which was 5.87 mg KOH/g. The previous study [23] also revealed that the saponification values of raw, sun-dried and roasted peanut oil were 201, 195 and 170 mg KOH/g respectively, while our results showed a big difference between the saponification values of the raw and pre-heat treated samples, as shown in Table 1. In a previous analysis by Angaye et al. [11] comparing the physicochemical properties of peanut oil extracted from fresh seeds (without heating) and oil extracted from heat treated seeds, the refractive index and relative density were found to be completely in agreement with the values obtained in this study, whereas the peroxide value and viscosity showed a clear difference. Anyasor et al. [24] analyzed the acidic value of oil seeds in three different geographic regions, the highest acid value was 4.49 mg KOH/g and peroxide 1.15 meq O_2/kg. Musa et al. [25] performed another analysis on the physicochemical properties of commercial peanut oils marketed in various Nigerian states. The highest acidity value was 6.83 mg KOH /g, which is comparable to the result obtained for raw oil, and the lowest saponification value in that analysis corresponded to our highest saponification value, which is for the sample heated for two hours. The relative density and refractive index of all samples were found to be in great accordance with a number of previous studies, while other physicochemical parameters showed slight differences [10, 26].

Table 1. Physicochemical Characteristics of the raw and heat-treated peanut oil.

Property	Raw oil	Heated oil for 30 min	Heated oil for 60 min	Heated oil for 120 min
Refractive index	1.467	1.467	1.468	1.468
Acid value (mg KOH/g)	5.91	5.87	5.87	5.88
Peroxide value (meq O_2/kg)	2.87	3.27	3.95	2.57
Saponification V. mg KOH/g	115.47	156.19	172.73	184.68
Color	< 1.5	< 2	< 3	< 3.5
Density (g/cm^3)	0.905	0.903	0.9026	0.9019

Kinematic viscosity mm^2/s	41.343	40.847	41.09	39.70
Water content %	0.0014	0.0017	0.0015	0.0012
Flash point °C	202	204	206	208

3.2 Characteristics of the biodiesel

Biodiesel's fuel properties are generally influenced by the type of feedstock used and their fatty acid content. The biodiesel produced from heat-treated peanut oil using the following conditions: 1:3, 1:4, 1:5 and 1:6 molar ratio of oil to ethanol, 0.5, 1, 1.5 and 2 wt% catalyst loading, 45 – 60°C reaction temperature and reaction time 30, 60, 90 and 120 min. were examined for physical and chemical properties. The results are presented in Table 2. Its characteristics have been compared with ASTM D 6751 [19] biodiesel standard and ASTM D975 [20] fossil diesel standard to confirm its acceptability as a fuel in diesel engines as shown in Table 3. The density of raw and heat-treated peanut methyl ester was found in the range of 806 – 810 kg/m^3. According to ASTM D975 [20], the fuel density should not exceed 950 kg/m^3. Density is an important parameter for assessing fuel quality as it gives an indication of the delay between fuel injection and combustion inside the engine (ignition quality), as well as the amount of energy per unit mass (specific energy). The high fuel density would increase fuel consumption and, as a result, nitrogen oxide emissions [27]. Kinematic viscosity is associated with the flow resistance of liquids caused by internal friction force between two layers of fluid flowing over each other. Therefore, it is a crucial property in deciding the quality of a biodiesel as a fuel. Owing to low fuel atomization, tube blockage, and potential deposits formation in the engine, high viscosity can reduce the consistency of the liquid fuel, resulting in inadequate combustion and extreme fouling [28]. The kinematic viscosity of biodiesel derived from peanut oil was determined to be 1.260, 1.240, 1.245 and 1.255 mm^2/s respectively. When compared to ASTM D6751 [19], it was observed that the viscosity was below the acceptable level, although it was lower than the acceptable limit for fossil diesel ASTM D975 [20] as shown in Table 3, it can be concluded that the ethyl esters of peanut oil have a lower propensity to form deposits in engines and that their gasoline is more desirable in terms of viscous properties. The flash point is the lowest temperature at which fuel vapors burn when an ignition source is used. As a consequence, it's a metric for a sample's tendency for forming a flammable mixture with air. The flash points obtained from peanuts biodiesel were 62, 65, 68 and 70.1°C respectively and all fell within the standard limit range set by ASTM D6751. 1:6 ratio of oil and ethanol has a higher flash point compared to other ratios, but all of them are significantly high and this result leads to their safer handling and storage. The cloud point and flow point of biodiesel have consequences for its use in cold weather conditions. The cloud point is the most common metric for determining a fuel's tendency to crystallize and the cloud point for peanut oil showed the same results at – 28°C. This indicates that peanut oil has a low tendency to form cloudy crystals easily at cold temperatures. Biodiesel also has the potential to undergo hydrolysis in the presence of water. Water-contaminated fuel can cause engine wear and malfunction. The water content of the biodiesel derived from peanut oil was 0.03% by volume. This value falls in the range (0.02 – 0.05) specified by ASTM standards. And all samples did not contain any traces of bottom sediments and water. The presence of sulfur in the atmosphere has harmful impacts on human health and the climate. It has long been understood that biodiesel is sulfur-free, which is a major improvement over fossil fuel, and the results obtained confirmed this fact. The sulfur content for all samples was 304.6, 355.0, 411.7, and 544.7 ppm respectively and all are within the ASTM D6751 [19] standard (500 maximum).

Table 2. Characteristics of biodiesel produced from Heat-treated Peanut oil.

	ethanol to oil ratio			
Test	3:1	4:1	5:1	6:1
Density (kg/m^3)	810	806	807	809
Kinematic viscosity @ 40°C (mm^2/s)	1.260	1.240	1.245	1.255

Sulphur content (ppm)	304.6	355.0	411.7	544.7
Flash point (°C)	62	65	68	70.1
Water content (%)	0.03	0.03	0.03	0.03
Color	0.5	0.5	0.5	0.5
Carbon residue (%)	0.02	0.02	0.02	0.02

Table 3. Test methods for ASTM D675 and standard (B100 biodiesel).

Property	Result	ASTM Test Method	ASTM Standard for biodiesel (ASTM D6751)	ASTM Standard for petrol diesel (ASTM D975)
Flash point °C	62 - 70.1	D - 93	130 min.	340 - 358 K
Water & Sediment % vol	0.03	D - 2709	0.050 (max.)	0.02 - 0.05
Kinematic Viscosity mm^2/s	1.240 - 1.260	D - 445	1.9 - 6.0	1.9 - 4.1
Sulphur (ppm)	304.6 - 544.7	D-5453	500 (max)	0.35 - 0.55%
Copper strip corrosion	1a	D-130	No. 3 (max.)	No. 3 (max.)
Cloud Point	– 28	D-2500	Report to customer	40 max oF
Carbon residue %	0.02	D-4530	0.050	0.35 - 0.40
Acid number (mgKOH/g)	0	D-664	0.80 (max.)	
Color	0.02	D-877	Not required	
Density kg/m^3	806 – 810	D-287		950 (max)

Source: Abdulkareem et al. (2012) [21]

4. Conclusion

Biodiesel is a sustainable and alternative fuel to petroleum-based diesel that has the ability to expand production to meet increasing demand. Its raw materials are renewable and plentiful. However, non-edible and waste vegetable oils are used in the manufacture of biodiesel due to food competition. In the present work, biodiesel from heat-treated peanut oil have been synthesized by acid-catalyzed transesterification with ethanol at different reaction conditions. The physicochemical properties of the oil and biodiesel were determined and compared with accepted standards i.e. ASTM Standard. Law viscosity, high flash point, law water and Sediment contents, law cloud point and law sulphur content are the most important measured characteristics of the biodiesel produced in this study

Acknowledgments

The authors are grateful to the International University of Africa, and to the Sudanese Thermal Generation Company, Sudan for providing the laboratory facilities necessary to conduct this research work.

References

[1] Chen R, Qin Z, Han J, Wang M, Taheripour F, Tyner W, et al. Life cycle energy and greenhouse gas emission effects of biodiesel in the United States with induced land use change impacts. *Bioresource Technology*. 2018; **251**:249-58.

[2] Abed KA, El Morsi AK, Sayed MM, Shaib AAE, Gad MS. Effect of waste cooking-oil biodiesel on performance and exhaust emissions of a diesel engine. *Egyptian Journal of Petroleum*. 2018; **27**(4):985-9.

[3] Mohd Noor CW, Noor MM, Mamat R. Biodiesel as alternative fuel for marine diesel engine applications: A review. *Renewable and Sustainable Energy Reviews*. 2018; **94**:127-42.

[4] Silva Filho SCd, Miranda AC, Silva TAF, Calarge FA, Souza RRd, Santana JCC, et al. Environmental and techno-economic considerations on biodiesel production from waste frying oil in São Paulo city. *Journal of Cleaner Production*. 2018; **183**:1034-42.
[5] Ge JC, Kim HY, Yoon SK, Choi NJ. Reducing volatile organic compound emissions from diesel engines using canola oil biodiesel fuel and blends. *Fuel*. 2018; **218**:266-74.
[6] Mizik T, Gyarmati G. Economic and Sustainability of Biodiesel Production—A Systematic Literature Review. *Clean Technologies*. 2021; **3**(1):19-36.
[7] Vicente G, Martínez M, Aracil J. Integrated biodiesel production: a comparison of different homogeneous catalysts systems. *Bioresource Technology*. 2004; **92**(3):297-305.
[8] Canakci M, Sanli H. Biodiesel production from various feedstocks and their effects on the fuel properties. *Journal of Industrial Microbiology and Biotechnology*. 2008; **35**(5):431-41.
[9] Kulkarni MG, Dalai AK. Waste cooking oil an economical source for biodiesel: A review. *Industrial & Engineering Chemistry Research*. 2006; **45**(9):2901-13.
[10] Gebremedhin, G., Shimelis, A., Tarekegn, B., & Cernak, M.(2018). Physicochemical and functional properties of Ethiopian (Roba Variety) peanut (*Arachishypogaea L.*) for industrial use. *Austin J Nutri Food Sci*. 2018; **6**(3): 1107.
[11] Angaye S. Maduelosi, NJ, & Amadi, C. Effect of heat on the physicochemical properties of groundnut oil. *International Journal of Science and Research (IJSR)*. 2015; **4**(1):1278-80.
[12] Mandloi S, Radadia B, Visavadia M, Vaghela A. A review of chemical characteristics (acid value and iodine value) of peanut oil. *Weekly Science Research Journal*. 2014; **1**(30):1-4.
[13] Standard A. D4052. Standard test method for density, relative density, and API gravity of liquids by digital density meter. 2011 ASTM Annual Book of Standards. 2018.
[14] Standard A. D1747. Standard test method for refractive index of viscous materials. ASTM International, West Conshohocken, PA, USA. 2019.
[15] Standard A. D445. Standard test method for kinematic viscosity of transparent and opaque liquids (and calculation of dynamic viscosity). ASTM International, West Conshohocken, PA, USA. 2019.
[16] AOCS O. AOCS Official Method Cd 8b-90: Peroxide value acetic acid-isooctane method, official methods and recommended practices of the AOCS. American Oil Chemists Society Press, Champaign. AOCS Press; 2011.
[17] AOCS O. AOCS Official Method Cd 3-25: Saponification value of fats and oils, official methods and recommended practices of the AOCS. American Oil Chemists Society Press, Champaign. AOCS Press; 2017.
[18] AOCS A. Official Method Cd 3d-63. Acid value of fats and oils, official methods and recommended practices of the AOCS. American Oil Chemists Society Press, Champaign. AOCS Press; 2017.
[19] ASTM D-6751-03a. Standard specification for biodiesel fuel blend stock (B100) for middle distillate fuels. ASTM International. 2003, 1-6.
[20] ASTM D975-20c. Standard specification for diesel fuel. ASTM International. 2020.
[21] Abdulkareem A, Jimoh A, Afolabi A, Odigure J, Patience D. Production and characterization of biofuel from non-edible oils: an alternative energy sources to petrol diesel: IntechOpen; 2012.
[22] Innovation AI. Ethanol industry outlook. Report Renewable Fuels Association. 2012.
[23] Ayoola P, Adeyeye A. Effect of heating on the chemical composition and physico–chemical properties of Arachis hypogea groundnut seed flour and oil. *Pakistan Journal of Nutrition*. 2010; **9**(8):751-4.
[24] Anyasor G, Ogunwenmo K, Oyelana O, Ajayi D, Dangana J. Chemical analyses of groundnut (Arachis hypogaea) oil. *Pakistan Journal of Nutrition*. 2009; **8**(3):269-72.
[25] Musa M, Sulaiman A, Bello I, Itumoh J, Bello K, Bello A, et al. Physicochemical properties of some commercial groundnut oil products sold in Sokoto metropolis. Northwest Nigeria. *Journal of Biological Science and Bioconservation*. 2012; **4**:17-24.

[26] Zahran HA, Tawfeuk HZ. Physicochemical properties of new peanut (Arachis hypogaea L.) varieties. *OCL*. 2019; **26:**19.
[27] Kataria J, Mohapatra SK, Kundu K. Biodiesel production from waste cooking oil using heterogeneous catalysts and its operational characteristics on variable compression ratio CI engine. *Journal of the Energy Institute*. 2019; **92**(2):275-87.
[28] Allen CAW, Watts KC, Ackman RG. Predicting the surface tension of biodiesel fuels from their fatty acid composition. *Journal of the American Oil Chemists' Society*. 1999; **76**(3):317-23.
[29] Priya IV, Sastry SV, Sahoo A. Acid Catalysis of Waste Cooking Oils for Biodiesel Production. *Iranian (Iranica) Journal of Energy & Environment*. 2019; **10**(2):91-5.

Journal of Physics: Conference Series **2063** (2021) 012028 doi:10.1088/1742-6596/2063/1/012028

Isolation, Preparation and Characterization of Polylactic Acid Film Reinforced with Nano silica

Zainab A Ali[1,] Alaa Karem Niamah[1] and Widad Salih Hannosh[2*]

[1] Food Science department, University of Basrah, Basrah, Iraq

[2] Chemistry department, University of Basrah, Basrah, Iraq

*Corresponding author:whidadhanoosh@yahoo.com

Abstract: Lactic acid was isolated by using microbial fermentation in whey media which was carried out by local strain *Lactobacillus paraplantarum* then polymerized using the acidic medium's polycondensation method. FTIR and H-NMR were used to characterize the isolated lactic acid monomer and polyalctic acid (PLA), and the results confirmed the chemical structures of the isolated lactic acid and PLA. GPC techniques were used to determine the molecular weight and molecular weight distribution of the prepared PLA; the result showed that the Polydispersity index (PDI) was 2.51. The Biodegradable composite films of poly lactic acid (PLA)/Nano silica powder were prepared by the composite film casting method using dichloromethane as solvent. In contrast, Nano silica was synthesis from hydrolysis of tetra ethoxysilane and was loaded in PLA in 1 to5 wt. %. The films were subjected to a tensile strength study. Thermogravimetric analysis (TGA) and differential thermal analysis (DSC) were used to evaluate PLA. Also, water absorption of the prepared composites was studied, and the result showed that the thermal stability and water absorption of these prepared films were increased with an increasing percentage of Nano silica, while the percentage of crystallinity of the PLA evaluated from DSC was 28 %.

1. Introduction

Nowadays, the environmental changings in the earth make hard issues for people's life in the all aspect, due to discharging of the huge quantities from waste and gases. The most discarded waste is remaining in the environmental for long time because it is not biodegradable [1, 2] .The plastic waste is occupied high percentage and it almost cannot recycled or it is not biodegradable. In addition to this obstacle, the major sources used in the plastic production is petrochemical or non-renewable fossil [3] Which they have several disadvantages, such that, diminishing and prices growth, production high amount of carbon dioxide in both the processing and the recycling, in addition to that, some toxic monomer used in the plastic production is pollutant and may be migrating to edible materials when used for packaging of food [4].

There are many schemes to overcame this drawback and reduces the effect of pollutants on the planetary and life. One of them by using ecofriendly and biodegradable polymers. These polymers included: Polyglycolide (PGA), Polyhydroxy butyrate (PHB), L-Polylactide (PLA) and Polyethylene glycol (PEG), [5].

Among these polymers, PLA is a bio-based, biodegradable polymer which can be prepared from lactic acid produced by microbial fermentation using renewable sources such as corn and sugarcane. It has found numerous applications in the medical and pharmaceutical fields and in the protection of goods and food [6]. Comparing with the others polymers, PLA has several advantages such as it consumes quantities of carbon dioxide during production [7]. It can be produced from renewable and benign resources, it is biodegradable, recyclable and compostable, and it's a good physical mechanical property and can be altered it through the manipulation of PLA architecture [8, 9]. Last three-decade, PLA has found numerous applications as a packing material and medical studies have shown that the amount of the lactic acid release from packaging material to food is lower than its level used in common food ingredients [10].

In this paper, we will use lactic acid produced by local strain *L. paraplantarum* bacteria to prepare PLA via polycondensation. In the present work, PLA used as a matrix and loaded it with 1 to 5 wt.% of Nanosilica and studied the effect of this filler loading on the tensile properties, thermal stability of PLA/ Nanosilica. ESM was used to characterize the silica powder prepared by the hydrolysis of tetraethoxysilane.

2. Experimental Part

2.1-Materials

Lactobacillus paraplantarum YL4 new local isolate was tested and recorded to be a local Iraqi isolate from Buffalo milk recorded under the name LP IQ MT622658, in European Nucleotide Archive (ENA), National Centre for Biotechnology Information (NCBI) and Gene Bank. MRS agar from Hi-media used for cultivation. Whey used as fermentation media, chloroform, calcium chloride, and hydrochloric acid solution (35%), potassium bromide was supplied from Fluka Comp. While dichloromethane, acetone, absolute ethanol, silicon tetrachloride, and dimethylformamide were supplied from Aldrich Comp. At the same time, solid Nano silica was synthesis in this study by the hydrolysis of precursor tetraethoxysilane monomer (prepared by silicone tetrachloride reaction with absolute ethanol). Ammonia solution acts as a catalyst, water, and ethanol, controlling the pH and temperature of the hydrolysis. The SEM technique detected the particle size of the silica.

2.2- Instruments.

2.2.1- Fourier Transform Infrared spectrum (FTIR). The Shimadzu FTIR 8400S device recorded the FTIR spectra of the lactic acid and the PLA. Each spectrum was recorded at the frequency range 400-4000 cm^{-1} using Sodium chloride cell for lactic acid, while for PLA, a Potassium bromide used. The KBr was previously oven-dried at 300°C to reduce the interference of water.

2.2.2- Gel permeation chromatography (GPC). The number average molecular weight (Mn) and weight average molecular weight (Mv) determined by Youngling ACME 9000 GPC/SEC in DMF solvent; With flow rate 1.00ml/min, and column set length 950 nm, Injection volume was 200 µl.

2.2.3 - Differential Scanning Calorimetry (DSC). DSC measurement was conducted on Shimadzu DSC-60. Dynamic scans were carried out at the temperature range of 25-350°C at a constant heating rate of 10°C/min, under a nitrogen atmosphere at a 20ml/min flux rate. About 10-15 mg of sample was used in aluminum crucible. On the other hand, the degree of crystallinity of PLA is calculated by using the following formula [11].

$$x_c = \frac{\Delta H_m}{\Delta H_f} x100\% \quad (1)$$

Where Xc is the degree of crystallinity

ΔH_m is the heat of fusion of the sample (from DSC curve)

ΔH_f is the heat of the fusion of 100% crystalline PLA.

Journal of Physics: Conference Series **2063** (2021) 012028 doi:10.1088/1742-6596/2063/1/012028

2.2.4 - Thermogravimetric Analysis (TGA). TGA measurement was evaluated on TGA Q50V20-13 build 39. The dynamic scan was measured in a temperature range (25-700) °C at a constant heating rate of 50 °C/min under a nitrogen atmosphere at a flow rate of 30 ml/min.

2.2.5 - Tensile strength and Elongation. The tensile strength and the elongation-to-cut ratio of the PLA film were measured using the Tensile Analyzer of the Polymer Research Center / the University of Basra, According to the American Society for Testing and Materials (ASTM) No. D-882.91 of 1996.

2.2.6 – Water absorption for PLA thin film. The specimens (dimension: 60 × 1 mm) were dried at 50°C in vacuum oven until a constant weight was attained, they were immersed in water in a thermostatic stainless-steel water bath at 30°C, 40°C and 50°C. Weight changes were recorded by periodic removal of the specimens from the water bath and weighing on a balance with a precision of 1 mg. The percentage change at any time t, (Mt) as a result of water absorption was determined by Equation 2, [12].

$$M_t = \frac{W_w - W_d}{W_d} \times 100\% \quad (2)$$

Where, Wd and Ww denote the weight of dry material (the initial weight) and weight of materials after exposure to water absorption, respectively. The percentage at maximum water absorption (Mm) was calculated as the average value of several consecutive measurements that showed no appreciable additional absorption.

2.2.7- Nuclear Magnetic Resonance (H-NMR). ^{1}H -NMR spectrum of isolated lactic acid was obtained on a Bruker spectrometer operating at 600 MHz, at room temperature. Deuterated dimethyl sulfoxide (DMSO)-d_6 were used as solvents, and the sample concentration was 10% (w/v).

2.3 – Synthesis methods

2.3.1. Synthesis of PLA. First *L. paraplantarum* isolate cultivated in MRS agar at 37 °C for 24-48 hrs. Which used to lactic acid production. Fermentation was carried out according to [13].The fermentation medium (whey) , pH adjusted to 6 , Inoculated with 1% inoculum with 8 $\times 10^7$ CFU / ml and incubated at 37 °C for 3 days. Lactic acid produced by bacteria was polymerized throughout condensation polymerization, carried out using the hydrochloric acid solution (35%) as a catalyst at 90-100 °C under nitrogen gas for 2-3 hrs. water was removed from time to time. Then the reaction mixture was cool to room temperature and washing several times with hot water to remove unreacted monomer, and remaining acid. The product was dissolved in chloroform, dried by adding calcium chloride and then filtrate. Finally, the polymer was precipitated by added methanol as white (sometimes slightly yellow) powder. Drying is carried out in an oven at 40°C under vacuum for 3 hrs. And then characterized by the different techniques.

2.3.2. Synthesis of PLA thin films. Firstly, The PLA is dried under an inert atmosphere at 40°C for 3-4 hours to remove any solvent traces. Secondly. A 20 wt. % solution of PLA in chloroform was spread onto a glass plate with a Gardner knife to obtain a film thickness of 50 ± 5 μm. The PLA film was kept overnight at room temperature to remove most of the solvent. Final drying was accomplished in a vacuum oven at 40°C for two days.

2.3.3. Synthesis of the PLA Nanocomposites. Before processing, the PLA granules were dried at 60°C for 24 hours in a vacuum oven to remove excess moisture. The PLA was dissolved in dichloromethane solvent and several percentages of nanosilica prepared in this study (1, 3 and 5) % was added to the solution of PLA and sonication for 2 hours at room temperature in order to get homogenies mixtures. Then the samples were casting in Petri dish and kept in a desiccator for the controlled the evaporation of the solvents for 2 days. Optically clear films with thickness ranging from 500 to 700 μm were obtained and subsequently dried at 80°C under vacuum for 2 days. The prepared films were kept in desiccator for further studies.

2.3.4. Film Thickness. The thickness estimated by take average of 5 readings using hand micrometer .

Journal of Physics: Conference Series **2063** (2021) 012028 doi:10.1088/1742-6596/2063/1/012028

3. Results and Discussion

3.1. Synthesis of PLA

PLA polymers produce by three routes from lactic acid, as shown in Figure 1.

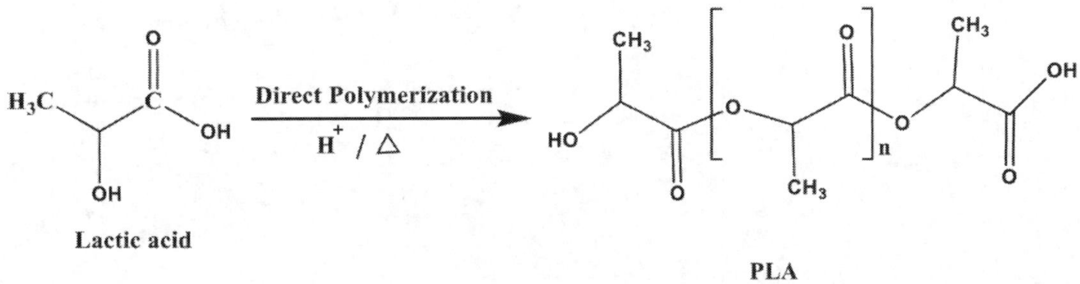

Figure 1. Scheme of polylactic polymerization

In this study , direct polycondensation of isolated lactic acid was used to obtained polylactic acid due to simple method .Here the condensation reaction obtained by the reaction of carboxylic groups and hydroxyl groups and the by product was water which was removed by distillation in order to get high yields polymers [14].Different catalyst were used to polymerize lactic acid and most of this research use antimony trioxide as catalyst [15]. Figure 2 shows the PLA powder.

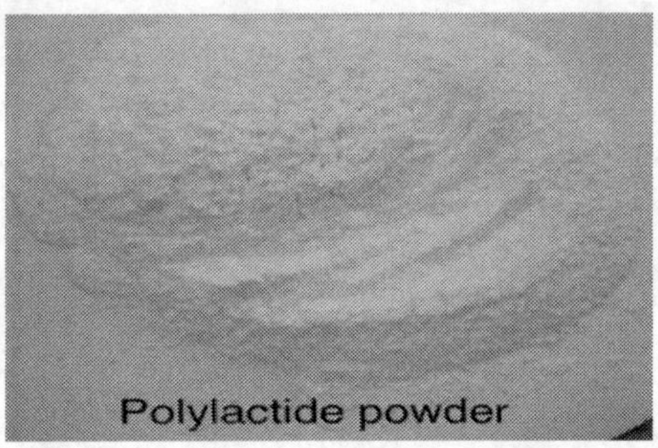

Figure 2. Photo of polymer powder prepared in this study

On the other hand, in this study, Nano silica was prepared through the hydrolysis of tetraethoxysilane monomer according to literature procedure [16] as shown below.

[Hydrolysis]

$$\equiv Si - OR + H_2O \Leftrightarrow \equiv Si - OH + ROH \quad (3)$$

[Alcohol condensation]

$$\equiv Si - OR + = Si - OH \Leftrightarrow \equiv Si - O - Si \equiv + ROH. \quad (4)$$

[Water condensation]

$$\equiv Si - OH + \equiv Si - OH \Leftrightarrow \equiv Si - O - Si \equiv + H_2O. \quad (5)$$

[Overall reaction]

$$Si(OR)_4 + 2H_2O \xrightarrow{OH-} SiO_2 \downarrow + 4ROH. \quad (6)$$

Where R is C_2H_5

Journal of Physics: Conference Series **2063** (2021) 012028 doi:10.1088/1742-6596/2063/1/012028

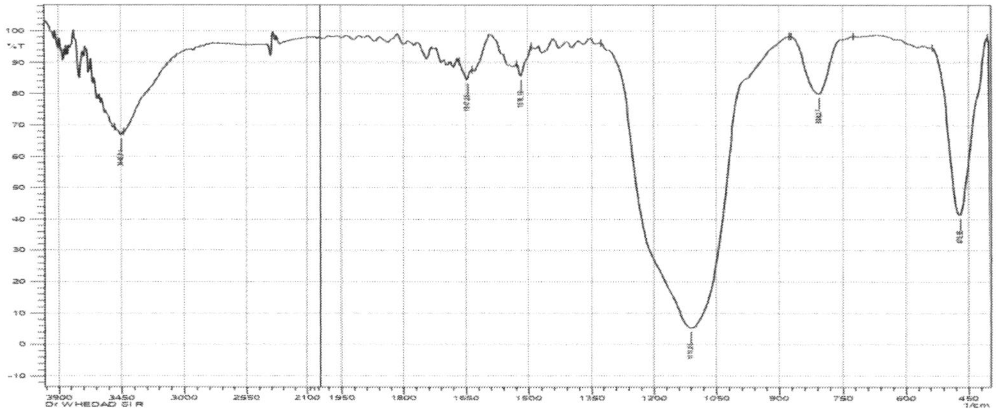

Figure 3. FTIR of the prepared silica

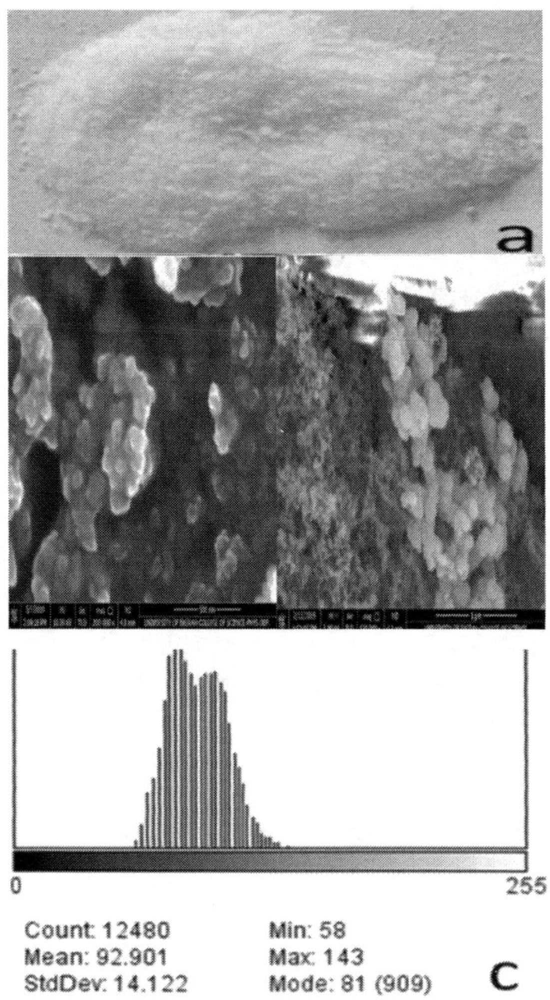

Figure 4. (a) Nano silica particle, (b) The SEM image of Nano silica powder prepared in this study and (c) The particle size distribution of Nanosilica

Figure 3. show the FTIR of the silica obtained from the hydrolysis process, the band at 3448 cm^{-1} indicate the presence of hydroxyl groups, while the characteristic band at 1111 cm^{-1} due to the Si-O-Si siloxane bond in their structures [17]. Figure 4. (a,b and c) Nano silica particle, SEM analysis of the products and the particle size distribution respectively.

3.2. Molecular weight of PLA:

GPC is an empirical method used to measure MMD and average molar masses of polymers [18]. Sometimes, the term SEC (size-exclusion chromatography) is used in the literature, but it is a more general concept for both GPC and GFC (gel filtration chromatography). In GPC, organic solvents are used as mobile phases, whereas in GFC, aqueous solvents are utilized. For the prepared PLA in this study, Figure 5 shows the molecular weight distribution and molecular weight distribution, While Table 1. show the values of the molecular weights calculated from the measurement.

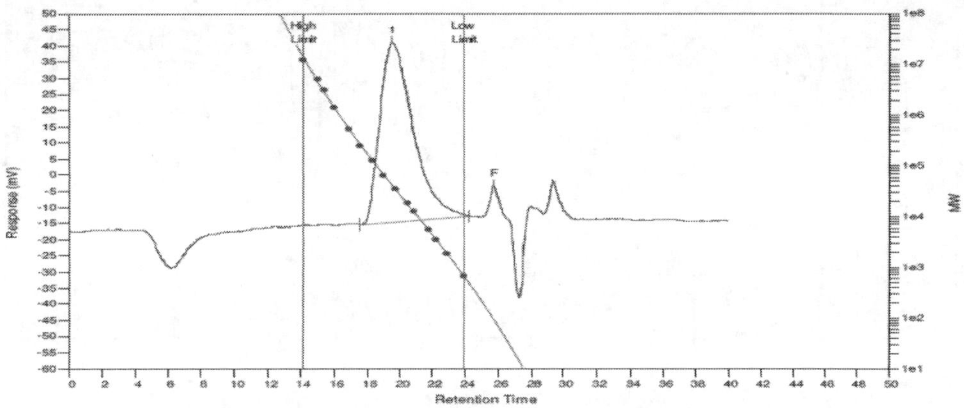

Figure 5, GPC Distribution curve of polylactic molecular weight

Table 1: The results of the molecular weights obtained from the measurement

Mn	Mw	Mz	Mz+1	Mv	PDI
15563	39193	63449	8449	35766	**2.5183**

Mn= number average molecular weight Mw=weight average molecular weight Mz=z average molecular weight. Mv=viscosity average molecular weight PDI= polydispersity index

From the results, we noted that the order of the different molecular weights according to their values is as follows:

$$M_n < M_w < M_z < M_v$$

This agreement to the most reported values of the molecular weight of polymers. With regard to the Polydispersity index, it value can be calculated from the equation below:

$$PDI = \frac{M_w}{M_n} \quad (7)$$

We note that the PDI value is 2.518, which means that the high molecular weight package is broad. These values depend on the type of polymerization process and its conditions. The research indicates that PDI values for additive polymers are in the range 1-1.5, while its values for condensing polymers are less than 2. When the number closes to the limits of 2 means, the high molecular weight is

Journal of Physics: Conference Series **2063** (2021) 012028 doi:10.1088/1742-6596/2063/1/012028

narrow. Thus the crystallization ratio is high. According to the prepared polymer, it gave a low percentage of crystallinity, which is around 28%, calculated from the DSC.

3.3. FTIR-analysis

The FTIR spectra were measured within a 400-4000 cm^{-1} range, for the isolated LA and PLA on sodium chloride and potassium bromide disk. The peak at 3510 cm-1 in the spectrum of the isolated LA as shown in Figure 6. Indicates to a hydroxyl group (OH), while the peaks (2993, 2947, 2885) cm^{-1} indicate the binding of C-H [19] .The peak at 1454 cm^{-1} is attributed to the CH3 group. The C-O-C within the region (1000-1300) cm-1 refers to the stretching vibration. The main peak, which is considered a clear indication of acid formation, appears at 1751 cm^{-1}, attributed to the carbonyl group as reported by [20].

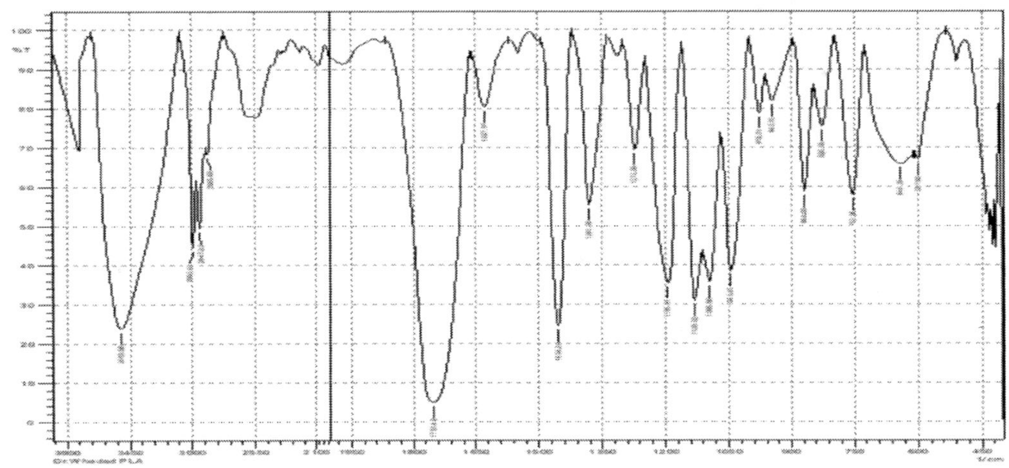

Figure 6, the FTIR of LA

On the other hand FTIR spectrum of PLA is shown in Figure 7. We noticed a decrease in the intensity of the beam belonging to the hydroxyl group 3506 cm^{-1}. This is evidence of the polymerization process. In addition to that, the presence of (2947, 2997) cm^{-1} peaks returns to the stretching C-H bond. The carbonyl group's peak in the acid 1751 cm^{-1} is shifted to 1759 cm^{-1}, with a stronger intensity in the polymer, which belongs to the ester group, which gives evidence of the polymerization of the acid. The peak at 1458 cm^{-1} refers to the CH_3 group. These results agreement with some researches [21, 22].

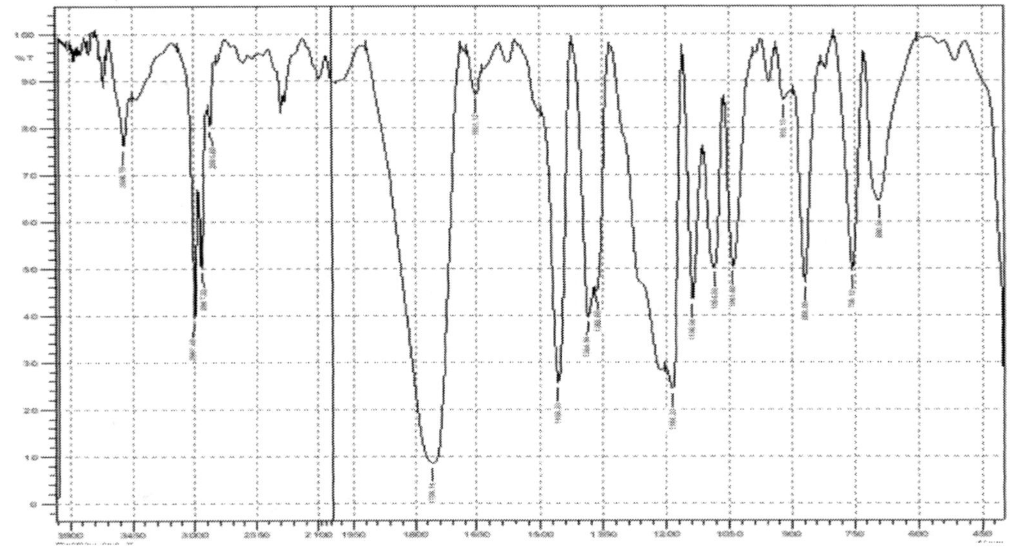

Figure 7, the FTIR of PLA

Journal of Physics: Conference Series **2063** (2021) 012028 doi:10.1088/1742-6596/2063/1/012028

3.4. H-NMR analysis

The 1H-NMR spectrum of isolated lactic acid is shown in Figure 8. The resonance signal at 1.3-1.6 (d, 3H, CH_3) is assigned to methyl protons ($-CH_3$), The resonance signalat ($\delta = 4.28$) corresponds to the methine linked to the OH group (23), while the peaks around 4.8 ppm due to deuterated solvent (D_2O) .

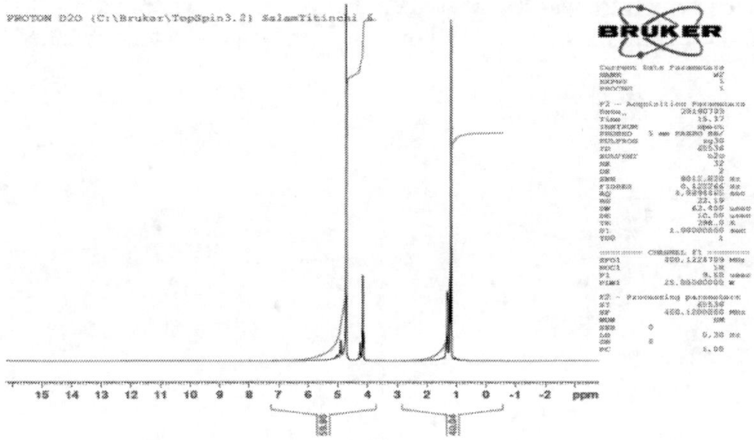

Figure 8. H-NMR spectrum of isolated lactic acid

3.5. DSC analysis

To investigate the thermal behavior of PLA, DSC measurements were carried out. PLA thermogram consists of a glass transition and melting endotherm peaks (Figure 9). The glass transition temperature, Tg, PLA, is 65.89 ◦C, and the specific heat capacity Cp, PLA 0.67 J/ (g ◦C). These values compare well with Tg = 61 ◦C and Cp ~ 0.55 J/ (g ◦C), as[24], While the endothermic peak at 158 ° C is attributed to the polymer's melting point [25]. On the other hand degree of crystallinity was determined from this analysis and was equal to 28.07, this mean that the prepared polymer has low degree of crystallinity.

$$X_c = \frac{26.28}{93.6} \times 100 = 28.07$$

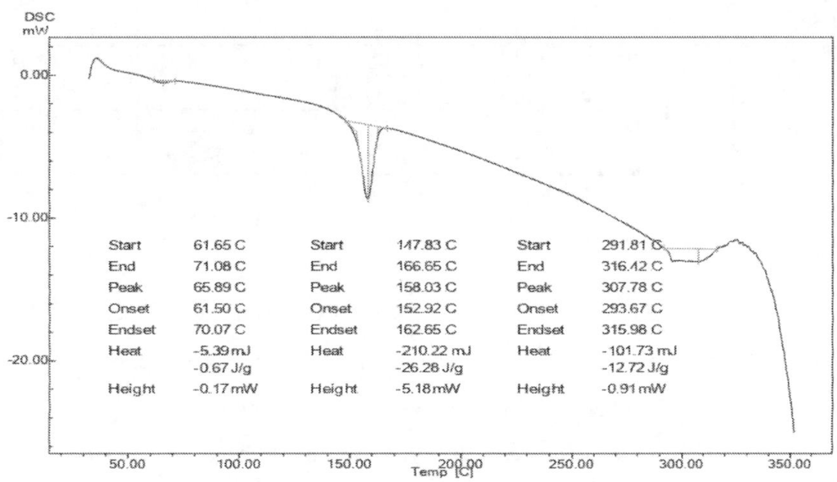

Figure 9. DSC thermogram of PLA

3.6. TGA analysis

TGA, or thermogravimetric analysis, has long been an accepted analysis method for polymers in industry . Its advantages are also easily seen from a forensic point of view, here TGA was carried out to investigate the importance of thermal degradation of PLA and PLA-composite. According to the TGA thermograms shown in Figures (10, 11), the degradation of PLA and PLA -Nano silica composite has one decomposition temperature higher than 400°C. In contrast, On the other hand, the decomposition rate of PLA- Nano silica composite was less than the other polymers, which means that the composite was more stable than the PLA alone. Table 2. Show the thermal parameters obtained from the TGA thermograms.

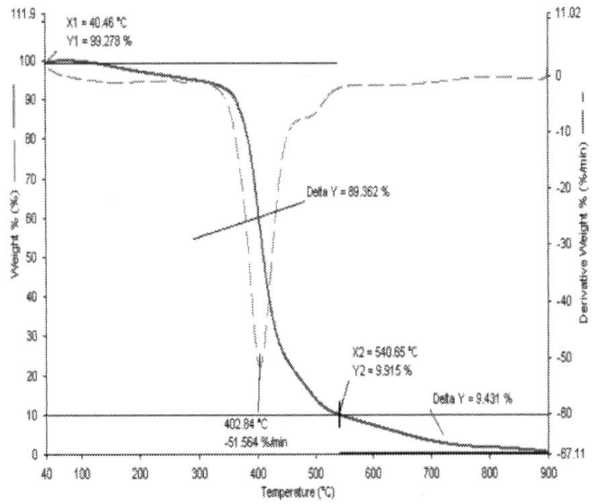

Figure 10. TGA thermogram of PLA

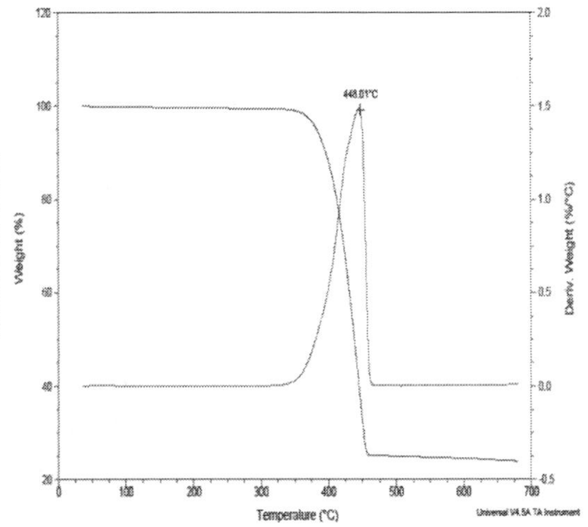

Figure 11. TGA thermogram of PLA+3%Nanosilica

Table 2. Thermal parameter of the prepared polymers

Sample No.	**Decomposition Temp.C^0**	**Temp. of 50% weight loss C^0**	**Char Content % At 600 C^0**	**Rate of Decomposition % / min.**
PLA	402	435	5	1.032
PLA-Nano silica	448	418	8	0.832

3.7. Tensile strength and Elongation of the PLA films

In order to assess the mechanical performance of PLA and PLA composite films, their tensile properties were studied. Figure 12. show the stress-strain curves of PLA Film. The effect of Nano silica loading on the tensile (strength, modulus, and elongation at break) are presented in Table 3., it can be observed that the tensile strength and modulus increased with filler content. This is probably because of better interfacial adhesion between the filler and the matrix by the van der Waals or induction interactions.

Journal of Physics: Conference Series **2063** (2021) 012028 doi:10.1088/1742-6596/2063/1/012028

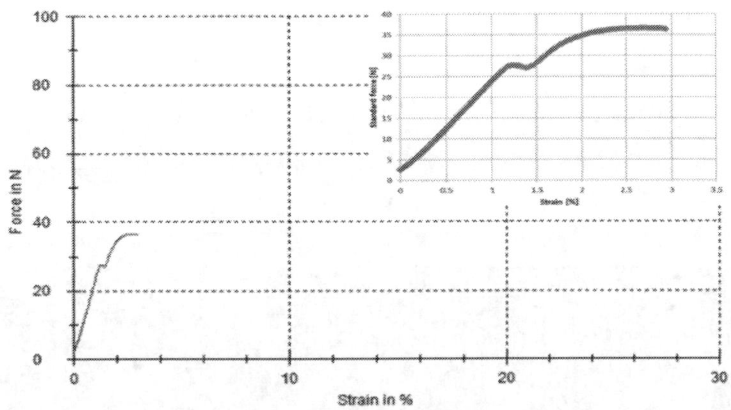

Figure 12. Stress-strain curve of PLA.

Table 3: Tensile properties of Nano silica-reinforced PLA as a function of silica content

Sample No.	Maximum Stress (Map)	Yong Modulus (Map)	Elongation at break %
PLA	19.8	986	2.9
PLA- 1% Nano silica	21.6	1011	2.7
PLA- 3% Nano silica	24.2	1045	2.6
PLA-5% Nano silica	31.8	1098	2.1

Table 4: Maximum water absorption of PLA Nanocomposites

Sample No.	**Mm value %**		
	25 C⁰	**35 C⁰**	**50 C⁰**
PLA alone	0.513	0.973	1.310
PLA-1 % Nano silica	0.721	1.427	1.732
PLA – 3% Nano silica	1.427	1.756	2.216
PLA-5% Nano silica	1.732	1.987	2.712

3.8. Water absorption

Table .4, shows the maximum water absorption (*Mm*), The *Mm* values of all PLA/silica are higher than that of unfilled PLA at immersion temperature of 30°C, 40°C and 50°C. This indicates that the presence of hydroxyl group in Nanosilica is prone to water absorption, which can be associated with its hydrophilicity. The *Mm* values of Nanosilica filled PLA (with and without impact modifiers) at 35°C and 50°C increased dramatically compared to the samples immersed at 30°C. It is suggested that at higher immersion temperatures (i.e., 35°C and 50°C), the swellable Nanosilica and the polymer chains became more flexible which create more free volume in PLA nanocomposites to enable more penetration sites for water molecules, and thus increase the *Mm* value. Figure 13.show different image of PLA.

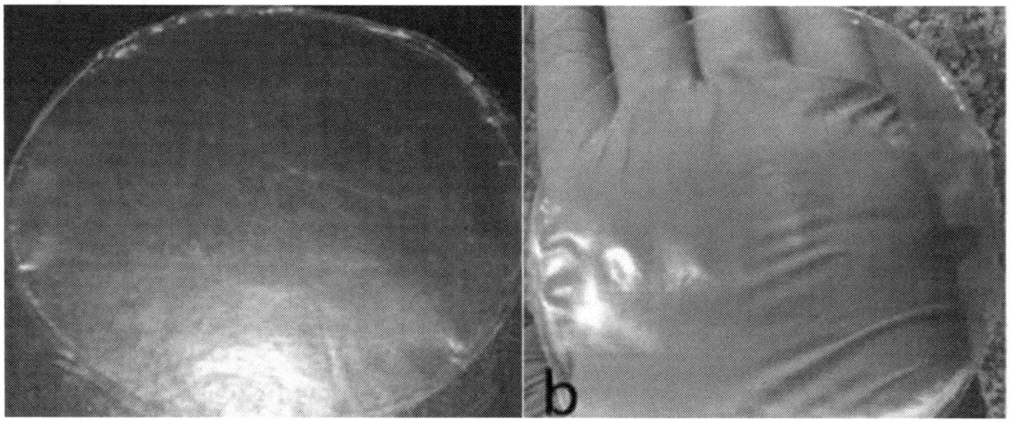

Figure.13: Image of, (a) PLA thin film alone, (b) PLA with Nanosilica film Conclusion

4. Conclusion

In this study lactic acid isolate and ,Then polymerized to Polylactic acid in acidic medium .FTIR and H-NMR used to characterized the product ,Thermal of PLA was evaluated using TGA and DSC techniques and from the result it has been found that the degree of crystallinity of PLA was 28% ,on the other hand composite thin film using Nanosilica was prepared and study of some mechanical properties the result shown that the tensile strength and modulus was increase with increasing percentage of Nanosilica.

References

[1] Singh AA, Sharma S, Srivastava M, Majumdar A. Modulating the properties of polylactic aci for packaging applications using biobased plasticizers and naturally obtained fillers 2020 Int. *J. Biol. Macromol.* **153** 1165.

[2] Ingrao C, Tricase C, Cholewa-Wójcik A, Kawecka A, Rana R, Siracusa V. Polylactic acid trays for fresh-food packaging: A Carbon Footprint assessment 2015 *Sci. Total Environ.* **537** 385.

[3] Collazo-Bigliardi S, Ortega-Toro R, Chiralt A. Using grafted poly (ε-caprolactone) for the compatibilization of thermoplastic starch-polylactic acid blends 2019 *React. Funct. Polym.* **142** 25.

[4] Chae Y, An Y-J. Current research trends on plastic pollution and ecological impacts on the soil ecosystem: A review 2018 *Environ. Pollut.* **240** 387.

[5] De Moura IG, de Sá AV, Abreu ASLM, Machado A V A, 2017 Food packaging, Elsevier;. p. 223-63, ch.7

[5] Panseri S, Martino P, Cagnardi P, Celano G, Tedesco D, Castrica M, Feasibility of biodegradable based packaging used for red meat storage during shelf-life: A pilot study. 2018

Food chemistry **249** 22.
[6] Zhou L, Zhao G, Jiang W. Mechanical properties of biodegradable polylactide/poly (ether-block-amide)/thermoplastic starch blends: Effect of the crosslinking of starch. 2016 *J. Appl. Polym. Sci.*133.
[7] Farah S, Anderson DG, Langer R. Physical and mechanical properties of PLA, and their functions in widespread applications—A comprehensive review 2016 *Adv. drug Deliv. Rev.* **107** 367.
[8] Öz A, Süfer Ö, Çelebi Sezer Y. Poly (lactic acid) films in food packaging systems. 2017 FSNT **2** 2574.
[9] Abd Alsaheb RA, Aladdin A, Othman NZ, Abd Malek R, Leng OM, Aziz R, Recent applications of polylactic acid in pharmaceutical and medical industries. 2015 *J. Chem. Pharm. Res.* **7** 51.
[10] Battegazzore D, Bocchini S, Frache A. Crystallization kinetics of poly (lactic acid)-talc composites. 2011 *Express Polym. Lett.* **5** 849.
[11] Tham WL, Poh BT, Ishak ZAM, Chow WS. Characterisation of water absorption of biodegradable poly (lactic acid)/halloysite nanotube nanocomposites at different temperatures. 2016 *J. Eng. Sci.* **12** 13.
[12] Reddy G, Altaf M, Naveena B, Venkateshwar M, Kumar EV. Amylolytic bacterial lactic acid fermentation—a review 2008 Biotech. Adv. **26** 22.
[13] Hamad K, Kaseem M, Yang H, Deri F, Ko Y. Properties and medical applications of polylactic acid: A review. 2015 Express Polym. Lett. **9** 435
[14] Mehta R, Kumar V, Bhunia H, Upadhyay S. Synthesis of poly (lactic acid): a review 2005 *J. Macromol. Sci., C: Polym. Rev.* **45** 325.
[15] Chruściel J, Ślusarski L. Synthesis of nanosilica by the sol-gel method and its activity toward polymers. 2003 *Mater. Sci.* **21** 461.
[16] Al-Oweini R, El-Rassy H. Synthesis and characterization by FTIR spectroscopy of silica aerogels prepared using several Si (OR) 4 and R″ Si (OR′) 3 precursors. 2009 *J. Mol. Struc.* **919** 140.
[17] Singh SK, Anthony P, Chowdhury A. High molecular weight poly (lactic acid) synthesized with apposite catalytic combination and longer time. 2018 *Orient. J. Chem.* **34** 1984.
[18] Păucean A, Vodnar D, Mureşan V, Fetea F, Ranga F, Man S, et al. Monitoring lactic acid concentrations by infrared spectroscopy: A new developed method for Lactobacillus fermenting media with potential food applications. 2017 *Acta Aliment.* **46** 420.
[19] Vodnar DC, Paucean A, DULF FV, Socaciu C. HPLC characterization of lactic acid formation and FTIR fingerprint of probiotic bacteria during fermentation processes. 2010 *Not Bot Horti Agrobo.* **38** 109.
[20] Mofokeng JP, Luyt A, Tábi T, Kovács J. Comparison of injection moulded, natural fibre-reinforced composites with PP and PLA as matrices. 2012 *J. Thermoplast. Compos. Mater.* **25** 927.
[21] Chieng BW, Ibrahim NA, Yunus WMZW, Hussein MZ, Then YY, Loo YY. Effects of graphene nanoplatelets and reduced graphene oxide on poly (lactic acid) and plasticized poly (lactic acid): a comparative study. 2014 *Polymers* **6** 2232.
[22] Kishida H, Jin F, Yan X, Moriya T, Enomoto H. Formation of lactic acid from glycolaldehyde by alkaline hydrothermal reaction. 2006 *Carbohydr. Res.* **341** 2619.
[23] Witzke DR, Narayan R, Kolstad JJ. Reversible kinetics and thermodynamics of the homopolymerization of L-lactide with 2-ethylhexanoic acid tin (II) salt 1997 *Macromolecules* **30** 7075.
[24] Day M, Nawaby A, Liao X. A DSC study of the crystallization behaviour of polylactic acid and its nanocomposites2006 *J.Therm. Anal. Calorim.* **86** 623
[25] Menczel JD, Prime RB. Thermal analysis of polymers. 2009.

Journal of Physics: Conference Series **2063** (2021) 012029 doi:10.1088/1742-6596/2063/1/012029

A mathematical optimization model for secondary settler

Zahir BAKIRI [1,2]

[1] Laboratoire de Génie des Procédés Chimiques (LGPC), Department of Chemical Engineering, Ferhat Abbas University of Setif-1, 19000, Algeria

[2] Department of Chemical Engineering, Salah Boubnider Constantine 3 University, Constantine, Algeria

Corresponding author: zahir.bakiri@univ-constantine3.dz

Abstract. The study of secondary settler modelling, which aims to establish the main model (one-dimensional-1D model), which is involved in some fundamental processes of the hydrodynamic behaviour of this liquid / solid separation unit and to engender variations of the sludge blanket height as a function of the operating parameters and maintaining of the municipal wastewater treatment plant of Setif. The objective of this research is focused on solid/liquid separation in the secondary settler by attempting a mathematical model that allows us to evaluate the sedimentation velocity as a function of the sludge settleability parameters.

1. Introduction

The activated sludge treatment process is the most used in Algeria and in the world to treat urban wastewater. The wastewater treatment plant of Setif city is a good example of this type of process. It should be noted that it was dimensioned for a capacity of 330,000 population equivalent to accommodate 66,000 m^3/d in dry weather. The activated sludge processes have a very high purification performance and reliability, especially for organic pollution. This type of treatment has several limitations, such as control the pollution degradation phenomena in the biological reactor and the separation problems in the secondary settler, which are strongly interconnected. The optimal operation of the wastewater treatment plant depends on all the parameters of the different processes. The activated sludge process is based on a highly active bacteriological culture mixed with wastewater, which results in the aerobic degradation of the organic matter in the aeration tank. The biomass formed in this system, which constitutes the solid phase, is eliminated by settling in the secondary settler. Biological treatment by activated sludge is generally based on a microbial culture maintained in suspension in an aerobic condition in the aeration tank fed by the wastewater to be treated [1]. The general organization of an activated sludge treatment plant is shown as follows (Figure 1). The final stage of treatment involves the simultaneous separation of sludge from the treated water by settling. Liu (2003) [2] assumes that settling is the last step in the treatment chain before the treated water is released into the natural environment. This settling is provided by the secondary settler. The degradation of organic matter carried out in the aeration tank, where microorganisms degrade organic matter by oxidation and the total suspended solids retained by bio-flocs [3,4]. It also includes the aeration system ensure oxygen, necessary for the growth of microorganisms, usually using high powerful turbines. The mixing equipment ensure that flocs are not settling and facilitate the homogeneous oxygenation of the aeration tank.

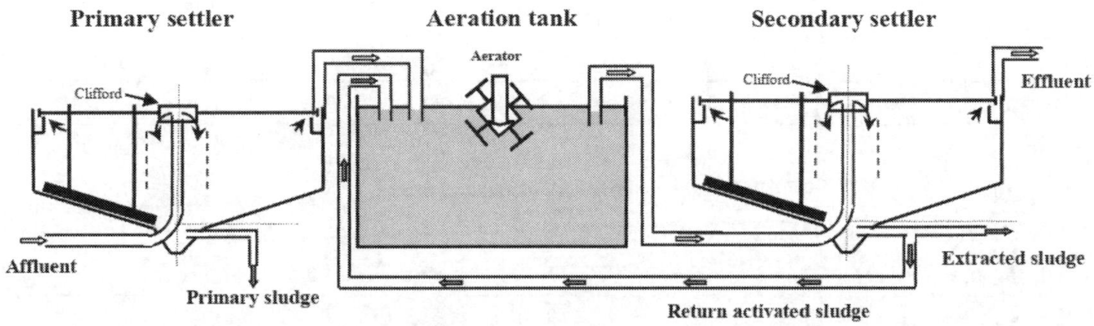

Figure 1. Activated sludge process

In this process, several types of hydraulic or biological dysfunctions may appear. Most often is the overflow of the sludge blanket to the receiving environment, due to disturbance of the hydraulic flow and excessive development of filamentous bacteria in the biological reactor. These disturbances generate difficulties settling sludge in the secondary settler. We specify that the excessive growth of filamentous bacteria, characterized by increasing the sludge volume index (> 200 mL / g), causes foaming by the presence of gas bubbles and thus leads to poor sludge settling and starting towards the receiving environment. On the other hand, modelling is one of the tools used to improve the knowledge processes and the best management of wastewater treatment plants (WWTP). It is essential to the improvement of the control strategies for wastewater treatment plants by optimizing the parameters of the different processes. For example, the increase of the suspended solids and the flux causes an important affecting the purification efficiency of the secondary settler because the residence time decreases, and the upward velocity of the suspended solids increases.

The modeling as an optimum operation of wastewater treatment plants via activated sludge is widely used for design, optimization and management of installations.

Many authors [5-12] have been particularly interested in two main objectives of dynamic modeling of wastewater treatment plants: the first is to optimize the operation of wastewater treatment plants (WWTP) and the second is to size installations and structures. The modelling of wastewater treatment plants is used to meet different objectives. It based on the operating parameters of the WWTP, such as residence time, aeration time, organic load, age of sludge. Modelling of WWTPs, in particular activated sludge plants, is a widely used tool to optimize the operation of the equipment. According to Gernaey et al. (2004) [13], modelling can be used for modifying the operating parameters of wastewater treatment plants to ensure economic treatment. It is also used to understand the performance of the aeration basin and the secondary settler.

The main objective of this work is the development of a mathematical model to simulate the variations of the height of the sludge blanket as a function of concentrations. An integrating the various operating parameters of the Setif wastewater treatment plant. This work focuses mainly on the hydrodynamic model of the one-dimensional secondary settler, based on the theory of the flow of solid particles and the continuity equation.

2. Average loads received by the treatment plant

Samples should be taken with greater accuracy using instantaneous sampling to provide qualitative information. We were particularly interested in the different elements that each have a specific role in the biological purification operations.

The average concentrations, the maximum recorded and the ratios characterizing the urban wastewater of the city of Setif are presented in Table 1.

Table 1. Parameters and ratios characterizing of the effluents coming in WWTP

Characteristics of the main parameters									
Parameters	**pH**	**BOD_5**	**COD**	**TSS**	**NTK**	**$N\text{-}NH_4$**	**$N\text{-}NO_2^-$**	**$N\text{-}NO_3^-$**	**Pt**
Units	--	mg/L	mg/L	mg/L	mg/L	mg/L	mg/L	mg/L	mg/L
Average	7.58	380	1020	475	71.3	58.4	5.8	7.48	5.5
Maximum registered	8.30	450	1070	665	84.5	65.7	7.2	8.40	7.1
Minimum registered	6.65	270	720	315	35.8	24.2	2.8	3.1	2.4

Ratios						
Ratios	COD/BOD_5	TSS/BOD_5	NTK/COD	COD/NTK	COD/ Pt	$N\text{-}NH_4/NTK$
Average	2,68	1,25	0,07	14,31	185,45	0,82
Maximum registered	3,96	2,46	0,12	29,89	445,83	1,84
Minimum registered	1,60	0,70	0,03	8,52	101,41	0,29

The results shown in Table 1 allowed us to conclude that the average concentrations of the pollution parameters are by those of the literature. In the general case, the main objective of wastewater treatment plants is to ensure the treatment of suspended solids (TSS), organic load (COD), ammonium (NH^{4+}), nitrate (NO^{3-}) and phosphorus. For the ratios, we observe the same average values as in the literature. The values of different forms of nitrogen and even for total phosphorus sometimes exceed the standards required by Executive Decrees No. 06-141 of April 19, 2006, and No. 10-23 of January 12, 2010.

3. Dynamic modelling of the secondary settler

The main design criteria for a solids separation unit from the liquid phase are mass loading, hydraulic inflow, peak flow rate, minimum depth, and sludge blanket [14]. Other criteria are defined in the literature, such as Daigger (1995) [15], Schuler and Hoon Jang (2007) [16], Zhang et al. (2006) [17], for conventional activated sludge for large plants designed to operate under optimal solid/liquid separation conditions, among these criteria, mixed liquor concentration, sludge volume index and recirculation ratio.

Our initial model is therefore a one-dimensional model based on solid particle flow theory and including dispersion.

The modelling as an optimum operation of wastewater treatment plants via activated sludge is widely used for design, optimization and management of installations. Many authors [18-25] have been particularly interested in two main objectives of dynamic modelling of wastewater treatment plants: the first is to optimize the operation of wastewater treatment plants and the second is to size installations and structures.

3.1. Numerical resolution

The hydraulic flow (Feed flux, Q_f) is divided into two flows: upward bulk flux (ascending clarification zone, Q_s) and and downward bulk flux (descending thickening zone, Q_r). We have added a graphic design (Figure 2) that will be useful for more detail.

Generally, three modes of sedimentation are considered: clarification, thickening, and compression. Because the last two zones are of the same nature, we assume only as a single zone. For this reason, the term thickening has been reserved for this manuscript.

Journal of Physics: Conference Series **2063** (2021) 012029 doi:10.1088/1742-6596/2063/1/012029

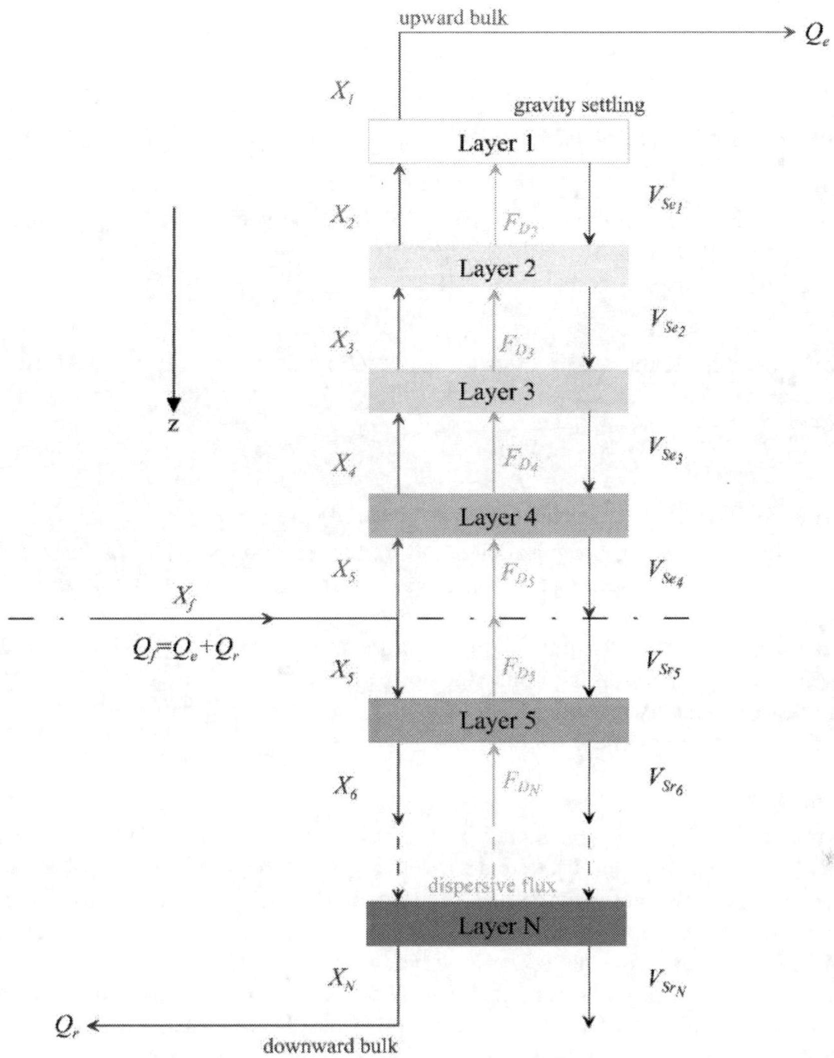

Figure 2. Stratification of secondary settler

3.2. Numerical resolution

Figure 2 shows the solid particle fluxes, sedimentation (F_s), hydraulic (F_h), and dispersive (F_D), in a layer of thickness ΔZ.

The balance equation (Figure 2) is as follows:

$$S F_s\big|_z + S F_h\big|_z + S F_D\big|_z = S F_s\big|_{z+dz} + S F_h\big|_{z+dz} + S F_D\big|_{z+dz} + V\frac{dX}{dt} + V(-r_X) \quad (1)$$

Where S is the section (m^2), V is the volume of settler, X is the total suspended solids (g/L), $(-r_X)$ is the rate of reaction, (g/L.min). After differentiation, Equation 1 can be represented as

$$dF_s + dF_h + dF_D + \frac{dX}{dt}dz + (-r_X)dz = 0 \quad (2)$$

Dividing by dz, Equation 2 takes the following form:

$$\frac{dF_s}{dz} + \frac{dF_h}{dz} + \frac{dF_D}{dz} + \frac{dX}{dt} + (-r_X) = 0 \quad (3)$$

After the replacement of the expressions of the fluxes, so

$$\frac{d(XV_s)}{dz}+\frac{d(V_h X)}{dz}+\frac{d\left(-D_s\frac{dX}{dz}\right)}{dz}+\frac{dX}{dt}+(-r_X)=0 \qquad (4)$$

Where $V_s = f(X)$ is the settling velocity (m/h), $V_h = Q/S$ is the hydraulic velocity (m/h), Q_f, Q_e and Q_r are the volumetric flow rates (f: feed, e: exit, and r: recycled) (L/h), D_s is the dispersion coefficient of solids in the thickening zone (m^2/h), z is the height of the secondary settler (m), The material balance in the thickening zone is as shown below:

$$\frac{d(XV_s)}{dz}+V_h\frac{dX}{dz}-D_s\frac{d^2X}{dz^2}+\frac{dX}{dt}+(-r_X)=0 \qquad (5)$$

Applying the same reasoning in the clarification zone, i.e. considering a layer and performing the mass balance with a relation to the biomass, we obtain the following equation (6):

$$-\frac{d(XV_s)}{dz}+V_h\frac{dX}{dz}-D_s'\frac{d^2X}{dz^2}+\frac{dX}{dt}+(-r_X)=0 \qquad (6)$$

D_s' is the dispersion coefficient in the clarification zone (m^2/h).

Equations (5) to (6) represent the dynamic model of the secondary settler, considering the axial dispersion of the N layers constituting the height (Z) of the settler as a function of the biomass concentration (X).

In steady-state, these non-linear differential equations are transformed into algebraic equations. It is solved by the iterative method, fixing the concentration at the surface of the settler and calculating the different concentrations of the layers constituting the two zones of clarification and thickening, which have been divided into five layers each.

4. Results and discussion

The sedimentation parameters must be affected by the pollution parameters and we have summarized the characteristics of the wastewater treated during the study period (Table 2). The pollutants are eliminated in the presence of microorganisms (organic matter) and by the growth of the bacterial culture (transformation metabolism of bacterial flocs), and by oxidation to CO_2 and H_2O which produces the energy necessary for the operation and production of new cellular materials. These parameters are strongly affected by filamentous bacteria and are manifested in two forms: foaming and filamentous expansion. The ability of sludge to be decanted is typically estimated by settling tests. The sludge index (Molhman index, MI) is influenced by the nature of the organic load. Several authors, such as Lee (1999) [10] and Hamilton et al. (1992) [22], include the dispersion term in their models and use a constant dispersion coefficient. In our case, we have also considered a constant dispersion coefficient for the settling process.

4.1.1. Sedimentation velocity and sludge index

An experimental example of relating sludge volume indice (SVI) to sedimentation velocity is summarized in Table 2. This table shows the physical and chemical characteristics of the sludge to be decanted (mixed liquor entering the secondary settler). They are microorganisms which are not the same from one station to another. It is very important to identify these characteristics to control the settling process.

Journal of Physics: Conference Series **2063** (2021) 012029 doi:10.1088/1742-6596/2063/1/012029

Table 2. Examples of experimental data for the sludge settleability.

Settlability (mL/L)	TSS (g/L)	SVI (mL/g)	V_S (m/h)
325	2.80	116	2.05
296	2.10	141	2.60
231	1.60	144	4.55
184	1.10	167	6.50
154	0.90	171	7.60

These low concentration data were obtained from the aeration tank of the wastewater treatment plant of the city of Setif. Several experiments were performed. During this period, the ability of sludge to settle and its qualities remained relatively homogeneous during these experiments. Note that the sludge indices (SVI) varied between 116 and 171 mL/g, with an average close to 140 mL/g.

Sub-models were developed, allowing us to estimate the settling velocity, by exploiting a large number of experimental results taken from the aeration tank of the Setif city wastewater treatment plant.

As a general practice, exponential functions are used to model the populations. According to the experimental results obtained, the variation of V_0 as a function of SVI is an increasing exponential function (same for population growth). In the literature, we find three types of models describing the variation in sedimentation velocity. The most widely used current models are usually basing on Vesilind [26] exponential model ($V_{S1} = k_1 e^{-n_1 X}$) and Yoshioka [27] power model ($V_{S2} = k_2 X^{-n_2}$). Other authors use the Cho et al. [28] function. This function is simply the expression of Vesilind divided by the biomass concentration ($V_{S3} = k_3 e^{-n_3 X} / X$).

Where k_i and n_i are the characteristic biomass parameters, X is the sludge concentration (total suspended solids), and V_{Si} is the settling velocities. There are many different types of expressions for settling velocity, as a function of sludge concentration that can be found in literature.

The proposed model for the average stabilized values of n=140 mL/g and V_o=2.6 m/h for WWTP of Setif. The initial conditions in these model's calibration are shown in Table 2. Note that our model is based on the Vesilind function.

This concentration represents the initial concentration (C_f), i.e. concentration of the mixed liquor in the aeration tank entering the secondary settler. It varies during the day and depending on the peak flow rate (low and medium load). This concentration is subjected at a settling in the secondary settler. It is concentrated at the bottom of this tank, and it is almost zero at the surface. For example, the concentration feeding the secondary settler is 5g/L. This concentration is almost zero at the top layer, and it can attend at 12 g/L at the bottom of the settler.

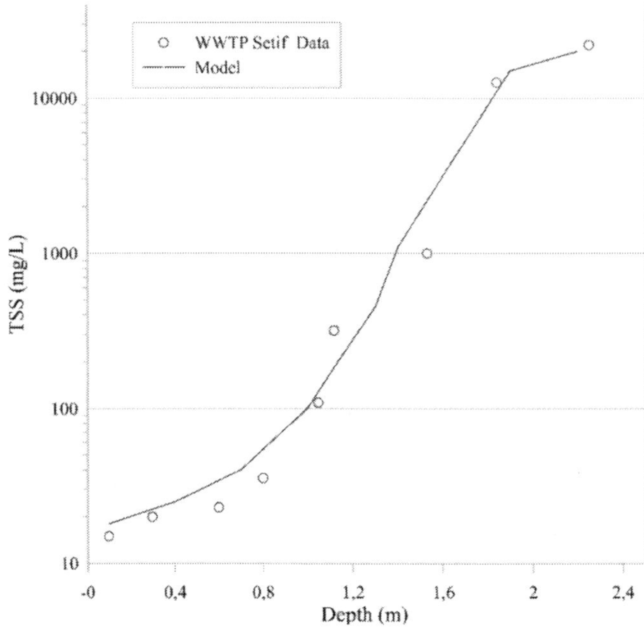

Figure 3. Comparison with our model and the experimental results of WWTP-Setif.

In order to validate the developed model, a comparison between the experimental results of WWTP-Setif and the predictions of the main decanter model is presented. The experimental results of WWTP were obtained in a settler with a cross section of 1661 m^2, a diameter of 46 m, a depth of 3 m, and an operating flow rate of 650 m^3/h. Figure 3 shows the variation of the sludge concentration as a function of the depth in secondary settler. We see from this figure that the model predictions are encouraging.

5. Conclusion

The model predictions developed is satisfactory, by comparison with the experimental results obtained in this work. This model is expressed as a function of the sludge settleability indices (SVI), whose range of variation is from 116 to 171 mL/g, obtained by performing more than 96 experiments. A one-dimensional dynamic model of the secondary clarifier has been developed, considering both clarification and thickening zones. In this model, we integrated the sedimentation velocity as a function of the sludge settleability index (SVI). The simulation of sludge concentrations and sludge blanket height also fits well with our experimental results. This simulation would be very complementary to the scheme adopted for the management and control of hydraulic surcharges.

References

[1] Katam, K., & Bhattacharyya, D. (2019). Simultaneous treatment of domestic wastewater and bio-lipid synthesis using immobilized and suspended cultures of microalgae and activated sludge. Journal of Industrial and Engineering Chemistry, 69, 295-303.

[2] Liu, Y. (2003). Chemically reduced excess sludge production in the activated sludge process. Chemosphere, 50(1), 1-7.

[3] Mohammadi, H., Sabzali, A., Gholami, M., Dehghanifard, E., & Mirzaei, R. (2012). Comparative study of SMBR and extended aeration activated sludge processes in the treatment of high-strength wastewaters. Desalination, 287, 109-115.

[4] Bakiri, Z., Chebli, D., & Nacef, S. (2012). Dynamic modelling of the secondary settler of a wastewater treatment via activated sludge to low-load. Energy Procedia, 18, 1-9.

[5] Chancelier J.Ph., Cohen De Lara M., Joannis C., Pacard, F. (1997), New insights in dynamic modelling of a secondary settler-II. Dynamical analysis, Wat. Res., Vol 31(8), pp 1857-1866.

[6] Diehu S., Jeppsson U. (1998), A model of the settler coupled to the biological reactor, War. Res., Vol32(2), pp 331-342.

[7] Diehl, S. (2000). On boundary conditions and solutions for ideal clarifier–thickener units. Chemical Engineering Journal, 80, 119-133.

[8] Marsilli-Libelli S. (1993), Dynamic modelling of sedimentation in the activated sludge, Civil. Eng. Syst., Vol 10, pp 207-224.

[9] Diehl S. (1997), Continuous Sedimentation of Multi-component Particles, Mathematical Methods in the Applied Science, Vol 20, pp1345-1364.

[10] Lee, T. T., F. Y. Wang, et R. B. Newell, 1999, Distributed parameter approach to the dynamics of complex biological processes, AIChE Journal; 45, 10: 2245-2268.

[11] Bakiri, Z., & Nacef, S. (2020). Development of an improved model for settling velocity and evaluation of the settleability characteristics. Water Environment Research, 92(7), 1089-1098.

[12] Wett B. (2002), A straight interpretation of the solids flux theory for a three-layer sedimentation model. Wat. Res.; 2949–2958.

[13] Germaey K.V., Vanrolleghem P.A., Lessard P., (2001), Modeling of a reactive primary clarifier, Water science and technology, Vol 43(7), pp 73-81.

[14] Guergachi, A. et G. Patry, 2003, Identification, verification and validation of process models in wastewater engineering:a critical review, Journal of Hydroinformatics; 5, 3: 181-188.

[15] Daigger, G. T., 1995, Development of refined clarifier operating diagrams using an updated settling characteristics database. Water Environ. Res.; 67: 95-100.

[16] Schuler, A. J et Hoon Jang, 2007, Density effects on activated sludge zone settling velocities, Water Research; 41:1814-1822.

[17] Zhang, D. J., Z. L. Li, P. L. Lu, T. Zhang et D. Y. Xua, (2006), A method for characterizing the complete settling process of activated sludge, Water Res.; 40,14: 2637-2644.

[18] Bajcar T., Gosar L., Sirok B., Steinman F., Rak G. (2010), Influence of flow field on sedimentation efficiency in a circular settling tank with peripheral inflow and central effluent, Chemical Engineering and Processing; 49 : 514–522.

[19] Bürger R., Diehl S., Nopens I. (2011), A consistent modelling methodology for secondary settling tanks in wastewater treatment, Water Research; 4 5: 2247 -2260.

[20] Gernaey K.V., Van Loosdrecht M.C.M., Henze M., Lind M.., B. Jørgensen S.B. (2004), Activated sludge wastewater treatment plant modelling and simulation: state of the art, Environmental Modelling & Software; 19 : 763–783.

[21] Gillot S., Choubert J-M., Racault Y., Héduit A. (2007), Etat de l'art de la modélisation du traitement biologique des eaux résiduaires, Tribune de l'eau ; 59(640) : 35-46.

[22] Hamilton J., Jain R., Antoniou P., Svoros S.A., Koopman B. (1992), Modelling and pilot-scale experimental verification for predenitrification process, Journal of Environmental Engineering; 118 : 38-55.

[23] Bakiri, Z., & Nacef, S. (2013). A simple model for secondary clarifier: application to wastewater treatment plant. Desalination and Water Treatment, 51(7-9), 1571-1576.

[24] Libelli, S., 1993, Dynamic modelling of sedimentation in the activated sludge, Civil. Eng. Syst. 10, 207-224.

[25] Lopez P. R. , Lavin A. G., Mahamud Lopez M. M., Bueno de las Heras J. L. (2008), Flow models for rectangular sedimentation tanks, Chemical Engineering and Processing; 47: 1705–1716.

[26] Vesilind P.A. (1968), Theoretical considerations: Design of prototype thickeners from batch settling tests, Wat. Sew. Wks. Vol 115(7), pp 302–307.

[27] Yoshioka N., Hotta Y., Tanaka S., Naito S. and Tsugami S. (1957) Continuous thickening of homogeneous flocculated slurries. Kagaku Kogaku 21, 66-74.

[28] Cho S.H., Colin F., Sardin M., Prost C. (1993), Settling velocity model of activated sludge, Water Research, Vol 27(7), pp 1237-1242.

Synthesis , characterization and spectral studies of mononuclear and binuclear Fe(II), Co (II), Ni (II), Zn(II) complexes with 2,4-dinitrophenylhydrazine dithiocarbamate

Enas Hazim Mohammed[1], Sahbaa Ali Ahmed[2a], Sariya Waleed Zedan[3]

[1]Department of general science College of basic education ,University of Mosul, Iraq.
[2,3]Department of Chemistry College of Science, University of Mosul, Iraq.
[a] Email: sahbaa-ali@uomosul.edu.iq

Abstract: New complexes of dithiocarbamate of the general formula [M(2,4- dinitrphenhydra - dtc)$_2$], [MM`(2,4- dinithrphenhydra-dtc)$_4$] where: M=Fe (II), Co(II), Ni(II),M`=Zn(II) ,2,4-dinitrphenhydra-dtc=2,4- di nitro phenyl hydrazine dithiocarbamate and [M(1,10- phen)$_3$][M`(2,4 –dinitrphenhydra-dtc)$_4$], [M(1,10- phen)$_3$][M`$_2$(2,4- dinitrphenhydra - dtc)$_6$] Where: M=Co(II), Ni(II), M`=Zn(II), (1,10- phen)= 1,10- phenanthroline monohydrate. Have been prepared and characterized by atomic absorption spectroscopy infrared, conductance measurements, electronic absorption spectra, nuclear magnetic resonance and magnetic susceptibility measurements. Magnetic moments and electronic spectra indicate that the complexes of the type [M(2,-dinitrphenhydra-dtc)$_2$], [MM`(2,4 -dinitrphenhydra - dtc)$_4$] are of square planner geometry while the complexes of the type [M(1,10-phen)$_3$][M`(2,4-dinitrphenhydra-dtc)$_4$] and [M(1,10 phen)$_3$][M`$_2$(2,4- dinitrphenhydra-dtc)$_6$] have octahedral geometry.

Keywords: Dithiocarbamate, Iron(II), Cobalt(II), Nickel(II), Zinc(II), Complexes.

1. Introduction

Dithiocarbamtes are valuable compounds due to their interesting chemistry and wide utilities and application, one of the most interesting aspects of dithiocarbamtes is their applications in coordination chemistry [1,2]. Most of aliphatic and aromatic dithiocarbamte synthesized until now may coordinate only through the dithio-group behaving as uni negative bidentate ligands in reactions with metal ions [3,4].

Dithiocarbamte complexes continue to attract attention due to their ability to stabilize transition metals in a variety of oxidation states and a developing interest in this area in the functionalization of the dithiocarbamate substituent [5,6].

Transition metal dithiocarbamate complexes find use in diverse application such as material science, medicine , agriculture [7] and possess interesting structural chemistry [8-10] that make their study continuously attractive and biological activity as antibacterial, cytostatic, antifungal and immunoregulatory most of these application are based on complexation properties of dithiocarbamate ligands with metal ions specially with transition metal ions [11,12].

In view of these interesting results , we have prepared a new dithiocarbamate ligand and its complexes with different metal ions and their nitrogenous adduct.

A number of dithiocarbamate adducts have been reported in literature [13] with various geometries such as square planar [14] octahedral [15].

Journal of Physics: Conference Series **2063** (2021) 012030 doi:10.1088/1742-6596/2063/1/012030

2. Experimental

Materials and Methods

All chemicals used in the present work were of analytical reagent (A.R) grade (B.D.H or fluka) chemical companies. Infrared spectra were recorded on Brucker 27. Alpha, A.T.R-diamond, lazer spectrophotometer in the 400- 4000 cm^{-1}range without using KBr discs. Conductivity measurements were carried out on a 10^{-3}M solution of the complexes in DMF using conductivity meter PCM_3 Jenway at ambient temperature. The electronic spectral were recorded on a Shimadzu Uv-visible spectra photometer uv-1650 pc for 10^{-3}M solution of complexes in DMF as solvent at 28c° using 1 cm quartz cells. Metal content were determined using 3000 atomic absorption. Melting points were recorded on an electro thermal 9300 apparatus. The magnetic measurement was carried out at 28c° on the solids by Sherwood scientific instrument.

Synthesis of sodium 2,4- dinitrophenyl hydrazine dithiocarbamate[6,16-18].

Aqueous solution of (4.0 gm, 0.1 mol) of sodium hydroxide was added to ethanolic solution of (19.8gm, 0.1mol) of 2,4- dinitrophenyl hydrazine with stirring. The mixture was cooled in an ice bath to this mixture (7.60 cm^3, 0.1mol) of carbon disulfide was added drop wise with continues stirring for 30 min in ice bath. The precipitate formed was extracted with (100ml) diethyl ether, filtered off washed with diethyl ether and dried under vacuum a yellow precipitate formed

A: synthesis of complex [M(2,4- dinitrphenhydra-dtc)$_2$] where :

M=Fe (II), Co(II), Ni(II) of $FeSO_4.7H_2O$ (0.278gm, 0.001 mol) or $Co(NO_3)_2$. $6H_2O$ (0.291 gm, 0.001 mol) or $Ni(NO_3)_2.6H_2O$ (
(80:20%) Acetone – water solution0.291gm, 0.001 mol) was added drop wise to ethanolic solution of sodium 2,4- dinitro phenyl hydrazine dithiocarbamate (0.592 gm, 0.002 mol) with stirring for 30 min until complete precipitation. The precipitate was filtered off washed with ethanol then with diethyl ether and dried under vacuum.

B: synthesis of complex [MM`(2,4-dinitrphenhydra-dtc)$_4$)] where :

M=Fe (II), Co(II), Ni(II), M`=Zn(II)

Aqueous solution of $ZnSO_4.7H_2O$ (0.2875 gm, 0.001 mol) was added drop wise with continuous stirring to ethanolic solution of sodium 2,4- dinitro phenyl hydrazine dithiocarbamate (1.184gm,0.004mol) to form Na_2[Zn(2,4- dinitrphenhydra-dtc)$_4$].

80:20% acetone- water solution of $FeSO_4.7H_2O$ (0.278gm, 0.001 mol) or $Co(NO_3)_2.6H_2O$ (0.291 gm , 0.001 mol) or $Ni(NO_3)_2.6H_2O$ (0.291 gm, 0.001 mol)was added drop wise with continuous stirring for 30 min until complete precipitation. The precipitate was filtered off, washed with ethanol then with diethyl ether and dried under vacuum.

C: synthesis of complex [M(1,10- phen)$_3$][M`(2,4-dinitrphenhydra-dtc)$_4$] where :

M=Co(II), Ni(II), M`=Zn(II)

Prepared by adding mixture of $Co(NO_3)_2.6H_2O$ (0.29gm, 0.001 mol) and 1,10 phenanthroline (0.7035gm,0.003 mol) or adding mixture of $Ni(NO_3)_2.6H_2O$ (0.29gm, 0.001 mol) and 1, 10 phenanthroline(0.7035 gm, 0.003 mol) which dissolved in 80:20 % acetone- water added to Na_2[Zn(2,4-dinitrphenhydra-dtc)$_4$] (1.2035 gm, 0.001 mol) which prepared in 1:B with continuous stirring for 30 min the mixture was leaved for 3 days until complete precipitation. The precipitate formed was filtered washed with acetone then dried under vacuum.

D: synthesis of complex [M(1,10- phen)$_3$][M`$_2$(2,4- dinitrphenhydra-dtc)$_6$] Where:

M=Co(II), Ni(II), M`=Zn(II)

Prepared by adding a mixture of $Co(NO_3)_2.6H_2O$ (0.29gm, 0.001 mol) and 1,10 phenanthroline monohydrate (0.7035gm , 0.003 mol) or adding mixture of $Ni(NO_3)_2.6H_2O$ (0.29gm, 0.001 mol) and 1,10 phenanthroline monohydrate (0.7035gm , 0.003 mol) which dissolved in 80:20% acetone-water solution to a mixture of $Na_2[Zn_2$(2,4-dinitrphenhydra-dtc)$_6$] which prepared of dissolving ZnSO4.7H2O(0.5750gm , 0.002mol) and sodium 2,4 –dinitro phenyl hydrazine dithiocarbamate (1.776 gm , 0.006 mol) drop wise with continuous stirring for 30 min, the precipitate formed was filtered, washed with acetone then dried under vacuum.

3. Results and discussion

The new ligand was prepared by the reaction of 2,4- dinitro phenyl hydrazine with sodium hydroxide and then added carbon disulphide. The complexes were prepared through direct reaction of the $FeSO_4.7H_2O$ or $Co(NO_3)_2.6H_2O$ or $Ni(NO_3)_2$6H2O with the above ligand in (1:2) molar ratio, other complexes were prepared through direct reaction of the $ZnSO_4.7H_2O$ with the $FeSO_4.7H_2O$ or $Co(NO_3)_2.6H_2O$ or $Ni(NO_3)_2.6H_2O$ with the above ligand.

The values of conductivity in dimethylformamide solution of the complexes ranged from (0.70→12.5) cm^2 ohm^{-1} mol^{-1} for the complex [M(2,4- dinitrphenhydra-dtc)$_2$] and complex [MM`(2,4-dinitrphenhydra-dtc)$_4$)] which are typical values for nonelectrolyte type [19]. While the values of conductivity of the complexes [M(1,10- phen)$_3$][M`(2,4-dinitrphenhydra-dtc)$_4$] [M(1,10- phen)$_3$][M`$_2$(2,4- dinitrphenhydra-dtc)$_6$] ranges from (74.4→84.3) cm^2. ohm^{-1}. mol^{-1} which are typical for electrolyte type as shown in table (1) [20].

3.1. Magnetic Susceptibility Measurements

The magnetic moments of the complexes were measured at 28 C°. The magnetic moments for Fe(II) , Co(II) and Ni(II) complexes (1→6) were (1.83→5.04) B.M. suggesting a square planner geometry as shown in table(1) [20].

The low value of magnetic moment for complex (5) is due to antiferromagnetic interaction. The magnetic moments of other complexes (7→10) were in the range (2.69→4.62) B.M in an octahedral geometry as shown in table (1) [21].

3.2. Infrared Spectral Studies

The important IR bands of the ligand and its complexes are listed in table (2). The stretching frequency of the ν (C---N) band for the dithiocarbamate was intermediate between the stretching frequencies associated with typical single and double-bonded carbon and nitrogen atoms [21]. The ν (C----N) and ν (C---S) of complex type [M(2,4-dinitrphenhydra-dtc)$_2$] and [MM`(2,4-dinitrphenhydra-dtc)$_4$)] were observed in the range of (1513-1603)$cm^{-1}$ and (1046-1077)$cm^{-1}$. The presence of only one band in the later region reports the bidentate coordination of the di thio ligand [22-25]. IR spectra of the complex type [M(2,4-dinitrphenhydra-dtc)$_2$] and [MM`(2,4-dinitrphenhydra-dtc)$_4$)] showed a new band at (407- 471) cm^{-1} which are evidence for the coordination of metal to sulfur ν (M-S) this behavior may be attributed to electron releasing of the amines which forces high electron density towards the sulfur atoms as shown in table (2).

Also the ν (M`- S) was observed in the range (401-414) $cm^{-1}$ for complex type [MM`(2,4-dinitrphenhydra-dtc)$_4$)]. The IR spectra of complex type [M(1,10 phen)$_3$][M`(2,4-dinitrphenhydra-dtc)$_4$] and [M(1,10- phen)$_3$][M`$_2$(2,4- dinitrphenhydra-dtc)$_6$] showed one band at (415-470) cm^{-1} which are evidence for the coordination of metal to sulfur ν (M`-S). in addition to the ν (M-N) was observed in the range(486-649) region [26] in the adduct as shown in table (2).

Journal of Physics: Conference Series **2063** (2021) 012030 doi:10.1088/1742-6596/2063/1/012030

Table 1. Analytical and some physical properties of the prepared complexes.

No	Compound	Color	M.P. C°	Molar conductivity	%M	% yield	μeff B.M.
	C7H5N4O4S2Na	light yellow	276	_	_	83.4	_
1.	[Fe(2,4- Dinitrphenhydra-dtc)$_2$]	Light Yellow	310	11.76	8.363 (9.27)	91.66	5.04
2.	[Co(2,4- Dinitrphenhydra-dtc)$_2$]	Light Brown	290	8.70	8.96 (9.74)	90	4.16
3.	[Ni(2,4- Dinitrphenhydra-dtc)$_2$]	Dark Brown	300	0.70	9.35 (9.73)	93.46	Diam
4.	[FeZn(2,4- Dinitrphenhydra-dtc)$_4$]	Light Yellow	254	9.38	4.24 (4.60)	61.85	3.51
5.	[CoZn(2,4- Dinitrphenhydra-dtc)$_4$]	Light Brown	256	12.5	4.46 (4.84)	52.67	1.83
6.	[NiZn(2,4- Dinitrphenhydra-dtc)$_4$]	Dark Yellow	285	10.6	4.22 (4.84)	53.08	Diam
7.	[Co(1,10- phen)$_3$]$^{+2}$[Zn(2,4-Dinitrphenhydra-dtc)$_4$]$^{-2}$	Dark Yellow	248	74.4	2.76 (3.07)	60.94	4.60
8.	[Ni(1,10- phen)$_3$]$^{+2}$[Zn(2,4-Dinitrphenhydra-dtc)$_4$]$^{-2}$	Light Yellow	260	78.8	2.64 (3.07)	5730	3.41
9.	[Co(1,10- phen)$_3$]$^{+2}$[Zn$_2$(2,4-Dinitrphenhydradtc)$_6$]$^{-2}$	Brown	190	82.7	2.04 (2.32)	70.27	2.69
10.	[Ni(1,10- phen)$_3$]$^{+2}$[Zn$_2$(2,4-Dinitrphenhydra-dtc)$_6$]$^{-2}$	Dark Brown	200	84.3	2.08 (2.32)	60.47	3.17

3.3. Electronic Spectral Studies

The UV-visible spectra of the ligand and its complexes of 10^{-3} M solution in DMF were recorded. The results were listed in table(2). The UV-visible spectrum of the Fe(II) complex (1) gives absorption bands at (21692)cm^{-1}, (32258)cm^{-1} regions. The UV- visible of Fe(II) complex(4)gives absorption bands at (21978)cm^{-1} and (33003)cm^{-1} regions which corresponds to (d-d) transition in square planner geometry as shown in table (2) [5, 27,28].

The Co(II) complex (2) exhibited an absorption band at (21505 cm^{-1}) regions which was assigned ($^2A_1g \rightarrow {}^2Eg$) transition and the Co (II) complex(5) exhibited an absorption band at (22371 cm^{-1}) regions which was assigned to ($^2A_1g \rightarrow {}^2Eg$) transition [29]. The Ni(II) complex (3) exhibited an absorption band at (27855 cm^{-1}) which was assigned to ($^1A_1g \rightarrow {}^1B_1g$) transition in square plannar geometry.

Journal of Physics: Conference Series **2063** (2021) 012030 doi:10.1088/1742-6596/2063/1/012030

The Ni(II)complex (6) show two absorption bands at (22124 cm^{-1}) and (32753 cm^{-1}) which were assigned to (^1A_1g→ ^1B_2g) and (^1A_1g→ ^1B_1g) transitions in square plannar geometry [30]. The complex (7) show an absorption band at (31153 cm^{-1}) region and the complex(9) show an absorption band at (23041 cm^{-1}) region which were assigned to ($^4T_1g(F)$→ ^4T_1g (P)) transition in an octahedral geometry of high spin [31-33].

The complex(8) show an absorption band at (34014 cm^{-1})region which was assigned to ($^3A_2g(F)$→ ^3T_1g (P)) transition and the complex(10) show two absorption bands at (22573 cm^{-1}) and (32501 cm^{-1}) region which were assigned to ($^3A_2g(F)$→ ^3T_1g (F)) and ($^3A_2g(F)$→ ^3T_1g (P)) transition in octahedral configuration as shown in table (2) [34].

Table 2. IR bands (cm^{-1}) and electronic data of ligand and prepared complexes

No	Compound	U.V band (lcm^{-1})	IR Spectra (cm^{-1})				
			ʋ(C-N)	ʋ (C-S)	ʋ (M-S)	ʋ (M'-S)	ʋ (M-N)
	C7H5N4O4S2Na	21.000 32.000	1512	1093	-	-	-
1.	[Fe(2,4- Dinitrphenhydra-dtc)$_2$]	21.692 32259	1516	1046	407	-	-
2.	[Co(2,4- Dinitrphenhyra-dtc)$_2$]	21505	1513	1051	411	-	-
3.	[Ni(2,4- Dinitrphenhydra-dtc)$_2$]	27855	1516	1077	403	-	-
4.	[FeZn(2,4- Dinitrphenhydra-dtc)$_4$]	21978 33003	1603	1043	460	401	-
5.	[CoZn(2,4- Dinitrphenhydra-dtc)$_4$]	22371 31153	1600	1059	462	413	-
6.	[NiZn(2,4- Dinitrphenhydra-dtc)$_4$]	22124 32573	1513	1058	471	414	-
7.	[Co(1,10- phen)$_3$]$^{+2}$[Zn(2,4-Dinitrphenhydra-dtc)$_4$]$^{-2}$	31153	1465	1116	-	470	510
8.	[Ni(1,10- phen)$_3$]$^{+2}$[Zn(2,4-Dinitrphenhydra-dtc)$_4$]$^{-2}$	34014	1453	1114	-	415	649
9.	[Co(1,10- phen)$_3$]$^{+2}$[Zn$_2$(2,4-Dinitrphenhydra-dtc)$_6$]$^{-2}$	23041	1455	1115	-	419	486
10.	[Ni(1,10- phen)$_3$]$^{+2}$[Zn$_2$(2,4-Dinitrphenhydra-dtc)$_6$]$^{-2}$	22573 32051	1457	1114	-	430	505

Journal of Physics: Conference Series **2063** (2021) 012030 doi:10.1088/1742-6596/2063/1/012030

3.4. The ^{1}H-NMR, Spectra for the Ligand L

^{1}H-NMR study of the ligand = 2,4- dinitrophenyl hydrazine dithiocarbmate The first peak was strong and observed in the region 2.45 because it was far from each of electron withdrawing groups which was agree with appeared theoretical measurements for the compound, in the in region 2.0 , (2) while the second peak has been showed in the region 4-2 in the N.M.R measurements, confirmed by the theoretical measurement of compound in the position 4.0 while the broad peak appeared around 9.1 -7.5 belonging to benzene cycle confirmed by the theoretical measurements and the interpretation of photons which connected with benzene cycle appeared as follow as. H5 appeared in the position, 8.97 because it is near from two withdrawing groups and it is confirmed by the theoretical measurement. H4 the other peak has been appeared in the region 8.2 because it is adjacent to one group of NO_2 and It is agree with the theoretical measurements 8.57.

The other peak in the region 7.5-7 belonging to the H3 and it is confirmed by approximately the theoretical and agree with and confirmed by published researches as shown in figure (1) [35-39].

4. Conclusion

In this work, the synthesis and characterization ligand and their complexes.. Physicochemical and spectroscopic methods have been put in place to verify bonding mode and over all complex structure. From These measurement results lead to the preparation of ten coordinate complexes and from above discussions the structure of complexes has been suggested which is complexes (1→6) a square plannar geometry and other complexes (7→10) were an octahedral geometry as shown in figure (2-5) .

Journal of Physics: Conference Series **2063** (2021) 012030 doi:10.1088/1742-6596/2063/1/012030

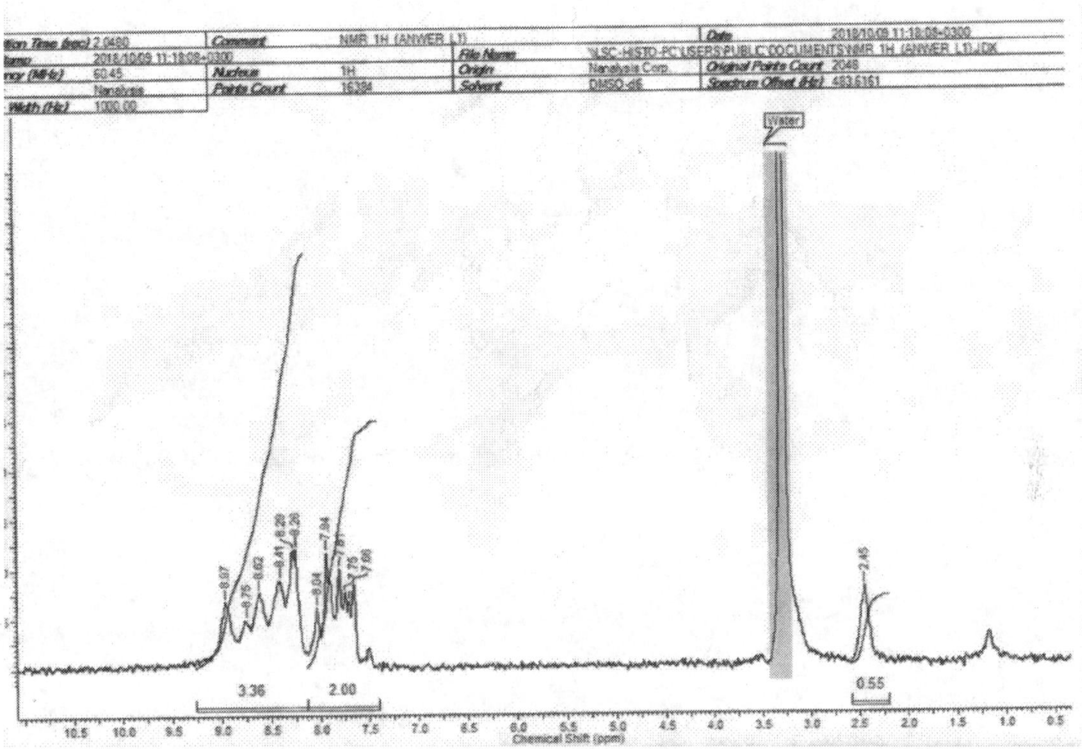

(a)

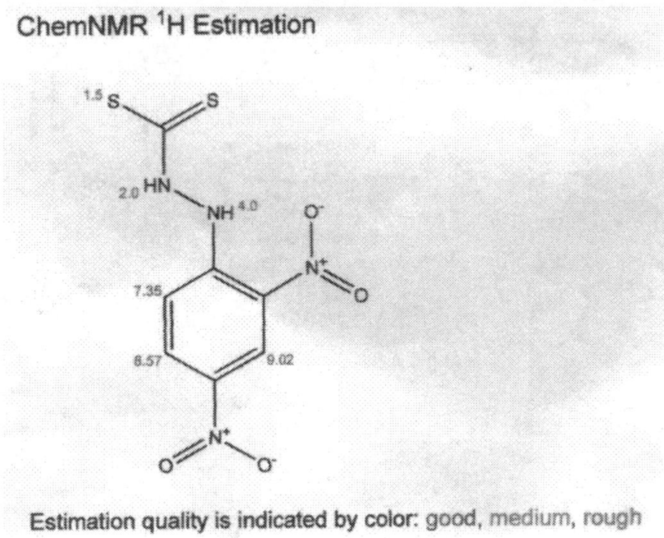

(b)

Journal of Physics: Conference Series **2063** (2021) 012030 doi:10.1088/1742-6596/2063/1/012030

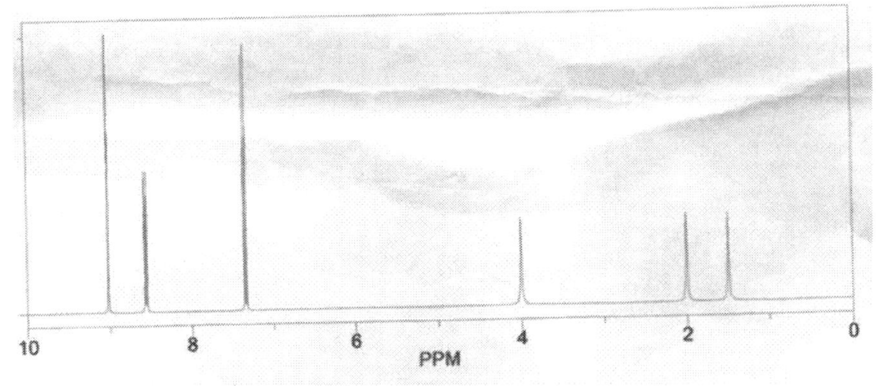

(c)

Figure 1(a,b,c). ^{1}H- NMR of the ligand Sodium -2,4- dinitrophenylhydrazine dithiocarbamate

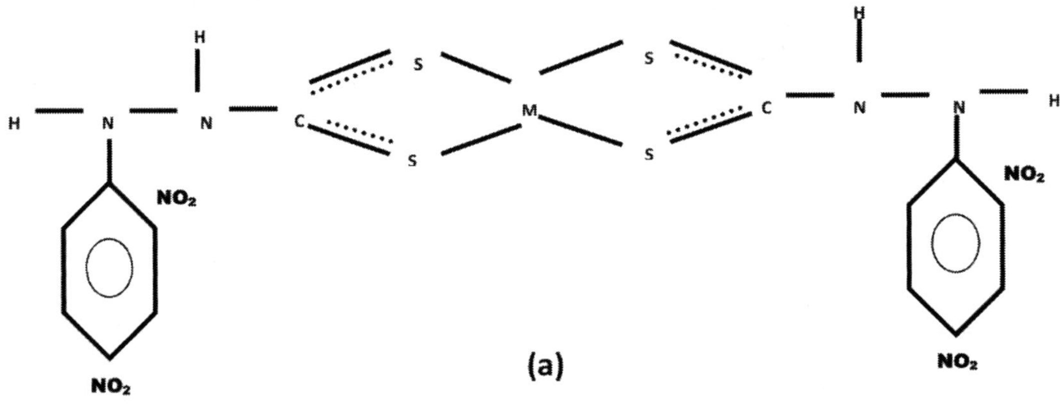

(a)

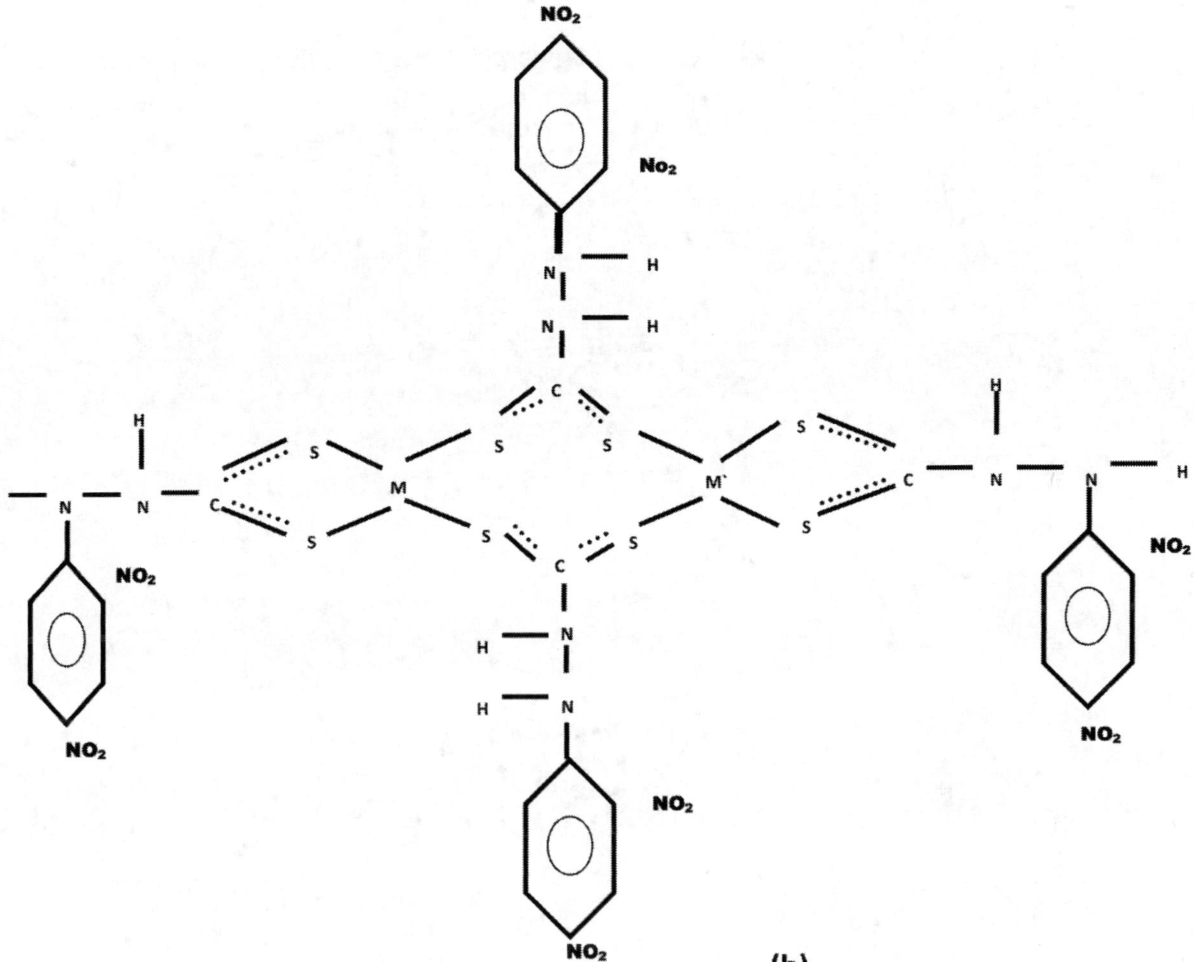

(b)

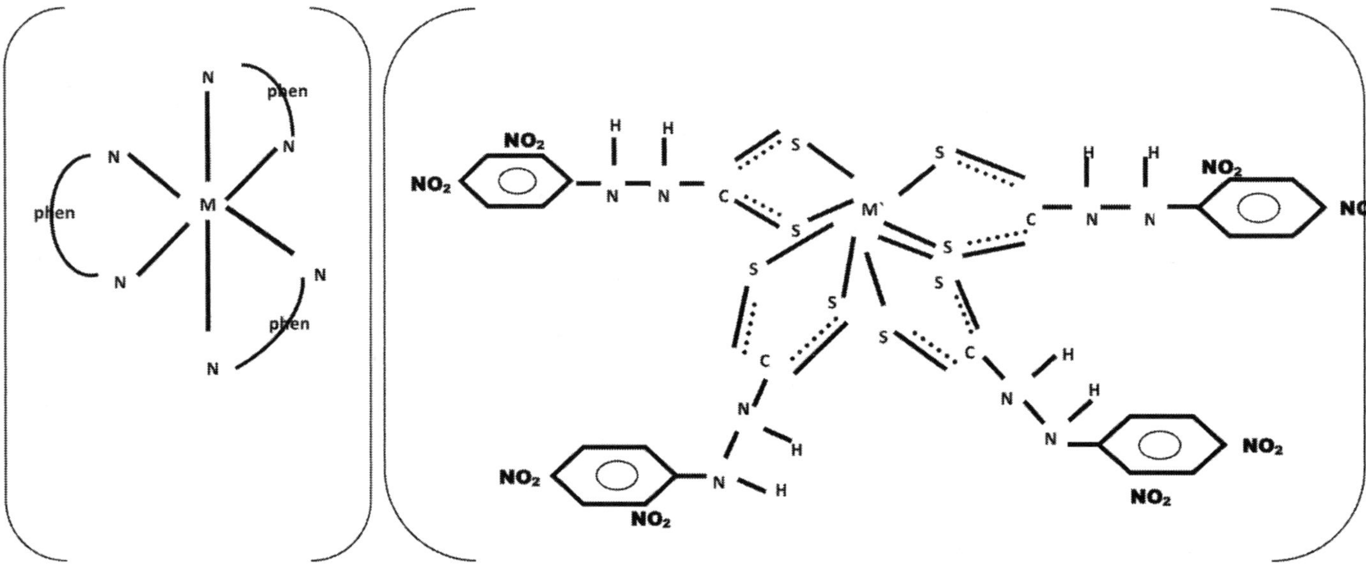

(c)

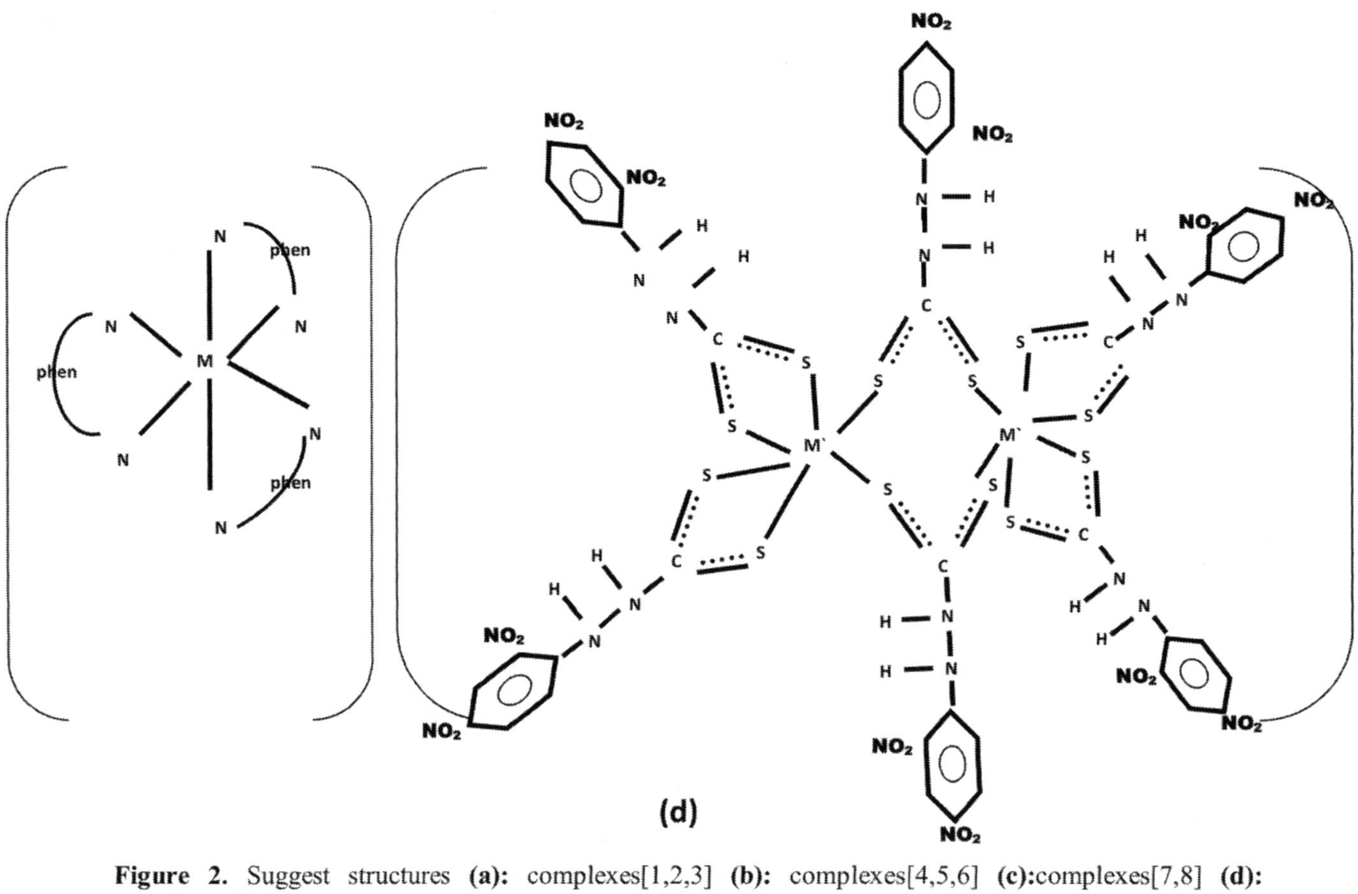

(d)

Figure 2. Suggest structures **(a):** complexes[1,2,3] **(b):** complexes[4,5,6] **(c):**complexes[7,8] **(d):** complexes [9,10].

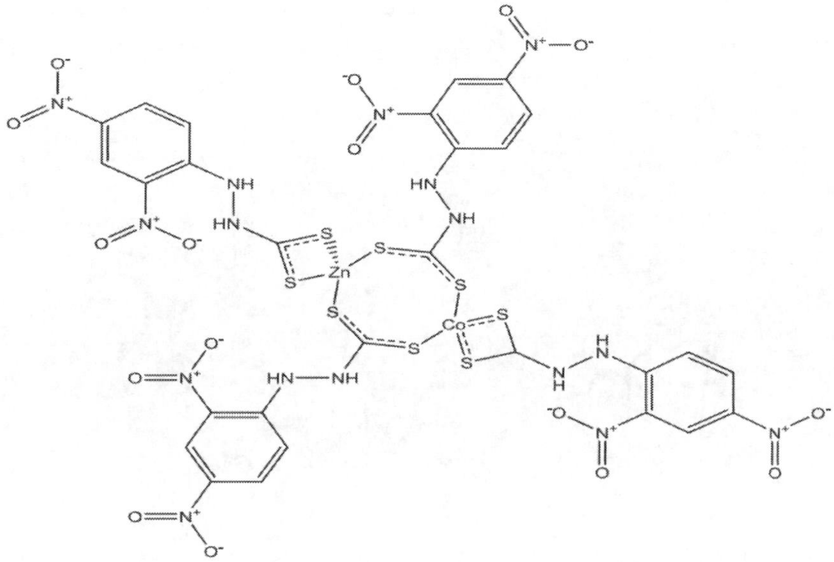

Figure 3. Suggest structure of complex (5)

Figure 4. Suggest structure of complex (5) by molecular mechanic

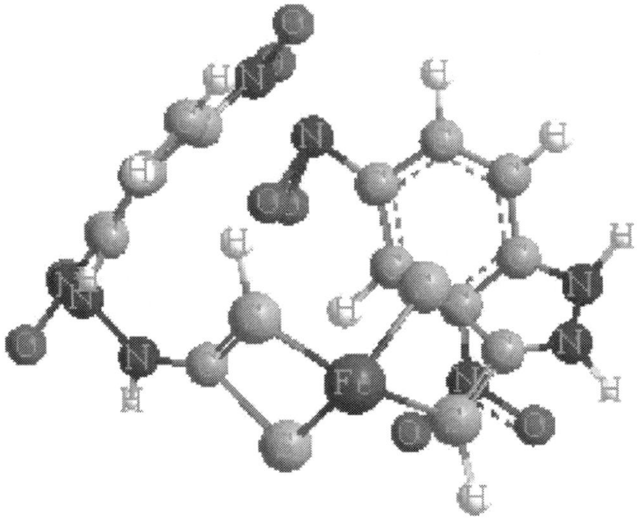

Figure 5. Suggest structure complex (1) by molecular mechanic

References

[1] Azim Z H, Katayoun M, Akram A and Vahid A 2011 *inorganical chemical Acta* 373 282-285.
[2] Ghufran Sh A, Kaiss R I and Mohammad F M 2020 *International Journal of Pharmaceutical Research* 12 1 1025-1032.
[3] Balakrishnan S, Duraisamy, S, Kasi M ,Kandasamy S, Sarkar R. and Kumarasamy A 2019 *Heliyon* 5 5 1687.
[4] Leka Z B, Leovac V M, Luk S, Sabo T J, Trifunovi S R and Szecsenyi K M 2006 *J.Therm Analy.calor* 83 3 687-691.
[5] Graeme H, Ebony J C,.Rainford R C R, and Brent l R 2009 *Inorganica chemical Acta* 362 1361-1364.
[6] Hogarth G , Richards E J C , Rainford B C R, kabir S E, Richards I , J.D.E.T , Wilton E and Zhang Q 2009 *Inorg chim Acta* 362 2020 – 2026.
[7] Shaheen F, Badshah A, Gielen M, Dusek M, Fejfarova K, Devos D and Mirza B 2007 *J.organomet. chem.* 6929 3019-3026.
[8] Onwundiwe D C and Ajibude P A 2011 *J.Chem crstallogr* 41 980-985.
[9] Al-Mukhtar S E and Aghwan M T 2013 *Rafidain journal of science* 24 4E 50-59.
[10] Hasan H A, Yousif E I , Al-Jeboori M J and Global E I Y 2018 *Journal of Global Pharma Technology* 10 03 875-882.
[11] Golcu A 2006 *Trans.Metal.Chemi* 31 405-412.
[12] Nami S A , Ullah I , Alam M , Lee D-U and Sarikavakli N 2016 *J. Of Photochemistry and Photobiology B: Biology* 160 392-399.
[13] Ekennia A C 2013 *Journal of Applied chemistry* Vol. 5 no. 2 PP . 36 – 39 ,.
[14] Onwaudiwe D C and Ajibade P A 2011 *International Journal of molecular sciences* Vol . 12 no 3 pp. 1964 – 1978.

[15] Kolawole G A and Osowole A A 2009 *Journal of coordination chemistry* Vol. 62 no 9 pp. 1437 – 1448.
[16] Bala R P and Venkatesh P 2015 *IJSR* 4 (2) 919-923.
[17] Ingle A D , Devghare H and Parase K 2013 *J. Chem. Pharm. Res.*, 5(7) 272-277.
[18] Kumar V, Singh V, Gupta A N , Manar K K , Drew M G and Singh N 2014 *J.Cryst. Eng. Comm.* 16 29 6765-6774.
[19] Geary W J 1971, The use if conductivity measurements in organic solvents for characterization of coordination compounds, *Coord. Chem. Rev* 7 8-122.
[20] Preti C and Tosi G 2007 J.lnorg.Nucl.Chem. 36 3725-3729.
[21] Srinivasan N, Valarmathi P, Thiramant S and Ciattin S 2010 *Trans.Met. Chem* 35 815-819.
[22] Serrano J L, Garcia L, Perez E, Sanchez G, Garcia J, Lopez G, Garcle G and Molins E 2003 *lnorg. Chim.Acta.* 355, 33-40.
[23] Bonati F and Ugo R 1967 *Journal of organomtallic chemistry* Vol . 10 no 2 pp 257- 268 . .
[24] Sebastian S , Sylvestre S, Sundaraganesan B , Amalanathan M , Ayyapan S, Oudayakuma K and Karthikeyan B 2013 *J.Spectrochimica Acta Part A: Mol. Bio. Spec.,* 107 167-178.
[25] Tekalea S U, Jadhava V B , Mundea S B and Pawar R B 2015 *J. Der Chemica Sinica,* 6 1 38-41.
[26] Raya I, Baba and Yamin B M 2006 *Malaysia.J.Analy. Sci,* 10 1 93-98.
[27] Abd el.wahed M G , Refat M S and El-Megharbel S M 2008 *Journal of molecular structure* Vol.888 no 1-3 pp. 416 – 429.
[28] Nicholls D 1973 The chemistry of iron. Cobalt and Nickel pergamon press Oxford, 1st Ed, pp,1037,1087,1088,1090,1091,1093, 1151, 1154.
[29] Bailar J C , Emeleus H J, Nyholm S R and Trotman- Dickenson A F 1973 Comprehensive inorganic chemistry 1st Ed, pergamon press, Oxford, 1151.
[30] Pandeya K B, T.S, waraich R Gaur C and Singh R P 1981 *J.lnorg Nucl, chem* 43 12 3159-3162.
[31] Martell A E 1971 Coordination chemistry van Nostrand Reinhold, New York, p.1.
[32] Singh N, Singh N K and Kaw C 1989 *Bull. Chem. Soc (JPN)* 62 3328-3333.
[33] Khan H, Badshah A , Rehman Z , Said M , Murtuza G , Shah A , Ahmed S , Butler I S and Fontaine F G 2012 *J. Polyhedron* 39 1 1-8.
[34] Al-Shemary R K , Al-Khazraji A M and Lateef A F S S M 2016 *Inter. J. Sci.and Res.* 5 5 1787-1793.
[35] Larkworthy L F, Murphy J M and Phillips D J 1968 *lonrg.chem* 7 1436-1443.
[36] Al- obaydee SZ 2011 Synthesis of some new Benzonaphthyridine derivativaties M.Sc. Thesis in organic Chemistry ,University of mousl , iraq
[37] Hassan A F 2015 *J. Chem. Mat.Res.* 7 4 p.9-15.
[38] Rivero I A , Espinoza K and Somanathan R 2004 *Molecules* 9 7 609-616.
[39] Tamizmani M , Kankanala R and Sivasankar C 2014 *Journal of Organometallic Chemistry* 763764 6-13.

Combined CFA-AFM Analytical approach for precipitation reaction regarding crystal growth building single and multiple monolayers based on surface area calculation with image surface roughness analysis (unseen surface)

Prof. Dr.Issam Mohammad Ali Shakir [1, 4], Prof. Dr. Nagham Shakir [2, 5]. Turkie and Dr. Hussein Fares Abd Al-razack [3, 6 *]
1, 2, 3 Department of chemistry, College of science, University of Baghda
4, 5, 6 Iraq, Baghdad, new Baghdad

*Husseinfares2003@gmail.com

Abstract. A reaction of Mebeverine hydrochloride (0.03mM) in pure form with sodium nitroprusside (0.07mM) to form OFF White precipitate. A constant feed was used to collect an enough amount in weight for AFM study . Continuous flow injection analysis was conducted as it is the aim of this study to combine FIA with AFM. To elucidate the study of surface morphology. Various parameters of AFM image surface roughness analysis were discussed in relation to the kind of precipitate formed. Skewness, kurtosis, peak-peak, ten peak height, fractal dimension, wavelength, core roughness depth, and reduced valley depth. All these with the four parameter mainly amplitude, hybrid, functional and spetial. Since no previous study of such was conducted; all usual mode of imaging was dealt with i.e., contact, non-contact and intermittent contact was defined . A non contact mode was used in this study. A detailed study of how the crystal growth buildup first mono layer (hypothetical based on the date obtained also how much each grain carry a concentration with the number that are on the first mono layer. Number of given samples of surface area calculation plus a demonstration of the hypothetically formed multi mono layers specially at high reactant concentration. The main aim of this project was the binding of AFM with FIA which is regarded as an new approach which might be a very useful knowledge for other researcher.

1. Introduction :

1.1. Introductory of AFM :

The imaging (Fig .no.1) , measuring and manipulating surfaces at the atomic scale is the aim of AFM and this study ,the main idea in the principle involved in the tip - sample interaction affect how the probe inter acts with the sample. If the probe experiences repulsive force the probe will be in contact mode otherwise as the probe moves further away from the surface dominate and the probe will be in non-contact mode.The primary imaging mode in AFM are available (Fig .no.2- 5) :

1- Contact mode when the probe surface separation is less than 0.5nm
2- Intermittent contact that occurs in the range of 0.5 and 2nm
3- Non-contact mode when the probe surface separation ranges from 0.1 to 10nm

The use of two techniques that never been used in the combined modes in explaining the topography of the precipitate formed that is measured as a turbid cloud causing either absorption or divergence or attenuation of incident photons released from white snow leds in a long distance and the signal is accumulated to cause a remarkable response or signal (any thing that serve to indicate assuming and trying in the design to avoid the noise (an electrical disturbances). The resultant peak (i.e., mountain with a pointed

Journal of Physics: Conference Series **2063** (2021) 012031 doi:10.1088/1742-6596/2063/1/012031

summit) within a given range i.e., the extent to which or the limit between which variable is possible. The instrument in use i.e., FIA , it has many uses for determination different species [1-8] .

The combination of this technique with Atomic Force microscopy (AFM) is a new trend in achieving a new approach for understanding the kind of formed precipitate and its properties. As for the author and the cited research article no preview record is available for such studies. Therefore the title of this chapter was chosen on this basis(ISDS-AFM) (Irradiation of Solo – Dual – System – AFM) [9] .

The imaging , measuring and manipulating surfaces at the atomic scale is the aim of AFM and this study the main idea in the principle involved in the tip-sample interaction affect how the probe inter acts with the sample. If the probe experiences repulsive force the probe will be in contact made otherwise as , the probe moves further away from the surface, attractive forces dominate and the probe will be in non-contact mode. Three primary imaging in AFM are available

1- Contact mode where the probe-surface separation is less than 0.5nm
1. Intermittent contact that occurs in the range of 0.5 and 2nm
2. Non-contact made where the probe be surface separation ranges from 0.1 to 10nm

Details of this study: the size of spotted area of imaging & the author will discuss four main item :

a) Amplitude parameter
b) Hybrid parameter
c) Functional parameter
d) Spatial parameter

Taking into account that number of particles, size represented by diameter for range of particulate assuming a sphere counting, skewness, kurtosis, roughness, ...etc.

While for continuous flow analysis (Fig. 6)where a known simple segment is injected via an injection valve through loading known precise volume measured by a sample loop and known volume through the available six port ; their function is to facilitate the repeatable precise measurements of volumes port lead 1-4 joined together loop sample (pre chosen volume of most probably the analyte) 2 inlet port , 3 outlet port , 5- excess (more than required), 6- sample injection port.

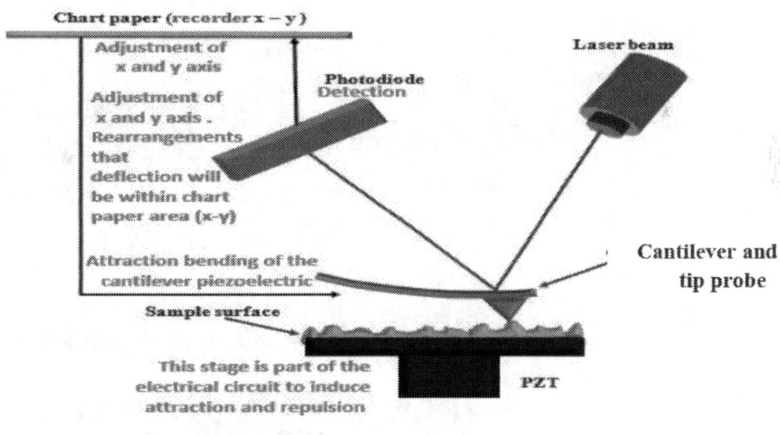

Figure 1: main item of Atomic Force Microscopy (in general)

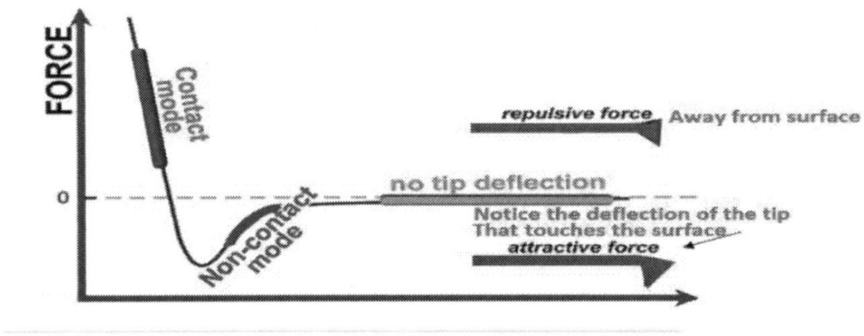

Figure 2: Tip-sample separation

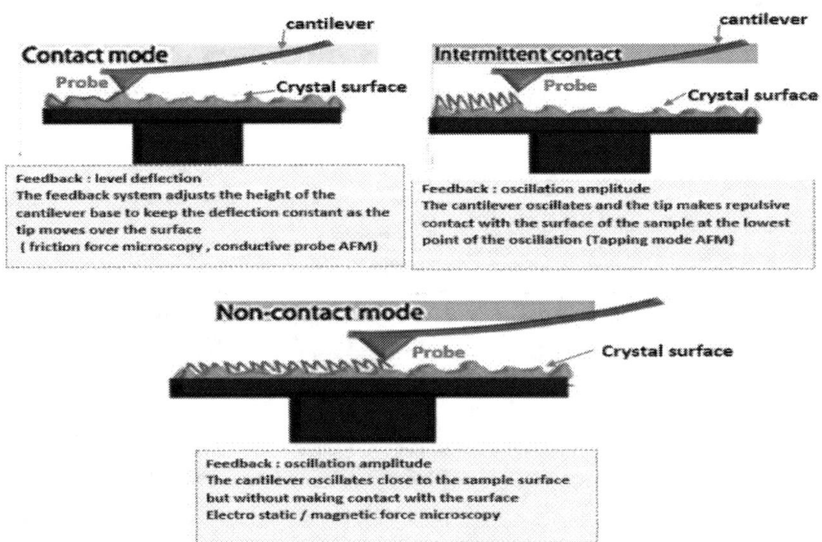

Figure 3: imaging mode in Atomic Force Microscopy

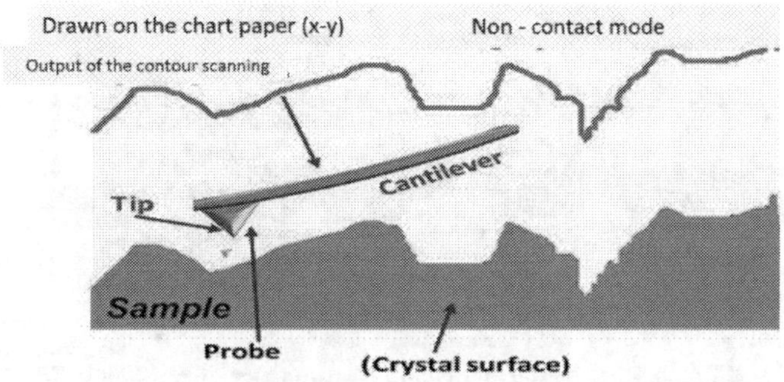

Figure 4: non- contact imaging profile in atomic force Microscopy

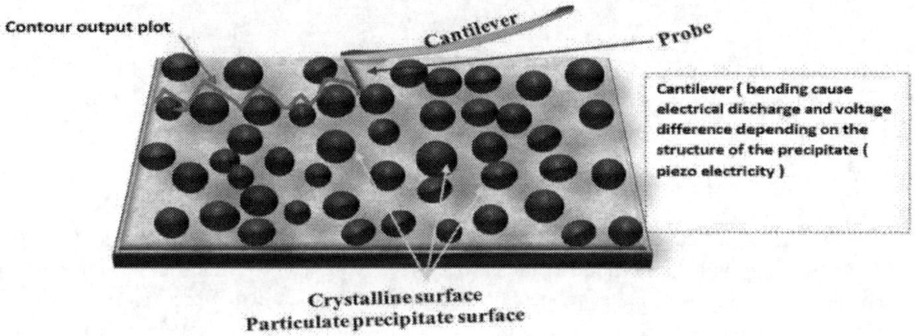

Figure 5: plot of scanned surface contour of the precipitate formed by the reaction of Mebeverine hydrochloride with sodium nitroprusside

2.Experimental

2.1 Chemicals

All chemicals were used of analytical-reagent grade while distilled water was used to prepare the solutions. A stock solution of **Mebeverine** HCl, 10 mmol.L^{-1} , $C_{25}H_{36}ClNO_5$) was prepared by dissolving ٠.٤٦٦ g in ١00 ml. A stock solution (50 mmol.L^{-1}) of **Sodium nitroprusside (SNP)**, $C_5FeN_6Na_2O$, (261.918 g/mol, BDH) was prepared by dissolving 3.2739 g in 250 ml D.W .

2.2Apparatu

The flow system consist of four parts as shown in figure 6 - Peristaltic pump – 4 channels (Switzerland) an Ismatic type ISM796 , A rotary 6- port injection valve (Teflon- chem inert) ,(IDEX corporation, USA , Electronic measuring [9] & readout system

2.3Methodology

Two lines manifold system (Fig. 6) was used .The sample is injected on a carrier stream line. The solution of which is propelled by the movement of a peristaltic pump of known flow rate (ml/min) to a manifold system.Fig.no.1 show a simplified diagram used in this study. Until the reaction product reaches the measurement of the formed reaction product where a signal of incident irradiation source is weakened. The obtained response is recorded via X- t potentiometric recorder or any available readout system . Various different designs were patented [10-15]. In this study many variable was dealt with leading to a fine conclusion and all result were discussed in a combined aspect. Sampling : collection of precipitate formed during the resulted reaction product from reactant Mebeverine hydrochloride with sodium nitroprusside (Fig.6)

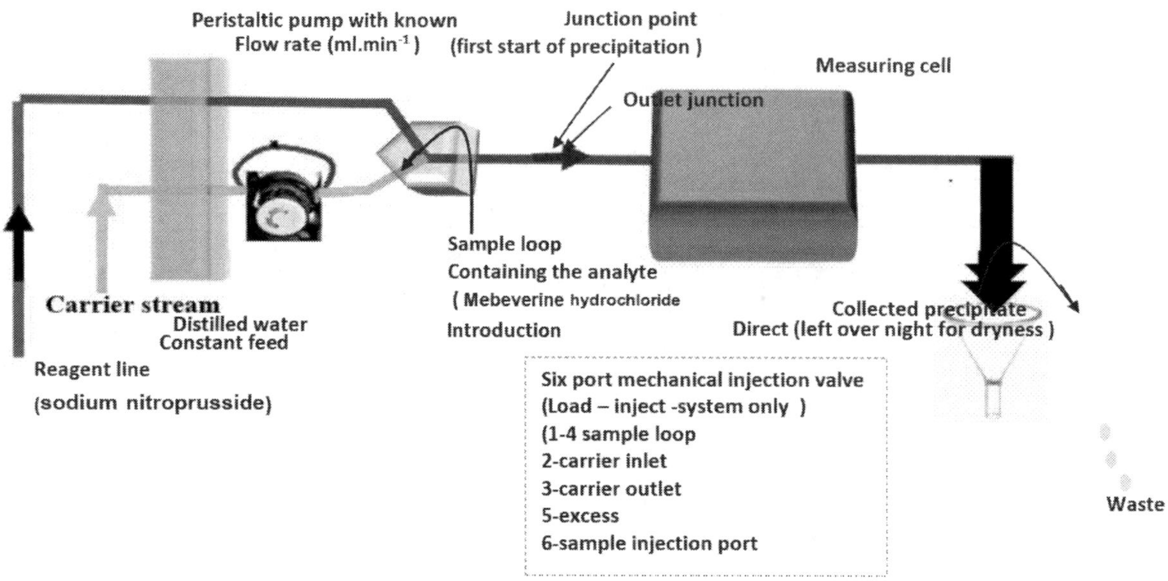

Figure 6: Flowgram of the manifold used in the determination of Mebeverine hydrochloride

The sampling is carried out in two ways (Fig.6) which should be followed :

First :Using an arbitrary concentration of reactants (mentioned above) until an enough amount of precipitate is formed on the filter paper placed in the funnel for separation of precipitate from solutions (carrier stream and reagent stream) .This collection will give you a constant feed of homogeneous precipitate .

Second :Using the same set up but the precipitate is collected during scatter ploted calibration graph build up .This can be named random samples because it will be collect low (at least repeated for n=3 measurement) as well as medium and high concentration . Therefore, it contains variable structural formation but at the end , it contain the same precipitate (what is needed is only few milligrams which quite enough for the study .

In both cases of sampling washing of formed precipitate on the funnel is washed from extra reagent and other mother liquid chemical used . The filter paper is left overnight covered gently to prevent any dust, when it is dried. The sample is ready for atomic force microscopy contour scan. The reactant of the intended reaction were not analysed by AFM because they of different population and can not be compared with the precipitate .as fineness (i.e., grain size will be completely different that the precipitate collected due to the policy of the manufacturing companies .Fig .no.7 shows the tip-sample interaction affect for scanned surface of the precipitate formed.

Journal of Physics: Conference Series **2063** (2021) 012031 doi:10.1088/1742-6596/2063/1/012031

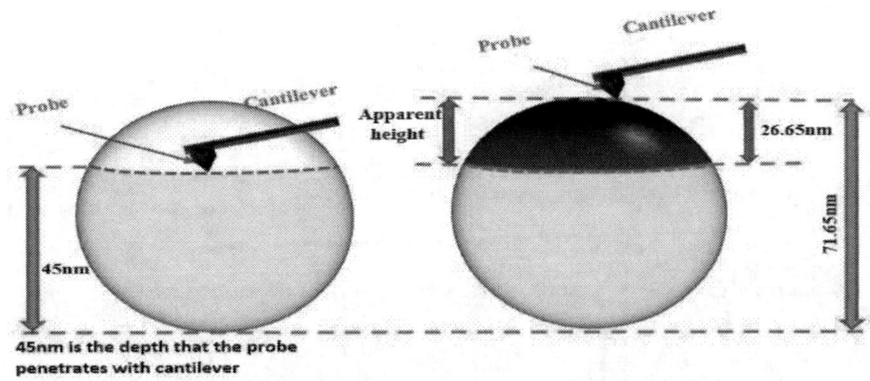

Figure 7: the difference between the probe scan (that gives depth) with the apparent height that the granule posses

3.Results & discussion :

This discussion addresses the principle feature of AFM in both surface morphology and surface growth characterization . Table 1 and Fig no. 8,9 shows : Grain No. : 382 & Avg.Diameter : 71.65. . Table 2 sum up the obtained data of diameter versus volume and Fig no. 10,11 shows the granularity cumulation distribution .

Discussing the image surface roughness analysis , it can be seen that Ssk (surface skewness) =0.00019 which mean that little or non existing of skewness i.e., the crystal growth is proceeding as a direct straight rise no bending which all depend on the nature of precipitate that is formed from homogeneous solution with homogeneous circular appearance on the image; it can be seen that it is circular no elongation (caused by bending or diverged from straight up rise crystal ; and since skewness is the degree of distortion from symmetrical bell curve or the normal distribution . It measure the lack of symmetry in data distribution. A symmetrical distribution will have a skewness of 0.0 which agree completely with Gaussian distribution , smoothed date with the chosen equation of r=0.9198 (**Fig . no. 9**) with percentage capital R^2 =84.61 i.e., the r value shows that the model chosen was a good choice which reflect the kind of granules that is of symmetrical shape while Fig no. 12 shows the image surface roughness analysis . The value of Sku (surface Curtosis) of 1.8 (Table 3) which approaches :

Leptokurtic (approach the value of 2 . As there are three kinds of kurtosis mainly (+) leptokurtic , (0) mesokutric (normal) and (-) platykutric. Also, it shows that there is little or no outlier crystal such as occluded or adsorbed within the structure of granules. Indication of great symmetry is quite evident. Kurtosis values can take -3 up to +3.

It also indicates of high purity of formed crystals. Even flow injection analysis is an on-line automated measurement method of analyte in this case a drug (**Mebeverine hydrochloride**). High purity is a necessity in drug analysis (to know what is measured and determined). Since the precipitated granules are mostly homogeneous which indicate the success of flow injection analysis in conducting the precipitation in high standard of analysis condition i.e., avoiding of interferences where ever they comes from as interfering material. The Sy(peak-peak) and Sx (ten point height) have the same value which also mean a symmetrical distribution between inter point distance at the x-axis which has the same y-axis height peak-peak also indicate the wavelength here it is 47.7nm which is the spacing between local peaks and valley with the consideration of their relative amplitude and individual frequencies

($\lambda_{\lambda} = 2\pi R_{\lambda}/\Delta_{\lambda}$) ($\lambda_{\lambda}$: varies according to the crystal type structure) , also ten point height gave 47.7nm for λ_{λ} which meant 5 repeatation of peak-valley that ensure regular formation of crystals in the growth process.

Core fluid retention index (1.49) while valley fluid retention index of 0.0692 indicate that the speed of precipitation was at high level that is retend during crystal growth buildup that enclose water molecules

The surface area of 47 agree quite well with the x-axis inter point distance of 47.7.

Journal of Physics: Conference Series **2063** (2021) 012031 doi:10.1088/1742-6596/2063/1/012031

Within spatial parameters high density noticed due to the close peaks. Fractional dimension of 2.59 is quite normal (which is a measure of how complicated a self-similar figure is in general it measures how many points lie in a given set .Also it captures the notion of how large a set is. If for example Fractial dimension = $\log 3^N / \log 2^N = N\log 3 / N\log 2 = 0.47712 / 0.30103 = 1.585$

Table no. 1 : Granularity cumulation distribution shows :
Grain No. : 382
Avg.Diameter : 71.65 , <= 10% Diameter : 40.00nm , <= 50% Diameter : 70.00nm , <= 90% Diameter : 100.00nm ,

Diameter (nm)<	Volume(%)	Cumulation(%)	Diameter (nm)<	Volume(%)	Cumulation(%)	Diameter (nm)<	Volume(%)	Cumulation(%)
35.00	2.62	2.62	65.00	9.69	40.05	95.00	5.76	84.55
40.00	5.76	8.38	70.00	7.07	47.12	100.00	4.97	89.53
45.00	4.97	13.35	75.00	7.85	54.97	105.00	4.71	94.24
50.00	5.50	18.85	80.00	7.59	62.57	110.00	3.14	97.38
55.00	5.24	24.08	85.00	7.59	70.16	115.00	1.83	99.21
60.00	6.28	30.37	90.00	8.64	78.80	120.00	0.79	100.00

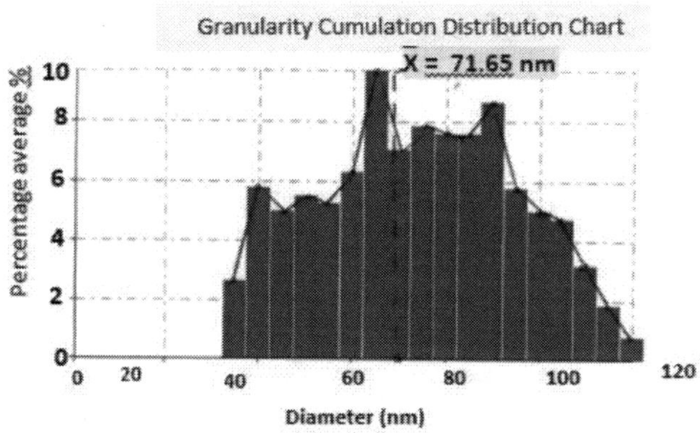

Figure 8: Granularity Cumulating Distribution Data report showing a Gaussian distribution

The chosen equation for Gaussian distribution will represent 84.61% of the all data used shown in table 1 and Figure 9 with high correlation coefficient of 0.9198 , limit of quantitation = 91.98% , i.e., quite useful in quantitative in research work already conducted

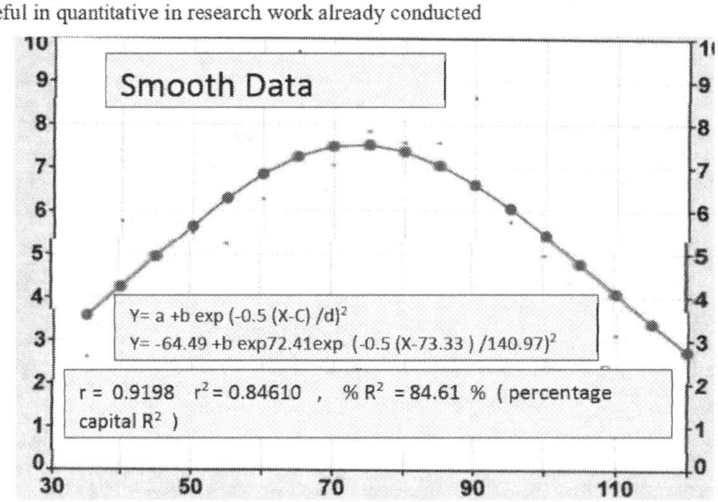

Figure 9: Shows : where X = granules diameter & y= % availability, r = 0.9198 r^2= 0.84610, % R^2 = 84.61 % (percentage capital R^2) , a= -64.49 , b= 72.41 , c= 73.33 , d= 140.97 using $Y= a + b \exp(-0.5 (X-C)/d)^2$ equation : $Y= -64.49 + b \exp 72.41 \exp(-0.5 (X-73.33)/140.97)^2$

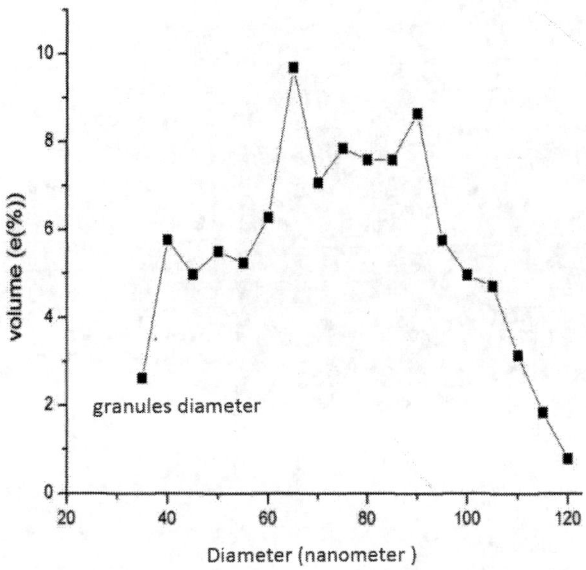

Figure 10: Variation of % volume versus granules diameter

Table 2: Distribution cumulative report of formed granules diameter

diameter nanometer	volume e(%)
35	2.62
40	5.76
45	4.97
50	5.5
55	5.24
60	6.28
65	9.69
70	7.07
75	7.85
80	7.59
85	7.59
90	8.64
95	5.76
100	4.97
105	4.71
110	3.14
115	1.83
120	0.79

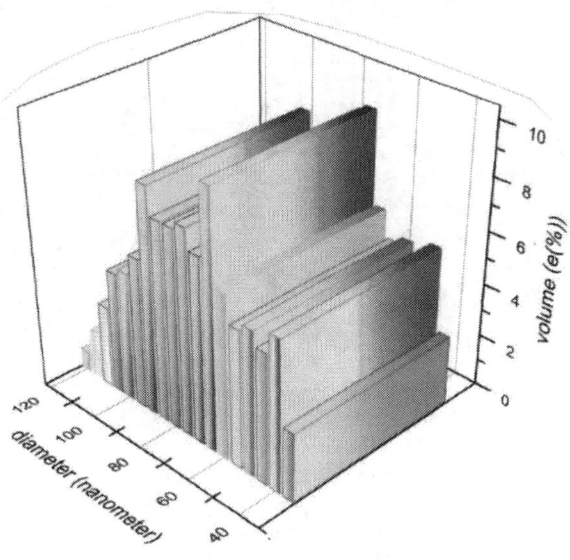

Figure 11: Granularity Accumulation distribution in 3 d forms showing variation of volume (V%) availability versus granules diameter (nm) . Showing that there is a shift toward a large diameter avoiding lower than 40 nm granules diameter restriction toward 40-120 nm is the range of available granules .

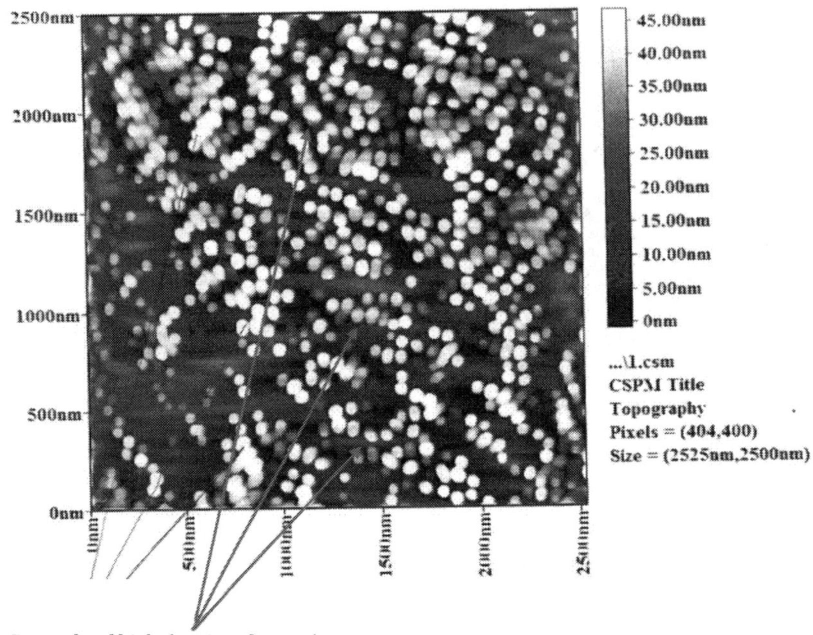

Figure 12: The imager surface roughness analysis shows a high density of granules . The probe was not able to see a higher depth of more than 45nm .Indicating that a close and rapid build up of granules even at constant steady flowing reactant reagents (Analyte plus reagents) .This fast grow of crystal indicate a colloidal form of precipitate .

Table 3: Imager surface roughness analysis by AFM

Amplitude parameters:	Value
Sa(Roughness Average)	11.9 nm
Sq(Root Mean Square)	13.8 nm
Ssk(Surface Skewness)	0.000195
Sku(Surface Kurtosis)	1.8
Sy(peak-peak)	47.7 nm
Sr(Ten Point Height)	47.7 nm
Hybrid Parameters:	
Ssc(Mean Summit Curvature)	-0.0336(1/ nm)
Sdq(Root Mean Square Slope)	1.16(1/ nm)
Sdr(Surface Area Ratio)	47
Functional Parameters:	
Sbi(Surface Bearing Index)	5.38
Sci(Core Fluide Retention Index)	1.49
Svi(Valley Fluid Retention Index)	0.0692
Spk(Reduced Summit Height)	5.13nm
Sk(Core Roughness Depth)	41.5nm
Svk(Reduced Valley Depth)	1.21nm
Spatial Parameters:	
Sds(Density of Summits)	214
Fractal Dimension	2.59

Journal of Physics: Conference Series **2063** (2021) 012031 doi:10.1088/1742-6596/2063/1/012031

In functional parameter the sk(core Roughness Depth =41.5 nm , while the reduced Valley depth = 1.21 nm .The difference is 40.29 nm which is very deep that the speed of precipitation is quite high that traps the water at early stages of precipitation leaving the valley trapped water to be near the surface of scanned area . As the total scanned depth is just little above 46.62 nm .Therefore the probe was not able to go beyond 45 nm . In Sa (Core Fluid Retention index of 1.49 compered to Svi (Valley Fluid Retention index = 0.0692 .The ratio is 21.53 i.e., the tendency is ≅ 22 time reserving power for water i.e., suggesting a gel precipitate . The roughness of 11.9 is the surface morphology of various variable peaks height giving rise to rough surface. (cast iron class A = 0.0048 inches (121920nm) while for copper 0.000059 inches (1498.6nm)) . Dryness of the precipitate will make the surface more rough , while a moist sample will fill up more Valleys . leaving a reflecting surface .

The calculation and discussion parameter of AFM as follow ::

Scanned surface area(fig . no. 13) = 2525nm x 2500 nm

= 6312500 nm^2

Assuming Spherical granules of average diameter of 71.65 nm

Surface area of a sphere (Single) = 4 πr^2

Single Sphere surface area = 4 x 3.14 x $(\frac{71.65}{2}\ nm)^2$

= 16119.88865 nm^2

Total number of spheres = 382 grains

Total surface area of all spheres = No. of grain x surface area of a single Sphere

= 382 x 16119.88865 nm^2

Total surface area of all spheres = 6157797.464 nm^2

Side walk area = Scanned area by the probe of AFM - Total surface area of all spheres

= 6312500 nm^2 - 6157797.464 nm^2

= 154702.5357 nm^2 (Side walk area)

A- If the value = -154702.5357 nm^2

The- 154702.5357 nm^2 represent the extra area remained after the formation of first ground monolayer .

Therefore : Remained surface area of spheres beyond the formation of first ground monolayer is equal to :

154702.5357 nm^2/ 16119.88865 nm^2

= 9.59699778

≅ 10 grains

10 grains (extra to the first ground mono layer i.e., start formation of second ground mono layer with this amount of grains) .

What remained in the first ground mono layer

Total number of grains – extra excess grains

382 - 10 = 372 grains

∴ First ground mono layer will have 372 granules

Second ground mono layer will have 10 granules

1 mole of molecules will have 6.02 x 10^{23} (it is a no. without unit)

1 mM of molecules ≅6.02 x 10^{20}

∴ 372 / 6.02 x 10^{23} = 62 x 10^{-23} mole

= 62 x 10^{-20} mM (each grain will have this concentration)

First ground monolayer have ; No. of gram x concentration of each

= 372grain x 62x10^{-20} → = 23064.0x10^{-20} mM concentration in mM (For the first ground monolayer)

The second hypothetical ground upper layer:

No.of grain x concentration of each grain:

10x 62 x 10^{-20} mM → 620 x10^{-20} mM

Journal of Physics: Conference Series **2063** (2021) 012031 doi:10.1088/1742-6596/2063/1/012031

Total captured concentration by both layer and have seen by the probe = what is available at first ground monolayer +what is available at the second ground monolayer

= 23064.0×10^{-20} mM + 620×10^{-20} mM

= 23684.0×10^{-20}mM

Now Total area of 372 grain – Total area of 10 grain will equal to TOTAL remained area as side walk

372x16119.88865 -10 x16119.88865 =

= 5835,400 nm^2

Remained area (side walk) = 5835400nm^2

Remained area / surface area of a single sphere (granules)

5835400nm^2 / 116119.88865nm^2 → 50.25 nm^2

That each granule will have a free area to move of 50.25 $nm^2 \cong 51$ nm^2

The 10 granules will occupy → 10 x51 = 510nm^2

(due to free space that the granules molecule will move within the second area)

6312500 nm^2 -510nm^2= 6311990 → 372 x51 → 18972.00nm^2

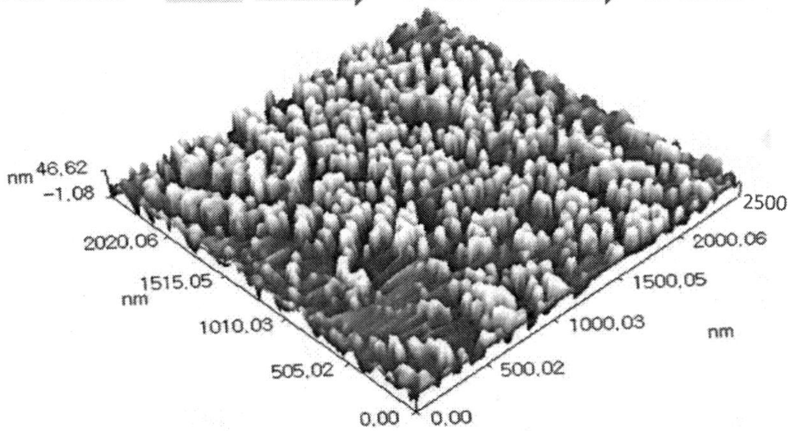

Figure 13: Image surface scan for the reaction of Mebeverine hydrochloride with sodium nitroprusside.

<= 10% diameter :40 nm (0.1 x 372 = 37.2 grains)
<= 50% diameter :70 nm (0.7 x 372 = 186 grains)
<= 90% diameter :100 nm (0.9 x 372 = 334.8 grains)
Since the combination of 37.2 + 186 + 334.8 = 558
∴ There is possibility that less than (<) will be considered here .
- Supersaturation is low and a spiral has developed , the growth rate should be proportional to the square of supersaturation , and when high , directly proportional to the supersaturation.
The formation of monolayer is lands that grow rapidly to the boundaries
Van Weimarn concern the variation in size and number of precipitate particles regarding concentration of the precipitating reagents. Van Weimarn postulated that a maximum would occur in the curve of particle size as a function of concentration of reactants and that average particle size would increase with time.
Van Weimarn relation→ Q-S/S
1- if =larg number means small particle size with large number of particles
2-if the results is small number mean large participate particulate with low number of particles Q/S -1
Q=total salt concentration
S=Molar solubility (increase with increasing the temperature)
Of course temperature of reaction will increase the solubility ,also PH value will be different at different temp. , reagent concentration , sample loop volume will play an important rule in obtaining a supersaturation at the outlet junction (c.f. Fig no.6), even it is not recommended .Teflon tubes are used as it has the nature of hydrophobicity (aid in water movements) .
It possible to use the dynamic range (analytical range) or working range i.e., calibration range and even linear range (linear dynamic range) but all depend on the analyst .
-When subtracting the area of scanned surface by the AFM from the total granules surface area .It is possible that the results of subtraction could be a positive value which mean that the concentration of formed

granules will occupy a first ground monolayers and there is still a space for more granules to build up this might happen only at low reactant concentration i.e, at lower part of the scatter point plot (at a glance of the calibration graph). While if the value of subtraction is a negative value this indicate that there is no room at first ground monolayer

monolayers causing to build up new second ground monolayer .This might happen at the average $\bar{x}$ & $\bar{y}$ i.e., the centroid value of the scatter plot of x (analyte concentration) versus obtained response (y) . Which expected to coincide with average formed precipitate (formed granules), Here at this point depending on the availability of excess extra granules that cover up the second ground monolayer (or even part of its surface area) .

+154702.5357 nm^2 at this stage there is enough space for the first ground monolayer to accommodate more granules . This extra available space can facilitate the presence of more granules ; concentration wise should help in building up more granules .The affected by the rate of flow rate , sample (analyte) concentration and reagent concentration and sample size (also temperature will play a good part in depicting the form of contour presented).

B- Scanned area by the probe equal to 6312500 nm^2 .

It will occupy the 382 granules by dividing total scanned area over the surface area of a single sphere (16119.88865 nm^2)

6312500 nm^2 / 6312500 nm^2 = 391 .596997786 ➡ 392 granules

Which mean an extra added granules number of 10 granules ; the first ground monolayer have a space of 10 more granules above what was seen by the probe (382 granules) .

$\therefore$ Therefore 392 x 51 nm^2 = 19992 nm^2

Scanned surface area by the probe = 6312500 nm^2

$\therefore$ 6312500 nm^2 - 19992 nm^2 = 6292508 nm^2

6292508 nm^2 is a free area not occupied by the granules at the used concentration of reactant concentration and parameters that was mentioned earlier (this will help in geometrical formation of the alleged crystal structure) i.e., support this by buildup as shown in Fig. …

This will conclude our detailed calculation .The area could be circular , or square ,or oblong .Both square and oblong are can calculate the length of each side .

(a x b)= area (nm^2) for oblong or (a x a)= area (nm^2) for square .

Fig. no.14shows a hypothetical representation of mono ground layer and second monolayer , Fig. no.15 : Samples of how surface area are calculated assuming of the presence of adsorbed ions and occluded ions , while fig 16& 17 shows the hypothetical build up monolayer forming a template for the basis of crystal growth .

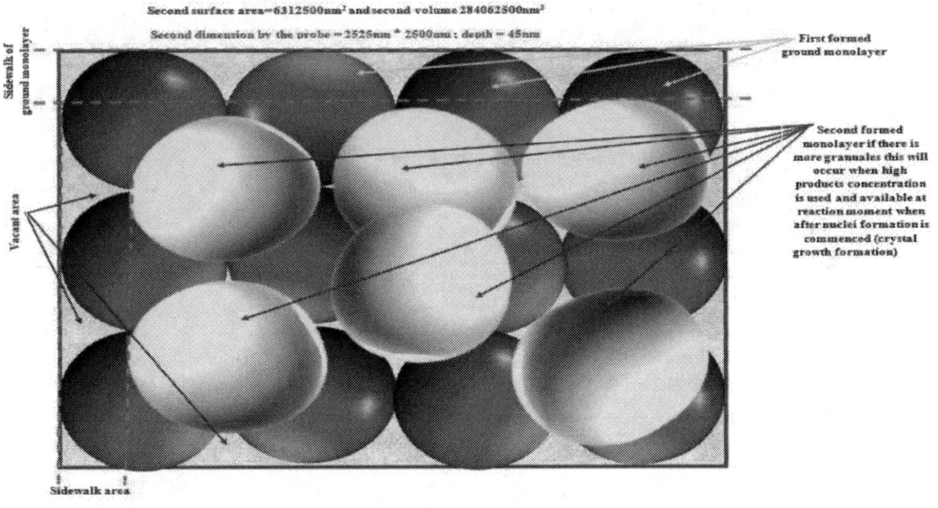

Figure 14: 3D- hypothetical representation of mono ground layer (brown) plus the formation of second monolayer (yellow) .

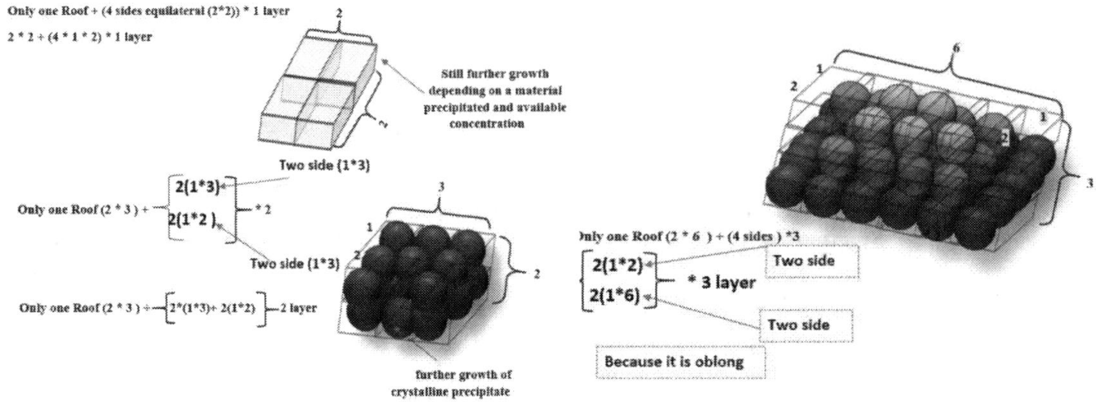

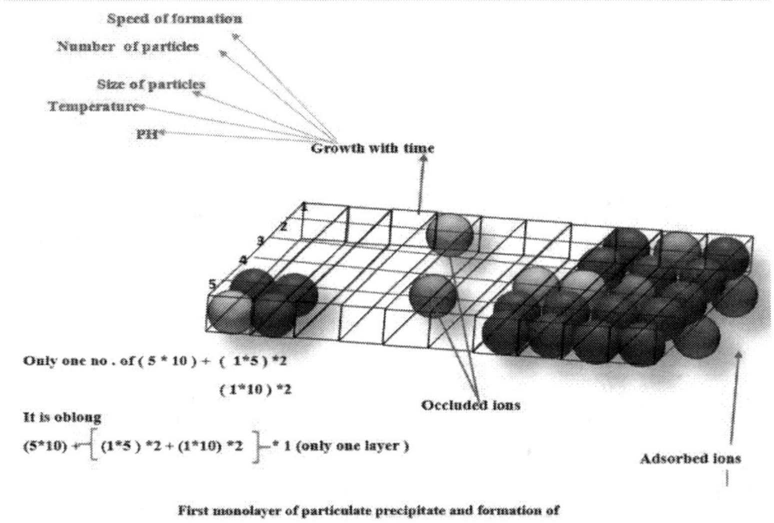

Figure 15: Samples of how surface area are calculated assuming of the presence of adsorbed ions and occluded ions (blue + orange)

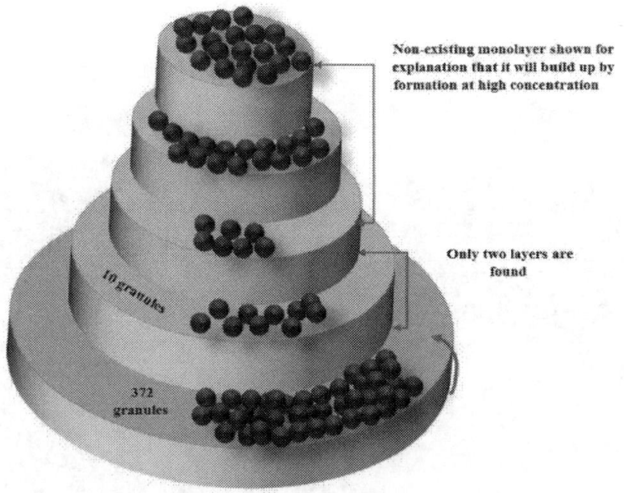

Total granules equal to 382 granules (see text). Assumption is made that layers are cylindrical to ease example of concentration.

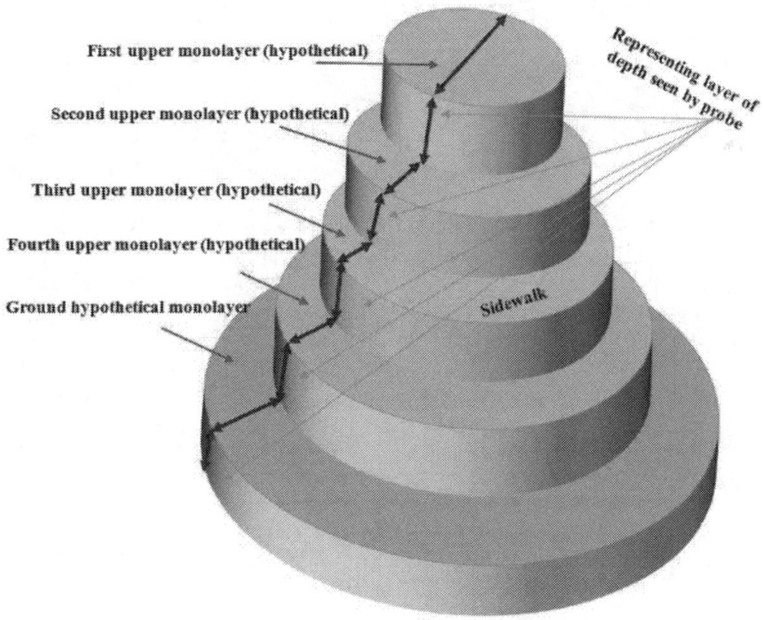

Figure 16 /A/ : Possible hypothetical illustration monolayers build up to form apparent peaks shown in fig. no.13

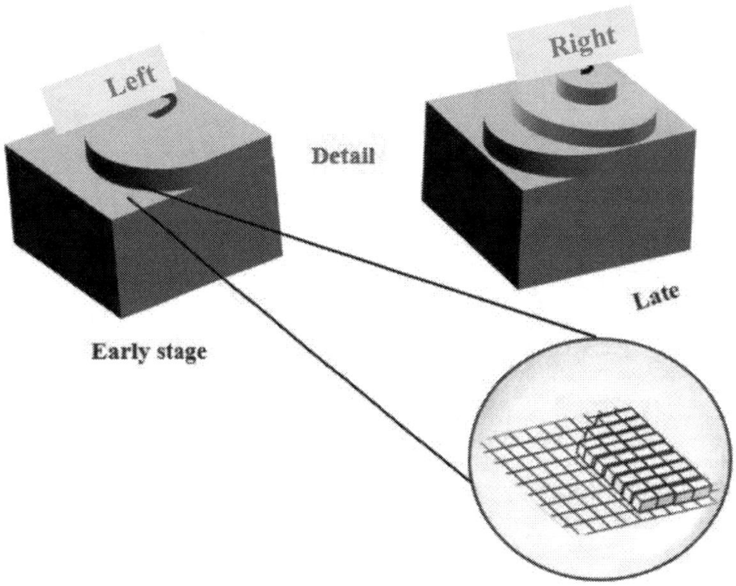

Figure 16/.B/: Schematic representation of crystal growth by screw dislocation. Left, an early stage; right, growth spiral at later stage; center, detail at the dislocation site , A similarity is shown in the given real photograph in figure no. 3 which is a hypothetically

References:

[1] IRQ. PATENT . No. 3613. Shakir ,I.M.A. and Al-Awadie ,Nagham .S.T., ISNAYAH-ST-CFI-Microphotometer S(3S_{BGR}X4-3(1D-Solar))-T(4S_{W}X4-2(1D-Solar)) ,2013 ,International classification ,G01J3/02,3.

[2] IRQ. PATENT . No. 3424. Shakir ,I.M.A. and Al-Awadie ,Nagham .S.T., Microphotometer Ayah 3SX3-3D-Solar ,2012 ,International classification ,G01N29/00,20.

[3] Issam M.A.S & Nagham S.T , 2005, New mode for the study of chemiluminescence reactions by tracking (distance or time) via multidetection points through optical fibers in a completely newly designed cell. Ph.D thesis, Baghdad University.

[4] Nagam S. T. Al-Awadi, Hussein F. Abd-Alrazach,2017.New mode for long path irradiation and two sided detection for the determination of ibuprofen in pure and pharmaceutical drugs using continuous flow feed via the use of ISNAG Fluorometer instrument, International journal of science and research , vol.6,pp.201-216.

[5] Issam M. 2014. New mode for the on-line determination of Amiloride in pure and pharmaceutical preparation using CFIA via the use of linear array from six LED (snow white) with one solar cell in a homemade AYAH 6SX1-T-1D solar CFI analyser. Iraqi journal of science. Vol.55(4B), pp:1708-1720.

[6] Issam M and Nagham S. 2011. Novel chemiluminometric-FIA method for the determination of DL-Histidine via loaded ion exchange resin with cobalt (II) ion. Tikrit journal of pure science. Vol.16 (1), pp:31-37.

[7] Nagham S. T. Al-Awadi, Hussein F. Abd-Alrazach,2018, Newly developed method for determination indomethacin using potassium hexacyanoferrate(III) by using ISNAG-Fluorimeter homemade via CFIA,Journal of global pharma technology, vol.10,pp.447-460.

[8] Issam M.A.S.2015. Promethazine-HCl determination using entrapped persulphate in water crystals by flow injection/stopped-flow technique and Ayah 3SX3-3D solar cell micro photometer, Iraqi journal of science, vol.56(1A),pp:25-.

[9] IRQ. PATENT . No. 5663. Shakir ,I.M.A. and Al-Awadie ,N.S.T., Dual and solo system for precipitated particulate in instrumental ON- Line analysis .2019, patent, International classification G01N21/01 .

[10] IRQ. PATENT . No. ٦١٠٠. Shakir ,I.M.A. and Al-Awadie ,N.S.T., Novel Multiple Continuous Flow Cells (hydrophilic & hydrophobic) works as a Solo Flow cell with Summed S/N responses in NAG- 4SX3 – 3 D instrument .20٢٠, patent, International classification G01N2021/032٥ , Iraqi classification 6 .

[11] IRQ. PATENT . No. 5604. Shakir ,I.M.A. and Al-Awadie ,N.S.T., Differential mass accumulation by the effect of neodymium in photometric .2018, patent, International classification B22F3/03 , H01F41/02 .

[12] IRQ. PATENT . No. 6091. Shakir ,I.M.A. and Al-Awadie ,N.S.T., ON – Line Analytical instrument for measurements of incident light attenuated and Diverged light ,NAG TUDI – WSLED – BLED -3 Solar , 2020, patent International classification G01N11/00 , , Iraqi classification 20 .

[13] IRQ. PATENT . No. 4859. Nagham .S.T., New System for Successive Irradiation and Spectral Signal Transfer with (Twelve) Optical Fiber using Continuous Flow Injection for Coloured and Turbid Solutions,2017 ,International classification , H01L31/0232 .

[14] IRQ. PATENT . No. 6136. Shakir ,I.M.A. and Al-Awadie ,N.S.T., Measure of incident retained photons excerted on precipitation reaction solutions and coloured solutions ,NAG - T . S_S . WSLED , 2020, patent, International classification , G01H13/00 , G01B15/025 Iraqi classification 4.

[15] IRQ. PATENT . No. 5776. Shakir ,I.M.A. and Al-Awadie , Nagham.S.T., Long distance chasing photometer for 300 mm length with 2mm path length to chase and to accumulate output resulted from Attenuated incident light 0-180° and the Diverged or Fluorescence light at 0-90° via two flow cells of 110 mm and 60 mm length (NAG – ADF-300-2) 2019, patent, International classification G01N2021/0328 .

AUTHOR INDEX

Institute of Physics
Dirac House, Temple Back
Bristol BS1 6BE UK

ISSN: 1742-6588
ISBN 978-1-7138-5034-2

9 781713 850342

CURRAN ASSOCIATES INC.
proceedings.com